JAN 1500
FEB 1900
MAR 1000
APR 1000
June 1500
JUy 300
Aug 500
Sept 1000
Sept 600
Oct 1000
Nov 1000
Dec 1000

JF 3400
M 4400
A 5400
J 6900
J 7200
A 7700
S 8700
S 9300
O 10,300   9300
N 11,300   10,300

J 11,300
F 9800
M 7900
A 6900
J 5900
J 4400
A 4100
S 3600
S 2600
O 2000
N 1000

13,300

Robert Lafore

Data Structures & Algorithms in Java

Second Edition

SAMS

800 East 96th Street, Indianapolis, Indiana 46240

Data Structures and Algorithms in Java, Second Edition

International Standard Book Number: 0-672-32453-9

Library of Congress Catalog Card Number: 2002106907

Printed in the United States of America

First Printing: December 2002

06   9   8   7

Trademarks

All terms mentioned in this book that are known to be trademarks or service marks have been appropriately capitalized. Sams Publishing cannot attest to the accuracy of this information. Use of a term in this book should not be regarded as affecting the validity of any trademark or service mark.

Warning and Disclaimer

Every effort has been made to make this book as complete and as accurate as possible, but no warranty or fitness is implied. The information provided is on an "as is" basis. The author and the publisher shall have neither liability nor responsibility to any person or entity with respect to any loss or damages arising from the information contained in this book.

Bulk Sales

Sams Publishing offers excellent discounts on this book when ordered in quantity for bulk purchases or special sales. For more information, please contact

U.S. Corporate and Government Sales
1-800-382-3419
corpsales@pearsontechgroup.com

For sales outside of the U.S., please contact

International Sales
international@pearsoned.com

Executive Editor
Michael Stephens

Acquisitions Editor
Carol Ackerman

Development Editor
Songlin Qiu

Managing Editor
Charlotte Clapp

Project Editor
Matt Purcell

Copy Editor
Chuck Hutchinson

Indexer
Johnna Dinse

Proofreader
Cindy Long

Technical Editor
Mike Kopack

Team Coordinator
Lynne Williams

Multimedia Developer
Dan Scherf

Interior Designer
Gary Adair

Cover Designer
Alan Clements

Production
Plan-it Publishing

Contents at a Glance

Table of Contents

Appendixes

About the Author

 **Robert Lafore** has degrees in Electrical Engineering and Mathematics, has worked as a systems analyst for the Lawrence Berkeley Laboratory, founded his own software company, and is a best-selling writer in the field of computer programming. Some of his current titles are *C++ Interactive Course* and *Object-Oriented Programming in C++*. Earlier best-selling titles include *Assembly Language Primer for the IBM PC and XT* and (back at the beginning of the computer revolution) *Soul of CP/M*.

Dedication

This book is dedicated to my readers, who have rewarded me over the years not only by buying my books, but with helpful suggestions and kind words. Thanks to you all.

Acknowledgments to the First Edition

My gratitude for the following people (and many others) cannot be fully expressed in this short acknowledgment. As always, Mitch Waite had the Java thing figured out before anyone else. He also let me bounce the applets off him until they did the job, and extracted the overall form of the project from a miasma of speculation. My editor, Kurt Stephan, found great reviewers, made sure everyone was on the same page, kept the ball rolling, and gently but firmly ensured that I did what I was supposed to do. Harry Henderson provided a skilled appraisal of the first draft, along with many valuable suggestions. Richard S. Wright, Jr., as technical editor, corrected numerous problems with his keen eye for detail. Jaime Niño, Ph.D., of the University of New Orleans, attempted to save me from myself and occasionally succeeded, but should bear no responsibility for my approach or coding details. Susan Walton has been a staunch and much-appreciated supporter in helping to convey the essence of the project to the non-technical. Carmela Carvajal was invaluable in extending our contacts with the academic world. Dan Scherf not only put the CD-ROM together, but was tireless in keeping me up to date on rapidly evolving software changes. Finally, Cecile Kaufman ably shepherded the book through its transition from the editing to the production process.

Acknowledgments to the Second Edition

My thanks to the following people at Sams Publishing for their competence, effort, and patience in the development of this second edition. Acquisitions Editor Carol Ackerman and Development Editor Songlin Qiu ably guided this edition through the complex production process. Project Editor Matt Purcell corrected a semi-infinite number of grammatical errors and made sure everything made sense. Tech Editor Mike Kopak reviewed the programs and saved me from several problems. Last but not least, Dan Scherf, an old friend from a previous era, provides skilled management of my code and applets on the Sams Web site.

We Want to Hear from You!

As the reader of this book, *you* are our most important critic and commentator. We value your opinion and want to know what we're doing right, what we could do better, what areas you'd like to see us publish in, and any other words of wisdom you're willing to pass our way.

As an executive editor for Sams Publishing, I welcome your comments. You can email or write me directly to let me know what you did or didn't like about this book—as well as what we can do to make our books better.

Please note that I cannot help you with technical problems related to the *topic* of this book. We do have a User Services group, however, where I will forward specific technical questions related to the book.

When you write, please be sure to include this book's title and author as well as your name, email address, and phone number. I will carefully review your comments and share them with the author and editors who worked on the book.

Email: feedback@samspublishing.com

Mail: Michael Stephens
Executive Editor
Sams Publishing
800 East 96th Street
Indianapolis, IN 46240 USA

For more information about this book or another Sams Publishing title, visit our Web site at www.samspublishing.com. Type the ISBN (excluding hyphens) or the title of a book in the Search field to find the page you're looking for.

Introduction

This introduction tells you briefly

- What's new in the Second Edition
- What this book is about
- Why it's different
- Who might want to read it
- What you need to know before you read it
- The software and equipment you need to use it
- How this book is organized

What's New in the Second Edition

This second edition of *Data Structures and Algorithms in Java* has been augmented to make it easier for the reader and for instructors using it as a text in computer science classes. Besides coverage of additional topics, we've added end-of-chapter questions, experiments, and programming projects.

Additional Topics

We've added a variety of interesting new topics to the book. Many provide a basis for programming projects. These new topics include

- Depth-first-search and game simulations
- The Josephus problem
- Huffman codes for data compression
- The Traveling Salesman problem
- Hamiltonian cycles
- The Knight's Tour puzzle
- Floyd's algorithm
- Warshall's algorithm
- 2-3 trees

- The knapsack problem
- Listing N things taken K at a time
- Folding-digits hash functions
- The radix sort

End-of-Chapter Questions

Short questions covering the key points of each chapter are included at the end of each chapter. The answers can be found in Appendix C, "Answers to Questions." These questions are intended as a self-test for readers, to ensure that they have understood the material.

Experiments

We include some suggested activities for the reader. These experiments often involve using the Workshop applets or example programs to examine certain features of an algorithm's operation, but some are pencil-and-paper or "thought experiments."

Programming Projects

Most importantly, we have included at the end of each chapter a number (usually five) of challenging programming projects. They cover a range of difficulty. The easiest are simple variations on the example programs. The most challenging are implementations of topics discussed in the text but for which there are no example programs. Solutions to the Programming Projects are not provided in this book, but see the adjacent note.

NOTE

It is expected that the programming projects will be useful for instructors looking for class assignments. To this end, qualified instructors can obtain suggested solutions to the programming projects in the form of source code and executable code. Contact the Sams Web site for information on Instructors Programs.

What This Book Is About

This book is about data structures and algorithms as used in computer programming. Data structures are ways in which data is arranged in your computer's memory (or stored on disk). Algorithms are the procedures a software program uses to manipulate the data in these structures.

Almost every computer program, even a simple one, uses data structures and algorithms. For example, consider a program that prints address labels. The program might use an array containing the addresses to be printed and a simple `for` loop to step through the array, printing each address.

The array in this example is a data structure, and the `for` loop, used for sequential access to the array, executes a simple algorithm. For uncomplicated programs with small amounts of data, such a simple approach might be all you need. However, for programs that handle even moderately large amounts of data, or which solve problems that are slightly out of the ordinary, more sophisticated techniques are necessary. Simply knowing the syntax of a computer language such as Java or C++ isn't enough.

This book is about what you need to know *after* you've learned a programming language. The material we cover here is typically taught in colleges and universities as a second-year course in computer science, after a student has mastered the fundamentals of programming.

What's Different About This Book

There are dozens of books on data structures and algorithms. What's different about this one? Three things:

- Our primary goal in writing this book is to make the topics we cover easy to understand.

- Demonstration programs called *Workshop applets* bring to life the topics we cover, showing you step by step, with "moving pictures," how data structures and algorithms work.

- The example code is written in Java, which is easier to understand than C, C++, or Pascal, the languages traditionally used to demonstrate computer science topics.

Let's look at these features in more detail.

Easy to Understand

Typical computer science textbooks are full of theory, mathematical formulas, and abstruse examples of computer code. This book, on the other hand, concentrates on simple explanations of techniques that can be applied to real-world problems. We avoid complex proofs and heavy math. There are lots of figures to augment the text.

Many books on data structures and algorithms include considerable material on software engineering. Software engineering is a body of study concerned with designing and implementing large and complex software projects.

However, it's our belief that data structures and algorithms are complicated enough without involving this additional discipline, so we have deliberately de-emphasized software engineering in this book. (We'll discuss the relationship of data structures and algorithms to software engineering in Chapter 1, "Overview.")

Of course, we do use an object-oriented approach, and we discuss various aspects of object-oriented design as we go along, including a mini-tutorial on OOP in Chapter 1. Our primary emphasis, however, is on the data structures and algorithms themselves.

Workshop Applets

From the Sams Web site you can download demonstration programs, in the form of Java applets, that cover the topics we discuss. These applets, which we call *Workshop applets*, will run on most Web browsers. (See Appendix A, "Running the Workshop Applets and Example Programs," for more details.) The Workshop applets create graphic images that show you in "slow motion" how an algorithm works.

For example, in one Workshop applet, each time you push a button, a bar chart shows you one step in the process of sorting the bars into ascending order. The values of variables used in the sorting algorithm are also shown, so you can see exactly how the computer code works when executing the algorithm. Text displayed in the picture explains what's happening.

Another applet models a binary tree. Arrows move up and down the tree, so you can follow the steps involved in inserting or deleting a node from the tree. There are more than 20 Workshop applets, at least one for each of the major topics in the book.

These Workshop applets make it far more obvious what a data structure really looks like, or what an algorithm is supposed to do, than a text description ever could. Of course, we provide a text description as well. The combination of Workshop applets, clear text, and illustrations should make things easy.

These Workshop applets are standalone graphics-based programs. You can use them as a learning tool that augments the material in the book. Note that they're not the same as the example code found in the text of the book, which we'll discuss next.

NOTE

The Workshop applets, in the form of Java .class files, are available on the Sams Web site at http://www.samspublishing.com/. Enter this book's ISBN (without the hyphens) in the Search box and click Search. When the book's title is displayed, click the title to go to a page where you can download the applets.

Java Example Code

The Java language is easier to understand (and write) than languages such as C and C++. The biggest reason for this is that Java doesn't use pointers. Some people are surprised that pointers aren't necessary for the creation of complex data structures and algorithms. In fact, eliminating pointers makes such code not only easier to write and to understand, but more secure and less prone to errors as well.

Java is a modern object-oriented language, which means we can use an object-oriented approach for the programming examples. This is important, because object-oriented programming (OOP) offers compelling advantages over the old-fashioned procedural approach, and is quickly supplanting it for serious program development. Don't be alarmed if you aren't familiar with OOP. It's not that hard to understand, especially in a pointer-free environment such as Java. We'll explain the basics of OOP in Chapter 1.

NOTE

Like the Workshop applets, the example programs (both source and executable files) can be downloaded from the Sams Web site.

Who This Book Is For

This book can be used as a text in a Data Structures and Algorithms course, typically taught in the second year of a computer science curriculum. However, it is also designed for professional programmers and for anyone else who needs to take the next step up from merely knowing a programming language. Because it's easy to understand, it is also appropriate as a supplemental text to a more formal course.

What You Need to Know Before You Read This Book

The only prerequisite for using this book is a knowledge of some programming language.

Although the example code is written in Java, you don't need to know Java to follow what's happening. Java is not hard to understand, and we've tried to keep the syntax as general as possible, avoiding baroque or Java-specific constructions whenever possible.

Of course, it won't hurt if you're already familiar with Java. Knowing C++ is essentially just as good, because Java syntax is based so closely on C++. The differences are minor as they apply to our example programs (except for the welcome elimination of pointers), and we'll discuss them in Chapter 1.

The Software You Need to Use This Book

To run the Workshop applets, you need a Web browser such as Microsoft Internet Explorer or Netscape Communicator. You can also use an applet viewer utility. Applet viewers are available with various Java development systems, including the free system from Sun Microsystems, which we'll discuss in Appendix A.

To run the example programs, you can use the MS-DOS utility in Microsoft Windows (called MS-DOS Prompt) or a similar text-oriented environment.

If you want to modify the source code for the example programs or write your own programs, you'll need a Java development system. Such systems are available commercially, or you can download an excellent basic system from Sun Microsystems, as described in Appendix A.

How This Book Is Organized

This section is intended for teachers and others who want a quick overview of the contents of the book. It assumes you're already familiar with the topics and terms involved in a study of data structures and algorithms.

The first two chapters are intended to ease the reader into data structures and algorithms as painlessly as possible.

Chapter 1, "Overview," presents an overview of the topics to be discussed and introduces a small number of terms that will be needed later on. For readers unfamiliar with object-oriented programming, it summarizes those aspects of this discipline that will be needed in the balance of the book, and for programmers who know C++ but not Java, the key differences between these languages are reviewed.

Chapter 2, "Arrays," focuses on arrays. However, there are two subtexts: the use of classes to encapsulate data storage structures and the class interface. Searching, insertion, and deletion in arrays and ordered arrays are covered. Linear searching and binary searching are explained. Workshop applets demonstrate these algorithms with unordered and ordered arrays.

In Chapter 3, "Simple Sorting," we introduce three simple (but slow) sorting techniques: the bubble sort, selection sort, and insertion sort. Each is demonstrated by a Workshop applet.

Chapter 4, "Stacks and Queues," covers three data structures that can be thought of as Abstract Data Types (ADTs): the stack, queue, and priority queue. These structures reappear later in the book, embedded in various algorithms. Each is demonstrated by a Workshop applet. The concept of ADTs is discussed.

Chapter 5, "Linked Lists," introduces linked lists, including doubly linked lists and double-ended lists. The use of references as "painless pointers" in Java is explained. A Workshop applet shows how insertion, searching, and deletion are carried out.

In Chapter 6, "Recursion," we explore recursion, one of the few chapter topics that is not a data structure. Many examples of recursion are given, including the Towers of Hanoi puzzle and the mergesort, which are demonstrated by Workshop applets.

Chapter 7, "Advanced Sorting," delves into some advanced sorting techniques: Shellsort and quicksort. Workshop applets demonstrate Shellsort, partitioning (the basis of quicksort), and two flavors of quicksort.

In Chapter 8, "Binary Trees," we begin our exploration of trees. This chapter covers the simplest popular tree structure: unbalanced binary search trees. A Workshop applet demonstrates insertion, deletion, and traversal of such trees.

Chapter 9, "Red-Black Trees," explains red-black trees, one of the most efficient balanced trees. The Workshop applet demonstrates the rotations and color switches necessary to balance the tree.

In Chapter 10, "2-3-4 Trees and External Storage," we cover 2-3-4 trees as an example of multiway trees. A Workshop applet shows how they work. We also discuss 2-3 trees and the relationship of 2-3-4 trees to B-trees, which are useful in storing external (disk) files.

Chapter 11, "Hash Tables," moves into a new field, hash tables. Workshop applets demonstrate several approaches: linear and quadratic probing, double hashing, and separate chaining. The hash-table approach to organizing external files is discussed.

In Chapter 12, "Heaps," we discuss the heap, a specialized tree used as an efficient implementation of a priority queue.

Chapters 13, "Graphs," and 14, "Weighted Graphs," deal with graphs, the first with unweighted graphs and simple searching algorithms, and the second with weighted graphs and more complex algorithms involving the minimum spanning trees and shortest paths.

In Chapter 15, "When to Use What," we summarize the various data structures described in earlier chapters, with special attention to which structure is appropriate in a given situation.

Appendix A, "Running the Workshop Applets and Example Programs," provides details on how to use these two kinds of software. It also tells how to use the Software Development Kit from Sun Microsystems, which can be used to modify the example programs and develop your own programs, and to run the applets and example programs.

Appendix B, "Further Reading," describes some books appropriate for further reading on data structures and other related topics.

Appendix C, "Answers to Questions," contains the answers to the end-of-chapter questions in the text.

Enjoy Yourself!

We hope we've made the learning process as painless as possible. Ideally, it should even be fun. Let us know if you think we've succeeded in reaching this ideal, or if not, where you think improvements might be made.

1

Overview

As you start this book, you may have some questions:

• What are data structures and algorithms?

• What good will it do me to know about them?

• Why can't I just use arrays and for loops to handle my data?

• When does it make sense to apply what I learn here?

This chapter attempts to answer these questions. We'll also introduce some terms you'll need to know and generally set the stage for the more detailed chapters to follow.

Next, for those of you who haven't yet been exposed to an object-oriented language, we'll briefly explain enough about OOP to get you started. Finally, for C++ programmers who don't know Java we'll point out some of the differences between these languages.

What Are Data Structures and Algorithms Good For?

The subject of this book is data structures and algorithms. A *data structure* is an arrangement of data in a computer's memory (or sometimes on a disk). Data structures include arrays, linked lists, stacks, binary trees, and hash tables, among others. *Algorithms* manipulate the data in these structures in various ways, such as searching for a particular data item and sorting the data.

What sorts of problems can you solve with a knowledge of these topics? As a rough approximation, we might divide the situations in which they're useful into three categories:

- Real-world data storage

- Programmer's tools

- Modeling

These are not hard-and-fast categories, but they may help give you a feeling for the usefulness of this book's subject matter. Let's look at them in more detail.

Real-World Data Storage

Many of the structures and techniques we'll discuss are concerned with how to handle real-world data storage. By real-world data, we mean data that describes physical entities external to the computer. As some examples, a personnel record describes an actual human being, an inventory record describes an existing car part or grocery item, and a financial transaction record describes, say, an actual check written to pay the electric bill.

A non-computer example of real-world data storage is a stack of 3-by-5 index cards. These cards can be used for a variety of purposes. If each card holds a person's name, address, and phone number, the result is an address book. If each card holds the name, location, and value of a household possession, the result is a home inventory.

Of course, index cards are not exactly state-of-the-art. Almost anything that was once done with index cards can now be done with a computer. Suppose you want to update your old index-card system to a computer program. You might find yourself with questions like these:

- How would you store the data in your computer's memory?

- Would your method work for a hundred file cards? A thousand? A million?

- Would your method permit quick insertion of new cards and deletion of old ones?

- Would it allow for fast searching for a specified card?

- Suppose you wanted to arrange the cards in alphabetical order. How would you sort them?

In this book, we will be discussing data structures that might be used in ways similar to a stack of index cards.

Of course, most programs are more complex than index cards. Imagine the database the Department of Motor Vehicles (or whatever it's called in your state) uses to keep track of drivers' licenses, or an airline reservations system that stores passenger and flight information. Such systems may include many data structures. Designing such complex systems requires the application of software engineering techniques, which we'll mention toward the end of this chapter.

Programmer's Tools

Not all data storage structures are used to store real-world data. Typically, real-world data is accessed more or less directly by a program's user. Some data storage structures, however, are not meant to be accessed by the user, but by the program itself. A programmer uses such structures as tools to facilitate some other operation. Stacks, queues, and priority queues are often used in this way. We'll see examples as we go along.

Real-World Modeling

Some data structures directly model real-world situations. The most important data structure of this type is the graph. You can use graphs to represent airline routes between cities or connections in an electric circuit or tasks in a project. We'll cover graphs in Chapter 13, "Graphs," and Chapter 14, "Weighted Graphs." Other data structures, such as stacks and queues, may also be used in simulations. A queue, for example, can model customers waiting in line at a bank or cars waiting at a toll booth.

Overview of Data Structures

Another way to look at data structures is to focus on their strengths and weaknesses. In this section we'll provide an overview, in the form of a table, of the major data storage structures we'll be discussing in this book. This is a bird's-eye view of a landscape that we'll be covering later at ground level, so don't be alarmed if the terms used are not familiar. Table 1.1 shows the advantages and disadvantages of the various data structures described in this book.

TABLE 1.1   Characteristics of Data Structures

Data Structure	Advantages	Disadvantages
Array	Quick insertion, very fast access if index known.	Slow search, slow deletion, fixed size.
Ordered array	Quicker search than unsorted array.	Slow insertion and deletion, fixed size.

TABLE 1.1   Continued

Data Structure	Advantages	Disadvantages
Stack	Provides last-in, first-out access.	Slow access to other items.
Queue	Provides first-in, first-out access.	Slow access to other items.
Linked list	Quick insertion, quick deletion.	Slow search.
Binary tree	Quick search, insertion, deletion (if tree remains balanced).	Deletion algorithm is complex.
Red-black tree	Quick search, insertion, deletion. Tree always balanced.	Complex.
2-3-4 tree	Quick search, insertion, deletion. Tree always balanced. Similar trees good for disk storage.	Complex.
Hash table	Very fast access if key known. Fast insertion.	Slow deletion, access slow if key not known, inefficient memory usage.
Heap	Fast insertion, deletion, access to largest item.	Slow access to other items.
Graph	Models real-world situations.	Some algorithms are slow and complex.

The data structures shown in Table 1.1, except the arrays, can be thought of as Abstract Data Types, or ADTs. We'll describe what this means in Chapter 5, "Linked Lists."

Overview of Algorithms

Many of the algorithms we'll discuss apply directly to specific data structures. For most data structures, you need to know how to

- Insert a new data item.
- Search for a specified item.
- Delete a specified item.

You may also need to know how to *iterate* through all the items in a data structure, visiting each one in turn so as to display it or perform some other action on it.

Another important algorithm category is *sorting*. There are many ways to sort data, and we devote Chapter 3, "Simple Sorting," and Chapter 7, "Advanced Sorting," to these algorithms.

The concept of *recursion* is important in designing certain algorithms. Recursion involves a method calling itself. We'll look at recursion in Chapter 6, "Recursion." (The term *method* is used in Java. In other languages, it is called a function, procedure, or subroutine.)

Some Definitions

Let's look at a few of the terms that we'll be using throughout this book.

Database

We'll use the term *database* to refer to all the data that will be dealt with in a particular situation. We'll assume that each item in a database has a similar format. As an example, if you create an address book using index cards, these cards constitute a database. The term *file* is sometimes used in this sense.

Record

Records are the units into which a database is divided. They provide a format for storing information. In the index card analogy, each card represents a record. A record includes all the information about some entity, in a situation in which there are many such entities. A record might correspond to a person in a personnel file, a car part in an auto supply inventory, or a recipe in a cookbook file.

Field

A record is usually divided into several *fields*. A field holds a particular kind of data. On an index card for an address book, a person's name, address, or telephone number is an individual field.

More sophisticated database programs use records with more fields. Figure 1.1 shows such a record, where each line represents a distinct field.

In Java (and other object-oriented languages), records are usually represented by *objects* of an appropriate class. Individual variables within an object represent data fields. Fields within a class object are called *fields* in Java (but *members* in some other languages such as C++).

```
Employee number:
Social security number:
Last name:
First name:
Street address:
City:
State:
Zip code:
Phone number:
Date of birth:
Date of first employment:
Salary:
```

FIGURE 1.1   A record with multiple fields.

Key

To search for a record within a database, you need to designate one of the record's fields as a *key* (or *search key*). You'll search for the record with a specific key. For instance, in an address book program, you might search in the name field of each record for the key "Brown." When you find the record with this key, you can access all its fields, not just the key. We might say that the key *unlocks* the entire record. You could search through the same file using the phone number field or the address field as the key. Any of the fields in Figure 1.1 could be used as a search key.

Object-Oriented Programming

This section is for those of you who haven't been exposed to object-oriented programming. However, caveat emptor. We cannot, in a few pages, do justice to all the innovative new ideas associated with OOP. Our goal is merely to make it possible for you to understand the example programs in the text.

If, after reading this section and examining some of the example code in the following chapters, you still find the whole OOP business as alien as quantum physics, you may need a more thorough exposure to OOP. See the reading list in Appendix B, "Further Reading," for suggestions.

Problems with Procedural Languages

OOP was invented because procedural languages, such as C, Pascal, and early versions of BASIC, were found to be inadequate for large and complex programs. Why was this?

There were two kinds of problems. One was the lack of correspondence between the program and the real world, and the other was the internal organization of the program.

Poor Modeling of the Real World

Conceptualizing a real-world problem using procedural languages is difficult. Methods carry out a task, while data stores information, but most real-world objects do both of these things. The thermostat on your furnace, for example, carries out tasks (turning the furnace on and off) but also stores information (the current temperature and the desired temperature).

If you wrote a thermostat control program in a procedural language, you might end up with two methods, `furnace_on()` and `furnace_off()`, but also two global variables, `currentTemp` (supplied by a thermometer) and `desiredTemp` (set by the user). However, these methods and variables wouldn't form any sort of programming unit; there would be no unit in the program you could call `thermostat`. The only such concept would be in the programmer's mind.

For large programs, which might contain hundreds of entities like thermostats, this procedural approach made things chaotic, error-prone, and sometimes impossible to implement at all. What was needed was a better match between things in the program and things in the outside world.

Crude Organizational Units

A more subtle, but related, problem had to do with a program's internal organization. Procedural programs were organized by dividing the code into methods. One difficulty with this kind of method-based organization was that it focused on methods at the expense of data. There weren't many options when it came to data. To simplify slightly, data could be local to a particular method, or it could be global—accessible to all methods. There was no way (at least not a flexible way) to specify that some methods could access a variable and others couldn't.

This inflexibility caused problems when several methods needed to access the same data. To be available to more than one method, such variables needed to be global, but global data could be accessed inadvertently by *any* method in the program. This lead to frequent programming errors. What was needed was a way to fine-tune data accessibility, allowing data to be available to methods with a need to access it, but hiding it from other methods.

Objects in a Nutshell

The idea of *objects* arose in the programming community as a solution to the problems with procedural languages.

Objects

Here's the amazing breakthrough that is the key to OOP: An object contains *both methods and variables*. A thermostat object, for example, would contain not only furnace_on() and furnace_off() methods, but also variables called currentTemp and desiredTemp. In Java, an object's variables such as these are called *fields*.

This new entity, the object, solves several problems simultaneously. Not only does an object in a program correspond more closely to an object in the real world, but it also solves the problem engendered by global data in the procedural model. The furnace_on() and furnace_off() methods can access currentTemp and desiredTemp. These variables are hidden from methods that are not part of thermostat, however, so they are less likely to be accidentally changed by a rogue method.

Classes

You might think that the idea of an object would be enough for one programming revolution, but there's more. Early on, it was realized that you might want to make several objects of the same type. Maybe you're writing a furnace control program for an entire apartment building, for example, and you need several dozen thermostat objects in your program. It seems a shame to go to the trouble of specifying each one separately. Thus, the idea of classes was born.

A *class* is a specification—a blueprint—for one or more objects. Here's how a thermostat class, for example, might look in Java:

```
class thermostat
    {
    private float currentTemp();
    private float desiredTemp();

    public void furnace_on()
       {
       // method body goes here
       }

    public void furnace_off()
       {
       // method body goes here
       }
    }  // end class thermostat
```

The Java keyword class introduces the class specification, followed by the name you want to give the class; here it's thermostat. Enclosed in curly brackets are the fields and methods that make up the class. We've left out the bodies of the methods; normally, each would have many lines of program code.

C programmers will recognize this syntax as similar to a structure, while C++ programmers will notice that it's very much like a class in C++, except that there's no semicolon at the end. (Why did we need the semicolon in C++ anyway?)

Creating Objects

Specifying a class doesn't create any objects of that class. (In the same way, specifying a structure in C doesn't create any variables.) To actually create objects in Java, you must use the keyword new. At the same time an object is created, you need to store a reference to it in a variable of suitable type—that is, the same type as the class.

What's a reference? We'll discuss references in more detail later. In the meantime, think of a reference as a name for an object. (It's actually the object's address, but you don't need to know that.)

Here's how we would create two references to type thermostat, create two new thermostat objects, and store references to them in these variables:

```
thermostat therm1, therm2;  // create two references

therm1 = new thermostat();  // create two objects and
therm2 = new thermostat();  // store references to them
```

Incidentally, creating an object is also called *instantiating* it, and an object is often referred to as an *instance* of a class.

Accessing Object Methods

After you specify a class and create some objects of that class, other parts of your program need to interact with these objects. How do they do that?

Typically, other parts of the program interact with an object's methods, not with its data (fields). For example, to tell the therm2 object to turn on the furnace, we would say

```
therm2.furnace_on();
```

The dot operator (.) associates an object with one of its methods (or occasionally with one of its fields).

At this point we've covered (rather telegraphically) several of the most important features of OOP. To summarize:

- Objects contain both methods and fields (data).

- A class is a specification for any number of objects.

- To create an object, you use the keyword new in conjunction with the class name.

- To invoke a method for a particular object, you use the dot operator.

These concepts are deep and far reaching. It's almost impossible to assimilate them the first time you see them, so don't worry if you feel a bit confused. As you see more classes and what they do, the mist should start to clear.

A Runnable Object-Oriented Program

Let's look at an object-oriented program that runs and generates actual output. It features a class called BankAccount that models a checking account at a bank. The program creates an account with an opening balance, displays the balance, makes a deposit and a withdrawal, and then displays the new balance. Listing 1.1 shows bank.java.

LISTING 1.1   The bank.java Program

```
// bank.java
// demonstrates basic OOP syntax
// to run this program: C>java BankApp
//////////////////////////////////////////////////////////////////
class BankAccount
   {
   private double balance;                    // account balance

   public BankAccount(double openingBalance) // constructor
      {
      balance = openingBalance;
      }

   public void deposit(double amount)        // makes deposit
      {
      balance = balance + amount;
      }

   public void withdraw(double amount)       // makes withdrawal
      {
      balance = balance - amount;
      }

   public void display()                      // displays balance
      {
      System.out.println("balance=" + balance);
      }
   } // end class BankAccount
//////////////////////////////////////////////////////////////////
```

LISTING 1.1   Continued

```
class BankApp
    {
    public static void main(String[] args)
        {
        BankAccount ba1 = new BankAccount(100.00); // create acct

        System.out.print("Before transactions, ");
        ba1.display();                             // display balance

        ba1.deposit(74.35);                        // make deposit
        ba1.withdraw(20.00);                       // make withdrawal

        System.out.print("After transactions, ");
        ba1.display();                             // display balance
        } // end main()
    } // end class BankApp
```

Here's the output from this program:

```
Before transactions, balance=100
After transactions, balance=154.35
```

There are two classes in bank.java. The first one, BankAccount, contains the fields and methods for our bank account. We'll examine it in detail in a moment. The second class, BankApp, plays a special role.

The BankApp **Class**

To execute the program in Listing 1.1 from an MS-DOS prompt, you type java BankApp following the C: prompt:

```
C:\>java BankApp
```

This command tells the java interpreter to look in the BankApp class for the method called main(). Every Java application must have a main() method; execution of the program starts at the beginning of main(), as you can see in Listing 1.1. (You don't need to worry yet about the String[] args argument in main().)

The main() method creates an object of class BankAccount, initialized to a value of 100.00, which is the opening balance, with this statement:

```
BankAccount ba1 = new BankAccount(100.00); // create acct
```

The System.out.print() method displays the string used as its argument, Before
transactions:, and the account displays its balance with this statement:

ba1.display();

The program then makes a deposit to, and a withdrawal from, the account:

ba1.deposit(74.35);
ba1.withdraw(20.00);

Finally, the program displays the new account balance and terminates.

The BankAccount Class
The only data field in the BankAccount class is the amount of money in the account,
called balance. There are three methods. The deposit() method adds an amount to
the balance, withdrawal() subtracts an amount, and display() displays the balance.

Constructors
The BankAccount class also features a *constructor*, which is a special method that's
called automatically whenever a new object is created. A constructor always has
exactly the same name as the class, so this one is called BankAccount(). This
constructor has one argument, which is used to set the opening balance when the
account is created.

A constructor allows a new object to be initialized in a convenient way. Without the
constructor in this program, you would have needed an additional call to deposit()
to put the opening balance in the account.

Public and Private
Notice the keywords public and private in the BankAccount class. These keywords
are *access modifiers* and determine which methods can access a method or field. The
balance field is preceded by private. A field or method that is private can be
accessed only by methods that are part of the same class. Thus, balance cannot be
accessed by statements in main() because main() is not a method in BankAccount.

All the methods in BankAccount have the access modifier public, however, so they
can be accessed by methods in other classes. That's why statements in main() can
call deposit(), withdrawal(), and display().

Data fields in a class are typically made private and methods are made public. This
protects the data; it can't be accidentally modified by methods of other classes. Any
outside entity that needs to access data in a class must do so using a method of the
same class. Data is like a queen bee, kept hidden in the middle of the hive, fed and
cared for by worker-bee methods.

Inheritance and Polymorphism

We'll briefly mention two other key features of object-oriented programming: inheritance and polymorphism.

Inheritance is the creation of one class, called the *extended* or derived class, from another class called the *base* class. The extended class has all the features of the base class, plus some additional features. For example, a `secretary` class might be derived from a more general `employee` class and include a field called `typingSpeed` that the `employee` class lacked.

In Java, inheritance is also called *subclassing*. The base class may be called the *superclass*, and the extended class may be called the *subclass*.

Inheritance enables you to easily add features to an existing class and is an important aid in the design of programs with many related classes. Inheritance thus makes it easy to reuse classes for a slightly different purpose, a key benefit of OOP.

Polymorphism involves treating objects of different classes in the same way. For polymorphism to work, these different classes must be derived from the same base class. In practice, polymorphism usually involves a method call that actually executes different methods for objects of different classes.

For example, a call to `display()` for a `secretary` object would invoke a display method in the `secretary` class, while the exact same call for a `manager` object would invoke a different display method in the `manager` class. Polymorphism simplifies and clarifies program design and coding.

For those not familiar with them, inheritance and polymorphism involve significant additional complexity. To keep the focus on data structures and algorithms, we have avoided these features in our example programs. Inheritance and polymorphism are important and powerful aspects of OOP but are not necessary for the explanation of data structures and algorithms.

Software Engineering

In recent years, it has become fashionable to begin a book on data structures and algorithms with a chapter on software engineering. We don't follow that approach, but let's briefly examine software engineering and see how it fits into the topics we discuss in this book.

Software engineering is the study of ways to create large and complex computer programs, involving many programmers. It focuses on the overall design of the programs and on the creation of that design from the needs of the end users. Software engineering is concerned with the life cycle of a software project, which includes specification, design, verification, coding, testing, production, and maintenance.

It's not clear that mixing software engineering on one hand and data structures and algorithms on the other actually helps the student understand either topic. Software engineering is rather abstract and is difficult to grasp until you've been involved yourself in a large project. The use of data structures and algorithms, on the other hand, is a nuts-and-bolts discipline concerned with the details of coding and data storage.

Accordingly, we focus on the essentials of data structures and algorithms. How do they really work? What structure or algorithm is best in a particular situation? What do they look like translated into Java code? As we noted, our intent is to make the material as easy to understand as possible. For further reading, we mention some books on software engineering in Appendix B.

Java for C++ Programmers

If you're a C++ programmer who has not yet encountered Java, you might want to read this section. We'll mention several ways that Java differs from C++.

This section is not intended to be a primer on Java. We don't even cover all the differences between C++ and Java. We're interested in only a few Java features that might make it hard for C++ programmers to figure out what's going on in the example programs.

No Pointers

The biggest difference between C++ and Java is that Java doesn't use pointers. To a C++ programmer, not using pointers may at first seem quite amazing. How can you get along without pointers?

Throughout this book we'll use pointer-free code to build complex data structures. You'll see that this approach is not only possible, but actually easier than using C++ pointers.

Actually, Java only does away with *explicit* pointers. Pointers, in the form of memory addresses, are still there, under the surface. It's sometimes said that, in Java, *everything* is a pointer. This statement is not completely true, but it's close. Let's look at the details.

References

Java treats primitive data types (such as `int`, `float`, and `double`) differently than objects. Look at these two statements:

```
int intVar;       // an int variable called intVar
BankAccount bc1;  // reference to a BankAccount object
```

In the first statement, a memory location called `intVar` actually holds a numerical value such as 127 (assuming such a value has been placed there). However, the memory location `bc1` does not hold the data of a `BankAccount` object. Instead, it contains the *address* of a `BankAccount` object that is actually stored elsewhere in memory. The name `bc1` is a *reference to* this object; it's not the object itself.

Actually, `bc1` won't hold a reference if it has not been assigned an object at some prior point in the program. Before being assigned an object, it holds a reference to a special object called `null`. In the same way, `intVar` won't hold a numerical value if it's never been assigned one. The compiler will complain if you try to use a variable that has never been assigned a value.

In C++, the statement

```
BankAccount bc1;
```

actually creates an object; it sets aside enough memory to hold all the object's data. In Java, all this statement creates is a place to put an object's memory address. You can think of a reference as a pointer with the syntax of an ordinary variable. (C++ has reference variables, but they must be explicitly specified with the & symbol.)

Assignment
It follows that the assignment operator (=) operates differently with Java objects than with C++ objects. In C++, the statement

```
bc2 = bc1;
```

copies all the data from an object called `bc1` into a different object called `bc2`. Following this statement, there are two objects with the same data. In Java, on the other hand, this same assignment statement copies the memory address that `bc1` refers to into `bc2`. Both `bc1` and `bc2` now refer to exactly the same object; they are references to it.

This can get you into trouble if you're not clear what the assignment operator does. Following the assignment statement shown above, the statement

```
bc1.withdraw(21.00);
```

and the statement

```
bc2.withdraw(21.00);
```

both withdraw $21 from *the same bank account object.*

Suppose you actually want to copy data from one object to another. In this case you must make sure you have two separate objects to begin with and then copy each field separately. The equal sign won't do the job.

The new **Operator**

Any object in Java must be created using new. However, in Java, new returns a reference, not a pointer as in C++. Thus, pointers aren't necessary to use new. Here's one way to create an object:

```
BankAccount ba1;
ba1 = new BankAccount();
```

Eliminating pointers makes for a more secure system. As a programmer, you can't find out the actual address of ba1, so you can't accidentally corrupt it. However, you probably don't need to know it, unless you're planning something wicked.

How do you release memory that you've acquired from the system with new and no longer need? In C++, you use delete. In Java, you don't need to worry about releasing memory. Java periodically looks through each block of memory that was obtained with new to see if valid references to it still exist. If there are no such references, the block is returned to the free memory store. This process is called *garbage collection*.

In C++ almost every programmer at one time or another forgets to delete memory blocks, causing "memory leaks" that consume system resources, leading to bad performance and even crashing the system. Memory leaks can't happen in Java (or at least hardly ever).

Arguments

In C++, pointers are often used to pass objects to functions to avoid the overhead of copying a large object. In Java, objects are always passed as references. This approach also avoids copying the object:

```
void method1()
   {
   BankAccount ba1 = new BankAccount(350.00);
   method2(ba1);
   }


void method2(BankAccount acct)
   {
   }
```

In this code, the references ba1 and acct both refer to the same object. In C++ acct would be a separate object, copied from ba1.

Primitive data types, on the other hand, are always passed by value. That is, a new variable is created in the method and the value of the argument is copied into it.

Equality and Identity

In Java, if you're talking about primitive types, the equality operator (==) will tell you whether two variables have the same value:

```
int intVar1 = 27;
int intVar2 = intVar1;
if(intVar1 == intVar2)
   System.out.println("They're equal");
```

This is the same as the syntax in C and C++, but in Java, because relational operators use references, they work differently with objects. The equality operator, when applied to objects, tells you whether two references are identical—that is, whether they refer to the same object:

```
carPart cp1 = new carPart("fender");
carPart cp2 = cp1;
if(cp1 == cp2)
   System.out.println("They're Identical");
```

In C++ this operator would tell you if two objects contained the same data. If you want to see whether two objects contain the same data in Java, you must use the equals() method of the Object class:

```
carPart cp1 = new carPart("fender");
carPart cp2 = cp1;
if( cp1.equals(cp2) )
   System.out.println("They're equal");
```

This technique works because all objects in Java are implicitly derived from the Object class.

Overloaded Operators

This point is easy: There are no overloaded operators in Java. In C++, you can redefine +, *, =, and most other operators so that they behave differently for objects of a particular class. No such redefinition is possible in Java. Use a named method instead, such as add() or whatever.

Primitive Variable Types

The primitive or built-in variable types in Java are shown in Table 1.2.

TABLE 1.2   Primitive Data Types

Name	Size in Bits	Range of Values
boolean	1	true or false
byte	8	–128 to +127
char	16	'\u0000' to '\uFFFF'
short	16	–32,768 to +32,767
int	32	–2,147,483,648 to +2,147,483,647
long	64	–9,223,372,036,854,775,808 to +9,223,372,036,854,775,807
float	32	Approximately 10^{-38} to 10^{+38}; 7 significant digits
double	64	Approximately 10^{-308} to 10^{+308}; 15 significant digits

Unlike C and C++, which use integers for true/false values, boolean is a distinct type in Java.

Type char is unsigned, and uses two bytes to accommodate the Unicode character representation scheme, which can handle international characters.

The int type varies in size in C and C++, depending on the specific computer platform; in Java an int is always 32 bits.

Literals of type float use the suffix F (for example, 3.14159F); literals of type double need no suffix. Literals of type long use suffix L (as in 45L); literals of the other integer types need no suffix.

Java is more strongly typed than C and C++; many conversions that were automatic in those languages require an explicit cast in Java.

All types not shown in Table 1.2, such as String, are classes.

Input/Output

There have been changes to input/output as Java has evolved. For the console-mode applications we'll be using as example programs in this book, some clunky-looking but effective constructions are available for input and output. They're quite different from the workhorse cout and cin approaches in C++ and printf() and scanf() in C.

Older versions of the Java Software Development Kit (SDK) required the line

```
import java.io.*;
```

at the beginning of the source file for all input/output routines. Now this line is needed only for input.

Output

You can send any primitive type (numbers and characters), and String objects as well, to the display with these statements:

```
System.out.print(var);     // displays var, no linefeed
System.out.println(var);   // displays var, then starts new line
```

The print() method leaves the cursor on the same line; println() moves it to the beginning of the next line.

In older versions of the SDK, a System.out.print() statement did not actually write anything to the screen. It had to be followed by a System.out.println()or System.out.flush() statement to display the entire buffer. Now it displays immediately.

You can use several variables, separated by plus signs, in the argument. Suppose in this statement the value of ans is 33:

```
System.out.println("The answer is " + ans);
```

Then the output will be

```
The answer is 33
```

Inputting a String

Input is considerably more involved than output. In general, you want to read any input as a String object. If you're actually inputting something else, say a character or number, you then convert the String object to the desired type.

As we noted, any program that uses input must include the statement

```
import java.io.*;
```

at the beginning of the program. Without this statement, the compiler will not recognize such entities as IOException and InputStreamReader.

String input is fairly baroque. Here's a method that returns a string entered by the user:

```
public static String getString() throws IOException
    {
    InputStreamReader isr = new InputStreamReader(System.in);
    BufferedReader br = new BufferedReader(isr);
    String s = br.readLine();
    return s;
    }
```

This method returns a String object, which is composed of characters typed on the keyboard and terminated with the Enter key. The details of the InputStreamReader and BufferedReader classes need not concern us here.

Besides importing java.io.*, you'll need to add throws IOException to all input methods, as shown in the preceding code. In fact, you'll need to add throws IOException to any method, such as main(), that calls any of the input methods.

Inputting a Character

Suppose you want your program's user to enter a character. (By *enter*, we mean typing something and pressing the Enter key.) The user may enter a single character or (incorrectly) more than one. Therefore, the safest way to read a character involves reading a String and picking off its first character with the charAt() method:

```
public static char getChar() throws IOException
    {
    String s = getString();
    return s.charAt(0);
    }
```

The charAt() method of the String class returns a character at the specified position in the String object; here we get the first character, which is number 0. This approach prevents extraneous characters being left in the input buffer. Such characters can cause problems with subsequent input.

Inputting Integers

To read numbers, you make a String object as shown before and convert it to the type you want using a conversion method. Here's a method, getInt(), that converts input into type int and returns it:

```
public int getInt() throws IOException
    {
    String s = getString();
    return Integer.parseInt(s);
    }
```

The parseInt() method of class Integer converts the string to type int. A similar routine, parseLong(), can be used to convert type long.

In older versions of the SDK, you needed to use the line

```
import java.lang.Integer;
```

at the beginning of any program that used parseInt(), but this convention is no longer necessary.

For simplicity, we don't show any error-checking in the input routines in the example programs. The user must type appropriate input, or an exception will occur. With the code shown here the exception will cause the program to terminate. In a serious program you should analyze the input string before attempting to convert it and should also catch any exceptions and process them appropriately.

Inputting Floating-Point Numbers

Types `float` and `double` can be handled in somewhat the same way as integers, but the conversion process is more complex. Here's how you read a number of type `double`:

```
public int getDouble() throws IOException
   {
   String s = getString();
   Double aDub = Double.valueOf(s);
   return aDub.doubleValue();
   }
```

The `String` is first converted to an object of type `Double` (uppercase *D*), which is a "wrapper" class for type `double`. A method of `Double` called `doubleValue()` then converts the object to type `double`.

For type `float`, there's an equivalent `Float` class, which has equivalent `valueOf()` and `floatValue()` methods.

Java Library Data Structures

The `java.util` package contains data structures, such as `Vector` (an extensible array), `Stack`, `Dictionary`, and `Hashtable`. In this book we'll usually ignore these built-in classes. We're interested in teaching fundamentals, not the details of a particular implementation. However, occasionally we'll find some of these structures useful. You must use the line

```
import java.util.*;
```

before you can use objects of these classes.

Although we don't focus on them, such class libraries, whether those that come with Java or others available from third-party developers, can offer a rich source of versatile, debugged storage classes. This book should equip you with the knowledge to know what sort of data structure you need and the fundamentals of how it works. Then you can decide whether you should write your own classes or use someone else's.

Summary

- A data structure is the organization of data in a computer's memory or in a disk file.

- The correct choice of data structure allows major improvements in program efficiency.

- Examples of data structures are arrays, stacks, and linked lists.

- An algorithm is a procedure for carrying out a particular task.

- In Java, an algorithm is usually implemented by a class method.

- Many of the data structures and algorithms described in this book are most often used to build databases.

- Some data structures are used as programmer's tools: They help execute an algorithm.

- Other data structures model real-world situations, such as telephone lines running between cities.

- A database is a unit of data storage composed of many similar records.

- A record often represents a real-world object, such as an employee or a car part.

- A record is divided into fields. Each field stores one characteristic of the object described by the record.

- A key is a field in a record that's used to carry out some operation on the data. For example, personnel records might be sorted by a LastName field.

- A database can be searched for all records whose key field has a certain value. This value is called a search key.

Questions

These questions are intended as a self-test for readers. Answers to the questions may be found in Appendix C.

1. In many data structures you can _____ a single record, _____ it, and _____ it.

2. Rearranging the contents of a data structure into a certain order is called _____ .

3. In a database, a field is

    a. a specific data item.

    b. a specific object.

    c. part of a record.

    d. part of an algorithm.

4. The field used when searching for a particular record is the _____ .

5. In object-oriented programming, an object

    a. is a class.

    b. may contain data and methods.

    c. is a program.

    d. may contain classes.

6. A class

    a. is a blueprint for many objects.

    b. represents a specific real-world object.

    c. will hold specific values in its fields.

    d. specifies the type of a method.

7. In Java, a class specification

    a. creates objects.

    b. requires the keyword new.

    c. creates references.

    d. none of the above.

8. When an object wants to do something, it uses a _____ .

9. In Java, accessing an object's methods requires the _____ operator.

10. In Java, boolean and byte are _____ .

(There are no experiments or programming projects for Chapter 1.)

2
Arrays

The array is the most commonly used data storage structure; it's built into most programming languages. Because arrays are so well known, they offer a convenient jumping-off place for introducing data structures and for seeing how object-oriented programming and data structures relate to one another. In this chapter we'll introduce arrays in Java and demonstrate a home-made array class.

We'll also examine a special kind of array, the ordered array, in which the data is stored in ascending (or descending) key order. This arrangement makes possible a fast way of searching for a data item: the binary search.

We'll start the chapter with a Java Workshop applet that shows insertion, searching, and deletion in an array. Then we'll show some sample Java code that carries out these same operations.

Later we'll examine ordered arrays, again starting with a Workshop applet. This applet will demonstrate a binary search. At the end of the chapter we'll talk about Big O notation, the most widely used measure of algorithm efficiency.

The Array Workshop Applet

Suppose you're coaching kids-league baseball, and you want to keep track of which players are present at the practice field. What you need is an attendance-monitoring program for your laptop—a program that maintains a database of the players who have shown up for practice. You can use a simple data structure to hold this data. There are several actions you would like to be able to perform:

- Insert a player into the data structure when the player arrives at the field.

- Check to see whether a particular player is present, by searching for the player's number in the structure.

- Delete a player from the data structure when that player goes home.

These three operations—insertion, searching, and deletion—will be the fundamental ones in most of the data storage structures we'll study in this book.

We'll often begin the discussion of a particular data structure by demonstrating it with a Workshop applet. This approach will give you a feeling for what the structure and its algorithms do, before we launch into a detailed explanation and demonstrate sample code. The Workshop applet called Array shows how an array can be used to implement insertion, searching, and deletion.

Now start up the Array Workshop applet, as described in Appendix A, "Running the Workshop Applets and Example Programs," with

```
C:\>appletviewer Array.html
```

Figure 2.1 shows the resulting array with 20 elements, 10 of which have data items in them. You can think of these items as representing your baseball players. Imagine that each player has been issued a team shirt with the player's number on the back. To make things visually interesting, the shirts come in a variety of colors. You can see each player's number and shirt color in the array.

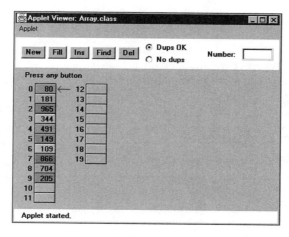

FIGURE 2.1   The Array Workshop applet.

This applet demonstrates the three fundamental procedures mentioned earlier:

- The Ins button inserts a new data item.
- The Find button searches for specified data item.
- The Del button deletes a specified data item.

Using the New button, you can create a new array of a size you specify. You can fill this array with as many data items as you want using the Fill button. Fill creates a set of items and randomly assigns them numbers and colors. The numbers are in the range 0 to 999. You can't create an array of more than 60 cells, and you can't, of course, fill more data items than there are array cells.

Also, when you create a new array, you'll need to decide whether duplicate items will be allowed; we'll return to this question in a moment. The default value is no duplicates, so the No Dups radio button is initially selected to indicate this setting.

Insertion

Start with the default arrangement of 20 cells and 10 data items, and the No Dups button selected. You insert a baseball player's number into the array when the player arrives at the practice field, having been dropped off by a parent. To insert a new item, press the Ins button once. You'll be prompted to enter the value of the item:

```
Enter key of item to insert
```

Type a number, say 678, into the text field in the upper-right corner of the applet. (Yes, it is hard to get three digits on the back of a kid's shirt.) Press Ins again and the applet will confirm your choice:

```
Will insert item with key 678
```

A final press of the button will cause a data item, consisting of this value and a random color, to appear in the first empty cell in the array. The prompt will say something like

```
Inserted item with key 678 at index 10
```

Each button press in a Workshop applet corresponds to a step that an algorithm carries out. The more steps required, the longer the algorithm takes. In the Array Workshop applet the insertion process is very fast, requiring only a single step. This is true because a new item is always inserted in the first vacant cell in the array, and the algorithm knows this location because it knows how many items are already in the array. The new item is simply inserted in the next available space. Searching and deletion, however, are not so fast.

In no-duplicates mode you're on your honor not to insert an item with the same key as an existing item. If you do, the applet displays an error message, but it won't prevent the insertion. The assumption is that you won't make this mistake.

Searching

To begin a search, click the Find button. You'll be prompted for the key number of the person you're looking for. Pick a number that appears on an item somewhere in the middle of the array. Type in the number and repeatedly press the Find button. At each button press, one step in the algorithm is carried out. You'll see the red arrow start at cell 0 and move methodically down the cells, examining a new one each time you press the button. The index number in the message

```
Checking next cell, index = 2
```

will change as you go along. When you reach the specified item, you'll see the message

```
Have found item with key 505
```

or whatever key value you typed in. Assuming duplicates are not allowed, the search will terminate as soon as an item with the specified key value is found.

If you have selected a key number that is not in the array, the applet will examine every occupied cell in the array before telling you that it can't find that item.

Notice that (again assuming duplicates are not allowed) the search algorithm must look through an average of half the data items to find a specified item. Items close to the beginning of the array will be found sooner, and those toward the end will be found later. If N is the number of items, the average number of steps needed to find an item is N/2. In the worst-case scenario, the specified item is in the last occupied cell, and N steps will be required to find it.

As we noted, the time an algorithm takes to execute is proportional to the number of steps, so searching takes much longer on the average (N/2 steps) than insertion (one step).

Deletion

To delete an item, you must first find it. After you type in the number of the item to be deleted, repeated button presses will cause the arrow to move, step by step, down the array until the item is located. The next button press deletes the item, and the cell becomes empty. (Strictly speaking, this step isn't necessary because we're going to copy over this cell anyway, but deleting the item makes it clearer what's happening.)

Implicit in the deletion algorithm is the assumption that *holes* are not allowed in the array. A hole is one or more empty cells that have filled cells above them (at higher index numbers). If holes are allowed, all the algorithms become more complicated because they must check to see whether a cell is empty before examining its contents. Also, the algorithms become less efficient because they must waste time looking at unoccupied cells. For these reasons, occupied cells must be arranged contiguously: no holes allowed.

Therefore, after locating the specified item and deleting it, the applet must shift the contents of each subsequent cell down one space to fill in the hole. Figure 2.2 shows an example.

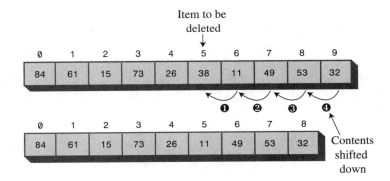

FIGURE 2.2   Deleting an item.

If the item in cell 5 (38, in Figure 2.2) is deleted, the item in 6 shifts into 5, the item in 7 shifts into 6, and so on to the last occupied cell. During the deletion process, when the item is located, the applet shifts down the contents of the higher-indexed cells as you continue to press the Del button.

A deletion requires (assuming no duplicates are allowed) searching through an average of N/2 elements and then moving the remaining elements (an average of N/2 moves) to fill up the resulting hole. This is N steps in all.

The Duplicates Issue

When you design a data storage structure, you need to decide whether items with duplicate keys will be allowed. If you're working with a personnel file and the key is an employee number, duplicates don't make much sense; there's no point in assigning the same number to two employees. On the other hand, if the key value is last names, then there's a distinct possibility several employees will have the same key value, so duplicates should be allowed.

Of course, for the baseball players, duplicate numbers should not be allowed. Keeping track of the players would be hard if more than one wore the same number.

The Array Workshop applet lets you select either option. When you use New to create a new array, you're prompted to specify both its size and whether duplicates are permitted. Use the radio buttons Dups OK or No Dups to make this selection.

If you're writing a data storage program in which duplicates are not allowed, you may need to guard against human error during an insertion by checking all the data items in the array to ensure that none of them already has the same key value as the item being inserted. This check is inefficient, however, and increases the number of steps required for an insertion from one to N. For this reason, our applet does not perform this check.

Searching with Duplicates

Allowing duplicates complicates the search algorithm, as we noted. Even if it finds a match, it must continue looking for possible additional matches until the last occupied cell. At least, this is one approach; you could also stop after the first match. How you proceed depends on whether the question is "Find me everyone with blue eyes" or "Find me someone with blue eyes."

When the Dups OK button is selected, the applet takes the first approach, finding all items matching the search key. This approach always requires N steps because the algorithm must go all the way to the last occupied cell.

Insertion with Duplicates

Insertion is the same with duplicates allowed as when they're not: A single step inserts the new item. But remember, if duplicates are not allowed, and there's a possibility the user will attempt to input the same key twice, you may need to check every existing item before doing an insertion.

Deletion with Duplicates

Deletion may be more complicated when duplicates are allowed, depending on exactly how "deletion" is defined. If it means to delete only the first item with a specified value, then, on the average, only N/2 comparisons and N/2 moves are necessary. This is the same as when no duplicates are allowed.

If, however, deletion means to delete *every* item with a specified key value, the same operation may require multiple deletions. Such an operation will require checking N cells and (probably) moving more than N/2 cells. The average depends on how the duplicates are distributed throughout the array.

The applet assumes this second meaning and deletes multiple items with the same key. This is complicated because each time an item is deleted, subsequent items must be shifted farther. For example, if three items are deleted, then items beyond the last deletion will need to be shifted three spaces. To see how this operation works, set the applet to Dups OK and insert three or four items with the same key. Then try deleting them.

Table 2.1 shows the average number of comparisons and moves for the three operations, first where no duplicates are allowed and then where they are allowed. N is the number of items in the array. Inserting a new item counts as one move.

TABLE 2.1   Duplicates OK Versus No Duplicates

	No Duplicates	Duplicates OK
Search	N/2 comparisons	N comparisons
Insertion	No comparisons, one move	No comparisons, one move
Deletion	N/2 comparisons, N/2 moves	N comparisons, more than N/2 moves

You can explore these possibilities with the Array Workshop applet.

The difference between N and N/2 is not usually considered very significant, except when you're fine-tuning a program. Of more importance, as we'll discuss toward the end of this chapter, is whether an operation takes one step, N steps, log(N) steps, or N^2 steps.

Not Too Swift

One of the significant things to notice when you're using the Array applet is the slow and methodical nature of the algorithms. With the exception of insertion, the algorithms involve stepping through some or all of the cells in the array. Different data structures offer much faster (but more complex) algorithms. We'll see one, the binary search on an ordered array, later in this chapter, and others throughout this book.

The Basics of Arrays in Java

The preceding section showed graphically the primary algorithms used for arrays. Now we'll see how to write programs to carry out these algorithms, but we first want to cover a few of the fundamentals of arrays in Java.

If you're a Java expert, you can skip ahead to the next section, but even C and C++ programmers should stick around. Arrays in Java use syntax similar to that in C and C++ (and not that different from other languages), but there are nevertheless some unique aspects to the Java approach.

Creating an Array

As we noted in Chapter 1, "Overview," there are two kinds of data in Java: primitive types (such as `int` and `double`) and objects. In many programming languages (even object-oriented ones such as C++), arrays are primitive types, but in Java they're treated as objects. Accordingly, you must use the `new` operator to create an array:

```
int[] intArray;            // defines a reference to an array
intArray = new int[100];   // creates the array, and
                           // sets intArray to refer to it
```

Or you can use the equivalent single-statement approach:

```
int[] intArray = new int[100];
```

The [] operator is the sign to the compiler we're naming an array object and not an ordinary variable. You can also use an alternative syntax for this operator, placing it after the name instead of the type:

```
int intArray[] = new int[100];  // alternative syntax
```

However, placing the [] after the `int` makes it clear that the [] is part of the type, not the name.

Because an array is an object, its name—`intArray` in the preceding code—is a reference to an array; it's not the array itself. The array is stored at an address elsewhere in memory, and `intArray` holds only this address.

Arrays have a `length` field, which you can use to find the size (the number of elements) of an array:

```
int arrayLength = intArray.length;   // find array size
```

As in most programming languages, you can't change the size of an array after it's been created.

Accessing Array Elements

Array elements are accessed using an index number in square brackets. This is similar to how other languages work:

```
temp = intArray[3];  // get contents of fourth element of array
intArray[7] = 66;    // insert 66 into the eighth cell
```

Remember that in Java, as in C and C++, the first element is numbered 0, so that the indices in an array of 10 elements run from 0 to 9.

If you use an index that's less than 0 or greater than the size of the array less 1, you'll get the Array Index Out of Bounds runtime error.

Initialization

Unless you specify otherwise, an array of integers is automatically initialized to 0 when it's created. Unlike C++, this is true even of arrays defined within a method (function). Say you create an array of objects like this:

```
autoData[] carArray = new autoData[4000];
```

Until the array elements are given explicit values, they contain the special null object. If you attempt to access an array element that contains null, you'll get the runtime error Null Pointer Assignment. The moral is to make sure you assign something to an element before attempting to access it.

You can initialize an array of a primitive type to something besides 0 using this syntax:

```
int[] intArray = { 0, 3, 6, 9, 12, 15, 18, 21, 24, 27 };
```

Perhaps surprisingly, this single statement takes the place of both the reference declaration and the use of new to create the array. The numbers within the curly brackets are called the *initialization list*. The size of the array is determined by the number of values in this list.

An Array Example

Let's look at some example programs that show how an array can be used. We'll start with an old-fashioned procedural version and then show the equivalent object-oriented approach. Listing 2.1 shows the old-fashioned version, called array.java.

LISTING 2.1   The array.java Program

```
// array.java
// demonstrates Java arrays
// to run this program: C>java arrayApp
/////////////////////////////////////////////////////////////////
class ArrayApp
   {
   public static void main(String[] args)
      {
      long[] arr;                  // reference to array
      arr = new long[100];         // make array
      int nElems = 0;              // number of items
```

LISTING 2.1   Continued

```
    int j;                      // loop counter
    long searchKey;             // key of item to search for
//------------------------------------------------------------
    arr[0] = 77;                // insert 10 items
    arr[1] = 99;
    arr[2] = 44;
    arr[3] = 55;
    arr[4] = 22;
    arr[5] = 88;
    arr[6] = 11;
    arr[7] = 00;
    arr[8] = 66;
    arr[9] = 33;
    nElems = 10;                // now 10 items in array
//------------------------------------------------------------
    for(j=0; j<nElems; j++)     // display items
       System.out.print(arr[j] + " ");
    System.out.println("");
//------------------------------------------------------------
    searchKey = 66;             // find item with key 66
    for(j=0; j<nElems; j++)         // for each element,
       if(arr[j] == searchKey)      // found item?
          break;                    // yes, exit before end
    if(j == nElems)                 // at the end?
       System.out.println("Can't find " + searchKey); // yes
    else
       System.out.println("Found " + searchKey);      // no
//------------------------------------------------------------
    searchKey = 55;             // delete item with key 55
    for(j=0; j<nElems; j++)         // look for it
    if(arr[j] == searchKey)
       break;
    for(int k=j; k<nElems-1; k++)      // move higher ones down
       arr[k] = arr[k+1];
    nElems--;                       // decrement size
//------------------------------------------------------------
    for(j=0; j<nElems; j++)     // display items
       System.out.print( arr[j] + " ");
    System.out.println("");
    }  // end main()
  }  // end class ArrayApp
```

In this program, we create an array called `arr`, place 10 data items (kids' numbers) in it, search for the item with value 66 (the shortstop, Louisa), display all the items, remove the item with value 55 (Freddy, who had a dentist appointment), and then display the remaining 9 items. The output of the program looks like this:

```
77 99 44 55 22 88 11 0 66 33
Found 66
77 99 44 22 88 11 0 66 33
```

The data we're storing in this array is type `long`. We use `long` to make it clearer that this is data; type `int` is used for index values. We've chosen a primitive type to simplify the coding. Generally, the items stored in a data structure consist of several fields, so they are represented by objects rather than primitive types. We'll see such an example toward the end of this chapter.

Insertion
Inserting an item into the array is easy; we use the normal array syntax:

```
arr[0] = 77;
```

We also keep track of how many items we've inserted into the array with the `nElems` variable.

Searching
The `searchKey` variable holds the value we're looking for. To search for an item, we step through the array, comparing `searchKey` with each element. If the loop variable `j` reaches the last occupied cell with no match being found, the value isn't in the array. Appropriate messages are displayed: `Found 66` or `Can't find 27`.

Deletion
Deletion begins with a search for the specified item. For simplicity, we assume (perhaps rashly) that the item is present. When we find it, we move all the items with higher index values down one element to fill in the "hole" left by the deleted element, and we decrement `nElems`. In a real program, we would also take appropriate action if the item to be deleted could not be found.

Display
Displaying all the elements is straightforward: We step through the array, accessing each one with `arr[j]` and displaying it.

Program Organization
The organization of `array.java` leaves something to be desired. The program has only one class, `ArrayApp`, and this class has only one method, `main()`. `array.java` is essentially an old-fashioned procedural program. Let's see if we can make it easier to understand (among other benefits) by making it more object oriented.

We're going to provide a gradual introduction to an object-oriented approach, using two steps. In the first, we'll separate the data storage structure (the array) from the rest of the program. The remaining part of the program will become a *user* of the structure. In the second step, we'll improve the communication between the storage structure and its user.

Dividing a Program into Classes

The array.java program in Listing 2.1 essentially consists of one big method. We can reap many benefits by dividing the program into classes. What classes? The data storage structure itself is one candidate, and the part of the program that uses this data structure is another. By dividing the program into these two classes, we can clarify the functionality of the program, making it easier to design and understand (and in real programs to modify and maintain).

In array.java we used an array as a data storage structure, but we treated it simply as a language element. Now we'll encapsulate the array in a class, called LowArray. We'll also provide class methods by which objects of other classes (the LowArrayApp class in this case) can access the array. These methods allow communication between LowArray and LowArrayApp.

Our first design of the LowArray class won't be entirely successful, but it will demonstrate the need for a better approach. The lowArray.java program in Listing 2.2 shows how it looks.

LISTING 2.2   The lowArray.java Program

```
// lowArray.java
// demonstrates array class with low-level interface
// to run this program: C>java LowArrayApp
//////////////////////////////////////////////////////////////
class LowArray
    {
    private long[] a;                // ref to array a
//-------------------------------------------------------------
    public LowArray(int size)        // constructor
        { a = new long[size]; }      // create array
//-------------------------------------------------------------
public void setElem(int index, long value)    // set value
        { a[index] = value; }
//-------------------------------------------------------------
    public long getElem(int index)            // get value
        { return a[index]; }
//-------------------------------------------------------------
```

LISTING 2.2   Continued

```
    }  // end class LowArray
/////////////////////////////////////////////////////////////
class LowArrayApp
    {
    public static void main(String[] args)
        {
        LowArray arr;                    // reference
        arr = new LowArray(100);         // create LowArray object
        int nElems = 0;                  // number of items in array
        int j;                           // loop variable

        arr.setElem(0, 77);              // insert 10 items
        arr.setElem(1, 99);
        arr.setElem(2, 44);
        arr.setElem(3, 55);
        arr.setElem(4, 22);
        arr.setElem(5, 88);
        arr.setElem(6, 11);
        arr.setElem(7, 00);
        arr.setElem(8, 66);
        arr.setElem(9, 33);
        nElems = 10;                     // now 10 items in array

        for(j=0; j<nElems; j++)          // display items
            System.out.print(arr.getElem(j) + " ");
        System.out.println("");

        int searchKey = 26;              // search for data item
        for(j=0; j<nElems; j++)              // for each element,
            if(arr.getElem(j) == searchKey) // found item?
                break;
        if(j == nElems)                      // no
            System.out.println("Can't find " + searchKey);
        else                             // yes
            System.out.println("Found " + searchKey);

                                     // delete value 55
        for(j=0; j<nElems; j++)              // look for it
        if(arr.getElem(j) == 55)
            break;
        for(int k=j; k<nElems; k++)          // higher ones down
```

LISTING 2.2   Continued

```
        arr.setElem(k, arr.getElem(k+1) );
    nElems--;                          // decrement size

    for(j=0; j<nElems; j++)      // display items
        System.out.print( arr.getElem(j) + " ");
    System.out.println("");
    }  // end main()
 }  // end class LowArrayApp
//////////////////////////////////////////////////////////////////
```

The output from the lowArray.java program is similar to that from array.java, except that we try to find a non-existent key value (26) before deleting the item with the key value 55:

```
77 99 44 55 22 88 11 0 66 33
Can't find 26
77 99 44 22 88 11 0 66 33
```

Classes LowArray and LowArrayApp

In lowArray.java, we essentially wrap the class LowArray around an ordinary Java array. The array is hidden from the outside world inside the class; it's private, so only LowArray class methods can access it. There are three LowArray methods: setElem() and getElem(), which insert and retrieve an element, respectively; and a constructor, which creates an empty array of a specified size.

Another class, LowArrayApp, creates an object of the LowArray class and uses it to store and manipulate data. Think of LowArray as a tool and LowArrayApp as a user of the tool. We've divided the program into two classes with clearly defined roles. This is a valuable first step in making a program object oriented.

A class used to store data objects, as is LowArray in the lowArray.java program, is sometimes called a *container class*. Typically, a container class not only stores the data but also provides methods for accessing the data and perhaps also sorting it and performing other complex actions on it.

Class Interfaces

We've seen how a program can be divided into separate classes. How do these classes interact with each other? Communication between classes and the division of responsibility between them are important aspects of object-oriented programming.

This point is especially true when a class may have many different users. Typically, a class can be used over and over by different users (or the same user) for different purposes. For example, someone might use the LowArray class in some other program to store the serial numbers of his traveler's checks. The class can handle this task just as well as it can store the numbers of baseball players.

If a class is used by many different programmers, the class should be designed so that it's easy to use. The way that a class user relates to the class is called the class *interface*. Because class fields are typically private, when we talk about the interface, we usually mean the class methods—what they do and what their arguments are. By calling these methods, a class user interacts with an object of the class. One of the important advantages conferred by object-oriented programming is that a class interface can be designed to be as convenient and efficient as possible. Figure 2.3 is a fanciful interpretation of the LowArray interface.

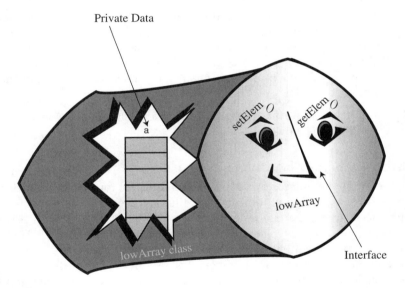

FIGURE 2.3   The LowArray interface.

Not So Convenient

The interface to the LowArray class in lowArray.java is not particularly convenient. The methods setElem() and getElem() operate on a low conceptual level, performing exactly the same tasks as the [ ] operator in an ordinary Java array. The class user, represented by the main() method in the LowArrayApp class, ends up having to carry out the same low-level operations it did in the non-class version of an array in the array.java program. The only difference was that it related to setElem() and getElem() instead of the [ ] operator. It's not clear that this approach is an improvement.

Also notice that there's no convenient way to display the contents of the array. Somewhat crudely, the LowArrayApp class simply uses a for loop and the getElem() method for this purpose. We could avoid repeated code by writing a separate method for LowArrayApp that it could call to display the array contents, but is it really the responsibility of the LowArrayApp class to provide this method?

Thus, lowArray.java demonstrates how to divide a program into classes, but it really doesn't buy us too much in practical terms. Let's see how to redistribute responsibilities between the classes to obtain more of the advantages of OOP.

Who's Responsible for What?

In the lowArray.java program, the main() routine in the LowArrayApp class, the user of the data storage structure, must keep track of the indices to the array. For some users of an array, who need random access to array elements and don't mind keeping track of the index numbers, this arrangement might make sense. For example, sorting an array, as we'll see in the next chapter, can make efficient use of this direct hands-on approach.

In a typical program, however, the user of the data storage device won't find access to the array indices to be helpful or relevant.

The highArray.java Example

Out next example program shows an improved interface for the storage structure class, called HighArray. Using this interface, the class user (the HighArrayApp class) no longer needs to think about index numbers. The setElem() and getElem() methods are gone; they're replaced by insert(), find(), and delete(). These new methods don't require an index number as an argument because the class takes responsibility for handling index numbers. The user of the class (HighArrayApp) is free to concentrate on the *what* instead of the *how*—what's going to be inserted, deleted, and accessed, instead of exactly how these activities are carried out.

Figure 2.4 shows the HighArray interface, and Listing 2.3 shows the highArray.java program.

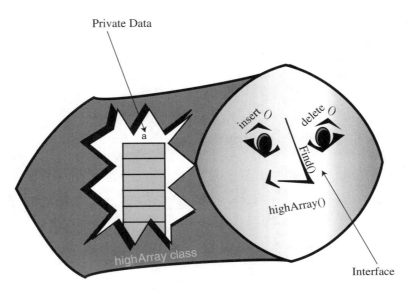

FIGURE 2.4   The HighArray interface.

LISTING 2.3   The highArray.java Program

```
// highArray.java
// demonstrates array class with high-level interface
// to run this program: C>java HighArrayApp
//////////////////////////////////////////////////////////////////
class HighArray
   {
   private long[] a;                // ref to array a
   private int nElems;              // number of data items
   //--------------------------------------------------------------
   public HighArray(int max)        // constructor
      {
      a = new long[max];              // create the array
      nElems = 0;                     // no items yet
      }
   //--------------------------------------------------------------
   public boolean find(long searchKey)
      {                              // find specified value
      int j;
      for(j=0; j<nElems; j++)          // for each element,
         if(a[j] == searchKey)         // found item?
```

LISTING 2.3   Continued

```
            break;                       // exit loop before end
      if(j == nElems)                    // gone to end?
         return false;                   // yes, can't find it
      else
         return true;                    // no, found it
      }  // end find()
//------------------------------------------------------------
   public void insert(long value)    // put element into array
      {
      a[nElems] = value;               // insert it
      nElems++;                        // increment size
      }
//------------------------------------------------------------
   public boolean delete(long value)
      {
      int j;
      for(j=0; j<nElems; j++)          // look for it
         if( value == a[j] )
            break;
      if(j==nElems)                    // can't find it
         return false;
      else                             // found it
         {
         for(int k=j; k<nElems; k++) // move higher ones down
            a[k] = a[k+1];
         nElems--;                     // decrement size
         return true;
         }
      }  // end delete()
//------------------------------------------------------------
   public void display()             // displays array contents
      {
      for(int j=0; j<nElems; j++)       // for each element,
         System.out.print(a[j] + " ");  // display it
      System.out.println("");
      }
//------------------------------------------------------------
   }  // end class HighArray
////////////////////////////////////////////////////////////////
class HighArrayApp
   {
   public static void main(String[] args)
```

LISTING 2.3   Continued

```
{
int maxSize = 100;              // array size
HighArray arr;                  // reference to array
arr = new HighArray(maxSize);   // create the array

arr.insert(77);                 // insert 10 items
arr.insert(99);
arr.insert(44);
arr.insert(55);
arr.insert(22);
arr.insert(88);
arr.insert(11);
arr.insert(00);
arr.insert(66);
arr.insert(33);

arr.display();                  // display items

int searchKey = 35;             // search for item
if( arr.find(searchKey) )
   System.out.println("Found " + searchKey);
else
   System.out.println("Can't find " + searchKey);

arr.delete(00);                 // delete 3 items
arr.delete(55);
arr.delete(99);

arr.display();                  // display items again
} // end main()
} // end class HighArrayApp
/////////////////////////////////////////////////////////////////
```

The HighArray class is now wrapped around the array. In main(), we create an array of this class and carry out almost the same operations as in the lowArray.java program: We insert 10 items, search for an item—one that isn't there—and display the array contents. Because deleting is so easy, we delete 3 items (0, 55, and 99) instead of 1 and finally display the contents again. Here's the output:

```
77 99 44 55 22 88 11 0 66 33
Can't find 35
77 44 22 88 11 66 33
```

Notice how short and simple `main()` is. The details that had to be handled by `main()` in `lowArray.java` are now handled by `HighArray` class methods.

In the `HighArray` class, the `find()` method looks through the array for the item whose key value was passed to it as an argument. It returns `true` or `false`, depending on whether it finds the item.

The `insert()` method places a new data item in the next available space in the array. A field called `nElems` keeps track of the number of array cells that are actually filled with data items. The `main()` method no longer needs to worry about how many items are in the array.

The `delete()` method searches for the element whose key value was passed to it as an argument and, when it finds that element, shifts all the elements in higher index cells down one cell, thus writing over the deleted value; it then decrements `nElems`.

We've also included a `display()` method, which displays all the values stored in the array.

The User's Life Made Easier

In `lowArray.java` (Listing 2.2), the code in `main()` to search for an item required eight lines; in `highArray.java`, it requires only one. The class user, the `HighArrayApp` class, need not worry about index numbers or any other array details. Amazingly, the class user doesn't even need to know *what kind of data structure* the `HighArray` class is using to store the data. The structure is hidden behind the interface. In fact, in the next section, we'll see the same interface used with a somewhat different data structure.

Abstraction

The process of separating the *how* from the *what*—how an operation is performed inside a class, as opposed to what's visible to the class user—is called *abstraction*. Abstraction is an important aspect of software engineering. By abstracting class functionality, we make it easier to design a program because we don't need to think about implementation details at too early a stage in the design process.

The Ordered Workshop Applet

Imagine an array in which the data items are arranged in order of ascending key values—that is, with the smallest value at index 0, and each cell holding a value larger than the cell below. Such an array is called an *ordered array*.

When we insert an item into this array, the correct location must be found for the insertion: just above a smaller value and just below a larger one. Then all the larger values must be moved up to make room.

Why would we want to arrange data in order? One advantage is that we can speed up search times dramatically using a *binary search*.

Start the Ordered Workshop applet, using the procedure described in Chapter 1. You'll see an array; it's similar to the one in the Array Workshop applet, but the data is ordered. Figure 2.5 shows this applet.

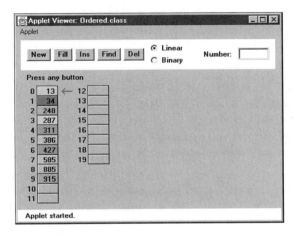

FIGURE 2.5   The Ordered Workshop applet.

In the ordered array we've chosen not to allow duplicates. As we saw earlier, this decision speeds up searching somewhat but slows down insertion.

Linear Search

Two search algorithms are available for the Ordered Workshop applet: linear and binary. Linear search is the default. Linear searches operate in much the same way as the searches in the unordered array in the Array applet: The red arrow steps along, looking for a match. The difference is that in the ordered array, the search quits if an item with a larger key is found.

Try out a linear search. Make sure the Linear radio button is selected. Then use the Find button to search for a non-existent value that, if it were present, would fit somewhere in the middle of the array. In Figure 2.5, this number might be 400. You'll see that the search terminates when the first item larger than 400 is reached; it's 427 in the figure. The algorithm knows there's no point looking further.

Try out the Ins and Del buttons as well. Use Ins to insert an item with a key value that will go somewhere in the middle of the existing items. You'll see that insertion requires moving all the items with key values larger than the item being inserted.

Use the Del button to delete an item from the middle of the array. Deletion works much the same as it did in the Array applet, shifting items with higher index numbers down to fill in the hole left by the deletion. In the ordered array, however, the deletion algorithm can quit partway through if it doesn't find the item, just as the search routine can.

Binary Search

The payoff for using an ordered array comes when we use a binary search. This kind of search is much faster than a linear search, especially for large arrays.

The Guess-a-Number Game

Binary search uses the same approach you did as a kid (if you were smart) to guess a number in the well-known children's guessing game. In this game, a friend asks you to guess a number she's thinking of between 1 and 100. When you guess a number, she'll tell you one of three things: Your guess is larger than the number she's thinking of, it's smaller, or you guessed correctly.

To find the number in the fewest guesses, you should always start by guessing 50. If your friend says your guess is too low, you deduce the number is between 51 and 100, so your next guess should be 75 (halfway between 51 and 100). If she says it's too high, you deduce the number is between 1 and 49, so your next guess should be 25.

Each guess allows you to divide the range of possible values in half. Finally, the range is only one number long, and that's the answer.

Notice how few guesses are required to find the number. If you used a linear search, guessing first 1, then 2, then 3, and so on, finding the number would take you, on the average, 50 guesses. In a binary search each guess divides the range of possible values in half, so the number of guesses required is far fewer. Table 2.2 shows a game session when the number to be guessed is 33.

TABLE 2.2   Guessing a Number

Step Number	Number Guessed	Result	Range of Possible Values
0			1–100
1	50	Too high	1–49
2	25	Too low	26–49
3	37	Too high	26–36
4	31	Too low	32–36
5	34	Too high	32–33
6	32	Too low	33–33
7	33	Correct	

The correct number is identified in only seven guesses. This is the maximum. You might get lucky and guess the number before you've worked your way all the way down to a range of one. This would happen if the number to be guessed was 50, for example, or 34.

Binary Search in the Ordered Workshop Applet

To perform a binary search with the Ordered Workshop applet, you must use the New button to create a new array. After the first press, you'll be asked to specify the size of the array (maximum 60) and which kind of searching scheme you want: linear or binary. Choose binary by clicking the Binary radio button. After the array is created, use the Fill button to fill it with data items. When prompted, type the amount (not more than the size of the array). A few more presses fills in all the items.

When the array is filled, pick one of the values in the array and see how you can use the Find button to locate it. After a few preliminary presses, you'll see the red arrow pointing to the algorithm's current guess, and you'll see the range shown by a vertical blue line adjacent to the appropriate cells. Figure 2.6 depicts the situation when the range is the entire array.

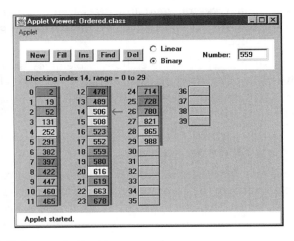

FIGURE 2.6   Initial range in the binary search.

At each press of the Find button, the range is halved and a new guess is chosen in the middle of the range. Figure 2.7 shows the next step in the process.

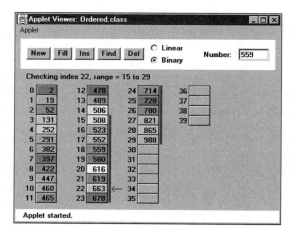

FIGURE 2.7   Range in step 2 of the binary search.

Even with a maximum array size of 60 items, a half-dozen button presses suffices to locate any item.

Try using the binary search with different array sizes. Can you figure out how many steps are necessary before you run the applet? We'll return to this question in the last section of this chapter.

Notice that the insertion and deletion operations also employ the binary search (when it's selected). The place where an item should be inserted is found with a binary search, as is an item to be deleted. In this applet, items with duplicate keys are not permitted.

Java Code for an Ordered Array

Let's examine some Java code that implements an ordered array. We'll use the OrdArray class to encapsulate the array and its algorithms. The heart of this class is the find() method, which uses a binary search to locate a specified data item. We'll examine this method in detail before showing the complete program.

Binary Search with the find() Method

The find() method searches for a specified item by repeatedly dividing in half the range of array elements to be considered. The method looks like this:

```
public int find(long searchKey)
   {
   int lowerBound = 0;
   int upperBound = nElems-1;
```

```
    int curIn;

while(true)
   {
   curIn = (lowerBound + upperBound ) / 2;
   if(a[curIn]==searchKey)
      return curIn;              // found it
   else if(lowerBound > upperBound)
      return nElems;             // can't find it
   else                          // divide range
      {
      if(a[curIn] < searchKey)
         lowerBound = curIn + 1; // it's in upper half
      else
         upperBound = curIn - 1; // it's in lower half
      } // end else divide range
   } // end while
} // end find()
```

The method begins by setting the lowerBound and upperBound variables to the first and last occupied cells in the array. Setting these variables specifies the range where the item we're looking for, searchKey, may be found. Then, within the while loop, the current index, curIn, is set to the middle of this range.

If we're lucky, curIn may already be pointing to the desired item, so we first check if this is true. If it is, we've found the item, so we return with its index, curIn.

Each time through the loop we divide the range in half. Eventually, the range will get so small that it can't be divided any more. We check for this in the next statement: If lowerBound is greater than upperBound, the range has ceased to exist. (When lowerBound equals upperBound, the range is one and we need one more pass through the loop.) We can't continue the search without a valid range, but we haven't found the desired item, so we return nElems, the total number of items. This isn't a valid index because the last filled cell in the array is nElems-1. The class user interprets this value to mean that the item wasn't found.

If curIn is not pointing at the desired item, and the range is still big enough, we're ready to divide the range in half. We compare the value at the current index, a[curIn], which is in the middle of the range, with the value to be found, searchKey.

If searchKey is larger, we know we should look in the upper half of the range. Accordingly, we move lowerBound up to curIn. Actually, we move it one cell beyond curIn because we've already checked curIn itself at the beginning of the loop.

If searchKey is smaller than a[curIn], we know we should look in the lower half of the range. So we move upperBound down to one cell below curIn. Figure 2.8 shows how the range is altered in these two situations.

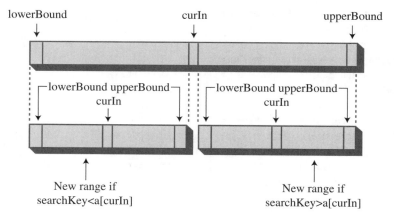

FIGURE 2.8   Dividing the range in a binary search.

The OrdArray Class

In general, the orderedArray.java program is similar to highArray.java (Listing 2.3). The main difference is that find() uses a binary search, as we've seen.

We could have used a binary search to locate the position where a new item will be inserted. This operation involves a variation on the find() routine, but for simplicity we retain the linear search in insert(). The speed penalty may not be important because, as we've seen, an average of half the items must be moved anyway when an insertion is performed, so insertion will not be very fast even if we locate the item with a binary search. However, for the last ounce of speed, you could change the initial part of insert() to a binary search (as is done in the Ordered Workshop applet). Similarly, the delete() method could call find() to figure out the location of the item to be deleted.

The OrdArray class includes a new size() method, which returns the number of data items currently in the array. This information is helpful for the class user, main(), when it calls find(). If find() returns nElems, which main() can discover with size(), then the search was unsuccessful. Listing 2.4 shows the complete listing for the orderedArray.java program.

LISTING 2.4   The orderedArray.java Program
_____

```java
// orderedArray.java
// demonstrates ordered array class
// to run this program: C>java OrderedApp
////////////////////////////////////////////////////////////////
class OrdArray
   {
   private long[] a;                   // ref to array a
   private int nElems;                 // number of data items
   //-----------------------------------------------------------
   public OrdArray(int max)            // constructor
      {
      a = new long[max];               // create array
      nElems = 0;
      }
   //-----------------------------------------------------------
   public int size()
      { return nElems; }
   //-----------------------------------------------------------
   public int find(long searchKey)
      {
      int lowerBound = 0;
      int upperBound = nElems-1;
      int curIn;

      while(true)
         {
         curIn = (lowerBound + upperBound ) / 2;
         if(a[curIn]==searchKey)
            return curIn;              // found it
         else if(lowerBound > upperBound)
            return nElems;             // can't find it
         else                          // divide range
            {
            if(a[curIn] < searchKey)
               lowerBound = curIn + 1; // it's in upper half
            else
               upperBound = curIn - 1; // it's in lower half
            }  // end else divide range
         }  // end while
      }  // end find()
   //-----------------------------------------------------------
```

LISTING 2.4   Continued

```
public void insert(long value)    // put element into array
   {
   int j;
   for(j=0; j<nElems; j++)        // find where it goes
      if(a[j] > value)            // (linear search)
         break;
   for(int k=nElems; k>j; k--)    // move bigger ones up
      a[k] = a[k-1];
   a[j] = value;                  // insert it
   nElems++;                      // increment size
   }  // end insert()
//--------------------------------------------------------------
public boolean delete(long value)
   {
   int j = find(value);
   if(j==nElems)                  // can't find it
      return false;
   else                           // found it
      {
      for(int k=j; k<nElems; k++) // move bigger ones down
         a[k] = a[k+1];
      nElems--;                   // decrement size
      return true;
      }
   }  // end delete()
//--------------------------------------------------------------
public void display()             // displays array contents
   {
   for(int j=0; j<nElems; j++)    // for each element,
      System.out.print(a[j] + " "); // display it
   System.out.println("");
   }
//--------------------------------------------------------------
   }  // end class OrdArray
////////////////////////////////////////////////////////////////
class OrderedApp
   {
   public static void main(String[] args)
      {
      int maxSize = 100;          // array size
      OrdArray arr;               // reference to array
```

LISTING 2.4   Continued

```
    arr = new OrdArray(maxSize);    // create the array

    arr.insert(77);                 // insert 10 items
    arr.insert(99);
    arr.insert(44);
    arr.insert(55);
    arr.insert(22);
    arr.insert(88);
    arr.insert(11);
    arr.insert(00);
    arr.insert(66);
    arr.insert(33);

    int searchKey = 55;             // search for item
    if( arr.find(searchKey) != arr.size() )
        System.out.println("Found " + searchKey);
    else
        System.out.println("Can't find " + searchKey);

    arr.display();                  // display items

    arr.delete(00);                 // delete 3 items
    arr.delete(55);
    arr.delete(99);

    arr.display();                  // display items again
    }  // end main()
  }  // end class OrderedApp
//////////////////////////////////////////////////////////////////
```

Advantages of Ordered Arrays

What have we gained by using an ordered array? The major advantage is that search times are much faster than in an unordered array. The disadvantage is that insertion takes longer because all the data items with a higher key value must be moved up to make room. Deletions are slow in both ordered and unordered arrays because items must be moved down to fill the hole left by the deleted item.

Ordered arrays are therefore useful in situations in which searches are frequent, but insertions and deletions are not. An ordered array might be appropriate for a database of company employees, for example. Hiring new employees and laying off existing ones would probably be infrequent occurrences compared with accessing an existing employee's record for information, or updating it to reflect changes in salary, address, and so on.

A retail store inventory, on the other hand, would not be a good candidate for an ordered array because the frequent insertions and deletions, as items arrived in the store and were sold, would run slowly.

Logarithms

In this section we'll explain how logarithms are used to calculate the number of steps necessary in a binary search. If you're a math major, you can probably skip this section. If math makes you break out in a rash, you can also skip it, except for taking a long hard look at Table 2.3.

We've seen that a binary search provides a significant speed increase over a linear search. In the number-guessing game, with a range from 1 to 100, a maximum of seven guesses is needed to identify any number using a binary search; just as in an array of 100 records, seven comparisons are needed to find a record with a specified key value. How about other ranges? Table 2.3 shows some representative ranges and the number of comparisons needed for a binary search.

TABLE 2.3   Comparisons Needed in Binary Search

Range	Comparisons Needed
10	4
100	7
1,000	10
10,000	14
100,000	17
1,000,000	20
10,000,000	24
100,000,000	27
1,000,000,000	30

Notice the differences between binary search times and linear search times. For very small numbers of items, the difference isn't dramatic. Searching 10 items would take an average of five comparisons with a linear search (N/2) and a maximum of four comparisons with a binary search. But the more items there are, the bigger the difference. With 100 items, there are 50 comparisons in a linear search, but only 7 in a binary search. For 1,000 items, the numbers are 500 versus 10, and for 1,000,000 items, they're 500,000 versus 20. We can conclude that for all but very small arrays, the binary search is greatly superior.

The Equation

You can verify the results of Table 2.3 by repeatedly dividing a range (from the first column) in half until it's too small to divide further. The number of divisions this process requires is the number of comparisons shown in the second column.

Repeatedly dividing the range by two is an algorithmic approach to finding the number of comparisons. You might wonder if you could also find the number using a simple equation. Of course, there is such an equation, and it's worth exploring here because it pops up from time to time in the study of data structures. This formula involves logarithms. (Don't panic yet.)

The numbers in Table 2.3 leave out some interesting data. They don't answer such questions as, What is the exact size of the maximum range that can be searched in five steps? To solve this problem, we must create a similar table, but one that starts at the beginning, with a range of one, and works up from there by multiplying the range by two each time. Table 2.4 shows how this looks for the first seven steps.

TABLE 2.4   Powers of Two

Step s, same as $\log_2(r)$	Range r	Range Expressed as Power of 2 (2^s)
0	1	2^0
1	2	2^1
2	4	2^2
3	8	2^3
4	16	2^4
5	32	2^5
6	64	2^6
7	128	2^7
8	256	2^8
9	512	2^9
10	1024	2^{10}

For our original problem with a range of 100, we can see that 6 steps don't produce a range quite big enough (64), while 7 steps cover it handily (128). Thus, the 7 steps that are shown for 100 items in Table 2.3 are correct, as are the 10 steps for a range of 1000.

Doubling the range each time creates a series that's the same as raising two to a power, as shown in the third column of Table 2.4. We can express this power as a formula. If s represents steps (the number of times you multiply by two—that is, the power to which two is raised) and r represents the range, then the equation is

$r = 2^s$

If you know s, the number of steps, this tells you r, the range. For example, if s is 6, the range is 2^6, or 64.

The Opposite of Raising Two to a Power

Our original question was the opposite of the one just described: Given the range, we want to know how many comparisons are required to complete a search. That is, given r, we want an equation that gives us s.

The inverse of raising something to a power is called a logarithm. Here's the formula we want, expressed with a logarithm:

$$s = \log_2(r)$$

This equation says that the number of steps (comparisons) is equal to the logarithm to the base 2 of the range. What's a logarithm? The base 2 logarithm of a number r is the number of times you must multiply two by itself to get r. In Table 2.4, we show that the numbers in the first column, s, are equal to $\log_2(r)$.

How do you find the logarithm of a number without doing a lot of dividing? Pocket calculators and most computer languages have a log function. It is usually log to the base 10, but you can convert easily to base 2 by multiplying by 3.322. For example, $\log_{10}(100) = 2$, so $\log_2(100) = 2$ times 3.322, or 6.644. Rounded up to the whole number 7, this is what appears in the column to the right of 100 in Table 2.4.

In any case, the point here isn't to calculate logarithms. It's more important to understand the relationship between a number and its logarithm. Look again at Table 2.3, which compares the number of items and the number of steps needed to find a particular item. Every time you multiply the number of items (the range) by a factor of 10, you add only three or four steps (actually 3.322, before rounding off to whole numbers) to the number needed to find a particular element. This is true because, as a number grows larger, its logarithm doesn't grow nearly as fast. We'll compare this logarithmic growth rate with that of other mathematical functions when we talk about Big O notation later in this chapter.

Storing Objects

In the Java examples we've shown so far, we've stored primitive variables of type long in our data structures. Storing such variables simplifies the program examples, but it's not representative of how you use data storage structures in the real world. Usually, the data items (records) you want to store are combinations of many fields. For a personnel record, you would store last name, first name, age, Social Security number, and so forth. For a stamp collection, you would store the name of the country that issued the stamp, its catalog number, condition, current value, and so on.

In our next Java example, we'll show how objects, rather than variables of primitive types, can be stored.

The Person Class

In Java, a data record is usually represented by a class object. Let's examine a typical class used for storing personnel data. Here's the code for the Person class:

```
class Person
   {
   private String lastName;
   private String firstName;
   private int age;
   //-------------------------------------------------------------
   public Person(String last, String first, int a)
      {                                 // constructor
      lastName = last;
      firstName = first;
      age = a;
      }
   //-------------------------------------------------------------
   public void displayPerson()
      {
      System.out.print("   Last name: " + lastName);
      System.out.print(", First name: " + firstName);
      System.out.println(", Age: " + age);
      }
   //-------------------------------------------------------------
   public String getLast()           // get last name
      { return lastName; }
   }  // end class Person
```

We show only three variables in this class, for a person's last name, first name, and age. Of course, records for most applications would contain many additional fields.

A constructor enables a new Person object to be created and its fields initialized. The displayPerson() method displays a Person object's data, and the getLast() method returns the Person's last name; this is the key field used for searches.

The classDataArray.java Program

The program that makes use of the Person class is similar to the highArray.java program (Listing 2.3) that stored items of type long. Only a few changes are necessary to adapt that program to handle Person objects. Here are the major changes:

- The type of the array a is changed to Person.

- The key field (the last name) is now a String object, so comparisons require the equals() method rather than the == operator. The getLast() method of Person obtains the last name of a Person object, and equals() does the comparison:

```
if( a[j].getLast().equals(searchName) )   // found item?
```

- The insert() method creates a new Person object and inserts it in the array, instead of inserting a long value.

The main() method has been modified slightly, mostly to handle the increased quantity of output. We still insert 10 items, display them, search for 1 item, delete 3 items, and display them all again. Listing 2.5 shows the complete classDataArray.java program.

LISTING 2.5   The classDataArray.java Program

```
// classDataArray.java
// data items as class objects
// to run this program: C>java ClassDataApp
//////////////////////////////////////////////////////////////////
class Person
   {
   private String lastName;
   private String firstName;
   private int age;
//--------------------------------------------------------------
   public Person(String last, String first, int a)
      {                                    // constructor
      lastName = last;
      firstName = first;
      age = a;
      }
//--------------------------------------------------------------
   public void displayPerson()
      {
      System.out.print("   Last name: " + lastName);
      System.out.print(", First name: " + firstName);
      System.out.println(", Age: " + age);
      }
//--------------------------------------------------------------
```

LISTING 2.5   Continued

```
   public String getLast()          // get last name
      { return lastName; }
   }  // end class Person
/////////////////////////////////////////////////////////////////
class ClassDataArray
   {
   private Person[] a;               // reference to array
   private int nElems;               // number of data items

   public ClassDataArray(int max)    // constructor
      {
      a = new Person[max];              // create the array
      nElems = 0;                       // no items yet
      }
//-------------------------------------------------------------
   public Person find(String searchName)
      {                                 // find specified value
      int j;
      for(j=0; j<nElems; j++)           // for each element,
         if( a[j].getLast().equals(searchName) )  // found item?
            break;                      // exit loop before end
      if(j == nElems)                   // gone to end?
         return null;                   // yes, can't find it
      else
         return a[j];                   // no, found it
      }  // end find()
//-------------------------------------------------------------
   // put person into array
   public void insert(String last, String first, int age)
      {
      a[nElems] = new Person(last, first, age);
      nElems++;                         // increment size
      }
//-------------------------------------------------------------
   public boolean delete(String searchName)
      {                                 // delete person from array
      int j;
      for(j=0; j<nElems; j++)           // look for it
         if( a[j].getLast().equals(searchName) )
            break;
      if(j==nElems)                     // can't find it
```

LISTING 2.5   Continued

```
            return false;
         else                           // found it
            {
            for(int k=j; k<nElems; k++)     // shift down
               a[k] = a[k+1];
            nElems--;                    // decrement size
            return true;
            }
         } // end delete()
//-------------------------------------------------------------
    public void displayA()              // displays array contents
         {
         for(int j=0; j<nElems; j++)      // for each element,
            a[j].displayPerson();         // display it
         }
//-------------------------------------------------------------
    } // end class ClassDataArray
////////////////////////////////////////////////////////////////
class ClassDataApp
    {
    public static void main(String[] args)
         {
         int maxSize = 100;              // array size
         ClassDataArray arr;             // reference to array
         arr = new ClassDataArray(maxSize);  // create the array
                                         // insert 10 items
         arr.insert("Evans", "Patty", 24);
         arr.insert("Smith", "Lorraine", 37);
         arr.insert("Yee", "Tom", 43);
         arr.insert("Adams", "Henry", 63);
         arr.insert("Hashimoto", "Sato", 21);
         arr.insert("Stimson", "Henry", 29);
         arr.insert("Velasquez", "Jose", 72);
         arr.insert("Lamarque", "Henry", 54);
         arr.insert("Vang", "Minh", 22);
         arr.insert("Creswell", "Lucinda", 18);

         arr.displayA();                 // display items

         String searchKey = "Stimson";  // search for item
         Person found;
```

LISTING 2.5   Continued

```
      found=arr.find(searchKey);
      if(found != null)
         {
         System.out.print("Found ");
         found.displayPerson();
         }
      else
         System.out.println("Can't find " + searchKey);

      System.out.println("Deleting Smith, Yee, and Creswell");
      arr.delete("Smith");            // delete 3 items
      arr.delete("Yee");
      arr.delete("Creswell");

      arr.displayA();                 // display items again
      }  // end main()
   }  // end class ClassDataApp
//////////////////////////////////////////////////////////////////
```

Here's the output of this program:

```
   Last name: Evans, First name: Patty, Age: 24
   Last name: Smith, First name: Lorraine, Age: 37
   Last name: Yee, First name: Tom, Age: 43
   Last name: Adams, First name: Henry, Age: 63
   Last name: Hashimoto, First name: Sato, Age: 21
   Last name: Stimson, First name: Henry, Age: 29
   Last name: Velasquez, First name: Jose, Age: 72
   Last name: Lamarque, First name: Henry, Age: 54
   Last name: Vang, First name: Minh, Age: 22
   Last name: Creswell, First name: Lucinda, Age: 18
Found    Last name: Stimson, First name: Henry, Age: 29
Deleting Smith, Yee, and Creswell
   Last name: Evans, First name: Patty, Age: 24
   Last name: Adams, First name: Henry, Age: 63
   Last name: Hashimoto, First name: Sato, Age: 21
   Last name: Stimson, First name: Henry, Age: 29
   Last name: Velasquez, First name: Jose, Age: 72
   Last name: Lamarque, First name: Henry, Age: 54
   Last name: Vang, First name: Minh, Age: 22
```

The `classDataArray.java` program shows that class objects can be handled by data storage structures in much the same way as primitive types. (Note that a serious program using the last name as a key would need to account for duplicate last names, which would complicate the programming as discussed earlier.)

Big O Notation

Automobiles are divided by size into several categories: subcompacts, compacts, midsize, and so on. These categories provide a quick idea what size car you're talking about, without needing to mention actual dimensions. Similarly, it's useful to have a shorthand way to say how efficient a computer algorithm is. In computer science, this rough measure is called "Big O" notation.

You might think that in comparing algorithms you would say things like "Algorithm A is twice as fast as algorithm B," but in fact this sort of statement isn't too meaningful. Why not? Because the proportion can change radically as the number of items changes. Perhaps you increase the number of items by 50%, and now A is three times as fast as B. Or you have half as many items, and A and B are now equal. What you need is a comparison that tells how an algorithm's speed is related to the number of items. Let's see how this looks for the algorithms we've seen so far.

Insertion in an Unordered Array: Constant

Insertion into an unordered array is the only algorithm we've seen that doesn't depend on how many items are in the array. The new item is always placed in the next available position, at `a[nElems]`, and `nElems` is then incremented. Insertion requires the same amount of time no matter how big N—the number of items in the array—is. We can say that the time, T, to insert an item into an unsorted array is a constant K:

$$T = K$$

In a real situation, the actual time (in microseconds or whatever) required by the insertion is related to the speed of the microprocessor, how efficiently the compiler has generated the program code, and other factors. The constant K in the preceding equation is used to account for all such factors. To find out what K is in a real situation, you need to measure how long an insertion took. (Software exists for this very purpose.) K would then be equal to that time.

Linear Search: Proportional to N

We've seen that, in a linear search of items in an array, the number of comparisons that must be made to find a specified item is, on the average, half of the total number of items. Thus, if N is the total number of items, the search time T is proportional to half of N:

$$T = K * N / 2$$

As with insertions, discovering the value of K in this equation would require timing a search for some (probably large) value of N and then using the resulting value of T to calculate K. When you know K, you can calculate T for any other value of N.

For a handier formula, we could lump the 2 into the K. Our new K is equal to the old K divided by 2. Now we have

T = K * N

This equation says that average linear search times are proportional to the size of the array. If an array is twice as big, searching it will take twice as long.

Binary Search: Proportional to log(N)

Similarly, we can concoct a formula relating T and N for a binary search:

$T = K * \log_2(N)$

As we saw earlier, the time is proportional to the base 2 logarithm of N. Actually, because any logarithm is related to any other logarithm by a constant (3.322 to go from base 2 to base 10), we can lump this constant into K as well. Then we don't need to specify the base:

T = K * log(N)

Don't Need the Constant

Big O notation looks like the formulas just described, but it dispenses with the constant K. When comparing algorithms, you don't really care about the particular microprocessor chip or compiler; all you want to compare is how T changes for different values of N, not what the actual numbers are. Therefore, the constant isn't needed.

Big O notation uses the uppercase letter O, which you can think of as meaning "order of." In Big O notation, we would say that a linear search takes O(N) time, and a binary search takes O(log N) time. Insertion into an unordered array takes O(1), or constant time. (That's the numeral 1 in the parentheses.)

Table 2.5 summarizes the running times of the algorithms we've discussed so far.

TABLE 2.5   Running Times in Big O Notation

Algorithm	Running Time in Big O Notation
Linear search	O(N)
Binary search	O(log N)
Insertion in unordered array	O(1)
Insertion in ordered array	O(N)
Deletion in unordered array	O(N)
Deletion in ordered array	O(N)

Figure 2.9 graphs some Big O relationships between time and number of items. Based on this graph, we might rate the various Big O values (very subjectively) like this: O(1) is excellent, O(log N) is good, O(N) is fair, and O(N²) is poor. O(N²) occurs in the bubble sort and also in certain graph algorithms that we'll look at later in this book.

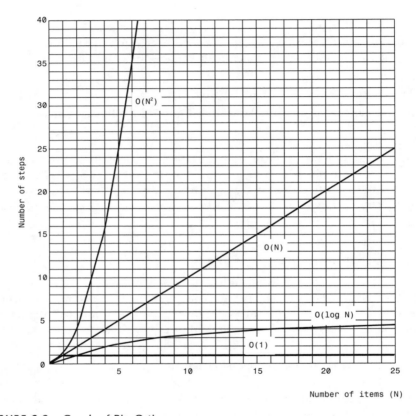

FIGURE 2.9   Graph of Big O times.

The idea in Big O notation isn't to give actual figures for running times but to convey how the running times are affected by the number of items. This is the most meaningful way to compare algorithms, except perhaps actually measuring running times in a real installation.

Why Not Use Arrays for Everything?

Arrays seem to get the job done, so why not use them for all data storage? We've already seen some of their disadvantages. In an unordered array you can insert items quickly, in O(1) time, but searching takes slow O(N) time. In an ordered array you can search quickly, in O(logN) time, but insertion takes O(N) time. For both kinds of arrays, deletion takes O(N) time because half the items (on the average) must be moved to fill in the hole.

It would be nice if there were data structures that could do everything—insertion, deletion, and searching—quickly, ideally in O(1) time, but if not that, then in O(logN) time. In the chapters ahead, we'll see how closely this ideal can be approached, and the price that must be paid in complexity.

Another problem with arrays is that their size is fixed when they are first created with new. Usually, when the program first starts, you don't know exactly how many items will be placed in the array later, so you guess how big it should be. If your guess is too large, you'll waste memory by having cells in the array that are never filled. If your guess is too small, you'll overflow the array, causing at best a message to the program's user, and at worst a program crash.

Other data structures are more flexible and can expand to hold the number of items inserted in them. The linked list, discussed in Chapter 5, "Linked Lists," is such a structure.

We should mention that Java includes a class called Vector that acts much like an array but is expandable. This added capability comes at the expense of some loss of efficiency.

You might want to try creating your own vector class. If the class user is about to overflow the internal array in this class, the insertion algorithm creates a new array of larger size, copies the old array contents to the new array, and then inserts the new item. This whole process would be invisible to the class user.

Summary

- Arrays in Java are objects, created with the new operator.

- Unordered arrays offer fast insertion but slow searching and deletion.

- Wrapping an array in a class protects the array from being inadvertently altered.

- A class interface is composed of the methods (and occasionally fields) that the class user can access.

- A class interface can be designed to make things simple for the class user.

- A binary search can be applied to an ordered array.

- The logarithm to the base B of a number A is (roughly) the number of times you can divide A by B before the result is less than 1.

- Linear searches require time proportional to the number of items in an array.

- Binary searches require time proportional to the logarithm of the number of items.

- Big O notation provides a convenient way to compare the speed of algorithms.

- An algorithm that runs in O(1) time is the best, O(log N) is good, O(N) is fair, and O(N^2) is pretty bad.

Questions

These questions are intended as a self-test for readers. Answers may be found in Appendix C.

1. Inserting an item into an unordered array

   a. takes time proportional to the size of the array.

   b. requires multiple comparisons.

   c. requires shifting other items to make room.

   d. takes the same time no matter how many items there are.

2. True or False: When you delete an item from an unordered array, in most cases you shift other items to fill in the gap.

3. In an unordered array, allowing duplicates

   a. increases times for all operations.

   b. increases search times in some situations.

   c. always increases insertion times.

   d. sometimes decreases insertion times.

4. True or False: In an unordered array, it's generally faster to find out an item is not in the array than to find out it is.

5. Creating an array in Java requires using the keyword _____ .

6. If class A is going to use class B for something, then

   a. class A's methods should be easy to understand.

   b. it's preferable if class B communicates with the program's user.

   c. the more complex operations should be placed in class A.

   d. the more work that class B can do, the better.

7. When class A is using class B for something, the methods and fields class A can access in class B are called class B's _____.

8. Ordered arrays, compared with unordered arrays, are

   a. much quicker at deletion.

   b. quicker at insertion.

   c. quicker to create.

   d. quicker at searching.

9. A logarithm is the inverse of _____ .

10. The base 10 logarithm of 1,000 is _____ .

11. The maximum number of elements that must be examined to complete a binary search in an array of 200 elements is

   a. 200.

   b. 8.

   c. 1.

   d. 13.

12. The base 2 logarithm of 64 is _____ .

13. True or False: The base 2 logarithm of 100 is 2.

14. Big O notation tells

   a. how the speed of an algorithm relates to the number of items.

   b. the running time of an algorithm for a given size data structure.

   c. the running time of an algorithm for a given number of items.

   d. how the size of a data structure relates to the number of items.

15. O(1) means a process operates in _____ time.

16. Either variables of primitive types or _____ can be placed in an array.

Experiments

Carrying out these experiments will help to provide insights into the topics covered in the chapter. No programming is involved.

1. Use the Array Workshop applet to insert, search for, and delete items. Make sure you can predict what it's going to do. Do this both when duplicates are allowed and when they're not.

2. Make sure you can predict in advance what range the Ordered Workshop applet will select at each step.

3. In an array holding an even number of data items, there is no middle item. Which item does the binary search algorithm examine first? Use the Ordered Workshop applet to find out.

Programming Projects

Writing programs to solve the Programming Projects helps to solidify your understanding of the material and demonstrates how the chapter's concepts are applied. (As noted in the Introduction, qualified instructors may obtain completed solutions to the Programming Projects on the publisher's Web site.)

2.1   To the `HighArray` class in the `highArray.java` program (Listing 2.3), add a method called `getMax()` that returns the value of the highest key in the array, or –1 if the array is empty. Add some code in `main()` to exercise this method. You can assume all the keys are positive numbers.

2.2   Modify the method in Programming Project 2.1 so that the item with the highest key is not only returned by the method, but also removed from the array. Call the method `removeMax()`.

2.3   The `removeMax()` method in Programming Project 2.2 suggests a way to sort the contents of an array by key value. Implement a sorting scheme that does not require modifying the `HighArray` class, but only the code in `main()`. You'll need a second array, which will end up inversely sorted. (This scheme is a rather crude variant of the selection sort in Chapter 3, "Simple Sorting.")

2.4   Modify the `orderedArray.java` program (Listing 2.4) so that the `insert()` and `delete()` routines, as well as `find()`, use a binary search, as suggested in the text.

2.5   Add a `merge()` method to the `OrdArray` class in the `orderedArray.java` program (Listing 2.4) so that you can merge two ordered source arrays into an ordered destination array. Write code in `main()` that inserts some random numbers into the two source arrays, invokes `merge()`, and displays the contents of the resulting destination array. The source arrays may hold different numbers of data items. In your algorithm you will need to compare the keys of the source arrays, picking the smallest one to copy to the destination. You'll also need to handle the situation when one source array exhausts its contents before the other.

2.6   Write a `noDups()` method for the `HighArray` class of the `highArray.java` program (Listing 2.3). This method should remove all duplicates from the array. That is, if three items with the key 17 appear in the array, `noDups()` should remove two of them. Don't worry about maintaining the order of the items. One approach is to first compare every item with all the other items and overwrite any duplicates with a `null` (or a distinctive value that isn't used for real keys). Then remove all the `null`s. Of course, the array size will be reduced.

3

Simple Sorting

As soon as you create a significant database, you'll probably think of reasons to sort it in various ways. You need to arrange names in alphabetical order, students by grade, customers by ZIP code, home sales by price, cities in order of increasing population, countries by GNP, stars by magnitude, and so on.

Sorting data may also be a preliminary step to searching it. As we saw in Chapter 2, "Arrays," a binary search, which can be applied only to sorted data, is much faster than a linear search.

Because sorting is so important and potentially so time-consuming, it has been the subject of extensive research in computer science, and some very sophisticated methods have been developed. In this chapter we'll look at three of the simpler algorithms: the bubble sort, the selection sort, and the insertion sort. Each is demonstrated with its own Workshop applet. In Chapter 7, "Advanced Sorting," we'll look at more sophisticated approaches: Shellsort and quicksort.

The techniques described in this chapter, while unsophisticated and comparatively slow, are nevertheless worth examining. Besides being easier to understand, they are actually better in some circumstances than the more sophisticated algorithms. The insertion sort, for example, is preferable to quicksort for small files and for almost-sorted files. In fact, an insertion sort is commonly used as a part of a quicksort implementation.

The example programs in this chapter build on the array classes we developed in the preceding chapter. The sorting algorithms are implemented as methods of similar array classes.

Be sure to try out the Workshop applets included in this chapter. They are more effective in explaining how the sorting algorithms work than prose and static pictures could ever be.

How Would You Do It?

Imagine that your kids-league baseball team (mentioned in Chapter 1, "Overview") is lined up on the field, as shown in Figure 3.1. The regulation nine players, plus an extra, have shown up for practice. You want to arrange the players in order of increasing height (with the shortest player on the left) for the team picture. How would you go about this sorting process?

FIGURE 3.1   The unordered baseball team.

As a human being, you have advantages over a computer program. You can see all the kids at once, and you can pick out the tallest kid almost instantly. You don't need to laboriously measure and compare everyone. Also, the kids don't need to occupy particular places. They can jostle each other, push each other a little to make room, and stand behind or in front of each other. After some ad hoc rearranging, you would have no trouble in lining up all the kids, as shown in Figure 3.2.

FIGURE 3.2   The ordered baseball team.

A computer program isn't able to glance over the data in this way. It can compare only two players at one time because that's how the comparison operators work. This tunnel vision on the part of algorithms will be a recurring theme. Things may seem simple to us humans, but the algorithm can't see the big picture and must, therefore, concentrate on the details and follow some simple rules.

The three algorithms in this chapter all involve two steps, executed over and over until the data is sorted:

1. Compare two items.

2. Swap two items, or copy one item.

However, each algorithm handles the details in a different way.

Bubble Sort

The bubble sort is notoriously slow, but it's conceptually the simplest of the sorting algorithms and for that reason is a good beginning for our exploration of sorting techniques.

Bubble Sort on the Baseball Players

Imagine that you're near-sighted (like a computer program) so that you can see only two of the baseball players at the same time, if they're next to each other and if you stand very close to them. Given this impediment, how would you sort them? Let's assume there are N players, and the positions they're standing in are numbered from 0 on the left to N-1 on the right.

The bubble sort routine works like this: You start at the left end of the line and compare the two kids in positions 0 and 1. If the one on the left (in 0) is taller, you swap them. If the one on the right is taller, you don't do anything. Then you move over one position and compare the kids in positions 1 and 2. Again, if the one on the left is taller, you swap them. This sorting process is shown in Figure 3.3.

Here are the rules you're following:

1. Compare two players.

2. If the one on the left is taller, swap them.

3. Move one position right.

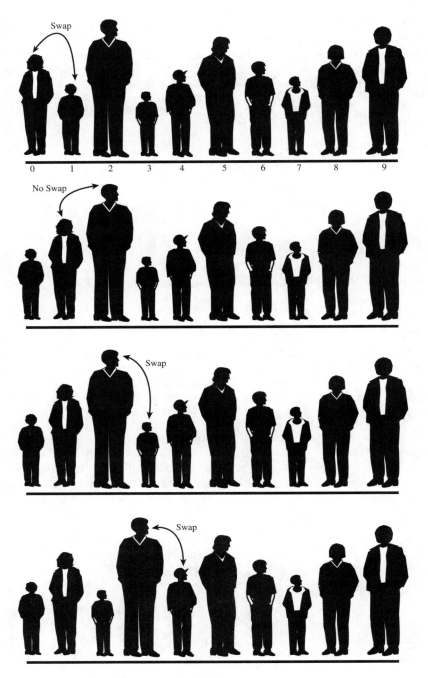

FIGURE 3.3 Bubble sort: the beginning of the first pass.

You continue down the line this way until you reach the right end. You have by no means finished sorting the kids, but you do know that the tallest kid is on the right. This must be true because, as soon as you encounter the tallest kid, you'll end up swapping him (or her) every time you compare two kids, until eventually he (or she) will reach the right end of the line. This is why it's called the bubble sort: As the algorithm progresses, the biggest items "bubble up" to the top end of the array. Figure 3.4 shows the baseball players at the end of the first pass.

Sorted

FIGURE 3.4   Bubble sort: the end of the first pass.

After this first pass through all the data, you've made N-1 comparisons and somewhere between 0 and N-1 swaps, depending on the initial arrangement of the players. The item at the end of the array is sorted and won't be moved again.

Now you go back and start another pass from the left end of the line. Again, you go toward the right, comparing and swapping when appropriate. However, this time you can stop one player short of the end of the line, at position N-2, because you know the last position, at N-1, already contains the tallest player. This rule could be stated as:

4. When you reach the first sorted player, start over at the left end of the line.

You continue this process until all the players are in order. Describing this process is much harder than demonstrating it, so let's watch the BubbleSort Workshop applet at work.

The BubbleSort Workshop Applet

Start the BubbleSort Workshop applet. You'll see something that looks like a bar graph, with the bar heights randomly arranged, as shown in Figure 3.5.

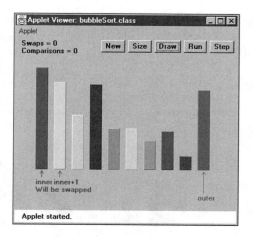

FIGURE 3.5   The BubbleSort Workshop applet.

The Run Button

This Workshop applet contains a two-speed graph: You can either let it run by itself, or you can single-step through the process. To get a quick idea what happens, click the Run button. The algorithm will bubble-sort the bars. When it finishes, in 10 seconds or so, the bars will be sorted, as shown in Figure 3.6.

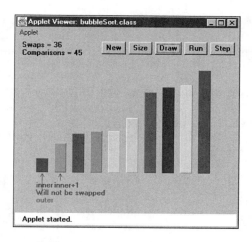

FIGURE 3.6   After the bubble sort.

The New Button

To do another sort, press the New button. New creates a new set of bars and initial-izes the sorting routine. Repeated presses of New toggle between two arrangements of bars: a random order, as shown in Figure 3.5, and an inverse ordering where the bars are sorted backward. This inverse ordering provides an extra challenge for many sorting algorithms.

The Step Button

The real payoff for using the BubbleSort Workshop applet comes when you single-step through a sort. You can see exactly how the algorithm carries out each step.

Start by creating a new randomly arranged graph with New. You'll see three arrows pointing at different bars. Two arrows, labeled inner and inner+1, are side by side on the left. Another arrow, outer, starts on the far right. (The names are chosen to correspond to the inner and outer loop variables in the nested loops used in the algorithm.)

Click once on the Step button. You'll see the inner and the inner+1 arrows move together one position to the right, swapping the bars if appropriate. These arrows correspond to the two players you compared, and possibly swapped, in the baseball scenario.

A message under the arrows tells you whether the contents of inner and inner+1 will be swapped, but you know this just from comparing the bars: If the taller one is on the left, they'll be swapped. Messages at the top of the graph tell you how many swaps and comparisons have been carried out so far. (A complete sort of 10 bars requires 45 comparisons and, on the average, about 22 swaps.)

Continue pressing Step. Each time inner and inner+1 finish going all the way from 0 to outer, the outer pointer moves one position to the left. At all times during the sorting process, all the bars to the right of outer are sorted; those to the left of (and at) outer are not.

The Size Button

The Size button toggles between 10 bars and 100 bars. Figure 3.7 shows what the 100 random bars look like.

You probably don't want to single-step through the sorting process for 100 bars, unless you're unusually patient. Press Run instead, and watch how the blue inner and inner+1 pointers seem to find the tallest unsorted bar and carry it down the row to the right, inserting it just to the left of the previously sorted bars.

Figure 3.8 shows the situation partway through the sorting process. The bars to the right of the red (longest) arrow are sorted. The bars to the left are beginning to look sorted, but much work remains to be done.

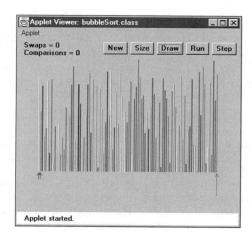

FIGURE 3.7   The BubbleSort applet with 100 bars.

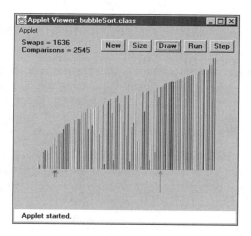

FIGURE 3.8   The 100 partly sorted bars.

If you started a sort with Run and the arrows are whizzing around, you can freeze the process at any point by pressing the Step button. You can then single-step to watch the details of the operation or press Run again to return to high-speed mode.

The Draw Button

Sometimes while running the sorting algorithm at full speed, the computer takes time off to perform some other task. This can result in some bars not being drawn. If this happens, you can press the Draw button to redraw all the bars. Doing so pauses the run, so you'll need to press the Run button again to continue.

You can press Draw at any time there seems to be a glitch in the display.

Java Code for a Bubble Sort

In the bubbleSort.java program, shown in Listing 3.1, a class called ArrayBub encapsulates an array a[], which holds variables of type long.

In a more serious program, the data would probably consist of objects, but we use a primitive type for simplicity. (We'll see how objects are sorted in the objectSort.java program in Listing 3.4.) Also, to reduce the size of the listing, we don't show find() and delete() methods with the ArrayBub class, although they would normally be part of a such a class.

LISTING 3.1   The bubbleSort.java Program

```java
// bubbleSort.java
// demonstrates bubble sort
// to run this program: C>java BubbleSortApp
////////////////////////////////////////////////////////////////
class ArrayBub
   {
   private long[] a;                // ref to array a
   private int nElems;              // number of data items
//--------------------------------------------------------------
   public ArrayBub(int max)         // constructor
      {
      a = new long[max];            // create the array
      nElems = 0;                   // no items yet
      }
//--------------------------------------------------------------
   public void insert(long value)   // put element into array
      {
      a[nElems] = value;            // insert it
      nElems++;                     // increment size
      }
//--------------------------------------------------------------
   public void display()            // displays array contents
      {
      for(int j=0; j<nElems; j++)   // for each element,
         System.out.print(a[j] + " ");  // display it
      System.out.println("");
      }
//--------------------------------------------------------------
   public void bubbleSort()
```

LISTING 3.1   Continued

```
      {
      int out, in;

      for(out=nElems-1; out>1; out--)    // outer loop (backward)
         for(in=0; in<out; in++)         // inner loop (forward)
            if( a[in] > a[in+1] )        // out of order?
               swap(in, in+1);           // swap them
      } // end bubbleSort()
//------------------------------------------------------------
   private void swap(int one, int two)
      {
      long temp = a[one];
      a[one] = a[two];
      a[two] = temp;
      }
//------------------------------------------------------------
   } // end class ArrayBub
////////////////////////////////////////////////////////////
class BubbleSortApp
   {
   public static void main(String[] args)
      {
      int maxSize = 100;              // array size
      ArrayBub arr;                   // reference to array
      arr = new ArrayBub(maxSize);    // create the array

      arr.insert(77);                 // insert 10 items
      arr.insert(99);
      arr.insert(44);
      arr.insert(55);
      arr.insert(22);
      arr.insert(88);
      arr.insert(11);
      arr.insert(00);
      arr.insert(66);
      arr.insert(33);

      arr.display();                  // display items

      arr.bubbleSort();               // bubble sort them
```

LISTING 3.1   Continued

```
    arr.display();                   // display them again
    }  // end main()
  }  // end class BubbleSortApp
//////////////////////////////////////////////////////////////
```

The constructor and the insert() and display() methods of this class are similar to those we've seen before. However, there's a new method: bubbleSort(). When this method is invoked from main(), the contents of the array are rearranged into sorted order.

The main() routine inserts 10 items into the array in random order, displays the array, calls bubbleSort() to sort it, and then displays it again. Here's the output:

```
77 99 44 55 22 88 11 0 66 33
0  11 22 33 44 55 66 77 88 99
```

The bubbleSort() method is only four lines long. Here it is, extracted from the listing:

```
public void bubbleSort()
    {
    int out, in;

    for(out=nElems-1; out>1; out--)    // outer loop (backward)
        for(in=0; in<out; in++)        // inner loop (forward)
            if( a[in] > a[in+1] )      // out of order?
                swap(in, in+1);        // swap them
    }  // end bubbleSort()
```

The idea is to put the smallest item at the beginning of the array (index 0) and the largest item at the end (index nElems-1). The loop counter out in the outer for loop starts at the end of the array, at nElems-1, and decrements itself each time through the loop. The items at indices greater than out are always completely sorted. The out variable moves left after each pass by in so that items that are already sorted are no longer involved in the algorithm.

The inner loop counter in starts at the beginning of the array and increments itself each cycle of the inner loop, exiting when it reaches out. Within the inner loop, the two array cells pointed to by in and in+1 are compared, and swapped if the one in in is larger than the one in in+1.

For clarity, we use a separate swap() method to carry out the swap. It simply exchanges the two values in the two array cells, using a temporary variable to hold the value of the first cell while the first cell takes on the value in the second and then setting the second cell to the temporary value. Actually, using a separate swap() method may not be a good idea in practice because the function call adds a small amount of overhead. If you're writing your own sorting routine, you may prefer to put the swap instructions in line to gain a slight increase in speed.

Invariants

In many algorithms there are conditions that remain unchanged as the algorithm proceeds. These conditions are called *invariants*. Recognizing invariants can be useful in understanding the algorithm. In certain situations they may also be helpful in debugging; you can repeatedly check that the invariant is true, and signal an error if it isn't.

In the bubbleSort.java program, the invariant is that the data items to the right of out are sorted. This remains true throughout the running of the algorithm. (On the first pass, nothing has been sorted yet, and there are no items to the right of out because it starts on the rightmost element.)

Efficiency of the Bubble Sort

As you can see by watching the BubbleSort Workshop applet with 10 bars, the inner and inner+1 arrows make nine comparisons on the first pass, eight on the second, and so on, down to one comparison on the last pass. For 10 items, this is

9 + 8 + 7 + 6 + 5 + 4 + 3 + 2 + 1 = 45

In general, where N is the number of items in the array, there are N-1 comparisons on the first pass, N-2 on the second, and so on. The formula for the sum of such a series is

$(N–1) + (N–2) + (N–3) + ... + 1 = N*(N–1)/2$

N*(N–1)/2 is 45 (10*9/2) when N is 10.

Thus, the algorithm makes about $N^2/2$ comparisons (ignoring the –1, which doesn't make much difference, especially if N is large).

There are fewer swaps than there are comparisons because two bars are swapped only if they need to be. If the data is random, a swap is necessary about half the time, so there will be about $N^2/4$ swaps. (Although in the worst case, with the initial data inversely sorted, a swap is necessary with every comparison.)

Both swaps and comparisons are proportional to N^2. Because constants don't count in Big O notation, we can ignore the 2 and the 4 and say that the bubble sort runs in $O(N^2)$ time. This is slow, as you can verify by running the BubbleSort Workshop applet with 100 bars.

Whenever you see one loop nested within another, such as those in the bubble sort and the other sorting algorithms in this chapter, you can suspect that an algorithm runs in $O(N^2)$ time. The outer loop executes N times, and the inner loop executes N (or perhaps N divided by some constant) times for each cycle of the outer loop. This means you're doing something approximately N*N or N^2 times.

Selection Sort

The selection sort improves on the bubble sort by reducing the number of swaps necessary from $O(N^2)$ to $O(N)$. Unfortunately, the number of comparisons remains $O(N^2)$. However, the selection sort can still offer a significant improvement for large records that must be physically moved around in memory, causing the swap time to be much more important than the comparison time. (Typically, this isn't the case in Java, where references are moved around, not entire objects.)

Selection Sort on the Baseball Players

Let's consider the baseball players again. In the selection sort, you can no longer compare only players standing next to each other. Thus, you'll need to remember a certain player's height; you can use a notebook to write it down. A magenta-colored towel will also come in handy.

A Brief Description

What's involved in the selection sort is making a pass through all the players and picking (or *selecting*, hence the name of the sort) the shortest one. This shortest player is then swapped with the player on the left end of the line, at position 0. Now the leftmost player is sorted and won't need to be moved again. Notice that in this algorithm the sorted players accumulate on the left (lower indices), whereas in the bubble sort they accumulated on the right.

The next time you pass down the row of players, you start at position 1, and, finding the minimum, swap with position 1. This process continues until all the players are sorted.

A More Detailed Description

In more detail, start at the left end of the line of players. Record the leftmost player's height in your notebook and throw the magenta towel on the ground in front of this person. Then compare the height of the next player to the right with the height in your notebook. If this player is shorter, cross out the height of the first player and record the second player's height instead. Also move the towel, placing it in front of this new "shortest" (for the time being) player. Continue down the row, comparing each player with the minimum. Change the minimum value in your notebook and move the towel whenever you find a shorter player. When you're done, the magenta towel will be in front of the shortest player.

Swap this shortest player with the player on the left end of the line. You've now sorted one player. You've made N-1 comparisons, but only one swap.

On the next pass, you do exactly the same thing, except that you can completely ignore the player on the left because this player has already been sorted. Thus, the algorithm starts the second pass at position 1, instead of 0. With each succeeding pass, one more player is sorted and placed on the left, and one less player needs to be considered when finding the new minimum. Figure 3.9 shows how this sort looks for the first three passes.

The SelectSort Workshop Applet

To see how the selection sort looks in action, try out the SelectSort Workshop applet. The buttons operate the same way as those in the BubbleSort applet. Use New to create a new array of 10 randomly arranged bars. The red arrow called outer starts on the left; it points to the leftmost unsorted bar. Gradually, it will move right as more bars are added to the sorted group on its left.

The magenta min arrow also starts out pointing to the leftmost bar; it will move to record the shortest bar found so far. (The magenta min arrow corresponds to the towel in the baseball analogy.) The blue inner arrow marks the bar currently being compared with the minimum.

As you repeatedly press Step, inner moves from left to right, examining each bar in turn and comparing it with the bar pointed to by min. If the inner bar is shorter, min jumps over to this new, shorter bar. When inner reaches the right end of the graph, min points to the shortest of the unsorted bars. This bar is then swapped with outer, the leftmost unsorted bar.

Figure 3.10 shows the situation midway through a sort. The bars to the left of outer are sorted, and inner has scanned from outer to the right end, looking for the short-est bar. The min arrow has recorded the position of this bar, which will be swapped with outer.

Use the Size button to switch to 100 bars, and sort a random arrangement. You'll see how the magenta min arrow hangs out with a perspective minimum value for a while and then jumps to a new one when the blue inner arrow finds a smaller candidate. The red outer arrow moves slowly but inexorably to the right, as the sorted bars accumulate to its left.

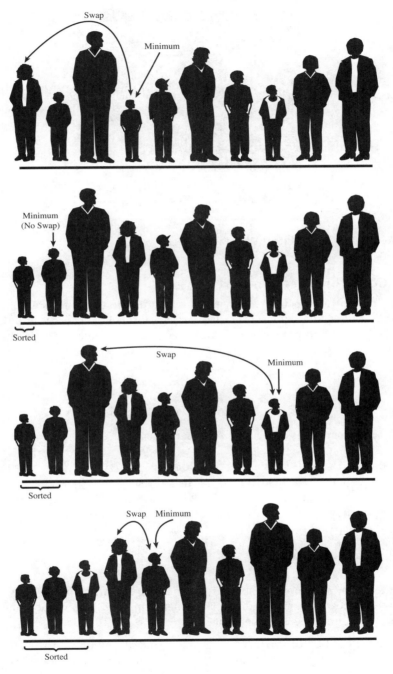

FIGURE 3.9   Selection sort on baseball players.

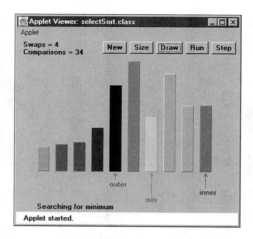

FIGURE 3.10   The SelectSort Workshop applet.

Java Code for Selection Sort

The listing for the selectSort.java program is similar to that for bubbleSort.java, except that the container class is called ArraySel instead of ArrayBub, and the bubbleSort() method has been replaced by selectSort(). Here's how this method looks:

```
public void selectionSort()
    {
    int out, in, min;

    for(out=0; out<nElems-1; out++)    // outer loop
        {
        min = out;                         // minimum
        for(in=out+1; in<nElems; in++) // inner loop
            if(a[in] < a[min] )            // if min greater,
                min = in;                  // we have a new min
        swap(out, min);                    // swap them
        }  // end for(out)
    }  // end selectionSort()
```

The outer loop, with loop variable out, starts at the beginning of the array (index 0) and proceeds toward higher indices. The inner loop, with loop variable in, begins at out and likewise proceeds to the right.

At each new position of in, the elements a[in] and a[min] are compared. If a[in] is smaller, then min is given the value of in. At the end of the inner loop, min points to the minimum value, and the array elements pointed to by out and min are swapped. Listing 3.2 shows the complete selectSort.java program.

LISTING 3.2   The selectSort.java Program

```java
// selectSort.java
// demonstrates selection sort
// to run this program: C>java SelectSortApp
////////////////////////////////////////////////////////////////
class ArraySel
    {
    private long[] a;               // ref to array a
    private int nElems;             // number of data items
//--------------------------------------------------------------
    public ArraySel(int max)        // constructor
        {
        a = new long[max];              // create the array
        nElems = 0;                     // no items yet
        }
//--------------------------------------------------------------
    public void insert(long value)  // put element into array
        {
        a[nElems] = value;              // insert it
        nElems++;                       // increment size
        }
//--------------------------------------------------------------
    public void display()           // displays array contents
        {
        for(int j=0; j<nElems; j++)     // for each element,
            System.out.print(a[j] + " "); // display it
        System.out.println("");
        }
//--------------------------------------------------------------
    public void selectionSort()
        {
        int out, in, min;

        for(out=0; out<nElems-1; out++)   // outer loop
            {
            min = out;                    // minimum
            for(in=out+1; in<nElems; in++) // inner loop
                if(a[in] < a[min] )        // if min greater,
                    min = in;              // we have a new min
```

LISTING 3.2   Continued

```
            swap(out, min);                 // swap them
            }  // end for(out)
        }  // end selectionSort()
//-------------------------------------------------------------
    private void swap(int one, int two)
        {
        long temp = a[one];
        a[one] = a[two];
        a[two] = temp;
        }
//-------------------------------------------------------------
    }  // end class ArraySel
////////////////////////////////////////////////////////////////
class SelectSortApp
    {
    public static void main(String[] args)
        {
        int maxSize = 100;          // array size
        ArraySel arr;               // reference to array
        arr = new ArraySel(maxSize); // create the array

        arr.insert(77);             // insert 10 items
        arr.insert(99);
        arr.insert(44);
        arr.insert(55);
        arr.insert(22);
        arr.insert(88);
        arr.insert(11);
        arr.insert(00);
        arr.insert(66);
        arr.insert(33);

        arr.display();              // display items

        arr.selectionSort();        // selection-sort them

        arr.display();              // display them again
        }  // end main()
    }  // end class SelectSortApp
////////////////////////////////////////////////////////////////
```

The output from `selectSort.java` is identical to that from `bubbleSort.java`:

```
77 99 44 55 22 88 11 0 66 33
0 11 22 33 44 55 66 77 88 99
```

Invariant

In the `selectSort.java` program, the data items with indices less than or equal to `out` are always sorted.

Efficiency of the Selection Sort

The selection sort performs the same number of comparisons as the bubble sort: N*(N-1)/2. For 10 data items, this is 45 comparisons. However, 10 items require fewer than 10 swaps. With 100 items, 4,950 comparisons are required, but fewer than 100 swaps. For large values of N, the comparison times will dominate, so we would have to say that the selection sort runs in $O(N^2)$ time, just as the bubble sort did. However, it is unquestionably faster because there are so few swaps. For smaller values of N, the selection sort may in fact be considerably faster, especially if the swap times are much larger than the comparison times.

Insertion Sort

In most cases the insertion sort is the best of the elementary sorts described in this chapter. It still executes in $O(N^2)$ time, but it's about twice as fast as the bubble sort and somewhat faster than the selection sort in normal situations. It's also not too complex, although it's slightly more involved than the bubble and selection sorts. It's often used as the final stage of more sophisticated sorts, such as quicksort.

Insertion Sort on the Baseball Players

To begin the insertion sort, start with your baseball players lined up in random order. (They wanted to play a game, but clearly there's no time for that.) It's easier to think about the insertion sort if we begin in the middle of the process, when the team is half sorted.

Partial Sorting

At this point there's an imaginary marker somewhere in the middle of the line. (Maybe you threw a red T-shirt on the ground in front of a player.) The players to the left of this marker are *partially sorted*. This means that they are sorted among themselves; each one is taller than the person to his or her left. However, the players aren't necessarily in their final positions because they may still need to be moved when previously unsorted players are inserted between them.

Note that partial sorting did not take place in the bubble sort and selection sort. In these algorithms a group of data items was completely sorted at any given time; in the insertion sort a group of items is only partially sorted.

The Marked Player

The player where the marker is, whom we'll call the "marked" player, and all the players on her right, are as yet unsorted. This is shown in Figure 3.11.a.

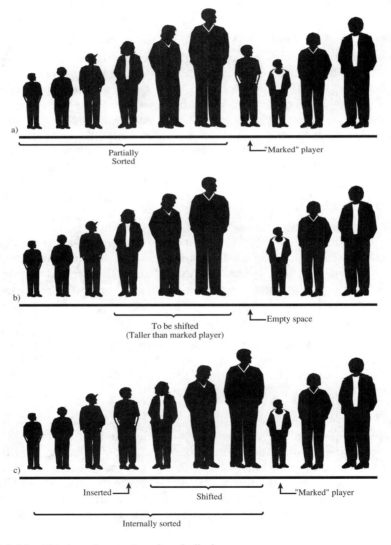

FIGURE 3.11   The insertion sort on baseball players.

What we're going to do is insert the marked player in the appropriate place in the (partially) sorted group. However, to do this, we'll need to shift some of the sorted players to the right to make room. To provide a space for this shift, we take the marked player out of line. (In the program this data item is stored in a temporary variable.) This step is shown in Figure 3.11.b.

Now we shift the sorted players to make room. The tallest sorted player moves into the marked player's spot, the next-tallest player into the tallest player's spot, and so on.

When does this shifting process stop? Imagine that you and the marked player are walking down the line to the left. At each position you shift another player to the right, but you also compare the marked player with the player about to be shifted. The shifting process stops when you've shifted the last player that's taller than the marked player. The last shift opens up the space where the marked player, when inserted, will be in sorted order. This step is shown in Figure 3.11.c.

Now the partially sorted group is one player bigger, and the unsorted group is one player smaller. The marker T-shirt is moved one space to the right, so it's again in front of the leftmost unsorted player. This process is repeated until all the unsorted players have been inserted (hence the name *insertion* sort) into the appropriate place in the partially sorted group.

The InsertSort Workshop Applet

Use the InsertSort Workshop applet to demonstrate the insertion sort. Unlike the other sorting applets, it's probably more instructive to begin with 100 random bars rather than 10.

Sorting 100 Bars

Change to 100 bars with the Size button, and click Run to watch the bars sort themselves before your very eyes. You'll see that the short red outer arrow marks the dividing line between the partially sorted bars to the left and the unsorted bars to the right. The blue inner arrow keeps starting from outer and zipping to the left, looking for the proper place to insert the marked bar. Figure 3.12 shows how this process looks when about half the bars are partially sorted.

The marked bar is stored in the temporary variable pointed to by the magenta arrow at the right end of the graph, but the contents of this variable are replaced so often that it's hard to see what's there (unless you slow down to single-step mode).

Sorting 10 Bars

To get down to the details, use Size to switch to 10 bars. (If necessary, use New to make sure they're in random order.)

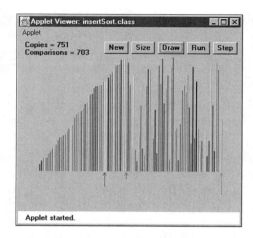

FIGURE 3.12   The InsertSort Workshop applet with 100 bars.

At the beginning, `inner` and `outer` point to the second bar from the left (array index 1), and the first message is `Will copy outer to temp`. This will make room for the shift. (There's no arrow for `inner-1`, but of course it's always one bar to the left of `inner`.)

Click the Step button. The bar at `outer` will be copied to `temp`. We say that items are copied from a source to a destination. When performing a copy, the applet removes the bar from the source location, leaving a blank. This is slightly misleading because in a real Java program the reference in the source would remain there. However, blanking the source makes it easier to see what's happening.

What happens next depends on whether the first two bars are already in order (smaller on the left). If they are, you'll see the message `Have compared inner-1 and temp, no copy necessary`.

If the first two bars are not in order, the message is `Have compared inner-1 and temp, will copy inner-1 to inner`. This is the shift that's necessary to make room for the value in `temp` to be reinserted. There's only one such shift on this first pass; more shifts will be necessary on subsequent passes. The situation is shown in Figure 3.13.

On the next click, you'll see the copy take place from `inner-1` to `inner`. Also, the `inner` arrow moves one space left. The new message is `Now inner is 0, so no copy necessary`. The shifting process is complete.

No matter which of the first two bars was shorter, the next click will show you `Will copy temp to inner`. This will happen, but if the first two bars were initially in order, you won't be able to tell a copy was performed because `temp` and `inner` hold the same bar. Copying data over the top of the same data may seem inefficient, but the algorithm runs faster if it doesn't check for this possibility, which happens comparatively infrequently.

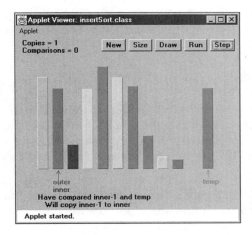

FIGURE 3.13   The InsertSort Workshop applet with 10 bars.

Now the first two bars are partially sorted (sorted with respect to each other), and the `outer` arrow moves one space right, to the third bar (index 2). The process repeats, with the `Will copy outer to temp` message. On this pass through the sorted data, there may be no shifts, one shift, or two shifts, depending on where the third bar fits among the first two.

Continue to single-step the sorting process. Again, you can see what's happening more easily after the process has run long enough to provide some sorted bars on the left. Then you can see how just enough shifts take place to make room for the reinsertion of the bar from `temp` into its proper place.

Java Code for Insertion Sort

Here's the method that carries out the insertion sort, extracted from the `insertSort.java` program:

```
public void insertionSort()
   {
   int in, out;

   for(out=1; out<nElems; out++)      // out is dividing line
      {
      long temp = a[out];        // remove marked item
      in = out;                  // start shifts at out
      while(in>0 && a[in-1] >= temp) // until one is smaller,
         {
         a[in] = a[in-1];             // shift item right,
```

```
        --in;                       // go left one position
      }
    a[in] = temp;                   // insert marked item
    }  // end for
  }  // end insertionSort()
```

In the outer for loop, out starts at 1 and moves right. It marks the leftmost unsorted data. In the inner while loop, in starts at out and moves left, until either temp is smaller than the array element there, or it can't go left any further. Each pass through the while loop shifts another sorted element one space right.

It may be hard to see the relation between the steps in the InsertSort Workshop applet and the code, so Figure 3.14 is an activity diagram of the insertionSort() method, with the corresponding messages from the InsertSort Workshop applet. Listing 3.3 shows the complete insertSort.java program.

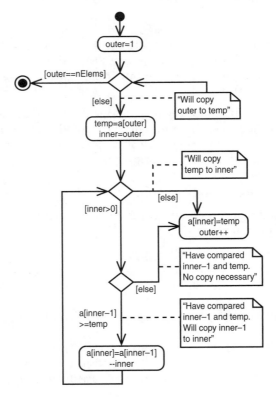

FIGURE 3.14   Activity diagram for insertSort().

LISTING 3.3   The insertSort.java Program

```
// insertSort.java
// demonstrates insertion sort
// to run this program: C>java InsertSortApp
//--------------------------------------------------------------
class ArrayIns
    {
    private long[] a;                 // ref to array a
    private int nElems;               // number of data items
//--------------------------------------------------------------
    public ArrayIns(int max)          // constructor
        {
        a = new long[max];               // create the array
        nElems = 0;                      // no items yet
        }
//--------------------------------------------------------------
    public void insert(long value)    // put element into array
        {
        a[nElems] = value;            // insert it
        nElems++;                     // increment size
        }
//--------------------------------------------------------------
    public void display()             // displays array contents
        {
        for(int j=0; j<nElems; j++)      // for each element,
            System.out.print(a[j] + " "); // display it
        System.out.println("");
        }
//--------------------------------------------------------------
    public void insertionSort()
        {
        int in, out;

        for(out=1; out<nElems; out++)    // out is dividing line
            {
            long temp = a[out];          // remove marked item
            in = out;                    // start shifts at out
            while(in>0 && a[in-1] >= temp) // until one is smaller,
                {
                a[in] = a[in-1];            // shift item to right
                --in;                       // go left one position
                }
```

LISTING 3.3   Continued

```
            a[in] = temp;                // insert marked item
        }  // end for
    }  // end insertionSort()
//----------------------------------------------------------------
    }  // end class ArrayIns
////////////////////////////////////////////////////////////////
class InsertSortApp
    {
    public static void main(String[] args)
        {
        int maxSize = 100;           // array size
        ArrayIns arr;                // reference to array
        arr = new ArrayIns(maxSize); // create the array

        arr.insert(77);              // insert 10 items
        arr.insert(99);
        arr.insert(44);
        arr.insert(55);
        arr.insert(22);
        arr.insert(88);
        arr.insert(11);
        arr.insert(00);
        arr.insert(66);
        arr.insert(33);

        arr.display();               // display items

        arr.insertionSort();         // insertion-sort them

        arr.display();               // display them again
        }  // end main()
    }  // end class InsertSortApp
////////////////////////////////////////////////////////////////
```

Here's the output from the insertSort.java program; it's the same as that from the other programs in this chapter:

```
77 99 44 55 22 88 11 0 66 33
0 11 22 33 44 55 66 77 88 99
```

Invariants in the Insertion Sort

At the end of each pass, following the insertion of the item from `temp`, the data items with smaller indices than `outer` are partially sorted.

Efficiency of the Insertion Sort

How many comparisons and copies does this algorithm require? On the first pass, it compares a maximum of one item. On the second pass, it's a maximum of two items, and so on, up to a maximum of N-1 comparisons on the last pass. This is

$$1 + 2 + 3 + \ldots + N\text{-}1 = N^*(N\text{-}1)/2$$

However, because on each pass an average of only half of the maximum number of items are actually compared before the insertion point is found, we can divide by 2, which gives

$$N^*(N\text{-}1)/4$$

The number of copies is approximately the same as the number of comparisons. However, a copy isn't as time-consuming as a swap, so for random data this algorithm runs twice as fast as the bubble sort and faster than the selection sort.

In any case, like the other sort routines in this chapter, the insertion sort runs in $O(N^2)$ time for random data.

For data that is already sorted or almost sorted, the insertion sort does much better. When data is in order, the condition in the `while` loop is never true, so it becomes a simple statement in the outer loop, which executes N-1 times. In this case the algorithm runs in O(N) time. If the data is almost sorted, insertion sort runs in almost O(N) time, which makes it a simple and efficient way to order a file that is only slightly out of order.

However, for data arranged in inverse sorted order, every possible comparison and shift is carried out, so the insertion sort runs no faster than the bubble sort. You can check this using the reverse-sorted data option (toggled with New) in the InsertSort Workshop applet.

Sorting Objects

For simplicity we've applied the sorting algorithms we've looked at thus far to a primitive data type: `long`. However, sorting routines will more likely be applied to objects than primitive types. Accordingly, we show a Java program in Listing 3.4, `objectSort.java`, that sorts an array of `Person` objects (last seen in the `classDataArray.java` program in Chapter 2).

Java Code for Sorting Objects

The algorithm used in our Java program is the insertion sort from the preceding section. The Person objects are sorted on lastName; this is the key field. The objectSort.java program is shown in Listing 3.4.

LISTING 3.4   The objectSort.java Program

```java
// objectSort.java
// demonstrates sorting objects (uses insertion sort)
// to run this program: C>java ObjectSortApp
//////////////////////////////////////////////////////////////////
class Person
    {
    private String lastName;
    private String firstName;
    private int age;
    //--------------------------------------------------------------
    public Person(String last, String first, int a)
        {                              // constructor
        lastName = last;
        firstName = first;
        age = a;
        }
    //--------------------------------------------------------------
    public void displayPerson()
        {
        System.out.print("   Last name: " + lastName);
        System.out.print(", First name: " + firstName);
        System.out.println(", Age: " + age);
        }
    //--------------------------------------------------------------
    public String getLast()            // get last name
        { return lastName; }
    } // end class Person
//////////////////////////////////////////////////////////////////
class ArrayInOb
    {
    private Person[] a;                // ref to array a
    private int nElems;                // number of data items
//--------------------------------------------------------------
    public ArrayInOb(int max)          // constructor
        {
```

LISTING 3.4   Continued

```
    a = new Person[max];           // create the array
    nElems = 0;                    // no items yet
    }
//------------------------------------------------------------
                               // put person into array
  public void insert(String last, String first, int age)
    {
    a[nElems] = new Person(last, first, age);
    nElems++;                      // increment size
    }
//------------------------------------------------------------
  public void display()            // displays array contents
    {
    for(int j=0; j<nElems; j++)    // for each element,
       a[j].displayPerson();       // display it
    System.out.println("");
    }
//------------------------------------------------------------
  public void insertionSort()
    {
    int in, out;

    for(out=1; out<nElems; out++)  // out is dividing line
       {
       Person temp = a[out];       // remove marked person
       in = out;                   // start shifting at out

       while(in>0 &&               // until smaller one found,
            a[in-1].getLast().compareTo(temp.getLast())>0)
         {
         a[in] = a[in-1];          // shift item to the right
         --in;                     // go left one position
         }
       a[in] = temp;               // insert marked item
       }  // end for
    }  // end insertionSort()
//------------------------------------------------------------
  }  // end class ArrayInOb
//////////////////////////////////////////////////////////////
class ObjectSortApp
  {
```

LISTING 3.4   Continued

```
public static void main(String[] args)
   {
   int maxSize = 100;            // array size
   ArrayInOb arr;                // reference to array
   arr = new ArrayInOb(maxSize); // create the array

   arr.insert("Evans", "Patty", 24);
   arr.insert("Smith", "Doc", 59);
   arr.insert("Smith", "Lorraine", 37);
   arr.insert("Smith", "Paul", 37);
   arr.insert("Yee", "Tom", 43);
   arr.insert("Hashimoto", "Sato", 21);
   arr.insert("Stimson", "Henry", 29);
   arr.insert("Velasquez", "Jose", 72);
   arr.insert("Vang", "Minh", 22);
   arr.insert("Creswell", "Lucinda", 18);

   System.out.println("Before sorting:");
   arr.display();                // display items

   arr.insertionSort();          // insertion-sort them

   System.out.println("After sorting:");
   arr.display();                // display them again
   } // end main()
 } // end class ObjectSortApp
//////////////////////////////////////////////////////////////
```

Here's the output of this program:

```
Before sorting:
   Last name: Evans, First name: Patty, Age: 24
   Last name: Smith, First name: Doc, Age: 59
   Last name: Smith, First name: Lorraine, Age: 37
   Last name: Smith, First name: Paul, Age: 37
   Last name: Yee, First name: Tom, Age: 43
   Last name: Hashimoto, First name: Sato, Age: 21
   Last name: Stimson, First name: Henry, Age: 29
   Last name: Velasquez, First name: Jose, Age: 72
   Last name: Vang, First name: Minh, Age: 22
   Last name: Creswell, First name: Lucinda, Age: 18
```

```
After sorting:
   Last name: Creswell, First name: Lucinda, Age: 18
   Last name: Evans, First name: Patty, Age: 24
   Last name: Hashimoto, First name: Sato, Age: 21
   Last name: Smith, First name: Doc, Age: 59
   Last name: Smith, First name: Lorraine, Age: 37
   Last name: Smith, First name: Paul, Age: 37
   Last name: Stimson, First name: Henry, Age: 29
   Last name: Vang, First name: Minh, Age: 22
   Last name: Velasquez, First name: Jose, Age: 72
   Last name: Yee, First name: Tom, Age: 43
```

Lexicographical Comparisons

The insertSort() method in objectSort.java is similar to that in insertSort.java, but it has been adapted to compare the lastName key values of records rather than the value of a primitive type.

We use the compareTo() method of the String class to perform the comparisons in the insertSort() method. Here's the expression that uses it:

```
a[in-1].getLast().compareTo(temp.getLast()) > 0
```

The compareTo() method returns different integer values depending on the lexicographical (that is, alphabetical) ordering of the String for which it's invoked and the String passed to it as an argument, as shown in Table 3.1.

TABLE 3.1 Operation of the compareTo() Method

s2.compareTo(s1)	Return Value
s1 < s2	< 0
s1 equals s2	0
s1 > s2	> 0

For example, if s1 is "cat" and s2 is "dog", the function will return a number less than 0. In the objectSort.java program, this method is used to compare the last name of a[in-1] with the last name of temp.

Stability

Sometimes it matters what happens to data items that have equal keys. For example, you may have employee data arranged alphabetically by last names. (That is, the last names were used as key values in the sort.) Now you want to sort the data by ZIP code, but you want all the items with the same ZIP code to continue to be sorted by last names. You want the algorithm to sort only what needs to be sorted, and leave everything else in its original order. Some sorting algorithms retain this secondary ordering; they're said to be *stable*.

All the algorithms in this chapter are stable. For example, notice the output of the objectSort.java program (Listing 3.4). Three persons have the last name of Smith. Initially, the order is Doc Smith, Lorraine Smith, and Paul Smith. After the sort, this ordering is preserved, despite the fact that the various Smith objects have been moved to new locations.

Comparing the Simple Sorts

There's probably no point in using the bubble sort, unless you don't have your algorithm book handy. The bubble sort is so simple that you can write it from memory. Even so, it's practical only if the amount of data is small. (For a discussion of what "small" means, see Chapter 15, "When to Use What.")

The selection sort minimizes the number of swaps, but the number of comparisons is still high. This sort might be useful when the amount of data is small and swapping data items is very time-consuming compared with comparing them.

The insertion sort is the most versatile of the three and is the best bet in most situations, assuming the amount of data is small or the data is almost sorted. For larger amounts of data, quicksort is generally considered the fastest approach; we'll examine quicksort in Chapter 7.

We've compared the sorting algorithms in terms of speed. Another consideration for any algorithm is how much memory space it needs. All three of the algorithms in this chapter carry out their sort *in place*, meaning that, besides the initial array, very little extra memory is required. All the sorts require an extra variable to store an item temporarily while it's being swapped.

You can recompile the example programs, such as bubbleSort.java, to sort larger amounts of data. By timing them for larger sorts, you can get an idea of the differences between them and the time required to sort different amounts of data on your particular system.

Summary

- The sorting algorithms in this chapter all assume an array as a data storage structure.

- Sorting involves comparing the keys of data items in the array and moving the items (actually, references to the items) around until they're in sorted order.

- All the algorithms in this chapter execute in $O(N^2)$ time. Nevertheless, some can be substantially faster than others.

- An invariant is a condition that remains unchanged while an algorithm runs.

- The bubble sort is the least efficient, but the simplest, sort.

- The insertion sort is the most commonly used of the $O(N^2)$ sorts described in this chapter.

- A sort is stable if the order of elements with the same key is retained.

- None of the sorts in this chapter require more than a single temporary variable, in addition to the original array.

Questions

These questions are intended as a self-test for readers. Answers may be found in Appendix C.

1. Computer sorting algorithms are more limited than humans in that

    a. humans are better at inventing new algorithms.

    b. computers can handle only a fixed amount of data.

    c. humans know what to sort, whereas computers need to be told.

    d. computers can compare only two things at a time.

2. The two basic operations in simple sorting are _____ items and _____ them (or sometimes _____ them).

3. True or False: The bubble sort always ends up comparing every item with every other item.

4. The bubble sort algorithm alternates between

    a. comparing and swapping.

    b. moving and copying.

    c. moving and comparing.

    d. copying and comparing.

5. True or False: If there are N items, the bubble sort makes exactly N*N comparisons.

6. In the selection sort,

    a. the largest keys accumulate on the left (low indices).

    b. a minimum key is repeatedly discovered.

    c. a number of items must be shifted to insert each item in its correctly sorted position.

    d. the sorted items accumulate on the right.

7. True or False: If, in a particular sorting situation, swaps take much longer than comparisons, the selection sort is about twice as fast as the bubble sort.

8. A copy is _____ times as fast as a swap.

9. What is the invariant in the selection sort?

10. In the insertion sort, the "marked player" described in the text corresponds to which variable in the insertSort.java program?

    a. in

    b. out

    c. temp

    d. a[out]

11. In the insertion sort, "partially sorted" means that

    a. some items are already sorted, but they may need to be moved.

    b. most items are in their final sorted positions, but a few still need to be sorted.

    c. only some of the items are sorted.

    d. group items are sorted among themselves, but items outside the group may need to be inserted in it.

12. Shifting a group of items left or right requires repeated _____.

13. In the insertion sort, after an item is inserted in the partially sorted group, it will

    a. never be moved again.

    b. never be shifted to the left.

    c. often be moved out of this group.

    d. find that its group is steadily shrinking.

14. The invariant in the insertion sort is that _____.

15. Stability might refer to

   a. items with secondary keys being excluded from a sort.

   b. keeping cities sorted by increasing population within each state, in a sort by state.

   c. keeping the same first names matched with the same last names.

   d. items keeping the same order of primary keys without regard to secondary keys.

Experiments

Carrying out these experiments will help to provide insights into the topics covered in the chapter. No programming is involved.

1. In bubbleSort.java (Listing 3.1) rewrite main() so it creates a large array and fills that array with data. You can use the following code to generate random numbers:

```
for(int j=0; j<maxSize; j++)          // fill array with
   {                                   // random numbers
   long n = (long)( java.lang.Math.random()*(maxSize-1) );
   arr.insert(n);
   }
```

Try inserting 10,000 items. Display the data before and after the sort. You'll see that scrolling the display takes a long time. Comment out the calls to display() so you can see how long the sort itself takes. The time will vary on different machines. Sorting 100,000 numbers will probably take less than 30 seconds. Pick an array size that takes about this long and time it. Then use the same array size to time selectSort.java (Listing 3.2) and insertSort.java (Listing 3.3). See how the speeds of these three sorts compare.

2. Devise some code to insert data in inversely sorted order (99,999, 99,998, 99,997, ...) into bubbleSort.java. Use the same amount of data as in Experiment 1. See how fast the sort runs compared with the random data in Experiment 1. Repeat this experiment with selectSort.java and insertSort.java.

3. Write code to insert data in already-sorted order (0, 1, 2, ...) into bubbleSort.java. See how fast the sort runs compared with Experiments 1 and 2. Repeat this experiment with selectSort.java and insertSort.java.

Programming Projects

Writing programs that solve the Programming Projects helps to solidify your under-standing of the material and demonstrates how the chapter's concepts are applied. (As noted in the Introduction, qualified instructors may obtain completed solutions to the Programming Projects on the publisher's Web site.)

3.1 In the bubbleSort.java program (Listing 3.1) and the BubbleSort Workshop applet, the in index always goes from left to right, finding the largest item and carrying it toward out on the right. Modify the bubbleSort() method so that it's bidirectional. This means the in index will first carry the largest item from left to right as before, but when it reaches out, it will reverse and carry the smallest item from right to left. You'll need two outer indexes, one on the right (the old out) and another on the left.

3.2 Add a method called median() to the ArrayIns class in the insertSort.java program (Listing 3.3). This method should return the median value in the array. (Recall that in a group of numbers half are larger than the median and half are smaller.) Do it the easy way.

3.3 To the insertSort.java program (Listing 3.3), add a method called noDups() that removes duplicates from a previously sorted array without disrupting the order. (You can use the insertionSort() method to sort the data, or you can simply use main() to insert the data in sorted order.) One can imagine schemes in which all the items from the place where a duplicate was discovered to the end of the array would be shifted down one space every time a duplicate was discovered, but this would lead to slow $O(N^2)$ time, at least when there were a lot of duplicates. In your algorithm, make sure no item is moved more than once, no matter how many duplicates there are. This will give you an algo-rithm with $O(N)$ time.

3.4 Another simple sort is the odd-even sort. The idea is to repeatedly make two passes through the array. On the first pass you look at all the pairs of items, a[j] and a[j+1], where j is odd (j = 1, 3, 5, ...). If their key values are out of order, you swap them. On the second pass you do the same for all the even values (j = 2, 4, 6, ...). You do these two passes repeatedly until the array is sorted. Replace the bubbleSort() method in bubbleSort.java (Listing 3.1) with an oddEvenSort() method. Make sure it works for varying amounts of data. You'll need to figure out how many times to do the two passes.

The odd-even sort is actually useful in a multiprocessing environment, where a separate processor can operate on each odd pair simultaneously and then on each even pair. Because the odd pairs are independent of each other, each pair can be checked—and swapped, if necessary—by a different processor. This makes for a very fast sort.

3.5 Modify the `insertionSort()` method in `insertSort.java` (Listing 3.3) so it counts the number of copies and the number of comparisons it makes during a sort and displays the totals. To count comparisons, you'll need to break up the double condition in the inner `while` loop. Use this program to measure the number of copies and comparisons for different amounts of inversely sorted data. Do the results verify $O(N^2)$ efficiency? Do the same for almost-sorted data (only a few items out of place). What can you deduce about the efficiency of this algorithm for almost-sorted data?

3.6 Here's an interesting way to remove duplicates from an array. The insertion sort uses a loop-within-a-loop algorithm that compares every item in the array with every other item. If you want to remove duplicates, this is one way to start. (See also Exercise 2.6 in Chapter 2.) Modify the `insertionSort()` method in the `insertSort.java` program so that it removes duplicates as it sorts. Here's one approach: When a duplicate is found, write over one of the duplicated items with a key value less than any normally used (such as –1, if all the normal keys are positive). Then the normal insertion sort algorithm, treating this new key like any other item, will put it at index 0. From now on the algorithm can ignore this item. The next duplicate will go at index 1, and so on. When the sort is finished, all the removed dups (now represented by –1 values) will be found at the beginning of the array. The array can then be resized and shifted down so it starts at 0.

4

Stacks and Queues

In this chapter we'll examine three data storage structures: the stack, the queue, and the priority queue. We'll begin by discussing how these structures differ from arrays; then we'll examine each one in turn. In the last section, we'll look at an operation in which the stack plays a significant role: parsing arithmetic expressions.

A Different Kind of Structure

There are significant differences between the data structures and algorithms we've seen in previous chapters and those we'll look at now. We'll discuss three of these differences before we examine the new structures in detail.

Programmer's Tools

Arrays—the data storage structure we've been examining thus far—as well as many other structures we'll encounter later in this book (linked lists, trees, and so on) are appropriate for the kind of data you might find in a database application. They're typically used for personnel records, inventories, financial data, and so on—data that corresponds to real-world objects or activities. These structures facilitate access to data: They make it easy to insert, delete, and search for particular items.

The structures and algorithms we'll examine in this chapter, on the other hand, are more often used as programmer's tools. They're primarily conceptual aids rather than full-fledged data storage devices. Their lifetime is typically shorter than that of the database-type structures. They are created and used to carry out a particular task during the operation of a program; when the task is completed, they're discarded.

Restricted Access

In an array, any item can be accessed, either immediately—if its index number is known—or by searching through a sequence of cells until it's found. In the data structures in this chapter, however, access is restricted: Only one item can be read or removed at a given time (unless you cheat).

The interface of these structures is designed to enforce this restricted access. Access to other items is (in theory) not allowed.

More Abstract

Stacks, queues, and priority queues are more abstract entities than arrays and many other data storage structures. They're defined primarily by their interface: the permissible operations that can be carried out on them. The underlying mechanism used to implement them is typically not visible to their user.

The underlying mechanism for a stack, for example, can be an array, as shown in this chapter, or it can be a linked list. The underlying mechanism for a priority queue can be an array or a special kind of tree called a *heap*. We'll return to the topic of one data structure being implemented by another when we discuss Abstract Data Types (ADTs) in Chapter 5, "Linked Lists."

Stacks

A stack allows access to only one data item: the last item inserted. If you remove this item, you can access the next-to-last item inserted, and so on. This capability is useful in many programming situations. In this section we'll see how a stack can be used to check whether parentheses, braces, and brackets are balanced in a computer program source file. At the end of this chapter, we'll see a stack playing a vital role in parsing (analyzing) arithmetic expressions such as 3*(4+5).

A stack is also a handy aid for algorithms applied to certain complex data structures. In Chapter 8, "Binary Trees," we'll see it used to help traverse the nodes of a tree. In Chapter 13, "Graphs," we'll apply it to searching the vertices of a graph (a technique that can be used to find your way out of a maze).

Most microprocessors use a stack-based architecture. When a method is called, its return address and arguments are pushed onto a stack, and when it returns, they're popped off. The stack operations are built into the microprocessor.

Some older pocket calculators used a stack-based architecture. Instead of entering arithmetic expressions using parentheses, you pushed intermediate results onto a stack. We'll learn more about this approach when we discuss parsing arithmetic expressions in the last section in this chapter.

The Postal Analogy

To understand the idea of a stack, consider an analogy provided by the U.S. Postal Service. Many people, when they get their mail, toss it onto a stack on the hall table or into an "in" basket at work. Then, when they have a spare moment, they process the accumulated mail from the top down. First, they open the letter on the top of the stack and take appropriate action—paying the bill, throwing it away, or whatever. After the first letter has been disposed of, they examine the next letter down, which is now the top of the stack, and deal with that. Eventually, they work their way down to the letter on the bottom of the stack (which is now the top). Figure 4.1 shows a stack of mail.

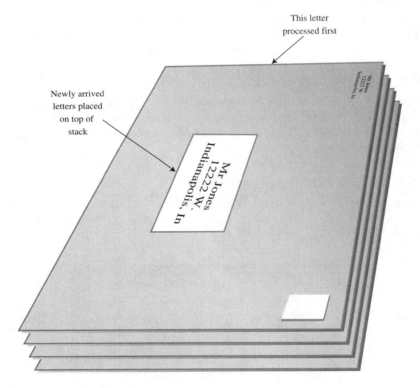

FIGURE 4.1   A stack of letters.

This "do the top one first" approach works all right as long as you can easily process all the mail in a reasonable time. If you can't, there's the danger that letters on the bottom of the stack won't be examined for months, and the bills they contain will become overdue.

Of course, many people don't rigorously follow this top-to-bottom approach. They may, for example, take the mail off the bottom of the stack, so as to process the oldest letter first. Or they might shuffle through the mail before they begin processing it and put higher-priority letters on top. In these cases, their mail system is no longer a stack in the computer-science sense of the word. If they take letters off the bottom, it's a queue; and if they prioritize it, it's a priority queue. We'll look at these possibilities later.

Another stack analogy is the tasks you perform during a typical workday. You're busy on a long-term project (A), but you're interrupted by a coworker asking you for temporary help with another project (B). While you're working on B, someone in accounting stops by for a meeting about travel expenses (C), and during this meeting you get an emergency call from someone in sales and spend a few minutes troubleshooting a bulky product (D). When you're done with call D, you resume meeting C; when you're done with C, you resume project B, and when you're done with B, you can (finally!) get back to project A. Lower-priority projects are "stacked up" waiting for you to return to them.

Placing a data item on the top of the stack is called *pushing* it. Removing it from the top of the stack is called *popping* it. These are the primary stack operations. A stack is said to be a Last-In-First-Out (LIFO) storage mechanism because the last item inserted is the first one to be removed.

The Stack Workshop Applet

Let's use the Stack Workshop applet to get an idea how stacks work. When you start up this applet, you'll see four buttons: New, Push, Pop, and Peek, as shown in Figure 4.2.

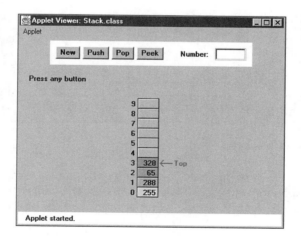

FIGURE 4.2   The Stack Workshop applet.

The Stack Workshop applet is based on an array, so you'll see an array of data items. Although it's based on an array, a stack restricts access, so you can't access elements using an index. In fact, the concept of a stack and the underlying data structure used to implement it are quite separate. As we noted earlier, stacks can also be implemented by other kinds of storage structures, such as linked lists.

The New Button

The stack in the Workshop applet starts off with four data items already inserted. If you want to start with an empty stack, the New button creates a new stack with no items. The next three buttons carry out the significant stack operations.

The Push Button

To insert a data item on the stack, use the button labeled Push. After the first press of this button, you'll be prompted to enter the key value of the item to be pushed. After you type the value into the text field, a few more presses will insert the item on the top of the stack.

A red arrow always points to the top of the stack—that is, the last item inserted. Notice how, during the insertion process, one step (button press) increments (moves up) the Top arrow, and the next step actually inserts the data item into the cell. If you reversed the order, you would overwrite the existing item at Top. When you're writing the code to implement a stack, it's important to keep in mind the order in which these two steps are executed.

If the stack is full and you try to push another item, you'll get the Can't insert: stack is full message. (Theoretically, an ADT stack doesn't become full, but the array implementing it does.)

The Pop Button

To remove a data item from the top of the stack, use the Pop button. The value popped appears in the Number text field; this corresponds to a pop() routine returning a value.

Again, notice the two steps involved: First, the item is removed from the cell pointed to by Top; then Top is decremented to point to the highest occupied cell. This is the reverse of the sequence used in the push operation.

The pop operation shows an item actually being removed from the array and the cell color becoming gray to show the item has been removed. This is a bit misleading, in that deleted items actually remain in the array until written over by new data. However, they cannot be accessed after the Top marker drops below their position, so conceptually they are gone, as the applet shows.

After you've popped the last item off the stack, the Top arrow points to –1, below the lowest cell. This position indicates that the stack is empty. If the stack is empty and you try to pop an item, you'll get the Can't pop: stack is empty message.

The Peek Button

Push and pop are the two primary stack operations. However, it's sometimes useful to be able to read the value from the top of the stack without removing it. The peek operation does this. By pushing the Peek button a few times, you'll see the value of the item at Top copied to the Number text field, but the item is not removed from the stack, which remains unchanged.

Notice that you can peek only at the top item. By design, all the other items are invisible to the stack user.

Stack Size

Stacks are typically small, temporary data structures, which is why we've shown a stack of only 10 cells. Of course, stacks in real programs may need a bit more room than this, but it's surprising how small a stack needs to be. A very long arithmetic expression, for example, can be parsed with a stack of only a dozen or so cells.

Java Code for a Stack

Let's examine a program, stack.java, that implements a stack using a class called StackX. Listing 4.1 contains this class and a short main() routine to exercise it.

LISTING 4.1   The stack.java Program

```
// stack.java
// demonstrates stacks
// to run this program: C>java StackApp
////////////////////////////////////////////////////////////
class StackX
   {
   private int maxSize;         // size of stack array
   private long[] stackArray;
   private int top;             // top of stack
//--------------------------------------------------------------
   public StackX(int s)          // constructor
      {
      maxSize = s;               // set array size
      stackArray = new long[maxSize];  // create array
      top = -1;                  // no items yet
      }
//--------------------------------------------------------------
   public void push(long j)      // put item on top of stack
      {
      stackArray[++top] = j;       // increment top, insert item
      }
```

LISTING 4.1   Continued

```
//--------------------------------------------------------------
   public long pop()              // take item from top of stack
      {
      return stackArray[top--];  // access item, decrement top
      }
//--------------------------------------------------------------
   public long peek()             // peek at top of stack
      {
      return stackArray[top];
      }
//--------------------------------------------------------------
   public boolean isEmpty()     // true if stack is empty
      {
      return (top == -1);
      }
//--------------------------------------------------------------
   public boolean isFull()      // true if stack is full
      {
      return (top == maxSize-1);
      }
//--------------------------------------------------------------
   } // end class StackX
////////////////////////////////////////////////////////////////
class StackApp
   {
   public static void main(String[] args)
      {
      StackX theStack = new StackX(10);  // make new stack
      theStack.push(20);                 // push items onto stack
      theStack.push(40);
      theStack.push(60);
      theStack.push(80);

      while( !theStack.isEmpty() )     // until it's empty,
         {                             // delete item from stack
         long value = theStack.pop();
         System.out.print(value);      // display it
         System.out.print(" ");
         } // end while
      System.out.println("");
      } // end main()
```

LISTING 4.1   Continued

```
    } // end class StackApp
//////////////////////////////////////////////////////////////
```

The `main()` method in the `StackApp` class creates a stack that can hold 10 items, pushes 4 items onto the stack, and then displays all the items by popping them off the stack until it's empty. Here's the output:

`80 60 40 20`

Notice how the order of the data is reversed. Because the last item pushed is the first one popped, the 80 appears first in the output.

This version of the `StackX` class holds data elements of type `long`. As noted in Chapter 3, "Simple Sorting," you can change this to any other type, including object types.

`StackX` **Class Methods**
The constructor creates a new stack of a size specified in its argument. The fields of the stack are made up of a variable to hold its maximum size (the size of the array), the array itself, and a variable `top`, which stores the index of the item on the top of the stack. (Note that we need to specify a stack size only because the stack is implemented using an array. If it had been implemented using a linked list, for example, the size specification would be unnecessary.)

The `push()` method increments `top` so it points to the space just above the previous top and stores a data item there. Notice again that `top` is incremented before the item is inserted.

The `pop()` method returns the value at `top` and then decrements `top`. This effectively removes the item from the stack; it's inaccessible, although the value remains in the array (until another item is pushed into the cell).

The `peek()` method simply returns the value at `top`, without changing the stack.

The `isEmpty()` and `isFull()` methods return `true` if the stack is empty or full, respectively. The `top` variable is at –1 if the stack is empty and `maxSize-1` if the stack is full.

Figure 4.3 shows how the stack class methods work.

Error Handling
There are different philosophies about how to handle stack errors. What happens if you try to push an item onto a stack that's already full or pop an item from a stack that's empty?

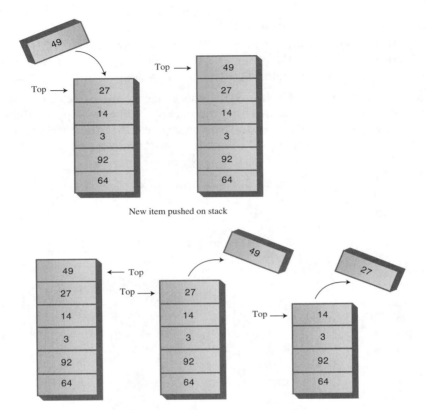

New item pushed on stack

Two items popped from stack

FIGURE 4.3 Operation of the StackX class methods.

We've left the responsibility for handling such errors up to the class user. The user should always check to be sure the stack is not full before inserting an item:

```
if( !theStack.isFull() )
   insert(item);
else
   System.out.print("Can't insert, stack is full");
```

In the interest of simplicity, we've left this code out of the main() routine (and anyway, in this simple program, we know the stack isn't full because it has just been initialized). We do include the check for an empty stack when main() calls pop().

Many stack classes check for these errors internally, in the push() and pop() methods. This is the preferred approach. In Java, a good solution for a stack class that discovers such errors is to throw an exception, which can then be caught and processed by the class user.

Stack Example 1: Reversing a Word

For our first example of using a stack, we'll examine a very simple task: reversing a word. When you run the program, it asks you to type in a word. When you press Enter, it displays the word with the letters in reverse order.

A stack is used to reverse the letters. First, the characters are extracted one by one from the input string and pushed onto the stack. Then they're popped off the stack and displayed. Because of its Last-In-First-Out characteristic, the stack reverses the order of the characters. Listing 4.2 shows the code for the reverse.java program.

LISTING 4.2   The reverse.java Program

```
// reverse.java
// stack used to reverse a string
// to run this program: C>java ReverseApp
import java.io.*;                    // for I/O
/////////////////////////////////////////////////////////////////
class StackX
   {
   private int maxSize;
   private char[] stackArray;
   private int top;
//-------------------------------------------------------------
   public StackX(int max)      // constructor
      {
      maxSize = max;
      stackArray = new char[maxSize];
      top = -1;
      }
//-------------------------------------------------------------
   public void push(char j)  // put item on top of stack
      {
      stackArray[++top] = j;
      }
//-------------------------------------------------------------
   public char pop()          // take item from top of stack
      {
      return stackArray[top--];
      }
//-------------------------------------------------------------
   public char peek()         // peek at top of stack
      {
      return stackArray[top];
```

LISTING 4.2   Continued

```
      }
//-------------------------------------------------------------
   public boolean isEmpty()  // true if stack is empty
      {
      return (top == -1);
      }
//-------------------------------------------------------------
   }  // end class StackX
////////////////////////////////////////////////////////////////
class Reverser
   {
   private String input;              // input string
   private String output;             // output string
//-------------------------------------------------------------
   public Reverser(String in)         // constructor
      { input = in; }
//-------------------------------------------------------------
   public String doRev()              // reverse the string
      {
      int stackSize = input.length();   // get max stack size
      StackX theStack = new StackX(stackSize);  // make stack

      for(int j=0; j<input.length(); j++)
         {
         char ch = input.charAt(j);     // get a char from input
         theStack.push(ch);             // push it
         }
      output = "";
      while( !theStack.isEmpty() )
         {
         char ch = theStack.pop();      // pop a char,
         output = output + ch;          // append to output
         }
      return output;
      }  // end doRev()
//-------------------------------------------------------------
   }  // end class Reverser
////////////////////////////////////////////////////////////////
class ReverseApp
   {
   public static void main(String[] args) throws IOException
```

LISTING 4.2   Continued

```
        {
        String input, output;
        while(true)
           {
           System.out.print("Enter a string: ");
           System.out.flush();
           input = getString();        // read a string from kbd
           if( input.equals("") )       // quit if [Enter]
              break;
                                        // make a Reverser
           Reverser theReverser = new Reverser(input);
           output = theReverser.doRev(); // use it
           System.out.println("Reversed: " + output);
           } // end while
        } // end main()
//-------------------------------------------------------------
    public static String getString() throws IOException
       {
       InputStreamReader isr = new InputStreamReader(System.in);
       BufferedReader br = new BufferedReader(isr);
       String s = br.readLine();
       return s;
       }
//-------------------------------------------------------------
    } // end class ReverseApp
//////////////////////////////////////////////////////////////
```

We've created a class `Reverser` to handle the reversing of the input string. Its key component is the method `doRev()`, which carries out the reversal, using a stack. The stack is created within `doRev()`, which sizes the stack according to the length of the input string.

In `main()` we get a string from the user, create a `Reverser` object with this string as an argument to the constructor, call this object's `doRev()` method, and display the return value, which is the reversed string. Here's some sample interaction with the program:

```
Enter a string: part
Reversed: trap
Enter a string:
```

Stack Example 2: Delimiter Matching

One common use for stacks is to parse certain kinds of text strings. Typically, the strings are lines of code in a computer language, and the programs parsing them are compilers.

To give the flavor of what's involved, we'll show a program that checks the delimiters in a line of text typed by the user. This text doesn't need to be a line of real Java code (although it could be), but it should use delimiters the same way Java does. The delimiters are the braces { and }, brackets [and], and parentheses (and). Each opening or left delimiter should be matched by a closing or right delimiter; that is, every { should be followed by a matching } and so on. Also, opening delimiters that occur later in the string should be closed before those occurring earlier. Here are some examples:

```
c[d]         // correct
a{b[c]d}e     // correct
a{b(c]d}e     // not correct; ] doesn't match (
a[b{c}d]e}    // not correct; nothing matches final }
a{b(c)        // not correct; nothing matches opening {
```

Opening Delimiters on the Stack

This delimiter-matching program works by reading characters from the string one at a time and placing opening delimiters when it finds them, on a stack. When it reads a closing delimiter from the input, it pops the opening delimiter from the top of the stack and attempts to match it with the closing delimiter. If they're not the same type (there's an opening brace but a closing parenthesis, for example), an error occurs. Also, if there is no opening delimiter on the stack to match a closing one, or if a delimiter has not been matched, an error occurs. A delimiter that hasn't been matched is discovered because it remains on the stack after all the characters in the string have been read.

Let's see what happens on the stack for a typical correct string:

```
a{b(c[d]e)f}
```

Table 4.1 shows how the stack looks as each character is read from this string. The entries in the second column show the stack contents, reading from the bottom of the stack on the left to the top on the right.

As the string is read, each opening delimiter is placed on the stack. Each closing delimiter read from the input is matched with the opening delimiter popped from the top of the stack. If they form a pair, all is well. Non-delimiter characters are not inserted on the stack; they're ignored.

TABLE 4.1   Stack Contents in Delimiter Matching

Character Read	Stack Contents
a	
{	{
b	{
(	{(
c	{(
[	{([
d	{([
]	{(
e	{(
)	{
f	{
}	

This approach works because pairs of delimiters that are opened last should be closed first. This matches the Last-In-First-Out property of the stack.

Java Code for brackets.java

The code for the parsing program, brackets.java, is shown in Listing 4.3. We've placed check(), the method that does the parsing, in a class called BracketChecker.

LISTING 4.3   The brackets.java Program

```java
// brackets.java
// stacks used to check matching brackets
// to run this program: C>java BracketsApp
import java.io.*;              // for I/O
////////////////////////////////////////////////////////////////
class StackX
   {
   private int maxSize;
   private char[] stackArray;
   private int top;
//--------------------------------------------------------------
   public StackX(int s)         // constructor
      {
      maxSize = s;
      stackArray = new char[maxSize];
      top = -1;
      }
//--------------------------------------------------------------
```

LISTING 4.3   Continued

```
   public void push(char j)   // put item on top of stack
       {
       stackArray[++top] = j;
       }
//------------------------------------------------------------
   public char pop()           // take item from top of stack
       {
       return stackArray[top--];
       }
//------------------------------------------------------------
   public char peek()          // peek at top of stack
       {
       return stackArray[top];
       }
//------------------------------------------------------------
   public boolean isEmpty()     // true if stack is empty
       {
       return (top == -1);
       }
//------------------------------------------------------------
   }  // end class StackX
////////////////////////////////////////////////////////////////
class BracketChecker
   {
   private String input;                 // input string
//------------------------------------------------------------
   public BracketChecker(String in)      // constructor
       { input = in; }
//------------------------------------------------------------
   public void check()
       {
       int stackSize = input.length();       // get max stack size
       StackX theStack = new StackX(stackSize);  // make stack

       for(int j=0; j<input.length(); j++)  // get chars in turn
          {
          char ch = input.charAt(j);         // get char
          switch(ch)
             {
             case '{':                        // opening symbols
             case '[':
```

LISTING 4.3   Continued

```
            case '(':
               theStack.push(ch);          // push them
               break;

            case '}':                       // closing symbols
            case ']':
            case ')':
               if( !theStack.isEmpty() )    // if stack not empty,
                  {
                  char chx = theStack.pop();   // pop and check
                  if( (ch=='}' && chx!='{') ||
                      (ch==']' && chx!='[') ||
                      (ch==')' && chx!='(') )
                     System.out.println("Error: "+ch+" at "+j);
                  }
               else                          // prematurely empty
                  System.out.println("Error: "+ch+" at "+j);
               break;
            default:     // no action on other characters
               break;
            } // end switch
         } // end for
      // at this point, all characters have been processed
      if( !theStack.isEmpty() )
         System.out.println("Error: missing right delimiter");
      } // end check()
//--------------------------------------------------------------
   } // end class BracketChecker
////////////////////////////////////////////////////////////////
class BracketsApp
   {
   public static void main(String[] args) throws IOException
      {
      String input;
      while(true)
         {
         System.out.print(
                    "Enter string containing delimiters: ");
         System.out.flush();
         input = getString();      // read a string from kbd
         if( input.equals("") )    // quit if [Enter]
```

LISTING 4.3   Continued

```
        break;
                                 // make a BracketChecker
        BracketChecker theChecker = new BracketChecker(input);
        theChecker.check();      // check brackets
        }  // end while
      }  // end main()
//----------------------------------------------------------
   public static String getString() throws IOException
      {
      InputStreamReader isr = new InputStreamReader(System.in);
      BufferedReader br = new BufferedReader(isr);
      String s = br.readLine();
      return s;
      }
//----------------------------------------------------------
   }  // end class BracketsApp
//////////////////////////////////////////////////////////////
```

The check() routine makes use of the StackX class from the reverse.java program (Listing 4.2). Notice how easy it is to reuse this class. All the code you need is in one place. This is one of the payoffs for object-oriented programming.

The main() routine in the BracketsApp class repeatedly reads a line of text from the user, creates a BracketChecker object with this text string as an argument, and then calls the check() method for this BracketChecker object. If it finds any errors, the check() method displays them; otherwise, the syntax of the delimiters is correct.

If it can, the check() method reports the character number where it discovered the error (starting at 0 on the left) and the incorrect character it found there. For example, for the input string

a{b(c]d}e

the output from check() will be

Error:] at 5

The Stack as a Conceptual Aid

Notice how convenient the stack is in the brackets.java program. You could have set up an array to do what the stack does, but you would have had to worry about keeping track of an index to the most recently added character, as well as other bookkeeping tasks. The stack is conceptually easier to use. By providing limited access to its contents, using the push() and pop() methods, the stack has made your program easier to understand and less error prone. (As carpenters will tell you, it's safer to use the right tool for the job.)

Efficiency of Stacks

Items can be both pushed and popped from the stack implemented in the `StackX` class in constant O(1) time. That is, the time is not dependent on how many items are in the stack and is therefore very quick. No comparisons or moves are necessary.

Queues

The word *queue* is British for *line* (the kind you wait in). In Britain, to "queue up" means to get in line. In computer science a queue is a data structure that is somewhat like a stack, except that in a queue the first item inserted is the first to be removed (First-In-First-Out, FIFO), while in a stack, as we've seen, the last item inserted is the first to be removed (LIFO). A queue works like the line at the movies: The first person to join the rear of the line is the first person to reach the front of the line and buy a ticket. The last person to line up is the last person to buy a ticket (or—if the show is sold out—to fail to buy a ticket). Figure 4.4 shows how such a queue looks.

People join the queue at the rear

People leave the queue at the front

FIGURE 4.4   A queue of people.

Queues are used as a programmer's tool as stacks are. We'll see an example where a queue helps search a graph in Chapter 13. They're also used to model real-world situations such as people waiting in line at a bank, airplanes waiting to take off, or data packets waiting to be transmitted over the Internet.

There are various queues quietly doing their job in your computer's (or the network's) operating system. There's a printer queue where print jobs wait for the printer to be available. A queue also stores keystroke data as you type at the keyboard. This way, if you're using a word processor but the computer is briefly doing something else when you hit a key, the keystroke won't be lost; it waits in the queue until the word processor has time to read it. Using a queue guarantees the keystrokes stay in order until they can be processed.

The Queue Workshop Applet

Let's use the Queue Workshop applet to get an idea how queues work. When you start up the applet, you'll see a queue with four items preinstalled, as shown in Figure 4.5.

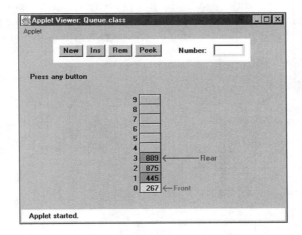

FIGURE 4.5   The Queue Workshop applet.

This applet demonstrates a queue based on an array. This is a common approach, although linked lists are also commonly used to implement queues.

The two basic queue operations are *inserting* an item, which is placed at the rear of the queue, and *removing* an item, which is taken from the front of the queue. This is similar to a person joining the rear of a line of movie-goers and, having arrived at the front of the line and purchased a ticket, removing herself from the front of the line.

The terms for insertion and removal in a stack are fairly standard; everyone says *push* and *pop*. Standardization hasn't progressed this far with queues. *Insert* is also called *put* or *add* or *enque*, while *remove* may be called *delete* or *get* or *deque*. The rear of the queue, where items are inserted, is also called the *back* or *tail* or *end*. The front, where items are removed, may also be called the *head*. We'll use the terms *insert*, *remove*, *front*, and *rear*.

The Insert Button

By repeatedly pressing the Ins button in the Queue Workshop applet, you can insert a new item. After the first press, you're prompted to enter a key value for a new item into the Number text field; this should be a number from 0 to 999. Subsequent presses will insert an item with this key at the rear of the queue and increment the Rear arrow so it points to the new item.

The Remove Button

Similarly, you can remove the item at the front of the queue using the Rem button. The item is removed, the item's value is stored in the Number field (corresponding to the remove() method returning a value), and the Front arrow is incremented. In the applet, the cell that held the deleted item is grayed to show it's gone. In a normal implementation, it would remain in memory but would not be accessible because Front had moved past it. The insert and remove operations are shown in Figure 4.6.

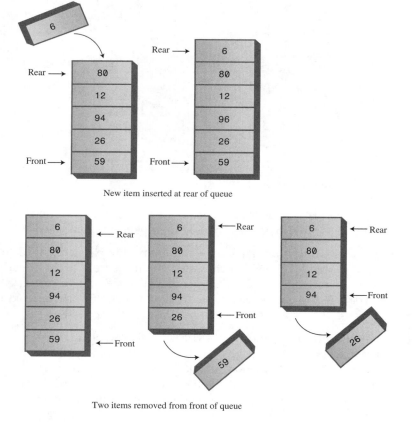

New item inserted at rear of queue

Two items removed from front of queue

FIGURE 4.6　Operation of the Queue class methods.

Unlike the situation in a stack, the items in a queue don't always extend all the way down to index 0 in the array. After some items are removed, Front will point at a cell with a higher index, as shown in Figure 4.7.

In Figure 4.7, notice that Front lies below Rear in the array; that is, Front has a lower index. As we'll see in a moment, this isn't always true.

The Peek Button
We show one other queue operation, peek. Peek finds the value of the item at the front of the queue without removing the item. (Like insert and remove, peek, when applied to a queue, is also called by a variety of other names.) If you press the Peek button, you'll see the value at Front transferred to the Number field. The queue is unchanged. This peek() method returns the value at the front of the queue. Some queue implementations have a rearPeek() and a frontPeek() method, but usually you want to know what you're about to remove, not what you just inserted.

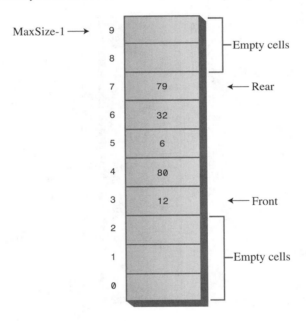

FIGURE 4.7   A queue with some items removed.

The New Button
If you want to start with an empty queue, you can use the New button to create one.

Empty and Full
If you try to remove an item when there are no more items in the queue, you'll get the Can't remove, queue is empty error message. If you try to insert an item when all the cells are already occupied, you'll get the Can't insert, queue is full message.

A Circular Queue

When you insert a new item in the queue in the Queue Workshop applet, the Front arrow moves upward, toward higher numbers in the array. When you remove an item, Rear also moves upward. Try these operations with the Workshop applet to convince yourself it's true. You may find the arrangement counter-intuitive, because the people in a line at the movies all move forward, toward the front, when a person leaves the line. We could move all the items in a queue whenever we deleted one, but that wouldn't be very efficient. Instead, we keep all the items in the same place and move the front and rear of the queue.

The trouble with this arrangement is that pretty soon the rear of the queue is at the end of the array (the highest index). Even if there are empty cells at the beginning of the array, because you've removed them with Rem, you still can't insert a new item because Rear can't go any further. Or can it? This situation is shown in Figure 4.8.

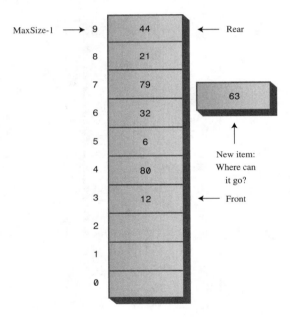

FIGURE 4.8   Rear arrow at the end of the array.

Wrapping Around

To avoid the problem of not being able to insert more items into the queue even when it's not full, the Front and Rear arrows *wrap around* to the beginning of the array. The result is a *circular queue* (sometimes called a *ring buffer*).

You can see how wraparound works with the Workshop applet. Insert enough items to bring the Rear arrow to the top of the array (index 9). Remove some items from the front of the array. Now insert another item. You'll see the Rear arrow wrap around from index 9 to index 0; the new item will be inserted there. This situation is shown in Figure 4.9.

Insert a few more items. The Rear arrow moves upward as you'd expect. Notice that after Rear has wrapped around, it's now below Front, the reverse of the original arrangement. You can call this a *broken sequence*: The items in the queue are in two different sequences in the array.

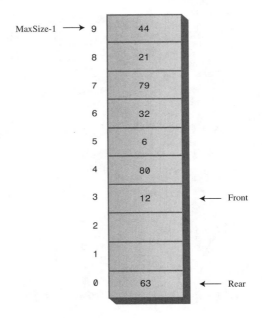

FIGURE 4.9   The Rear arrow wraps around.

Delete enough items so that the Front arrow also wraps around. Now you're back to the original arrangement, with Front below Rear. The items are in a single *contiguous sequence*.

Java Code for a Queue

The queue.java program features a Queue class with insert(), remove(), peek(), isFull(), isEmpty(), and size() methods.

The main() program creates a queue of five cells, inserts four items, removes three items, and inserts four more. The sixth insertion invokes the wraparound feature. All the items are then removed and displayed. The output looks like this:

```
40 50 60 70 80
```

Listing 4.4 shows the queue.java program.

LISTING 4.4   The queue.java Program

```java
// queue.java
// demonstrates queue
// to run this program: C>java QueueApp
////////////////////////////////////////////////////////////
class Queue
   {
   private int maxSize;
   private long[] queArray;
   private int front;
   private int rear;
   private int nItems;
//-------------------------------------------------------------
   public Queue(int s)          // constructor
      {
      maxSize = s;
      queArray = new long[maxSize];
      front = 0;
      rear = -1;
      nItems = 0;
      }
//-------------------------------------------------------------
   public void insert(long j)   // put item at rear of queue
      {
      if(rear == maxSize-1)              // deal with wraparound
         rear = -1;
      queArray[++rear] = j;              // increment rear and insert
      nItems++;                          // one more item
      }
//-------------------------------------------------------------
   public long remove()         // take item from front of queue
      {
      long temp = queArray[front++]; // get value and incr front
      if(front == maxSize)               // deal with wraparound
         front = 0;
      nItems--;                          // one less item
      return temp;
      }
//-------------------------------------------------------------
   public long peekFront()      // peek at front of queue
      {
```

LISTING 4.4   Continued

```
        return queArray[front];
        }
//-------------------------------------------------------------
    public boolean isEmpty()     // true if queue is empty
        {
        return (nItems==0);
        }
//-------------------------------------------------------------
    public boolean isFull()      // true if queue is full
        {
        return (nItems==maxSize);
        }
//-------------------------------------------------------------
    public int size()            // number of items in queue
        {
        return nItems;
        }
//-------------------------------------------------------------
    } // end class Queue
////////////////////////////////////////////////////////////////
class QueueApp
    {
    public static void main(String[] args)
        {
        Queue theQueue = new Queue(5);  // queue holds 5 items

        theQueue.insert(10);            // insert 4 items
        theQueue.insert(20);
        theQueue.insert(30);
        theQueue.insert(40);

        theQueue.remove();              // remove 3 items
        theQueue.remove();              //    (10, 20, 30)
        theQueue.remove();

        theQueue.insert(50);            // insert 4 more items
        theQueue.insert(60);            //    (wraps around)
        theQueue.insert(70);
        theQueue.insert(80);

        while( !theQueue.isEmpty() )    // remove and display
            {                           //    all items
```

LISTING 4.4   Continued

```
        long n = theQueue.remove();   // (40, 50, 60, 70, 80)
        System.out.print(n);
        System.out.print(" ");
        }
    System.out.println("");
    }  // end main()
  }  // end class QueueApp
```

We've chosen an approach in which Queue class fields include not only front and rear, but also the number of items currently in the queue: nItems. Some queue implementations don't use this field; we'll show this alternative later.

The insert() Method
The insert() method assumes that the queue is not full. We don't show it in main(), but normally you should call insert() only after calling isFull() and getting a return value of false. (It's usually preferable to place the check for fullness in the insert() routine and cause an exception to be thrown if an attempt was made to insert into a full queue.)

Normally, insertion involves incrementing rear and inserting at the cell rear now points to. However, if rear is at the top of the array, at maxSize-1, then it must wrap around to the bottom of the array before the insertion takes place. This is done by setting rear to –1, so when the increment occurs, rear will become 0, the bottom of the array. Finally, nItems is incremented.

The remove() Method
The remove() method assumes that the queue is not empty. You should call isEmpty() to ensure this is true before calling remove(), or build this error-checking into remove().

Removal always starts by obtaining the value at front and then incrementing front. However, if this puts front beyond the end of the array, it must then be wrapped around to 0. The return value is stored temporarily while this possibility is checked. Finally, nItems is decremented.

The peek() Method
The peek() method is straightforward: It returns the value at front. Some implementations allow peeking at the rear of the array as well; such routines are called something like peekFront() and peekRear() or just front() and rear().

The isEmpty(), isFull(), and size() Methods
The isEmpty(), isFull(), and size() methods all rely on the nItems field, respectively checking if it's 0, if it's maxSize, or returning its value.

Implementation Without an Item Count

The inclusion of the field nItems in the Queue class imposes a slight overhead on the insert() and remove() methods in that they must respectively increment and decrement this variable. This may not seem like an excessive penalty, but if you're dealing with huge numbers of insertions and deletions, it might influence performance.

Accordingly, some implementations of queues do without an item count and rely on the front and rear fields to figure out whether the queue is empty or full and how many items are in it. When this is done, the isEmpty(), isFull(), and size() routines become surprisingly complicated because the sequence of items may be either broken or contiguous, as we've seen.

Also, a strange problem arises. The front and rear pointers assume certain positions when the queue is full, but they can assume these exact same positions when the queue is empty. The queue can then appear to be full and empty at the same time.

This problem can be solved by making the array one cell larger than the maximum number of items that will be placed in it. Listing 4.5 shows a Queue class that implements this no-count approach. This class uses the no-count implementation.

LISTING 4.5   The Queue Class Without nItems

```
class Queue
   {
   private int maxSize;
   private long[] queArray;
   private int front;
   private int rear;
//----------------------------------------------------------------
   public Queue(int s)          // constructor
      {
      maxSize = s+1;                    // array is 1 cell larger
      queArray = new long[maxSize];  // than requested
      front = 0;
      rear = -1;
      }
//----------------------------------------------------------------
   public void insert(long j)  // put item at rear of queue
      {
      if(rear == maxSize-1)
         rear = -1;
      queArray[++rear] = j;
      }
//----------------------------------------------------------------
   public long remove()          // take item from front of queue
      {
```

LISTING 4.5   Continued

```
        long temp = queArray[front++];
        if(front == maxSize)
           front = 0;
        return temp;
        }
//-------------------------------------------------------------
    public long peek()            // peek at front of queue
        {
        return queArray[front];
        }
//-------------------------------------------------------------
    public boolean isEmpty()      // true if queue is empty
        {
        return ( rear+1==front || (front+maxSize-1==rear) );
        }
//-------------------------------------------------------------
    public boolean isFull()       // true if queue is full
        {
        return ( rear+2==front || (front+maxSize-2==rear) );
        }
//-------------------------------------------------------------
    public int size()             // (assumes queue not empty)
        {
        if(rear >= front)             // contiguous sequence
           return rear-front+1;
        else                          // broken sequence
           return (maxSize-front) + (rear+1);
        }
//-------------------------------------------------------------
    } // end class Queue
```

Notice the complexity of the isFull(), isEmpty(), and size() methods. This no-count approach is seldom needed in practice, so we'll refrain from discussing it in detail.

Efficiency of Queues

As with a stack, items can be inserted and removed from a queue in O(1) time.

Deques

A *deque* is a double-ended queue. You can insert items at either end and delete them from either end. The methods might be called `insertLeft()` and `insertRight()`, and `removeLeft()` and `removeRight()`.

If you restrict yourself to `insertLeft()` and `removeLeft()` (or their equivalents on the right), the deque acts like a stack. If you restrict yourself to `insertLeft()` and `removeRight()` (or the opposite pair), it acts like a queue.

A deque provides a more versatile data structure than either a stack or a queue and is sometimes used in container class libraries to serve both purposes. However, it's not used as often as stacks and queues, so we won't explore it further here.

Priority Queues

A priority queue is a more specialized data structure than a stack or a queue. However, it's a useful tool in a surprising number of situations. Like an ordinary queue, a priority queue has a front and a rear, and items are removed from the front. However, in a priority queue, items are ordered by key value so that the item with the lowest key (or in some implementations the highest key) is always at the front. Items are inserted in the proper position to maintain the order.

Here's how the mail sorting analogy applies to a priority queue. Every time the postman hands you a letter, you insert it into your pile of pending letters according to its priority. If it must be answered immediately (the phone company is about to disconnect your modem line), it goes on top, whereas if it can wait for a leisurely answer (a letter from your Aunt Mabel), it goes on the bottom. Letters with intermediate priorities are placed in the middle; the higher the priority, the higher their position in the pile. The top of the pile of letters corresponds to the front of the priority queue.

When you have time to answer your mail, you start by taking the letter off the top (the front of the queue), thus ensuring that the most important letters are answered first. This situation is shown in Figure 4.10.

Like stacks and queues, priority queues are often used as programmer's tools. We'll see one used in finding something called a minimum spanning tree for a graph, in Chapter 14, "Weighted Graphs."

Also, like ordinary queues, priority queues are used in various ways in certain computer systems. In a preemptive multitasking operating system, for example, programs may be placed in a priority queue so the highest-priority program is the next one to receive a time-slice that allows it to execute.

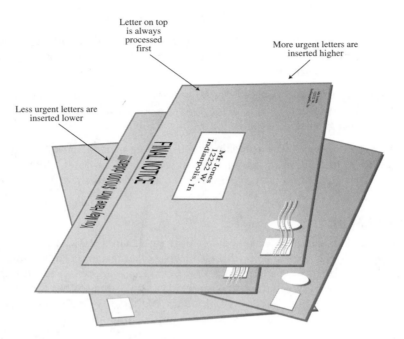

FIGURE 4.10 Letters in a priority queue.

In many situations you want access to the item with the lowest key value (which might represent the cheapest or shortest way to do something). Thus, the item with the smallest key has the highest priority. Somewhat arbitrarily, we'll assume that's the case in this discussion, although there are other situations in which the highest key has the highest priority.

Besides providing quick access to the item with the smallest key, you also want a priority queue to provide fairly quick insertion. For this reason, priority queues are, as we noted earlier, often implemented with a data structure called a heap. We'll look at heaps in Chapter 12, "Heaps." In this chapter, we'll show a priority queue implemented by a simple array. This implementation suffers from slow insertion, but it's simpler and is appropriate when the number of items isn't high or insertion speed isn't critical.

The PriorityQ Workshop Applet

The PriorityQ Workshop applet implements a priority queue with an array, in which the items are kept in sorted order. It's an *ascending-priority* queue, in which the item with smallest key has the highest priority and is accessed with remove(). (If the highest-key item were accessed, it would be a *descending-priority* queue.)

The minimum-key item is always at the top (highest index) in the array, and the largest item is always at index 0. Figure 4.11 shows the arrangement when the applet is started. Initially, there are five items in the queue.

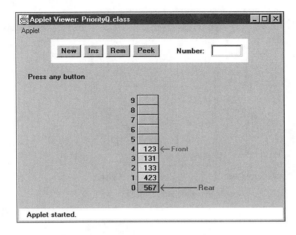

FIGURE 4.11 The PriorityQ Workshop applet.

The Insert Button
Try inserting an item. You'll be prompted to type the new item's key value into the Number field. Choose a number that will be inserted somewhere in the middle of the values already in the queue. For example, in Figure 4.11 you might choose 300. Then, as you repeatedly press Ins, you'll see that the items with smaller keys are shifted up to make room. A black arrow shows which item is being shifted. When the appropriate position is found, the new item is inserted into the newly created space.

Notice that there's no wraparound in this implementation of the priority queue. Insertion is slow of necessity because the proper in-order position must be found, but deletion is fast. A wraparound implementation wouldn't improve the situation. Note too that the Rear arrow never moves; it always points to index 0 at the bottom of the array.

The Delete Button
The item to be removed is always at the top of the array, so removal is quick and easy; the item is removed and the Front arrow moves down to point to the new top of the array. No shifting or comparisons are necessary.

In the PriorityQ Workshop applet, we show Front and Rear arrows to provide a comparison with an ordinary queue, but they're not really necessary. The algorithms know that the front of the queue is always at the top of the array at `nItems-1`, and they insert items in order, not at the rear. Figure 4.12 shows the operation of the `PriorityQ` class methods.

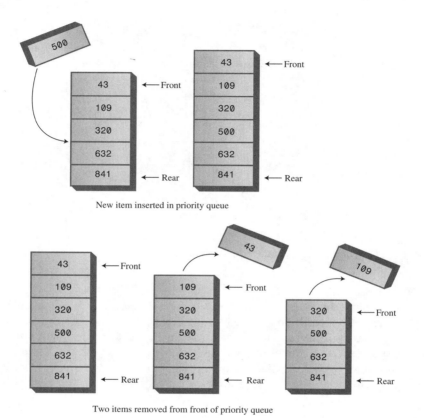

FIGURE 4.12 Operation of the `PriorityQ` class methods.

The Peek and New Buttons

You can peek at the minimum item (find its value without removing it) with the Peek button, and you can create a new, empty, priority queue with the New button.

Other Implementation Possibilities

The implementation shown in the PriorityQ Workshop applet isn't very efficient for insertion, which involves moving an average of half the items.

Another approach, which also uses an array, makes no attempt to keep the items in sorted order. New items are simply inserted at the top of the array. This makes insertion very quick, but unfortunately it makes deletion slow because the smallest item must be searched for. This approach requires examining all the items and shifting half of them, on the average, down to fill in the hole. In most situations the quick-deletion approach shown in the Workshop applet is preferred.

For small numbers of items, or situations in which speed isn't critical, implementing a priority queue with an array is satisfactory. For larger numbers of items, or when speed is critical, the heap is a better choice.

Java Code for a Priority Queue

The Java code for a simple array-based priority queue is shown in Listing 4.6.

LISTING 4.6   The `priorityQ.java` Program

```java
// priorityQ.java
// demonstrates priority queue
// to run this program: C>java PriorityQApp
////////////////////////////////////////////////////////////
class PriorityQ
   {
   // array in sorted order, from max at 0 to min at size-1
   private int maxSize;
   private long[] queArray;
   private int nItems;
//-------------------------------------------------------------
   public PriorityQ(int s)          // constructor
      {
      maxSize = s;
      queArray = new long[maxSize];
      nItems = 0;
      }
//-------------------------------------------------------------
   public void insert(long item)    // insert item
      {
      int j;

      if(nItems==0)                          // if no items,
         queArray[nItems++] = item;          // insert at 0
      else                                   // if items,
         {
         for(j=nItems-1; j>=0; j--)          // start at end,
            {
            if( item > queArray[j] )         // if new item larger,
               queArray[j+1] = queArray[j];  // shift upward
```

LISTING 4.6   Continued

```
            else                      // if smaller,
               break;                 // done shifting
            }  // end for
         queArray[j+1] = item;        // insert it
         nItems++;
         }  // end else (nItems > 0)
      }  // end insert()
//---------------------------------------------------------------
   public long remove()              // remove minimum item
      { return queArray[--nItems]; }
//---------------------------------------------------------------
   public long peekMin()             // peek at minimum item
      { return queArray[nItems-1]; }
//---------------------------------------------------------------
   public boolean isEmpty()          // true if queue is empty
      { return (nItems==0); }
//---------------------------------------------------------------
   public boolean isFull()           // true if queue is full
      { return (nItems == maxSize); }
//---------------------------------------------------------------
   }  // end class PriorityQ
////////////////////////////////////////////////////////////////
class PriorityQApp
   {
   public static void main(String[] args) throws IOException
      {
      PriorityQ thePQ = new PriorityQ(5);
      thePQ.insert(30);
      thePQ.insert(50);
      thePQ.insert(10);
      thePQ.insert(40);
      thePQ.insert(20);

      while( !thePQ.isEmpty() )
         {
         long item = thePQ.remove();
         System.out.print(item + " ");  // 10, 20, 30, 40, 50
         }  // end while
      System.out.println("");
      }  // end main()
//---------------------------------------------------------------
   }  // end class PriorityQApp
```

In main() we insert five items in random order, and then remove and display them. The smallest item is always removed first, so the output is

```
10, 20, 30, 40, 50
```

The insert() method checks whether there are any items; if not, it inserts one at index 0. Otherwise, it starts at the top of the array and shifts existing items upward until it finds the place where the new item should go. Then it inserts the item and increments nItems. Note that if there's any chance the priority queue is full, you should check for this possibility with isFull() before using insert().

The front and rear fields aren't necessary as they were in the Queue class because, as we noted, front is always at nItems-1 and rear is always at 0.

The remove() method is simplicity itself: It decrements nItems and returns the item from the top of the array. The peekMin() method is similar, except it doesn't decrement nItems. The isEmpty() and isFull() methods check if nItems is 0 or maxSize, respectively.

Efficiency of Priority Queues

In the priority-queue implementation we show here, insertion runs in O(N) time, while deletion takes O(1) time. We'll see how to improve insertion time with heaps in Chapter 12.

Parsing Arithmetic Expressions

So far in this chapter, we've introduced three different data storage structures. Let's shift gears now and focus on an important application for one of these structures. This application is *parsing* (that is, analyzing) arithmetic expressions such as 2+3 or 2*(3+4) or ((2+4)*7)+3*(9–5). The storage structure it uses is the stack. In the brackets.java program (Listing 4.3), we saw how a stack could be used to check whether delimiters were formatted correctly. Stacks are used in a similar, although more complicated, way for parsing arithmetic expressions.

In some sense this section should be considered optional. It's not a prerequisite to the rest of the book, and writing code to parse arithmetic expressions is probably not something you need to do every day, unless you are a compiler writer or are designing pocket calculators. Also, the coding details are more complex than any we've seen so far. However, seeing this important use of stacks is educational, and the issues raised are interesting in their own right.

As it turns out, it's fairly difficult, at least for a computer algorithm, to evaluate an arithmetic expression directly. It's easier for the algorithm to use a two-step process:

1. Transform the arithmetic expression into a different format, called postfix notation.

2. Evaluate the postfix expression.

Step 1 is a bit involved, but step 2 is easy. In any case, this two-step approach results in a simpler algorithm than trying to parse the arithmetic expression directly. Of course, for a human it's easier to parse the ordinary arithmetic expression. We'll return to the difference between the human and computer approaches in a moment.

Before we delve into the details of steps 1 and 2, we'll introduce postfix notation.

Postfix Notation

Everyday arithmetic expressions are written with an *operator* (+, −, *, or /) placed between two *operands* (numbers, or symbols that stand for numbers). This is called *infix* notation because the operator is written inside the operands. Thus, we say 2+2 and ½, or, using letters to stand for numbers, A+B and ⅘.

In postfix notation (which is also called Reverse Polish Notation, or RPN, because it was invented by a Polish mathematician), the operator *follows* the two operands. Thus, A+B becomes AB+, and ⅘ becomes AB/. More complex infix expressions can likewise be translated into postfix notation, as shown in Table 4.2. We'll explain how the postfix expressions are generated in a moment.

TABLE 4.2   Infix and Postfix Expressions

Infix	Postfix
A+B–C	AB+C–
A*B/C	AB*C/
A+B*C	ABC*+
A*B+C	AB*C+
A*(B+C)	ABC+*
A*B+C*D	AB*CD*+
(A+B)*(C–D)	AB+CD–*
((A+B)*C)–D	AB+C*D–
A+B*(C–D/(E+F))	ABCDEF+/–*+

Some computer languages also have an operator for raising a quantity to a power (typically, the ^ character), but we'll ignore that possibility in this discussion.

Besides infix and postfix, there's also a *prefix* notation, in which the operator is written before the operands: +AB instead of AB+. This notation is functionally similar to postfix but seldom used.

Translating Infix to Postfix

The next several pages are devoted to explaining how to translate an expression from infix notation into postfix. This algorithm is fairly involved, so don't worry if every detail isn't clear at first. If you get bogged down, you may want to skip ahead to the section "Evaluating Postfix Expressions." To understand how to create a postfix expression, you might find it helpful to see how a postfix expression is evaluated; for example, how the value 14 is extracted from the expression 234+*, which is the postfix equivalent of 2*(3+4). (Notice that in this discussion, for ease of writing, we restrict ourselves to expressions with single-digit numbers, although these expressions may evaluate to multidigit numbers.)

How Humans Evaluate Infix

How do you translate infix to postfix? Let's examine a slightly easier question first: How does a human evaluate a normal infix expression? Although, as we stated earlier, such evaluation is difficult for a computer, we humans do it fairly easily because of countless hours in Mr. Klemmer's math class. It's not hard for us to find the answer to 3+4+5, or 3*(4+5). By analyzing how we evaluate this expression, we can achieve some insight into the translation of such expressions into postfix.

Roughly speaking, when you "solve" an arithmetic expression, you follow rules something like this:

1. You read from left to right. (At least, we'll assume this is true. Sometimes people skip ahead, but for purposes of this discussion, you should assume you must read methodically, starting at the left.)

2. When you've read enough to evaluate two operands and an operator, you do the calculation and substitute the answer for these two operands and operator. (You may also need to solve other pending operations on the left, as we'll see later.)

3. You continue this process—going from left to right and evaluating when possible—until the end of the expression.

Tables 4.3, 4.4, and 4.5 show three examples of how simple infix expressions are evaluated. Later, in Tables 4.6, 4.7, and 4.8, we'll see how closely these evaluations mirror the process of translating infix to postfix.

To evaluate 3+4−5, you would carry out the steps shown in Table 4.3.

TABLE 4.3   Evaluating 3+4–5

Item Read	Expression Parsed So Far	Comments
3	3	
+	3+	
4	3+4	
–	7	When you see the –, you can evaluate 3+4.
	7–	
5	7–5	
End	2	When you reach the end of the expression, you can evaluate 7–5.

You can't evaluate the 3+4 until you see what operator follows the 4. If it's an * or /, you need to wait before applying the + sign until you've evaluated the * or /.

However, in this example the operator following the 4 is a –, which has the same precedence as a +, so when you see the –, you know you can evaluate 3+4, which is 7. The 7 then replaces the 3+4. You can evaluate the 7–5 when you arrive at the end of the expression.

Figure 4.13 shows this process in more detail. Notice how you go from left to right reading items from the input, and then, when you have enough information, you go from right to left, recalling previously examined input and evaluating each operand-operator-operand combination.

Because of precedence relationships, evaluating 3+4*5 is a bit more complicated, as shown in Table 4.4.

TABLE 4.4   Evaluating 3+4*5

Item Read	Expression Parsed So Far	Comments
3	3	
+	3+	
4	3+4	
*	3+4*	You can't evaluate 3+4 because * is higher precedence than +.
5	3+4*5	When you see the 5, you can evaluate 4*5.
	3+20	
End	23	When you see the end of the expression, you can evaluate 3+20.

Here you can't add the 3 until you know the result of 4*5. Why not? Because multiplication has a higher precedence than addition. In fact, both * and / have a higher precedence than + and –, so all multiplications and divisions must be carried out before any additions or subtractions (unless parentheses dictate otherwise; see the next example).

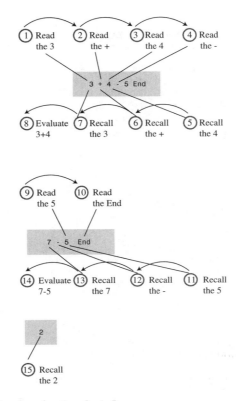

FIGURE 4.13 Details of evaluating 3+4–5.

Often you can evaluate as you go from left to right, as in the preceding example. However, you need to be sure, when you come to an operand-operator-operand combination such as A+B, that the operator on the right side of the B isn't one with a higher precedence than the +. If it does have a higher precedence, as in this example, you can't do the addition yet. However, after you've read the 5, the multiplication can be carried out because it has the highest priority; it doesn't matter whether a * or / follows the 5. However, you still can't do the addition until you've found out what's beyond the 5. When you find there's nothing beyond the 5 but the end of the expression, you can go ahead and do the addition. Figure 4.14 shows this process.

Parentheses are used to override the normal precedence of operators. Table 4.5 shows how you would evaluate 3*(4+5). Without the parentheses, you would do the multiplication first; with them, you do the addition first.

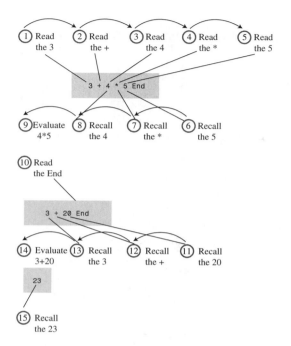

FIGURE 4.14 Details of evaluating 3+4*5.

TABLE 4.5   Evaluating 3*(4+5)

Item Read	Expression Parsed So Far	Comments
3	3	
*	3*	
(	3*(	
4	3*(4	You can't evaluate 3*4 because of the parenthesis.
+	3*(4+	
5	3*(4+5	You can't evaluate 4+5 yet.
)	3*(4+5)	When you see the), you can evaluate 4+5.
	3*9	After you've evaluated 4+5, you can evaluate 3*9.
	27	
End		Nothing left to evaluate.

Here we can't evaluate anything until we've reached the closing parenthesis. Multiplication has a higher or equal precedence compared to the other operators, so ordinarily we could carry out 3*4 as soon as we see the 4. However, parentheses have an even higher precedence than * and /. Accordingly, we must evaluate anything in parentheses before using the result as an operand in any other calculation. The closing parenthesis tells us we can go ahead and do the addition. We find that 4+5 is 9, and when we know this, we can evaluate 3*9 to obtain 27. Reaching the end of the expression is an anticlimax because there's nothing left to evaluate. This process is shown in Figure 4.15.

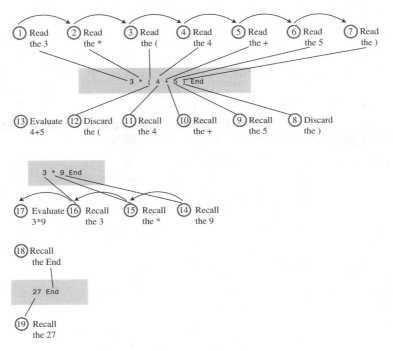

FIGURE 4.15 Details of evaluating 3*(4+5).

As we've seen, in evaluating an infix arithmetic expression, you go both forward and backward through the expression. You go forward (left to right) reading operands and operators. When you have enough information to apply an operator, you go backward, recalling two operands and an operator and carrying out the arithmetic.

Sometimes you must defer applying operators if they're followed by higher precedence operators or by parentheses. When this happens, you must apply the later, higher-precedence, operator first; then go backward (to the left) and apply earlier operators.

We could write an algorithm to carry out this kind of evaluation directly. However, as we noted, it's actually easier to translate into postfix notation first.

How Humans Translate Infix to Postfix

To translate infix to postfix notation, you follow a similar set of rules to those for evaluating infix. However, there are a few small changes. You don't do any arithmetic. The idea is not to evaluate the infix expression, but to rearrange the operators and operands into a different format: postfix notation. The resulting postfix expression will be evaluated later.

As before, you read the infix from left to right, looking at each character in turn. As you go along, you copy these operands and operators to the postfix output string. The trick is knowing when to copy what.

If the character in the infix string is an operand, you copy it immediately to the postfix string. That is, if you see an A in the infix, you write an A to the postfix. There's never any delay: You copy the operands as you get to them, no matter how long you must wait to copy their associated operators.

Knowing when to copy an operator is more complicated, but it's the same as the rule for evaluating infix expressions. Whenever you could have used the operator to evaluate part of the infix expression (if you were evaluating instead of translating to postfix), you instead copy it to the postfix string.

Table 4.6 shows how A+B–C is translated into postfix notation.

TABLE 4.6   Translating A+B–C into Postfix

Character Read from Infix Expression	Infix Expression Parsed So Far	Postfix Expression Written So Far	Comments
A	A	A	
+	A+	A	
B	A+B	AB	
–	A+B–	AB+	When you see the –, you can copy the + to the postfix string.
C	A+B–C	AB+C	
End	A+B–C	AB+C–	When you reach the end of the expression, you can copy the –.

Notice the similarity of this table to Table 4.3, which showed the evaluation of the infix expression 3+4–5. At each point where you would have done an evaluation in the earlier table, you instead simply write an operator to the postfix output.

Table 4.7 shows the translation of A+B*C to postfix. This evaluation is similar to Table 4.4, which covered the evaluation of 3+4*5.

TABLE 4.7 Translating A+B*C to Postfix

Character Read from Infix Expression	Infix Expression Parsed So Far	Postfix Expression Written So Far	Comments
A	A	A	
+	A+	A	
B	A+B	AB	
*	A+B*	AB	You can't copy the + because * is higher precedence than +.
C	A+B*C	ABC	When you see the C, you can copy the *.
	A+B*C	ABC*	
End	A+B*C	ABC*+	When you see the end of the expression, you can copy the +.

As the final example, Table 4.8 shows how A*(B+C) is translated to postfix. This process is similar to evaluating 3*(4+5) in Table 4.5. You can't write any postfix operators until you see the closing parenthesis in the input.

TABLE 4.8 Translating A*(B+C) into Postfix

Character Read from Infix Expression	Infix Expression Parsed so Far	Postfix Expression Written So Far	Comments
A	A	A	
*	A*	A	
(	A*(	A	
B	A*(B	AB	You can't copy * because of the parenthesis.
+	A*(B+	AB	
C	A*(B+C	ABC	You can't copy the + yet.
)	A*(B+C)	ABC+	When you see the), you can copy the +.
	A*(B+C)	ABC+*	After you've copied the +, you can copy the *.
End	A*(B+C)	ABC+*	Nothing left to copy.

As in the numerical evaluation process, you go both forward and backward through the infix expression to complete the translation to postfix. You can't write an operator to the output (postfix) string if it's followed by a higher-precedence operator or a left parenthesis. If it is, the higher-precedence operator or the operator in parentheses must be written to the postfix before the lower-priority operator.

Saving Operators on a Stack

You'll notice in both Table 4.7 and Table 4.8 that the order of the operators is reversed going from infix to postfix. Because the first operator can't be copied to the output until the second one has been copied, the operators were output to the postfix string in the opposite order they were read from the infix string. A longer example may make this operation clearer. Table 4.9 shows the translation to postfix of the infix expression A+B*(C–D). We include a column for stack contents, which we'll explain in a moment.

TABLE 4.9   Translating A+B*(C–D) to Postfix

Character Read from Infix Expression	Infix Expression Parsed So Far	Postfix Expression Written So Far	Stack Contents
A	A	A	
+	A+	A	+
B	A+B	AB	+
*	A+B*	AB	+*
(	A+B*(	AB	+*(
C	A+B*(C	ABC	+*(
–	A+B*(C–	ABC	+*(–
D	A+B*(C–D	ABCD	+*(–
)	A+B*(C–D)	ABCD–	+*(
	A+B*(C–D)	ABCD–	+*(
	A+B*(C–D)	ABCD–	+*
	A+B*(C–D)	ABCD–*	+
	A+B*(C–D)	ABCD–*+	

Here we see the order of the operands is +*– in the original infix expression, but the reverse order, –*+, in the final postfix expression. This happens because * has higher precedence than +, and –, because it's in parentheses, has higher precedence than *.

This order reversal suggests a stack might be a good place to store the operators while we're waiting to use them. The last column in Table 4.9 shows the stack contents at various stages in the translation process.

Popping items from the stack allows you to, in a sense, go backward (right to left) through the input string. You're not really examining the entire input string, only the operators and parentheses. They were pushed on the stack when reading the input, so now you can recall them in reverse order by popping them off the stack.

The operands (A, B, and so on) appear in the same order in infix and postfix, so you can write each one to the output as soon as you encounter it; they don't need to be stored on a stack.

Translation Rules

Let's make the rules for infix-to-postfix translation more explicit. You read items from the infix input string and take the actions shown in Table 4.10. These actions are described in pseudocode, a blend of Java and English.

In this table, the < and >= symbols refer to the operator precedence relationship, not numerical values. The opThis operator has just been read from the infix input, while the opTop operator has just been popped off the stack.

TABLE 4.10   Infix to Postfix Translation Rules

Item Read from Input (Infix)	Action
Operand	Write it to output (postfix)
Open parenthesis (	Push it on stack
Close parenthesis)	While stack not empty, repeat the following:
	Pop an item,
	If item is not (, write it to output
	Quit loop if item is (
Operator (opThis)	If stack empty,
	Push opThis
	Otherwise,
	While stack not empty, repeat:
	Pop an item,
	If item is (, push it, or
	If item is an operator (opTop), and
	If opTop < opThis, push opTop, or
	If opTop >= opThis, output opTop
	Quit loop if opTop < opThis or item is (
	Push opThis
No more items	While stack not empty,
	Pop item, output it.

Convincing yourself that these rules work may take some effort. Tables 4.11, 4.12, and 4.13 show how the rules apply to three example infix expressions. These tables are similar to Tables 4.6, 4.7, and 4.8, except that the relevant rules for each step have been added. Try creating similar tables by starting with other simple infix expressions and using the rules to translate some of them to postfix.

TABLE 4.11   Translation Rules Applied to A+B–C

Character Read from Infix	Infix Parsed So Far	Postfix Written So Far	Stack Contents	Rule
A	A	A		Write operand to output.
+	A+	A	+	If stack empty, push opThis.
B	A+B	AB	+	Write operand to output.
–	A+B–	AB		Stack not empty, so pop item.
	A+B–	AB+		opThis is –, opTop is +, opTop>=opThis, so output opTop.
	A+B–	AB+	–	Then push opThis.
C	A+B–C	AB+C	–	Write operand to output.
End	A+B–C	AB+C–		Pop leftover item, output it.

TABLE 4.12   Translation Rules Applied to A+B*C

Character Read From Infix	Infix Parsed So Far	Postfix Written So Far	Stack Contents	Rule
A	A	A		Write operand to postfix.
+	A+	A	+	If stack empty, push opThis.
B	A+B	AB	+	Write operand to output.
*	A+B*	AB	+	Stack not empty, so pop opTop.
	A+B*	AB	+	opThis is *, opTop is +, opTop<opThis, so push opTop.
	A+B*	AB	+*	Then push opThis.
C	A+B*C	ABC	+*	Write operand to output.
End	A+B*C	ABC*	+	Pop leftover item, output it.
	A+B*C	ABC*+		Pop leftover item, output it.

TABLE 4.13   Translation Rules Applied to A*(B+C)

Character Read From Infix	Infix Parsed So Far	Postfix Written So Far	Stack Contents	Rule
A	A	A		Write operand to postfix.
*	A*	A	*	If stack empty, push opThis.
(	A*(	A	*(	Push (on stack.

TABLE 4.13   Continued

Character Read From Infix	Infix Parsed So Far	Postfix Written So Far	Stack Contents	Rule
B	A*(B	AB	*(	Write operand to postfix.
+	A*(B+	AB	*	Stack not empty, so pop item.
	A*(B+	AB	*(	It's (, so push it.
	A*(B+	AB	*(+	Then push opThis.
C	A*(B+C	ABC	*(+	Write operand to postfix.
)	A*(B+C)	ABC+	*(	Pop item, write to output.
	A*(B+C)	ABC+	*	Quit popping if (.
End	A*(B+C)	ABC+*		Pop leftover item, output it.

Java Code to Convert Infix to Postfix

Listing 4.7 shows the infix.java program, which uses the rules from Table 4.10 to translate an infix expression to a postfix expression.

LISTING 4.7   The infix.java Program

```
// infix.java
// converts infix arithmetic expressions to postfix
// to run this program: C>java InfixApp
import java.io.*;            // for I/O
//////////////////////////////////////////////////////////////
class StackX
    {
    private int maxSize;
    private char[] stackArray;
    private int top;
//-------------------------------------------------------------
    public StackX(int s)          // constructor
       {
       maxSize = s;
       stackArray = new char[maxSize];
       top = -1;
       }
//-------------------------------------------------------------
    public void push(char j)  // put item on top of stack
       { stackArray[++top] = j; }
//-------------------------------------------------------------
    public char pop()             // take item from top of stack
```

LISTING 4.7   Continued

```
         { return stackArray[top--]; }
//-------------------------------------------------------------
   public char peek()           // peek at top of stack
      { return stackArray[top]; }
//-------------------------------------------------------------
   public boolean isEmpty()  // true if stack is empty
      { return (top == -1); }
//-------------------------------------------------------------
   public int size()            // return size
      { return top+1; }
//-------------------------------------------------------------
   public char peekN(int n)  // return item at index n
      { return stackArray[n]; }
//-------------------------------------------------------------
   public void displayStack(String s)
      {
      System.out.print(s);
      System.out.print("Stack (bottom-->top): ");
      for(int j=0; j<size(); j++)
         {
         System.out.print( peekN(j) );
         System.out.print(' ');
         }
      System.out.println("");
      }
//-------------------------------------------------------------
   }  // end class StackX
/////////////////////////////////////////////////////////////////
class InToPost                       // infix to postfix conversion
   {
   private StackX theStack;
   private String input;
   private String output = "";
//-------------------------------------------------------------
   public InToPost(String in)    // constructor
      {
      input = in;
      int stackSize = input.length();
      theStack = new StackX(stackSize);
      }
//-------------------------------------------------------------
```

LISTING 4.7   Continued

```
    public String doTrans()        // do translation to postfix
       {
       for(int j=0; j<input.length(); j++)
          {
          char ch = input.charAt(j);
          theStack.displayStack("For "+ch+" "); // *diagnostic*
          switch(ch)
             {
             case '+':                  // it's + or -
             case '-':
                gotOper(ch, 1);         // go pop operators
                break;                  //    (precedence 1)
             case '*':                  // it's * or /
             case '/':
                gotOper(ch, 2);         // go pop operators
                break;                  //    (precedence 2)
             case '(':                  // it's a left paren
                theStack.push(ch);      // push it
                break;
             case ')':                  // it's a right paren
                gotParen(ch);           // go pop operators
                break;
             default:                   // must be an operand
                output = output + ch; // write it to output
                break;
             }  // end switch
          }  // end for
       while( !theStack.isEmpty() )      // pop remaining opers
          {
          theStack.displayStack("While ");  // *diagnostic*
          output = output + theStack.pop(); // write to output
          }
       theStack.displayStack("End   ");     // *diagnostic*
       return output;                  // return postfix
       }  // end doTrans()
//--------------------------------------------------------------
    public  void gotOper(char opThis, int prec1)
       {                                 // got operator from input
       while( !theStack.isEmpty() )
          {
          char opTop = theStack.pop();
```

LISTING 4.7   Continued

```
          if( opTop == '(' )            // if it's a '('
             {
             theStack.push(opTop);      // restore '('
             break;
             }
          else                          // it's an operator
             {
             int prec2;                 // precedence of new op

             if(opTop=='+' || opTop=='-') // find new op prec
                prec2 = 1;
             else
                prec2 = 2;
             if(prec2 < prec1)          // if prec of new op less
                {                       //    than prec of old
                theStack.push(opTop);   // save newly-popped op
                break;
                }
             else                       // prec of new not less
                output = output + opTop; // than prec of old
             }  // end else (it's an operator)
          }  // end while
       theStack.push(opThis);           // push new operator
       }  // end gotOp()
//-------------------------------------------------------------
   public  void gotParen(char ch)
       {                                // got right paren from input
       while( !theStack.isEmpty() )
          {
          char chx = theStack.pop();
          if( chx == '(' )              // if popped '('
             break;                     // we're done
          else                          // if popped operator
             output = output + chx;     // output it
          }  // end while
       }  // end popOps()
//-------------------------------------------------------------
   }  // end class InToPost
/////////////////////////////////////////////////////////////
class InfixApp
   {
```

LISTING 4.7   Continued

```
public static void main(String[] args) throws IOException
   {
   String input, output;
   while(true)
      {
      System.out.print("Enter infix: ");
      System.out.flush();
      input = getString();       // read a string from kbd
      if( input.equals("") )     // quit if [Enter]
         break;
                                 // make a translator
      InToPost theTrans = new InToPost(input);
      output = theTrans.doTrans(); // do the translation
      System.out.println("Postfix is " + output + '\n');
      } // end while
   } // end main()
//--------------------------------------------------------------
   public static String getString() throws IOException
      {
      InputStreamReader isr = new InputStreamReader(System.in);
      BufferedReader br = new BufferedReader(isr);
      String s = br.readLine();
      return s;
      }
//--------------------------------------------------------------
   } // end class InfixApp
///////////////////////////////////////////////////////////////
```

The main() routine in the InfixApp class asks the user to enter an infix expression.
The input is read with the readString() utility method. The program creates an
InToPost object, initialized with the input string. Then it calls the doTrans() method
for this object to perform the translation. This method returns the postfix output
string, which is displayed.

The doTrans() method uses a switch statement to handle the various translation rules
shown in Table 4.10. It calls the gotOper() method when it reads an operator and the
gotParen() method when it reads a closing parenthesis,). These methods implement
the second two rules in the table, which are more complex than other rules.

We've included a displayStack() method to display the entire contents of the stack
in the StackX class. In theory, this isn't playing by the rules; you're supposed to
access the item only at the top. However, as a diagnostic aid, this routine is useful if

you want to see the contents of the stack at each stage of the translation. Here's some sample interaction with `infix.java`:

```
Enter infix: A*(B+C)-D/(E+F)
For A Stack (bottom-->top):
For * Stack (bottom-->top):
For ( Stack (bottom-->top): *
For B Stack (bottom-->top): * (
For + Stack (bottom-->top): * (
For C Stack (bottom-->top): * ( +
For ) Stack (bottom-->top): * ( +
For - Stack (bottom-->top): *
For D Stack (bottom-->top): -
For / Stack (bottom-->top): -
For ( Stack (bottom-->top): - /
For E Stack (bottom-->top): - / (
For + Stack (bottom-->top): - / (
For F Stack (bottom-->top): - / ( +
For ) Stack (bottom-->top): - / ( +
While Stack (bottom-->top): - /
While Stack (bottom-->top): -
End   Stack (bottom-->top):
Postfix is ABC+*DEF+/-
```

The output shows where the `displayStack()` method was called (from the `for` loop, the `while` loop, or at the end of the program) and, within the `for` loop, what character has just been read from the input string.

You can use single-digit numbers like 3 and 7 instead of symbols like A and B. They're all just characters to the program. For example:

```
Enter infix: 2+3*4
For 2 Stack (bottom-->top):
For + Stack (bottom-->top):
For 3 Stack (bottom-->top): +
For * Stack (bottom-->top): +
For 4 Stack (bottom-->top): + *
While Stack (bottom-->top): + *
While Stack (bottom-->top): +
End   Stack (bottom-->top):
Postfix is 234*+
```

Of course, in the postfix output, the 234 means the separate numbers 2, 3, and 4.

The infix.java program doesn't check the input for errors. If you type an incorrect infix expression, the program will provide erroneous output or crash and burn.

Experiment with this program. Start with some simple infix expressions, and see if you can predict what the postfix will be. Then run the program to verify your answer. Pretty soon, you'll be a postfix guru, much sought after at cocktail parties.

Evaluating Postfix Expressions

As you can see, converting infix expressions to postfix expressions is not trivial. Is all this trouble really necessary? Yes, the payoff comes when you evaluate a postfix expression. Before we show how simple the algorithm is, let's examine how a human might carry out such an evaluation.

How Humans Evaluate Postfix

Figure 4.16 shows how a human can evaluate a postfix expression using visual inspection and a pencil.

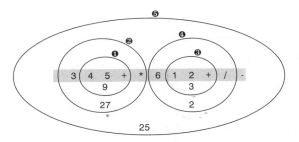

FIGURE 4.16 Visual approach to postfix evaluation of 345+*612+/−.

Start with the first operator on the left, and draw a circle around it and the two operands to its immediate left. Then apply the operator to these two operands—performing the actual arithmetic—and write down the result inside the circle. In the figure, evaluating 4+5 gives 9.

Now go to the next operator to the right, and draw a circle around it, the circle you already drew, and the operand to the left of that. Apply the operator to the previous circle and the new operand, and write the result in the new circle. Here 3*9 gives 27. Continue this process until all the operators have been applied: 1+2 is 3, and 6/3 is 2. The answer is the result in the largest circle: 27−2 is 25.

Rules for Postfix Evaluation

How do we write a program to reproduce this evaluation process? As you can see, each time you come to an operator, you apply it to the last two operands you've seen. This suggests that it might be appropriate to store the operands on a stack.

(This is the opposite of the infix-to-postfix translation algorithm, where *operators* were stored on the stack.) You can use the rules shown in Table 4.14 to evaluate postfix expressions.

TABLE 4.14   Evaluating a Postfix Expression

Item Read from Postfix Expression	Action
Operand	Push it onto the stack.
Operator	Pop the top two operands from the stack and apply the operator to them. Push the result.

When you're done, pop the stack to obtain the answer. That's all there is to it. This process is the computer equivalent of the human circle-drawing approach of Figure 4.16.

Java Code to Evaluate Postfix Expressions
In the infix-to-postfix translation, we used symbols (A, B, and so on) to stand for numbers. This approach worked because we weren't performing arithmetic operations on the operands but merely rewriting them in a different format.

Now we want to evaluate a postfix expression, which means carrying out the arithmetic and obtaining an answer. Thus, the input must consist of actual numbers. To simplify the coding, we've restricted the input to single-digit numbers.

Our program evaluates a postfix expression and outputs the result. Remember numbers are restricted to one digit. Here's some simple interaction:

```
Enter postfix: 57+
5 Stack (bottom-->top):
7 Stack (bottom-->top): 5
+ Stack (bottom-->top): 5 7
Evaluates to 12
```

You enter digits and operators, with no spaces. The program finds the numerical equivalent. Although the input is restricted to single-digit numbers, the results are not; it doesn't matter if something evaluates to numbers greater than 9. As in the infix.java program, we use the displayStack() method to show the stack contents at each step. Listing 4.8 shows the postfix.java program.

LISTING 4.8   The postfix.java Program

```
// postfix.java
// parses postfix arithmetic expressions
// to run this program: C>java PostfixApp
```

LISTING 4.8   Continued

```java
import java.io.*;              // for I/O
//////////////////////////////////////////////////////////////////
class StackX
   {
   private int maxSize;
   private int[] stackArray;
   private int top;
//-------------------------------------------------------------
   public StackX(int size)        // constructor
      {
      maxSize = size;
      stackArray = new int[maxSize];
      top = -1;
      }
//-------------------------------------------------------------
   public void push(int j)      // put item on top of stack
      { stackArray[++top] = j; }
//-------------------------------------------------------------
   public int pop()             // take item from top of stack
      { return stackArray[top--]; }
//-------------------------------------------------------------
   public int peek()            // peek at top of stack
      { return stackArray[top]; }
//-------------------------------------------------------------
   public boolean isEmpty()     // true if stack is empty
      { return (top == -1); }
//-------------------------------------------------------------
   public boolean isFull()      // true if stack is full
      { return (top == maxSize-1); }
//-------------------------------------------------------------
   public int size()            // return size
      { return top+1; }
//-------------------------------------------------------------
   public int peekN(int n)      // peek at index n
      { return stackArray[n]; }
//-------------------------------------------------------------
   public void displayStack(String s)
      {
      System.out.print(s);
      System.out.print("Stack (bottom-->top): ");
      for(int j=0; j<size(); j++)
```

LISTING 4.8   Continued

```
            {
            System.out.print( peekN(j) );
            System.out.print(' ');
            }
        System.out.println("");
        }
//-------------------------------------------------------------
    }  // end class StackX
/////////////////////////////////////////////////////////////////
class ParsePost
    {
    private StackX theStack;
    private String input;
//-------------------------------------------------------------
    public ParsePost(String s)
        { input = s; }
//-------------------------------------------------------------
    public int doParse()
        {
        theStack = new StackX(20);              // make new stack
        char ch;
        int j;
        int num1, num2, interAns;

        for(j=0; j<input.length(); j++)         // for each char,
            {
            ch = input.charAt(j);               // read from input
            theStack.displayStack(""+ch+" ");   // *diagnostic*
            if(ch >= '0' && ch <= '9')          // if it's a number
                theStack.push( (int)(ch-'0') ); //   push it
            else                                // it's an operator
                {
                num2 = theStack.pop();          // pop operands
                num1 = theStack.pop();
                switch(ch)                      // do arithmetic
                    {
                    case '+':
                        interAns = num1 + num2;
                        break;
                    case '-':
                        interAns = num1 - num2;
```

LISTING 4.8   Continued

```
                        break;
                  case '*':
                     interAns = num1 * num2;
                     break;
                  case '/':
                     interAns = num1 / num2;
                     break;
                  default:
                     interAns = 0;
                  }  // end switch
               theStack.push(interAns);        // push result
               }  // end else
            }  // end for
      interAns = theStack.pop();               // get answer
      return interAns;
      }  // end doParse()
   }  // end class ParsePost
//////////////////////////////////////////////////////////////////
class PostfixApp
   {
   public static void main(String[] args) throws IOException
      {
      String input;
      int output;

      while(true)
         {
         System.out.print("Enter postfix: ");
         System.out.flush();
         input = getString();         // read a string from kbd
         if( input.equals("") )       // quit if [Enter]
            break;
                                      // make a parser
         ParsePost aParser = new ParsePost(input);
         output = aParser.doParse();  // do the evaluation
         System.out.println("Evaluates to " + output);
         }  // end while
      }  // end main()
//------------------------------------------------------------------
   public static String getString() throws IOException
      {
```

LISTING 4.8   Continued

```
    InputStreamReader isr = new InputStreamReader(System.in);
    BufferedReader br = new BufferedReader(isr);
    String s = br.readLine();
    return s;
    }
//------------------------------------------------------------
  } // end class PostfixApp
////////////////////////////////////////////////////////////////
```

The main() method in the PostfixApp class gets the postfix string from the user and then creates a ParsePost object, initialized with this string. It then calls the doParse() method of ParsePost to carry out the evaluation.

The doParse() method reads through the input string character by character. If the character is a digit, it's pushed onto the stack. If it's an operator, it's applied immediately to the two operators on the top of the stack. (These operators are guaranteed to be on the stack already because the input string is in postfix notation.)

The result of the arithmetic operation is pushed onto the stack. After the last character (which must be an operator) is read and applied, the stack contains only one item, which is the answer to the entire expression.

Here's some interaction with more complex input: the postfix expression 345+*612+/–, which we showed a human evaluating in Figure 4.16. This expression corresponds to the infix 3*(4+5)–6/(1+2). (We saw an equivalent translation using letters instead of numbers in the previous section: A*(B+C)–D/(E+F) in infix is ABC+*DEF+/– in postfix.) Here's how the postfix is evaluated by the postfix.java program:

```
Enter postfix: 345+*612+/-
3 Stack (bottom-->top):
4 Stack (bottom-->top): 3
5 Stack (bottom-->top): 3 4
+ Stack (bottom-->top): 3 4 5
* Stack (bottom-->top): 3 9
6 Stack (bottom-->top): 27
1 Stack (bottom-->top): 27 6
2 Stack (bottom-->top): 27 6 1
+ Stack (bottom-->top): 27 6 1 2
/ Stack (bottom-->top): 27 6 3
- Stack (bottom-->top): 27 2
Evaluates to 25
```

As with the `infix.java` program (Listing 4.7), `postfix.java` doesn't check for input errors. If you type in a postfix expression that doesn't make sense, results are unpredictable.

Experiment with the program. Trying different postfix expressions and seeing how they're evaluated will give you an understanding of the process faster than reading about it.

Summary

- Stacks, queues, and priority queues are data structures usually used to simplify certain programming operations.

- In these data structures, only one data item can be accessed.

- A stack allows access to the last item inserted.

- The important stack operations are pushing (inserting) an item onto the top of the stack and popping (removing) the item that's on the top.

- A queue allows access to the first item that was inserted.

- The important queue operations are inserting an item at the rear of the queue and removing the item from the front of the queue.

- A queue can be implemented as a circular queue, which is based on an array in which the indices wrap around from the end of the array to the beginning.

- A priority queue allows access to the smallest (or sometimes the largest) item.

- The important priority queue operations are inserting an item in sorted order and removing the item with the smallest key.

- These data structures can be implemented with arrays or with other mechanisms such as linked lists.

- Ordinary arithmetic expressions are written in infix notation, so-called because the operator is written between the two operands.

- In postfix notation, the operator follows the two operands.

- Arithmetic expressions are typically evaluated by translating them to postfix notation and then evaluating the postfix expression.

- A stack is a useful tool both for translating an infix to a postfix expression and for evaluating a postfix expression.

Questions

These questions are intended as a self-test for readers. Answers may be found in Appendix C.

1. Suppose you push 10, 20, 30, and 40 onto the stack. Then you pop three items. Which one is left on the stack?

2. Which of the following is true?

   a. The pop operation on a stack is considerably simpler than the remove operation on a queue.

   b. The contents of a queue can wrap around, while those of a stack cannot.

   c. The top of a stack corresponds to the front of a queue.

   d. In both the stack and the queue, items removed in sequence are taken from increasingly high index cells in the array.

3. What do LIFO and FIFO mean?

4. True or False: A stack or a queue often serves as the underlying mechanism on which an ADT array is based.

5. Assume an array is numbered with index 0 on the left. A queue representing a line of movie-goers, with the first to arrive numbered 1, has the ticket window on the right. Then

   a. there is no numerical correspondence between the index numbers and the movie-goer numbers.

   b. the array index numbers and the movie-goer numbers increase in opposite left-right directions.

   c. the array index numbers correspond numerically to the locations in the line of movie-goers.

   d. the movie-goers and the items in the array move in the same direction.

6. As other items are inserted and removed, does a particular item in a queue move along the array from lower to higher indices, or higher to lower?

7. Suppose you insert 15, 25, 35, and 45 into a queue. Then you remove three items. Which one is left?

8. True or False: Pushing and popping items on a stack and inserting and removing items in a queue all take O(N) time.

9. A queue might be used to hold

   a. the items to be sorted in an insertion sort.

   b. reports of a variety of imminent attacks on the star ship Enterprise.

   c. keystrokes made by a computer user writing a letter.

   d. symbols in an algebraic expression being evaluated.

10. Inserting an item into a typical priority queue takes what big O time?

11. The term *priority* in a priority queue means that

    a. the highest priority items are inserted first.

    b. the programmer must prioritize access to the underlying array.

    c. the underlying array is sorted by the priority of the items.

    d. the lowest priority items are deleted first.

12. True or False: At least one of the methods in the priorityQ.java program (Listing 4.6) uses a linear search.

13. One difference between a priority queue and an ordered array is that

    a. the lowest-priority item cannot be extracted easily from the array as it can from the priority queue.

    b. the array must be ordered while the priority queue need not be.

    c. the highest priority item can be extracted easily from the priority queue but not from the array.

    d. All of the above.

14. Suppose you based a priority queue class on the OrdArray class in the orderedArray.java program (Listing 2.4) in Chapter 2, "Arrays." This will buy you binary search capability. If you wanted the best performance for your priority queue, would you need to modify the OrdArray class?

15. A priority queue might be used to hold

    a. passengers to be picked up by a taxi from different parts of the city.

    b. keystrokes made at a computer keyboard.

    c. squares on a chessboard in a game program.

    d. planets in a solar system simulation.

Experiments

Carrying out these experiments will help to provide insights into the topics covered in the chapter. No programming is involved.

1. Start with the initial configuration of the Queue Workshop applet. Alternately remove and insert items. (This way, you can reuse the deleted key value for the new item without typing it.) Notice how the group of four items crawls up to the top of the queue and then reappears at the bottom and keeps climbing.

2. Using the PriorityQ Workshop applet, figure out the positions of the Front and Rear arrows when the priority queue is full and when it is empty. Why can't a priority queue wrap around like an ordinary queue?

3. Think about how you remember the events in your life. Are there times when they seem to be stored in your brain in a stack? In a queue? In a priority queue?

Programming Projects

Writing programs that solve the Programming Projects helps to solidify your understanding of the material and demonstrates how the chapter's concepts are applied. (As noted in the Introduction, qualified instructors may obtain completed solutions to the Programming Projects on the publisher's Web site.)

4.1 Write a method for the Queue class in the queue.java program (Listing 4.4) that displays the contents of the queue. Note that this does not mean simply displaying the contents of the underlying array. You should show the queue contents from the first item inserted to the last, without indicating to the viewer whether the sequence is broken by wrapping around the end of the array. Be careful that one item and no items display properly, no matter where front and rear are.

4.2 Create a Deque class based on the discussion of deques (double-ended queues) in this chapter. It should include insertLeft(), insertRight(), removeLeft(), removeRight(), isEmpty(), and isFull() methods. It will need to support wrap-around at the end of the array, as queues do.

4.3 Write a program that implements a stack class that is based on the Deque class in Programming Project 4.2. This stack class should have the same methods and capabilities as the StackX class in the stack.java program (Listing 4.1).

4.4 The priority queue shown in Listing 4.6 features fast removal of the high-priority item but slow insertion of new items. Write a program with a revised PriorityQ class that has fast O(1) insertion time but slower removal of the high-priority item. Include a method that displays the contents of the priority queue, as suggested in Programming Project 4.1.

4.5 Queues are often used to simulate the flow of people, cars, airplanes, transactions, and so on. Write a program that models checkout lines at a supermarket, using the Queue class from the queue.java program (Listing 4.4). Several lines of customers should be displayed; you can use the display() method of Programming Project 4.1. You can add a new customer by pressing a key. You'll need to determine how the customer will decide which line to join. The checkers will take random amounts of time to process each customer (presumably depending on how many groceries the customer has). Once checked out, the customer is removed from the line. For simplicity, you can simulate the passing of time by pressing a key. Perhaps every keypress indicates the passage of one minute. (Java, of course, has more sophisticated ways to handle time.)

5

Linked Lists

In Chapter 2, "Arrays," we saw that arrays had certain disadvantages as data storage structures. In an unordered array, searching is slow, whereas in an ordered array, insertion is slow. In both kinds of arrays, deletion is slow. Also, the size of an array can't be changed after it's created.

In this chapter we'll look at a data storage structure that solves some of these problems: the *linked list*. Linked lists are probably the second most commonly used general-purpose storage structures after arrays.

The linked list is a versatile mechanism suitable for use in many kinds of general-purpose databases. It can also replace an array as the basis for other storage structures such as stacks and queues. In fact, you can use a linked list in many cases in which you use an array, unless you need frequent random access to individual items using an index.

Linked lists aren't the solution to all data storage problems, but they are surprisingly versatile and conceptually simpler than some other popular structures such as trees. We'll investigate their strengths and weaknesses as we go along.

In this chapter we'll look at simple linked lists, double-ended lists, sorted lists, doubly linked lists, and lists with iterators (an approach to random access to list elements). We'll also examine the idea of Abstract Data Types (ADTs), and see how stacks and queues can be viewed as ADTs and how they can be implemented as linked lists instead of arrays.

Links

In a linked list, each data item is embedded in a *link*. A link is an object of a class called something like Link. Because there are many similar links in a list, it makes sense to use a separate class for them, distinct from the linked list itself. Each `Link` object contains a reference (usually called `next`) to the next link in the list. A field in the list itself contains a reference to the first link. This relationship is shown in Figure 5.1.

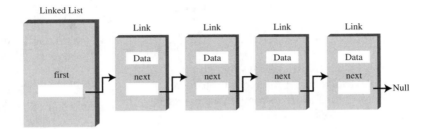

FIGURE 5.1 Links in a list.

Here's part of the definition of a class `Link`. It contains some data and a reference to the next link:

```
class Link
    {
    public int iData;     // data
    public double dData;  // data
    public Link next;     // reference to next link
    }
```

This kind of class definition is sometimes called *self-referential* because it contains a field—called `next` in this case—of the same type as itself.

We show only two data items in the link: an `int` and a `double`. In a typical application there would be many more. A personnel record, for example, might have name, address, Social Security number, title, salary, and many other fields. Often an object of a class that contains this data is used instead of the items:

```
class Link
    {
    public inventoryItem iI;  // object holding data
    public Link next;         // reference to next link
    }
```

References and Basic Types

You can easily get confused about references in the context of linked lists, so let's review how they work.

Being able to put a field of type Link inside the class definition of this same type may seem odd. Wouldn't the compiler be confused? How can it figure out how big to make a Link object if a link contains a link and the compiler doesn't already know how big a Link object is?

The answer is that in Java a Link object doesn't really contain another Link object, although it may look like it does. The next field of type Link is only a *reference to* another link, not an object.

A reference is a number that *refers to* an object. It's the object's address in the computer's memory, but you don't need to know its value; you just treat it as a magic number that tells you where the object is. In a given computer/operating system, all references, no matter what they refer to, are the same size. Thus, it's no problem for the compiler to figure out how big this field should be and thereby construct an entire Link object.

Note that in Java, primitive types such as int and double are stored quite differently than objects. Fields containing primitive types do not contain references, but actual numerical values like 7 or 3.14159. A variable definition like

```
double salary = 65000.00;
```

creates a space in memory and puts the number 65000.00 into this space. However, a reference to an object like

```
Link aLink = someLink;
```

puts a reference to an object of type Link, called someLink, into aLink. The someLink object itself is located elsewhere. It isn't moved, or even created, by this statement; it must have been created before. To create an object, you must always use new:

```
Link someLink = new Link();
```

Even the someLink field doesn't hold an object; it's still just a reference. The object is somewhere else in memory, as shown in Figure 5.2.

Other languages, such as C++, handle objects quite differently than Java. In C++ a field like

```
Link next;
```

actually contains an object of type Link. You can't write a self-referential class definition in C++ (although you can put a pointer to a Link in class Link; a pointer is similar to a reference). C++ programmers should keep in mind how Java handles objects; this usage may be counter-intuitive.

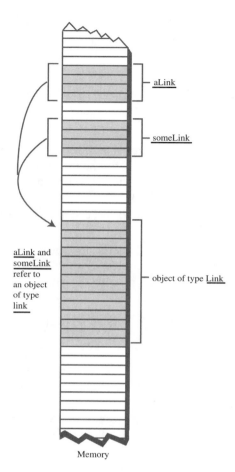

aLink

someLink

aLink and
someLink
refer to
an object
of type
link

object of type Link

Memory

FIGURE 5.2 Objects and references in memory.

Relationship, Not Position

Let's examine one of the major ways in which linked lists differ from arrays. In an array each item occupies a particular position. This position can be directly accessed using an index number. It's like a row of houses: You can find a particular house using its address.

In a list the only way to find a particular element is to follow along the chain of elements. It's more like human relations. Maybe you ask Harry where Bob is. Harry doesn't know, but he thinks Jane might know, so you go and ask Jane. Jane saw Bob leave the office with Sally, so you call Sally's cell phone. She dropped Bob off at

Peter's office, so...but you get the idea. You can't access a data item directly; you must use relationships between the items to locate it. You start with the first item, go to the second, then the third, until you find what you're looking for.

The LinkList Workshop Applet

The LinkList Workshop applet provides three list operations. You can insert a new data item, search for a data item with a specified key, and delete a data item with a specified key. These operations are the same ones we explored in the Array Workshop applet in Chapter 2; they're suitable for a general-purpose database application.

Figure 5.3 shows how the LinkList Workshop applet looks when it's started. Initially, there are 13 links on the list.

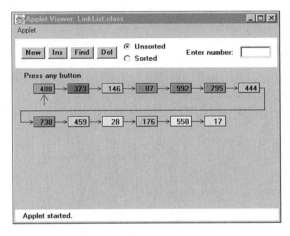

FIGURE 5.3 The LinkList Workshop applet.

The Insert Button

If you think 13 is an unlucky number, you can insert a new link. Press the Ins button, and you'll be prompted to enter a key value between 0 and 999. Subsequent presses will generate a link with this data in it, as shown in Figure 5.4.

In this version of a linked list, new links are always inserted at the beginning of the list. This is the simplest approach, although you can also insert links anywhere in the list, as we'll see later.

A final press on Ins will redraw the list so the newly inserted link lines up with the other links. This redrawing doesn't represent anything happening in the program itself, it just makes the display neater.

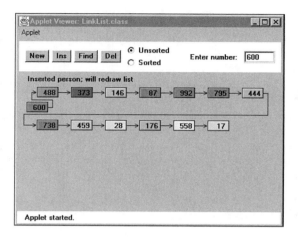

FIGURE 5.4 A new link being inserted.

The Find Button

The Find button allows you to find a link with a specified key value. When prompted, type in the value of an existing link, preferably one somewhere in the middle of the list. As you continue to press the button, you'll see the red arrow move along the list, looking for the link. A message informs you when the arrow finds the link. If you type a non-existent key value, the arrow will search all the way to the end of the list before reporting that the item can't be found.

The Delete Button

You can also delete a key with a specified value. Type in the value of an existing link and repeatedly press Del. Again, the arrow will move along the list, looking for the link. When the arrow finds the link, it simply removes that link and connects the arrow from the previous link straight across to the following link. This is how links are removed: The reference to the preceding link is changed to point to the following link.

A final keypress redraws the picture, but again redrawing just provides evenly spaced links for aesthetic reasons; the length of the arrows doesn't correspond to anything in the program.

> **NOTE**
>
> The LinkList Workshop applet can create both unsorted and sorted lists. Unsorted is the default. We'll show how to use the applet for sorted lists when we discuss them later in this chapter.

A Simple Linked List

Our first example program, linkList.java, demonstrates a simple linked list. The only operations allowed in this version of a list are

- Inserting an item at the beginning of the list

- Deleting the item at the beginning of the list

- Iterating through the list to display its contents

These operations are fairly easy to carry out, so we'll start with them. (As we'll see later, these operations are also all you need to use a linked list as the basis for a stack.)

Before we get to the complete linkList.java program, we'll look at some important parts of the Link and LinkList classes.

The Link Class

You've already seen the data part of the Link class. Here's the complete class definition:

```
class Link
   {
   public int iData;            // data item
   public double dData;         // data item
   public Link next;            // next link in list
// -------------------------------------------------------------
   public Link(int id, double dd) // constructor
      {
      iData = id;               // initialize data
      dData = dd;               // ('next' is automatically
      }                         //   set to null)
// -------------------------------------------------------------
   public void displayLink()    // display ourself
      {
      System.out.print("{" + iData + ", " + dData + "} ");
      }
   } // end class Link
```

In addition to the data, there's a constructor and a method, displayLink(), that displays the link's data in the format {22, 33.9}. Object purists would probably object to naming this method displayLink(), arguing that it should be simply display(). Using the shorter name would be in the spirit of polymorphism, but it makes the listing somewhat harder to understand when you see a statement like

```
current.display();
```

and you've forgotten whether `current` is a `Link` object, a `LinkList` object, or something else.

The constructor initializes the data. There's no need to initialize the `next` field because it's automatically set to `null` when it's created. (However, you could set it to `null` explicitly, for clarity.) The `null` value means it doesn't refer to anything, which is the situation until the link is connected to other links.

We've made the storage type of the `Link` fields (`iData` and so on) `public`. If they were `private`, we would need to provide public methods to access them, which would require extra code, thus making the listing longer and harder to read. Ideally, for security we would probably want to restrict `Link`-object access to methods of the `LinkList` class. However, without an inheritance relationship between these classes, that's not very convenient. We could use the default access specifier (no keyword) to give the data *package access* (access restricted to classes in the same directory), but that has no effect in these demo programs, which occupy only one directory anyway. The `public` specifier at least makes it clear that this data isn't private. In a more serious program you would probably want to make all the data fields in the `Link` class private.

The `LinkList` Class

The `LinkList` class contains only one data item: a reference to the first link on the list. This reference is called `first`. It's the only permanent information the list maintains about the location of any of the links. It finds the other links by following the chain of references from `first`, using each link's `next` field:

```
class LinkList
   {
   private Link first;          // ref to first link on list

// -------------------------------------------------------------
   public void LinkList()       // constructor
      {
      first = null;             // no items on list yet
      }
// -------------------------------------------------------------
   public boolean isEmpty()     // true if list is empty
      {
      return (first==null);
      }
```

```
// -----------------------------------------------------------
//   ...    other methods go here
   }
```

The constructor for LinkList sets first to null. This isn't really necessary because, as we noted, references are set to null automatically when they're created. However, the explicit constructor makes it clear that this is how first begins.

When first has the value null, we know there are no items on the list. If there were any items, first would contain a reference to the first one. The isEmpty() method uses this fact to determine whether the list is empty.

The insertFirst() **Method**

The insertFirst() method of LinkList inserts a new link at the beginning of the list. This is the easiest place to insert a link because first already points to the first link. To insert the new link, we need only set the next field in the newly created link to point to the old first link and then change first so it points to the newly created link. This situation is shown in Figure 5.5.

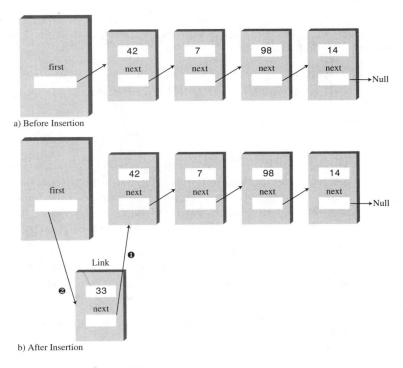

a) Before Insertion

b) After Insertion

FIGURE 5.5 Inserting a new link.

In `insertFirst()` we begin by creating the new link using the data passed as arguments. Then we change the link references as we just noted:

```
                                    // insert at start of list
public void insertFirst(int id, double dd)
   {                                // make new link
   Link newLink = new Link(id, dd);
   newLink.next = first;            // newLink --> old first
   first = newLink;                 // first --> newLink
   }
```

The `-->` arrows in the comments in the last two statements mean that a link (or the `first` field) connects to the next (downstream) link. (In doubly linked lists we'll see upstream connections as well, symbolized by `<--` arrows.) Compare these two statements with Figure 5.5. Make sure you understand how the statements cause the links to be changed, as shown in the figure. This kind of reference manipulation is the heart of linked-list algorithms.

The `deleteFirst()` Method

The `deleteFirst()` method is the reverse of `insertFirst()`. It disconnects the first link by rerouting `first` to point to the second link. This second link is found by looking at the `next` field in the first link:

```
public Link deleteFirst()         // delete first item
   {                              // (assumes list not empty)
   Link temp = first;             // save reference to link
   first = first.next;            // delete it: first-->old next
   return temp;                   // return deleted link
   }
```

The second statement is all you need to remove the first link from the list. We choose to also return the link, for the convenience of the user of the linked list, so we save it in `temp` before deleting it and return the value of `temp`. Figure 5.6 shows how `first` is rerouted to delete the object.

In C++ and similar languages, you would need to worry about deleting the link itself after it was disconnected from the list. It's in memory somewhere, but now nothing refers to it. What will become of it? In Java, the garbage collection process will destroy it at some point in the future; it's not your responsibility.

Notice that the `deleteFirst()` method assumes the list is not empty. Before calling it, your program should verify this fact with the `isEmpty()` method.

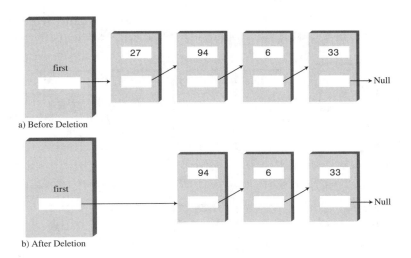

a) Before Deletion

b) After Deletion

FIGURE 5.6 Deleting a link.

The displayList() Method

To display the list, you start at first and follow the chain of references from link to link. A variable current points to (or technically *refers* to) each link in turn. It starts off pointing to first, which holds a reference to the first link. The statement

```
current = current.next;
```

changes current to point to the next link because that's what's in the next field in each link. Here's the entire displayList() method:

```
public void displayList()
   {
   System.out.print("List (first-->last): ");
   Link current = first;      // start at beginning of list
   while(current != null)     // until end of list,
      {
      current.displayLink();  // print data
      current = current.next; // move to next link
      }
   System.out.println("");
   }
```

The end of the list is indicated by the next field in the last link pointing to null rather than another link. How did this field get to be null? It started that way when the link was created and was never given any other value because it was always at the end of the list. The while loop uses this condition to terminate itself when it reaches the end of the list. Figure 5.7 shows how current steps along the list.

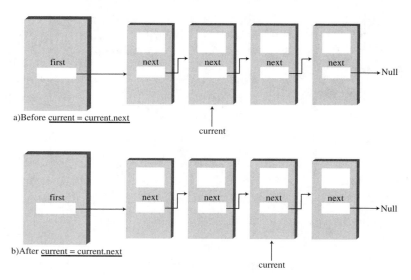

FIGURE 5.7 Stepping along the list.

At each link, the displayList() method calls the displayLink() method to display the data in the link.

The linkList.java **Program**

Listing 5.1 shows the complete linkList.java program. You've already seen all the components except the main() routine.

LISTING 5.1   The linkList.java Program

```java
// linkList.java
// demonstrates linked list
// to run this program: C>java LinkListApp
/////////////////////////////////////////////////////////////////
class Link
   {
   public int iData;               // data item (key)
   public double dData;            // data item
   public Link next;               // next link in list
// --------------------------------------------------------------
```

LISTING 5.1   Continued

```java
    public Link(int id, double dd) // constructor
       {
       iData = id;                 // initialize data
       dData = dd;                 // ('next' is automatically
       }                           //  set to null)
// ------------------------------------------------------------
    public void displayLink()      // display ourself
       {
       System.out.print("{" + iData + ", " + dData + "} ");
       }
    } // end class Link
////////////////////////////////////////////////////////////////
class LinkList
   {
   private Link first;             // ref to first link on list

// ------------------------------------------------------------
    public LinkList()              // constructor
       {
       first = null;               // no items on list yet
       }
// ------------------------------------------------------------
    public boolean isEmpty()       // true if list is empty
       {
       return (first==null);
       }
// ------------------------------------------------------------
                                   // insert at start of list
    public void insertFirst(int id, double dd)
       {                           // make new link
       Link newLink = new Link(id, dd);
       newLink.next = first;       // newLink --> old first
       first = newLink;            // first --> newLink
       }
// ------------------------------------------------------------
    public Link deleteFirst()      // delete first item
       {                           // (assumes list not empty)
       Link temp = first;          // save reference to link
       first = first.next;         // delete it: first-->old next
       return temp;                // return deleted link
       }
```

LISTING 5.1   Continued

```
// --------------------------------------------------------------
   public void displayList()
      {
      System.out.print("List (first-->last): ");
      Link current = first;        // start at beginning of list
      while(current != null)       // until end of list,
         {
         current.displayLink();    // print data
         current = current.next;   // move to next link
         }
      System.out.println("");
      }
// --------------------------------------------------------------
   } // end class LinkList
//////////////////////////////////////////////////////////////////
class LinkListApp
   {
   public static void main(String[] args)
      {
      LinkList theList = new LinkList();  // make new list

      theList.insertFirst(22, 2.99);      // insert four items
      theList.insertFirst(44, 4.99);
      theList.insertFirst(66, 6.99);
      theList.insertFirst(88, 8.99);

      theList.displayList();              // display list

      while( !theList.isEmpty() )         // until it's empty,
         {
         Link aLink = theList.deleteFirst();  // delete link
         System.out.print("Deleted ");        // display it
         aLink.displayLink();
         System.out.println("");
         }
      theList.displayList();              // display list
      } // end main()
   } // end class LinkListApp
//////////////////////////////////////////////////////////////////
```

In main() we create a new list, insert four new links into it with insertFirst(), and display it. Then, in the while loop, we remove the items one by one with deleteFirst() until the list is empty. The empty list is then displayed. Here's the output from linkList.java:

```
List (first-->last): {88, 8.99} {66, 6.99} {44, 4.99} {22, 2.99}
Deleted {88, 8.99}
Deleted {66, 6.99}
Deleted {44, 4.99}
Deleted {22, 2.99}
List (first-->last):
```

Finding and Deleting Specified Links

Our next example program adds methods to search a linked list for a data item with a specified key value and to delete an item with a specified key value. These, along with insertion at the start of the list, are the same operations carried out by the LinkList Workshop applet. The complete linkList2.java program is shown in Listing 5.2.

LISTING 5.2 The linkList2.java Program

```java
// linkList2.java
// demonstrates linked list
// to run this program: C>java LinkList2App
////////////////////////////////////////////////////////////////
class Link
   {
   public int iData;            // data item (key)
   public double dData;         // data item
   public Link next;            // next link in list
// -------------------------------------------------------------
   public Link(int id, double dd) // constructor
      {
      iData = id;
      dData = dd;
      }
// -------------------------------------------------------------
   public void displayLink()      // display ourself
      {
      System.out.print("{" + iData + ", " + dData + "} ");
      }
   }  // end class Link
```

LISTING 5.2   Continued

```
/////////////////////////////////////////////////////////////
class LinkList
   {
   private Link first;            // ref to first link on list
// ----------------------------------------------------------
   public LinkList()              // constructor
      {
      first = null;              // no links on list yet
      }
// ----------------------------------------------------------
   public void insertFirst(int id, double dd)
      {                          // make new link
      Link newLink = new Link(id, dd);
      newLink.next = first;      // it points to old first link
      first = newLink;           // now first points to this
      }
// ----------------------------------------------------------
   public Link find(int key)      // find link with given key
      {                          // (assumes non-empty list)
      Link current = first;             // start at 'first'
      while(current.iData != key)       // while no match,
         {
         if(current.next == null)       // if end of list,
            return null;                // didn't find it
         else                           // not end of list,
            current = current.next;     // go to next link
         }
      return current;                   // found it
      }
// ----------------------------------------------------------
   public Link delete(int key)    // delete link with given key
      {                          // (assumes non-empty list)
      Link current = first;             // search for link
      Link previous = first;
      while(current.iData != key)
         {
         if(current.next == null)
            return null;                // didn't find it
         else
            {
            previous = current;         // go to next link
```

LISTING 5.2 Continued

```
                current = current.next;
                }
            }                              // found it
        if(current == first)               // if first link,
            first = first.next;            //    change first
        else                               // otherwise,
            previous.next = current.next;  //    bypass it
        return current;
        }
// --------------------------------------------------------------
    public void displayList()      // display the list
        {
        System.out.print("List (first-->last): ");
        Link current = first;          // start at beginning of list
        while(current != null)         // until end of list,
            {
            current.displayLink();     // print data
            current = current.next;    // move to next link
            }
        System.out.println("");
        }
// --------------------------------------------------------------
    } // end class LinkList
////////////////////////////////////////////////////////////////
class LinkList2App
    {
    public static void main(String[] args)
        {
        LinkList theList = new LinkList();  // make list

        theList.insertFirst(22, 2.99);      // insert 4 items
        theList.insertFirst(44, 4.99);
        theList.insertFirst(66, 6.99);
        theList.insertFirst(88, 8.99);

        theList.displayList();              // display list

        Link f = theList.find(44);          // find item
        if( f != null)
            System.out.println("Found link with key " + f.iData);
        else
```

LISTING 5.2   Continued

```
            System.out.println("Can't find link");

      Link d = theList.delete(66);         // delete item
      if( d != null )
         System.out.println("Deleted link with key " + d.iData);
      else
         System.out.println("Can't delete link");

      theList.displayList();               // display list
      }  // end main()
   }  // end class LinkList2App
//////////////////////////////////////////////////////////////
```

The main() routine makes a list, inserts four items, and displays the resulting list. It then searches for the item with key 44, deletes the item with key 66, and displays the list again. Here's the output:

```
List (first-->last): {88, 8.99} {66, 6.99} {44, 4.99} {22, 2.99}
Found link with key 44
Deleted link with key 66
List (first-->last): {88, 8.99} {44, 4.99} {22, 2.99}
```

The find() Method

The find() method works much like the displayList() method in the linkList.java program. The reference current initially points to first and then steps its way along the links by setting itself repeatedly to current.next. At each link, find() checks whether that link's key is the one it's looking for. If the key is found, it returns with a reference to that link. If find() reaches the end of the list without finding the desired link, it returns null.

The delete() Method

The delete() method is similar to find() in the way it searches for the link to be deleted. However, it needs to maintain a reference not only to the current link (current), but to the link preceding the current link (previous). It does so because, if it deletes the current link, it must connect the preceding link to the following link, as shown in Figure 5.8. The only way to tell where the preceding link is located is to maintain a reference to it.

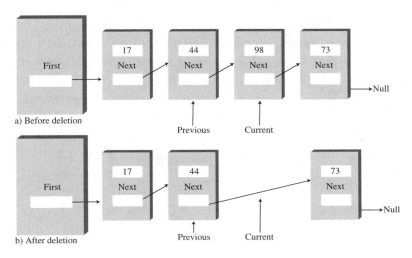

FIGURE 5.8 Deleting a specified link.

At each cycle through the while loop, just before current is set to current.next, previous is set to current. This keeps it pointing at the link preceding current.

To delete the current link once it's found, the next field of the previous link is set to the next link. A special case arises if the current link is the first link because the first link is pointed to by the LinkList's first field and not by another link. In this case the link is deleted by changing first to point to first.next, as we saw in the linkList.java program with the deleteFirst() method. Here's the code that covers these two possibilities:

```
                                    // found it
if(current == first)                // if first link,
   first = first.next;              //    change first
else                                // otherwise,
   previous.next = current.next;    //    bypass link
```

Other Methods

We've seen methods to insert and delete items at the start of a list, and to find a specified item and delete a specified item. You can imagine other useful list methods. For example, an insertAfter() method could find a link with a specified key value and insert a new link following it. We'll see such a method when we talk about list iterators at the end of this chapter.

Double-Ended Lists

A double-ended list is similar to an ordinary linked list, but it has one additional feature: a reference to the last link as well as to the first. Figure 5.9 shows such a list.

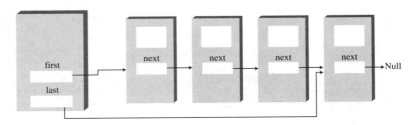

FIGURE 5.9 A double-ended list.

The reference to the last link permits you to insert a new link directly at the end of the list as well as at the beginning. Of course, you can insert a new link at the end of an ordinary single-ended list by iterating through the entire list until you reach the end, but this approach is inefficient.

Access to the end of the list as well as the beginning makes the double-ended list suitable for certain situations that a single-ended list can't handle efficiently. One such situation is implementing a queue; we'll see how this technique works in the next section.

Listing 5.3 contains the firstLastList.java program, which demonstrates a double-ended list. (Incidentally, don't confuse the double-ended list with the doubly linked list, which we'll explore later in this chapter.)

LISTING 5.3   The firstLastList.java Program

```java
// firstLastList.java
// demonstrates list with first and last references
// to run this program: C>java FirstLastApp
////////////////////////////////////////////////////////////////
class Link
   {
   public long dData;                // data item
   public Link next;                 // next link in list
// -------------------------------------------------------------
   public Link(long d)               // constructor
      { dData = d; }
// -------------------------------------------------------------
   public void displayLink()         // display this link
      { System.out.print(dData + " "); }
```

LISTING 5.3   Continued

```
// --------------------------------------------------------------
   }  // end class Link
////////////////////////////////////////////////////////////////
class FirstLastList
   {
   private Link first;              // ref to first link
   private Link last;               // ref to last link
// --------------------------------------------------------------
   public FirstLastList()           // constructor
      {
      first = null;                 // no links on list yet
      last = null;
      }
// --------------------------------------------------------------
   public boolean isEmpty()         // true if no links
      { return first==null; }
// --------------------------------------------------------------
   public void insertFirst(long dd) // insert at front of list
      {
      Link newLink = new Link(dd);  // make new link

      if( isEmpty() )               // if empty list,
         last = newLink;            // newLink <-- last
      newLink.next = first;         // newLink --> old first
      first = newLink;              // first --> newLink
      }
// --------------------------------------------------------------
   public void insertLast(long dd)  // insert at end of list
      {
      Link newLink = new Link(dd);  // make new link
      if( isEmpty() )               // if empty list,
         first = newLink;           // first --> newLink
      else
         last.next = newLink;       // old last --> newLink
      last = newLink;               // newLink <-- last
      }
// --------------------------------------------------------------
   public long deleteFirst()        // delete first link
      {                             // (assumes non-empty list)
      long temp = first.dData;
      if(first.next == null)        // if only one item
```

LISTING 5.3   Continued

```
        last = null;              // null <-- last
      first = first.next;         // first --> old next
      return temp;
      }
// --------------------------------------------------------------
   public void displayList()
      {
      System.out.print("List (first-->last): ");
      Link current = first;       // start at beginning
      while(current != null)      // until end of list,
         {
         current.displayLink();   // print data
         current = current.next;  // move to next link
         }
      System.out.println("");
      }
// --------------------------------------------------------------
   }  // end class FirstLastList
////////////////////////////////////////////////////////////////
class FirstLastApp
   {
   public static void main(String[] args)
      {                                  // make a new list
      FirstLastList theList = new FirstLastList();

      theList.insertFirst(22);    // insert at front
      theList.insertFirst(44);
      theList.insertFirst(66);

      theList.insertLast(11);     // insert at rear
      theList.insertLast(33);
      theList.insertLast(55);

      theList.displayList();      // display the list

      theList.deleteFirst();      // delete first two items
      theList.deleteFirst();

      theList.displayList();      // display again
      }  // end main()
   }  // end class FirstLastApp
////////////////////////////////////////////////////////////////
```

For simplicity, in this program we've reduced the number of data items in each link from two to one. This makes it easier to display the link contents. (Remember that in a serious program there would be many more data items, or a reference to another object containing many data items.)

This program inserts three items at the front of the list, inserts three more at the end, and displays the resulting list. It then deletes the first two items and displays the list again. Here's the output:

```
List (first-->last): 66 44 22 11 33 55
List (first-->last): 22 11 33 55
```

Notice how repeated insertions at the front of the list reverse the order of the items, while repeated insertions at the end preserve the order.

The double-ended list class is called the FirstLastList. As discussed, it has two data items, first and last, which point to the first item and the last item in the list. If there is only one item in the list, both first and last point to it, and if there are no items, they are both null.

The class has a new method, insertLast(), that inserts a new item at the end of the list. This process involves modifying last.next to point to the new link and then changing last to point to the new link, as shown in Figure 5.10.

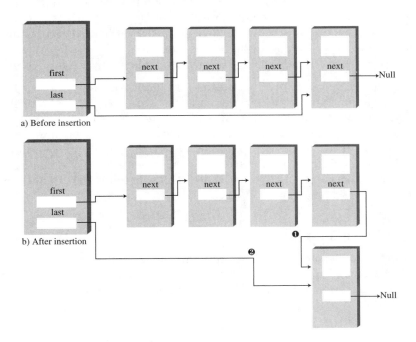

FIGURE 5.10 Insertion at the end of a list.

The insertion and deletion routines are similar to those in a single-ended list. However, both insertion routines must watch out for the special case when the list is empty prior to the insertion. That is, if isEmpty() is true, then insertFirst() must set last to the new link, and insertLast() must set first to the new link.

If inserting at the beginning with insertFirst(), first is set to point to the new link, although when inserting at the end with insertLast(), last is set to point to the new link. Deleting from the start of the list is also a special case if it's the last item on the list: last must be set to point to null in this case.

Unfortunately, making a list double-ended doesn't help you to delete the last link because there is still no reference to the next-to-last link, whose next field would need to be changed to null if the last link were deleted. To conveniently delete the last link, you would need a doubly linked list, which we'll look at soon. (Of course, you could also traverse the entire list to find the last link, but that's not very efficient.)

Linked-List Efficiency

Insertion and deletion at the beginning of a linked list are very fast. They involve changing only one or two references, which takes O(1) time.

Finding, deleting, or inserting next to a specific item requires searching through, on the average, half the items in the list. This requires O(N) comparisons. An array is also O(N) for these operations, but the linked list is nevertheless faster because nothing needs to be moved when an item is inserted or deleted. The increased efficiency can be significant, especially if a copy takes much longer than a comparison.

Of course, another important advantage of linked lists over arrays is that a linked list uses exactly as much memory as it needs and can expand to fill all of available memory. The size of an array is fixed when it's created; this usually leads to inefficiency because the array is too large, or to running out of room because the array is too small. Vectors, which are expandable arrays, may solve this problem to some extent, but they usually expand in fixed-sized increments (such as doubling the size of the array whenever it's about to overflow). This solution is still not as efficient a use of memory as a linked list.

Abstract Data Types

In this section we'll shift gears and discuss a topic that's more general than linked lists: Abstract Data Types (ADTs). What is an ADT? Roughly speaking, it's a way of looking at a data structure: focusing on what it does and ignoring how it does its job.

Stacks and queues are examples of ADTs. We've already seen that both stacks and queues can be implemented using arrays. Before we return to a discussion of ADTs, let's see how stacks and queues can be implemented using linked lists. This discussion will demonstrate the "abstract" nature of stacks and queues: how they can be considered separately from their implementation.

A Stack Implemented by a Linked List

When we created a stack in Chapter 4, "Stacks and Queues," we used an ordinary Java array to hold the stack's data. The stack's push() and pop() operations were actually carried out by array operations such as

```
arr[++top] = data;
```

and

```
data = arr[top--];
```

which insert data into, and take it out of, an array.

We can also use a linked list to hold a stack's data. In this case the push() and pop() operations would be carried out by operations like

```
theList.insertFirst(data)
```

and

```
data = theList.deleteFirst()
```

The user of the stack class calls push() and pop() to insert and delete items without knowing, or needing to know, whether the stack is implemented as an array or as a linked list. Listing 5.4 shows how a stack class called LinkStack can be implemented using the LinkList class instead of an array. (Object purists would argue that the name LinkStack should be simply Stack because users of this class shouldn't need to know that it's implemented as a list.)

LISTING 5.4   The linkStack.java Program

```
// linkStack.java
// demonstrates a stack implemented as a list
// to run this program: C>java LinkStackApp
////////////////////////////////////////////////////////////////
class Link
    {
    public long dData;              // data item
    public Link next;               // next link in list
```

LISTING 5.4   Continued

```java
// -----------------------------------------------------------------
   public Link(long dd)            // constructor
      { dData = dd; }
// -----------------------------------------------------------------
   public void displayLink()       // display ourself
      { System.out.print(dData + " "); }
   }  // end class Link
/////////////////////////////////////////////////////////////////////
class LinkList
   {
   private Link first;             // ref to first item on list
// -----------------------------------------------------------------
   public LinkList()               // constructor
      { first = null; }            // no items on list yet
// -----------------------------------------------------------------
   public boolean isEmpty()        // true if list is empty
      { return (first==null); }
// -----------------------------------------------------------------
   public void insertFirst(long dd) // insert at start of list
      {                            // make new link
      Link newLink = new Link(dd);
      newLink.next = first;        // newLink --> old first
      first = newLink;             // first --> newLink
      }
// -----------------------------------------------------------------
   public long deleteFirst()       // delete first item
      {                            // (assumes list not empty)
      Link temp = first;           // save reference to link
      first = first.next;          // delete it: first-->old next
      return temp.dData;           // return deleted link
      }
// -----------------------------------------------------------------
   public void displayList()
      {
      Link current = first;        // start at beginning of list
      while(current != null)       // until end of list,
         {
         current.displayLink();    // print data
         current = current.next;   // move to next link
         }
      System.out.println("");
```

LISTING 5.4   Continued

```
      }
// -------------------------------------------------------------
   }  // end class LinkList
///////////////////////////////////////////////////////////////
class LinkStack
   {
   private LinkList theList;
//-------------------------------------------------------------
   public LinkStack()             // constructor
      {
      theList = new LinkList();
      }
//-------------------------------------------------------------
   public void push(long j)     // put item on top of stack
      {
      theList.insertFirst(j);
      }
//-------------------------------------------------------------
   public long pop()             // take item from top of stack
      {
      return theList.deleteFirst();
      }
//-------------------------------------------------------------
   public boolean isEmpty()       // true if stack is empty
      {
      return ( theList.isEmpty() );
      }
//-------------------------------------------------------------
   public void displayStack()
      {
      System.out.print("Stack (top-->bottom): ");
      theList.displayList();
      }
//-------------------------------------------------------------
   }  // end class LinkStack
///////////////////////////////////////////////////////////////
class LinkStackApp
   {
   public static void main(String[] args)
      {
      LinkStack theStack = new LinkStack(); // make stack
```

LISTING 5.4   Continued

```
        theStack.push(20);                  // push items
        theStack.push(40);

        theStack.displayStack();            // display stack

        theStack.push(60);                  // push items
        theStack.push(80);

        theStack.displayStack();            // display stack

        theStack.pop();                     // pop items
        theStack.pop();

        theStack.displayStack();            // display stack
        }  // end main()
    }  // end class LinkStackApp
//////////////////////////////////////////////////////////////////
```

The main() routine creates a stack object, pushes two items on it, displays the stack, pushes two more items, and displays the stack again. Finally, it pops two items and displays the stack a third time. Here's the output:

```
Stack (top-->bottom): 40 20
Stack (top-->bottom): 80 60 40 20
Stack (top-->bottom): 40 20
```

Notice the overall organization of this program. The main() routine in the LinkStackApp class relates only to the LinkStack class. The LinkStack class relates only to the LinkList class. There's no communication between main() and the LinkList class.

More specifically, when a statement in main() calls the push() operation in the LinkStack class, this method in turn calls insertFirst() in the LinkList class to actually insert data. Similarly, pop() calls deleteFirst() to delete an item, and displayStack() calls displayList() to display the stack. To the class user, writing code in main(), there is no difference between using the list-based LinkStack class and using the array-based stack class from the stack.java program (Listing 4.1) in Chapter 4.

A Queue Implemented by a Linked List

Here's a similar example of an ADT implemented with a linked list. Listing 5.5 shows a queue implemented as a double-ended linked list.

LISTING 5.5   The `linkQueue.java` Program

```java
// linkQueue.java
// demonstrates queue implemented as double-ended list
// to run this program: C>java LinkQueueApp
////////////////////////////////////////////////////////////////
class Link
   {
   public long dData;                // data item
   public Link next;                 // next link in list
// -------------------------------------------------------------
   public Link(long d)               // constructor
      { dData = d; }
// -------------------------------------------------------------
   public void displayLink()         // display this link
      { System.out.print(dData + " "); }
// -------------------------------------------------------------
   }  // end class Link
////////////////////////////////////////////////////////////////
class FirstLastList
   {
   private Link first;               // ref to first item
   private Link last;                // ref to last item
// -------------------------------------------------------------
   public FirstLastList()            // constructor
      {
      first = null;                  // no items on list yet
      last = null;
      }
// -------------------------------------------------------------
   public boolean isEmpty()          // true if no links
      { return first==null; }
// -------------------------------------------------------------
   public void insertLast(long dd) // insert at end of list
      {
      Link newLink = new Link(dd);   // make new link
      if( isEmpty() )                // if empty list,
         first = newLink;            // first --> newLink
      else
         last.next = newLink;        // old last --> newLink
      last = newLink;                // newLink <-- last
      }
// -------------------------------------------------------------
```

LISTING 5.5   Continued

```
    public long deleteFirst()         // delete first link
        {                             // (assumes non-empty list)
        long temp = first.dData;
        if(first.next == null)        // if only one item
            last = null;              // null <-- last
        first = first.next;           // first --> old next
        return temp;
        }
//  --------------------------------------------------------------
    public void displayList()
        {
        Link current = first;         // start at beginning
        while(current != null)        // until end of list,
            {
            current.displayLink();    // print data
            current = current.next;   // move to next link
            }
        System.out.println("");
        }
//  --------------------------------------------------------------
    }  // end class FirstLastList
//////////////////////////////////////////////////////////////////
class LinkQueue
    {
    private FirstLastList theList;
//  --------------------------------------------------------------
    public LinkQueue()                // constructor
        { theList = new FirstLastList(); }  // make a 2-ended list
//  --------------------------------------------------------------
    public boolean isEmpty()          // true if queue is empty
        { return theList.isEmpty(); }
//  --------------------------------------------------------------
    public void insert(long j)        // insert, rear of queue
        { theList.insertLast(j); }
//  --------------------------------------------------------------
    public long remove()              // remove, front of queue
        { return theList.deleteFirst(); }
//  --------------------------------------------------------------
    public void displayQueue()
        {
        System.out.print("Queue (front-->rear): ");
```

LISTING 5.5   Continued

```
        theList.displayList();
        }
//------------------------------------------------------------
    }  // end class LinkQueue
/////////////////////////////////////////////////////////////
class LinkQueueApp
    {
    public static void main(String[] args)
        {
        LinkQueue theQueue = new LinkQueue();
        theQueue.insert(20);                  // insert items
        theQueue.insert(40);

        theQueue.displayQueue();              // display queue

        theQueue.insert(60);                  // insert items
        theQueue.insert(80);

        theQueue.displayQueue();              // display queue

        theQueue.remove();                    // remove items
        theQueue.remove();

        theQueue.displayQueue();              // display queue
        }  // end main()
/////////////////////////////////////////////////////////////
```

The program creates a queue, inserts two items, inserts two more items, and removes
two items; following each of these operations the queue is displayed. Here's the
output:

```
Queue (front-->rear): 20 40
Queue (front-->rear): 20 40 60 80
Queue (front-->rear): 60 80
```

Here the methods insert() and remove() in the LinkQueue class are implemented by
the insertLast() and deleteFirst() methods of the FirstLastList class. We've substituted a linked list for the array used to implement the queue in the queue.java
program (Listing 4.4) of Chapter 4.

The linkStack.java and linkQueue.java programs emphasize that stacks and queues
are conceptual entities, separate from their implementations. A stack can be implemented equally well by an array or by a linked list. What's important about a stack is

the push() and pop() operations and how they're used; it's not the underlying mechanism used to implement these operations.

When would you use a linked list as opposed to an array as the implementation of a stack or queue? One consideration is how accurately you can predict the amount of data the stack or queue will need to hold. If this isn't clear, the linked list gives you more flexibility than an array. Both are fast, so speed is probably not a major consideration.

Data Types and Abstraction

Where does the term *Abstract Data Type* come from? Let's look at the *data type* part of it first and then return to *abstract*.

Data Types

The phrase *data type* covers a lot of ground. It was first applied to built-in types such as int and double. This is probably what you first think of when you hear the term.

When you talk about a primitive type, you're actually referring to two things: a data item with certain characteristics and permissible operations on that data. For example, type int variables in Java can have whole-number values between −2,147,483,648 and +2,147,483,647, and the operators +, −, *, /, and so on can be applied to them. The data type's permissible operations are an inseparable part of its identity; understanding the type means understanding what operations can be performed on it.

With the advent of object-oriented programming, you could now create your own data types using classes. Some of these data types represent numerical quantities that are used in ways similar to primitive types. You can, for example, define a class for time (with fields for hours, minutes, seconds), a class for fractions (with numerator and denominator fields), and a class for extra-long numbers (characters in a string represent the digits). All these classes can be added and subtracted like int and double, except that in Java you must use methods with functional notation like add() and sub() rather than operators like + and −.

The phrase *data type* seems to fit naturally with such quantity-oriented classes. However, it is also applied to classes that don't have this quantitative aspect. In fact, *any* class represents a data type, in the sense that a class is made up of data (fields) and permissible operations on that data (methods).

By extension, when a data storage structure like a stack or queue is represented by a class, it too can be referred to as a data type. A stack is different in many ways from an int, but they are both defined as a certain arrangement of data and a set of operations on that data.

Abstraction

The word *abstract* means "considered apart from detailed specifications or implementation." An abstraction is the essence or important characteristics of something. The office of president, for example, is an abstraction, considered apart from the individual who happens to occupy that office. The powers and responsibilities of the office remain the same, while individual office-holders come and go.

In object-oriented programming, then, an Abstract Data Type is a class considered without regard to its implementation. It's a description of the data in the class (fields), a list of operations (methods) that can be carried out on that data, and instructions on how to use these operations. Specifically excluded are the details of how the methods carry out their tasks. As a class user, you're told what methods to call, how to call them, and the results you can expect, but not how they work.

The meaning of *Abstract Data Type* is further extended when it's applied to data structures such as stacks and queues. As with any class, it means the data and the operations that can be performed on it, but in this context even the fundamentals of how the data is stored become invisible to the user. Users not only don't know how the methods work, they also don't know what structure is used to store the data.

For the stack, the user knows that push() and pop() (and perhaps a few other methods) exist and how they work. The user doesn't (at least not usually) need to know how push() and pop() work, or whether data is stored in an array, a linked list, or some other data structure like a tree.

The Interface

An ADT specification is often called an *interface*. It's what the class user sees—usually its public methods. In a stack class, push() and pop() and similar methods form the interface.

ADT Lists

Now that we know what an Abstract Data Type is, we can mention another one: the *list*. A list (sometimes called a linear list) is a group of items arranged in a linear order. That is, they're lined up in a certain way, like beads on a string or houses on a street. Lists support certain fundamental operations. You can insert an item, delete an item, and usually read an item from a specified location (the third item, say).

Don't confuse the ADT list with the linked list we've been discussing in this chapter. A list is defined by its interface: the specific methods used to interact with it. This interface can be implemented by various structures, including arrays and linked lists. The list is an abstraction of such data structures.

ADTs as a Design Tool

The ADT concept is a useful aid in the software design process. If you need to store data, start by considering the operations that need to be performed on that data. Do you need access to the last item inserted? The first one? An item with a specified key? An item in a certain position? Answering such questions leads to the definition of an ADT. Only after the ADT is completely defined should you worry about the details of how to represent the data and how to code the methods that access the data.

By decoupling the specification of the ADT from the implementation details, you can simplify the design process. You also make it easier to change the implementation at some future time. If a user relates only to the ADT interface, you should be able to change the implementation without "breaking" the user's code.

Of course, once the ADT has been designed, the underlying data structure must be carefully chosen to make the specified operations as efficient as possible. If you need random access to element N, for example, the linked-list representation isn't so good because random access isn't an efficient operation for a linked list. You'd be better off with an array.

NOTE

Remember that the ADT concept is only a conceptual tool. Data storage structures are not divided cleanly into some that are ADTs and some that are used to implement ADTs. A linked list, for example, doesn't need to be wrapped in a list interface to be useful; it can act as an ADT on its own, or it can be used to implement another data type such as a queue. A linked list can be implemented using an array, and an array-type structure can be implemented using a linked list. What's an ADT and what's a more basic structure must be determined in a given context.

Sorted Lists

In the linked lists we've seen thus far, there was no requirement that data be stored in order. However, for certain applications it's useful to maintain the data in sorted order within the list. A list with this characteristic is called a *sorted list*.

In a sorted list, the items are arranged in sorted order by key value. Deletion is often limited to the smallest (or the largest) item in the list, which is at the start of the list, although sometimes find() and delete() methods, which search through the list for specified links, are used as well.

In general you can use a sorted list in most situations in which you use a sorted array. The advantages of a sorted list over a sorted array are speed of insertion (because elements don't need to be moved) and the fact that a list can expand to fill available memory, while an array is limited to a fixed size. However, a sorted list is somewhat more difficult to implement than a sorted array.

Later we'll look at one application for sorted lists: sorting data. A sorted list can also be used to implement a priority queue, although a heap (see Chapter 12, "Heaps") is a more common implementation.

The LinkList Workshop applet introduced at the beginning of this chapter demonstrates sorted as well as unsorted lists. To see how sorted lists work, use the New button to create a new list with about 20 links, and when prompted, click on the Sorted button. The result is a list with data in sorted order, as shown in Figure 5.11.

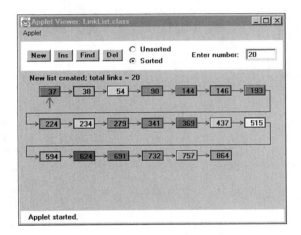

FIGURE 5.11 The LinkList Workshop applet with a sorted list.

Use the Ins button to insert a new item. Type in a value that will fall somewhere in the middle of the list. Watch as the algorithm traverses the links, looking for the appropriate insertion place. When it finds the correct location, it inserts the new link, as shown in Figure 5.12.

With the next press of Ins, the list will be redrawn to regularize its appearance. You can also find a specified link using the Find button and delete a specified link using the Del button.

Java Code to Insert an Item in a Sorted List

To insert an item in a sorted list, the algorithm must first search through the list until it finds the appropriate place to put the item: this is just before the first item that's larger, as shown in Figure 5.12.

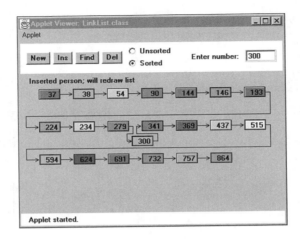

FIGURE 5.12 A newly inserted link.

When the algorithm finds where to put it, the item can be inserted in the usual way by changing next in the new link to point to the next link and changing next in the previous link to point to the new link. However, we need to consider some special cases: The link might need to be inserted at the beginning of the list, or it might need to go at the end. Let's look at the code:

```
public void insert(long key) // insert in order
   {
   Link newLink = new Link(key);    // make new link
   Link previous = null;            // start at first
   Link current = first;
                                    // until end of list,
   while(current != null && key > current.dData)
      {                             // or key > current,
      previous = current;
      current = current.next;       // go to next item
      }
   if(previous==null)              // at beginning of list
      first = newLink;             //    first --> newLink
   else                            // not at beginning
      previous.next = newLink;     //    old prev --> newLink
   newLink.next = current;         // newLink --> old current
   } // end insert()
```

We need to maintain a previous reference as we move along, so we can modify the previous link's next field to point to the new link. After creating the new link, we prepare to search for the insertion point by setting current to first in the usual way. We also set previous to null; this step is important because later we'll use this null value to determine whether we're still at the beginning of the list.

The while loop is similar to those we've used before to search for the insertion point, but there's an added condition. The loop terminates when the key of the link currently being examined (current.dData) is no longer smaller than the key of the link being inserted (key); this is the most usual case, where a key is inserted somewhere in the middle of the list.

However, the while loop also terminates if current is null. This happens at the end of the list (the next field of the last element is null), or if the list is empty to begin with (first is null).

When the while loop terminates, then, we may be at the beginning, the middle, or the end of the list, or the list may be empty.

If we're at the beginning, or the list is empty, previous will be null; so we set first to the new link. Otherwise, we're in the middle of the list, or at the end, and we set previous.next to the new link.

In any case we set the new link's next field to current. If we're at the end of the list, current is null, so the new link's next field is appropriately set to this value.

The sortedList.java **Program**

The sortedList.java example shown in Listing 5.6 presents a SortedList class with insert(), remove(), and displayList() methods. Only the insert() routine is different from its counterpart in non-sorted lists.

LISTING 5.6   The sortedList.java Program

```
// sortedList.java
// demonstrates sorted list
// to run this program: C>java SortedListApp
//////////////////////////////////////////////////////////////
class Link
    {
    public long dData;              // data item
    public Link next;               // next link in list
// -------------------------------------------------------------
    public Link(long dd)            // constructor
       { dData = dd; }
// -------------------------------------------------------------
    public void displayLink()       // display this link
       { System.out.print(dData + " "); }
```

LISTING 5.6   Continued

```
    }  // end class Link
/////////////////////////////////////////////////////////////////
class SortedList
    {
    private Link first;                 // ref to first item on list
// -------------------------------------------------------------
    public SortedList()                 // constructor
        { first = null; }
// -------------------------------------------------------------
    public boolean isEmpty()            // true if no links
        { return (first==null); }
// -------------------------------------------------------------
    public void insert(long key)        // insert, in order
        {
        Link newLink = new Link(key);   // make new link
        Link previous = null;           // start at first
        Link current = first;
                                        // until end of list,
        while(current != null && key > current.dData)
            {                           // or key > current,
            previous = current;
            current = current.next;     // go to next item
            }
        if(previous==null)              // at beginning of list
            first = newLink;            // first --> newLink
        else                            // not at beginning
            previous.next = newLink;    // old prev --> newLink
        newLink.next = current;         // newLink --> old current
        }  // end insert()
// -------------------------------------------------------------
    public Link remove()                // return & delete first link
        {                               // (assumes non-empty list)
        Link temp = first;              // save first
        first = first.next;             // delete first
        return temp;                    // return value
        }
// -------------------------------------------------------------
    public void displayList()
        {
        System.out.print("List (first-->last): ");
        Link current = first;           // start at beginning of list
```

LISTING 5.6   Continued

```
      while(current != null)      // until end of list,
         {
         current.displayLink();   // print data
         current = current.next;  // move to next link
         }
      System.out.println("");
      }
   }  // end class SortedList
/////////////////////////////////////////////////////////////////
class SortedListApp
   {
   public static void main(String[] args)
      {                              // create new list
      SortedList theSortedList = new SortedList();
      theSortedList.insert(20);    // insert 2 items
      theSortedList.insert(40);

      theSortedList.displayList(); // display list

      theSortedList.insert(10);    // insert 3 more items
      theSortedList.insert(30);
      theSortedList.insert(50);

      theSortedList.displayList(); // display list

      theSortedList.remove();      // remove an item

      theSortedList.displayList(); // display list
      }  // end main()
   }  // end class SortedListApp
/////////////////////////////////////////////////////////////////
```

In main() we insert two items with key values 20 and 40. Then we insert three more items, with values 10, 30, and 50. These values are inserted at the beginning of the list, in the middle, and at the end, showing that the insert() routine correctly handles these special cases. Finally, we remove one item, to show removal is always from the front of the list. After each change, the list is displayed. Here's the output from sortedList.java:

```
List (first-->last): 20 40
List (first-->last): 10 20 30 40 50
List (first-->last): 20 30 40 50
```

Efficiency of Sorted Linked Lists

Insertion and deletion of arbitrary items in the sorted linked list require O(N) comparisons (N/2 on the average) because the appropriate location must be found by stepping through the list. However, the minimum value can be found, or deleted, in O(1) time because it's at the beginning of the list. If an application frequently accesses the minimum item, and fast insertion isn't critical, then a sorted linked list is an effective choice. A priority queue might be implemented by a sorted linked list, for example.

List Insertion Sort

A sorted list can be used as a fairly efficient sorting mechanism. Suppose you have an array of unsorted data items. If you take the items from the array and insert them one by one into the sorted list, they'll be placed in sorted order automatically. If you then remove them from the list and put them back in the array, the array will be sorted.

This type of sort turns out to be substantially more efficient than the more usual insertion sort within an array, described in Chapter 3, "Simple Sorting," because fewer copies are necessary. It's still an O(N^2) process because inserting each item into the sorted list involves comparing a new item with an average of half the items already in the list, and there are N items to insert, resulting in about N^2/4 comparisons. However, each item is copied only twice: once from the array to the list and once from the list to the array. N*2 copies compares favorably with the insertion sort within an array, where there are about N^2 copies.

Listing 5.7 shows the listInsertionSort.java program, which starts with an array of unsorted items of type link, inserts them into a sorted list (using a constructor), and then removes them and places them back into the array.

LISTING 5.7   The listInsertionSort.java Program

```
// listInsertionSort.java
// demonstrates sorted list used for sorting
// to run this program: C>java ListInsertionSortApp
/////////////////////////////////////////////////////////////
class Link
   {
   public long dData;                // data item
   public Link next;                 // next link in list
// -------------------------------------------------------------
   public Link(long dd)              // constructor
      { dData = dd; }
// -------------------------------------------------------------
```

LISTING 5.7   Continued

```
    }  // end class Link
////////////////////////////////////////////////////////////////
class SortedList
    {
    private Link first;              // ref to first item on list
// --------------------------------------------------------------
    public SortedList()              // constructor (no args)
        { first = null; }            // initialize list
// --------------------------------------------------------------
    public SortedList(Link[] linkArr)  // constructor (array
        {                              // as argument)
        first = null;                  // initialize list
        for(int j=0; j<linkArr.length; j++)  // copy array
            insert( linkArr[j] );            // to list
        }
// --------------------------------------------------------------
    public void insert(Link k)       // insert (in order)
        {
        Link previous = null;            // start at first
        Link current = first;

                                         // until end of list,
        while(current != null && k.dData > current.dData)
            {                            // or key > current,
            previous = current;
            current = current.next;      // go to next item
            }
        if(previous==null)               // at beginning of list
            first = k;                   // first --> k
        else                             // not at beginning
            previous.next = k;           // old prev --> k
        k.next = current;                // k --> old current
        }  // end insert()
// --------------------------------------------------------------
    public Link remove()             // return & delete first link
        {                            // (assumes non-empty list)
        Link temp = first;               // save first
        first = first.next;              // delete first
        return temp;                     // return value
        }
// --------------------------------------------------------------
    }  // end class SortedList
```

LISTING 5.7   Continued

```
//////////////////////////////////////////////////////////////
class ListInsertionSortApp
   {
   public static void main(String[] args)
      {
      int size = 10;
                                    // create array of links
      Link[] linkArray = new Link[size];

      for(int j=0; j<size; j++)  // fill array with links
         {                               // random number
         int n = (int)(java.lang.Math.random()*99);
         Link newLink = new Link(n);  // make link
         linkArray[j] = newLink;       // put in array
         }
                                    // display array contents
      System.out.print("Unsorted array: ");
      for(int j=0; j<size; j++)
         System.out.print( linkArray[j].dData + " " );
      System.out.println("");
                                    // create new list
                                    // initialized with array
      SortedList theSortedList = new SortedList(linkArray);

      for(int j=0; j<size; j++)  // links from list to array
         linkArray[j] = theSortedList.remove();
                                    // display array contents
      System.out.print("Sorted Array:   ");
      for(int j=0; j<size; j++)
         System.out.print(linkArray[j].dData + " ");
      System.out.println("");
      }  // end main()
   }  // end class ListInsertionSortApp
//////////////////////////////////////////////////////////////
```

This program displays the values in the array before the sorting operation and again afterward. Here's some sample output:

```
Unsorted array: 59 69 41 56 84 15 86 81 37 35
Sorted array:   15 35 37 41 56 59 69 81 84 86
```

The output will be different each time because the initial values are generated randomly.

A new constructor for SortedList takes an array of Link objects as an argument and inserts the entire contents of this array into the newly created list. By doing so, it helps make things easier for the client (the main() routine).

We've also made a change to the insert() routine in this program. It now accepts a Link object as an argument, rather than a long. We do this so we can store Link objects in the array and insert them directly into the list. In the sortedList.java program (Listing 5.6), it was more convenient to have the insert() routine create each Link object, using the long value passed as an argument.

The downside of the list insertion sort, compared with an array-based insertion sort, is that it takes somewhat more than twice as much memory: The array and linked list must be in memory at the same time. However, if you have a sorted linked list class handy, the list insertion sort is a convenient way to sort arrays that aren't too large.

Doubly Linked Lists

Let's examine another variation on the linked list: the *doubly linked* list (not to be confused with the double-ended list). What's the advantage of a doubly linked list? A potential problem with ordinary linked lists is that it's difficult to traverse backward along the list. A statement like

```
current=current.next
```

steps conveniently to the next link, but there's no corresponding way to go to the previous link. Depending on the application, this limitation could pose problems.

For example, imagine a text editor in which a linked list is used to store the text. Each text line on the screen is stored as a String object embedded in a link. When the editor's user moves the cursor downward on the screen, the program steps to the next link to manipulate or display the new line. But what happens if the user moves the cursor upward? In an ordinary linked list, you would need to return current (or its equivalent) to the start of the list and then step all the way down again to the new current link. This isn't very efficient. You want to make a single step upward.

The doubly linked list provides this capability. It allows you to traverse backward as well as forward through the list. The secret is that each link has two references to other links instead of one. The first is to the next link, as in ordinary lists. The second is to the previous link. This type of list is shown in Figure 5.13.

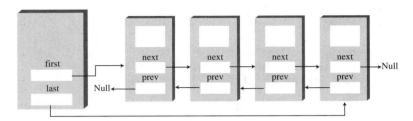

FIGURE 5.13 A doubly linked list.

The beginning of the specification for the Link class in a doubly linked list looks like this:

```
class Link
   {
   public long dData;              // data item
   public Link next;              // next link in list
   public link previous;          // previous link in list
   ...
   }
```

The downside of doubly linked lists is that every time you insert or delete a link you must deal with four links instead of two: two attachments to the previous link and two attachments to the following one. Also, of course, each link is a little bigger because of the extra reference.

A doubly linked list doesn't necessarily need to be a double-ended list (keeping a reference to the last element on the list) but creating it this way is useful, so we'll include it in our example.

We'll show the complete listing for the doublyLinked.java program soon, but first let's examine some of the methods in its doublyLinkedList class.

Traversal

Two display methods demonstrate traversal of a doubly linked list. The displayForward() method is the same as the displayList() method we've seen in ordinary linked lists. The displayBackward() method is similar but starts at the last element in the list and proceeds toward the start of the list, going to each element's previous field. This code fragment shows how this process works:

```
Link current = last;            // start at end
while(current != null)          // until start of list,
   current = current.previous;  // move to previous link
```

Incidentally, some people take the view that, because you can go either way equally easily on a doubly linked list, there is no preferred direction and therefore terms like previous and next are inappropriate. If you prefer, you can substitute direction-neutral terms such as left and right.

Insertion

We've included several insertion routines in the DoublyLinkedList class. The insertFirst() method inserts at the beginning of the list, insertLast() inserts at the end, and insertAfter() inserts following an element with a specified key.

Unless the list is empty, the insertFirst() routine changes the previous field in the old first link to point to the new link and changes the next field in the new link to point to the old first link. Finally, it sets first to point to the new link. This process is shown in Figure 5.14.

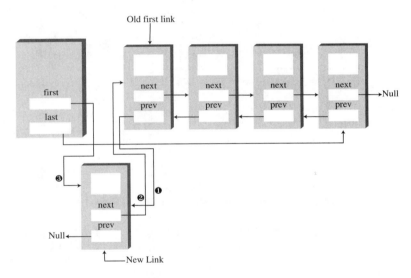

FIGURE 5.14 Insertion at the beginning.

If the list is empty, the last field must be changed instead of the first.previous field. Here's the code:

```
if( isEmpty() )                         // if empty list,
    last = newLink;                     // newLink <-- last
else
    first.previous = newLink;           // newLink <-- old first
newLink.next = first;                   // newLink --> old first
first = newLink;                        // first --> newLink
```

The insertLast() method is the same process applied to the end of the list; it's a mirror image of insertFirst().

The insertAfter() method inserts a new link following the link with a specified key value. It's a bit more complicated because four connections must be made. First, the link with the specified key value must be found. This procedure is handled the same way as the find() routine in the linkList2.java program (Listing 5.2). Then, assuming we're not at the end of the list, two connections must be made between the new link and the next link, and two more between current and the new link. This process is shown in Figure 5.15.

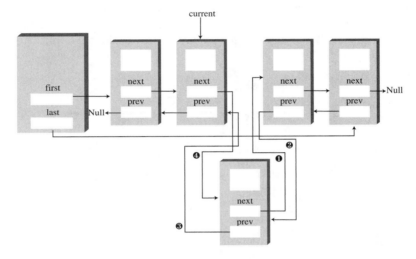

FIGURE 5.15 Insertion at an arbitrary location.

If the new link will be inserted at the end of the list, its next field must point to null, and last must point to the new link. Here's the insertAfter() code that deals with the links:

```
if(current==last)               // if last link,
   {
   newLink.next = null;         // newLink --> null
   last = newLink;              // newLink <-- last
   }
else                            // not last link,
   {
   newLink.next = current.next; // newLink --> old next
                                // newLink <-- old next
   current.next.previous = newLink;
```

```
   }
newLink.previous = current;     // old current <-- newLink
current.next = newLink;         // old current --> newLink
```

Perhaps you're unfamiliar with the use of two dot operators in the same expression. It's a natural extension of a single dot operator. The expression

```
current.next.previous
```

means the previous field of the link referred to by the next field in the link current.

Deletion

There are three deletion routines: deleteFirst(), deleteLast(), and deleteKey(). The first two are fairly straightforward. In deleteKey(), the key being deleted is current. Assuming the link to be deleted is neither the first nor the last one in the list, the next field of current.previous (the link before the one being deleted) is set to point to current.next (the link following the one being deleted), and the previous field of current.next is set to point to current.previous. This disconnects the current link from the list. Figure 5.16 shows how this disconnection looks, and the following two statements carry it out:

```
current.previous.next = current.next;
current.next.previous = current.previous;
```

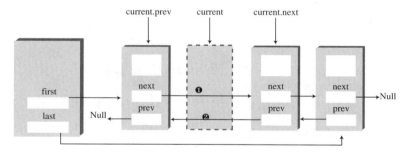

FIGURE 5.16 Deleting an arbitrary link.

Special cases arise if the link to be deleted is either the first or last in the list because first or last must be set to point to the next or the previous link. Here's the code from deleteKey() for dealing with link connections:

```
if(current==first)              // first item?
   first = current.next;        // first --> old next
else                            // not first
                                // old previous --> old next
```

```
            current.previous.next = current.next;

if(current==last)                  // last item?
   last = current.previous;        // old previous <-- last
else                               // not last
                                   // old previous <-- old next
   current.next.previous = current.previous;
```

The doublyLinked.java Program

Listing 5.8 shows the complete doublyLinked.java program, which includes all the
routines just discussed.

LISTING 5.8   The doublyLinked.java Program

```
// doublyLinked.java
// demonstrates doubly-linked list
// to run this program: C>java DoublyLinkedApp
////////////////////////////////////////////////////////////////
class Link
   {
   public long dData;                 // data item
   public Link next;                  // next link in list
   public Link previous;              // previous link in list
// -------------------------------------------------------------
   public Link(long d)                // constructor
      { dData = d; }
// -------------------------------------------------------------
   public void displayLink()          // display this link
      { System.out.print(dData + " "); }
// -------------------------------------------------------------
   }  // end class Link
////////////////////////////////////////////////////////////////
class DoublyLinkedList
   {
   private Link first;                // ref to first item
   private Link last;                 // ref to last item
// -------------------------------------------------------------
   public DoublyLinkedList()          // constructor
      {
      first = null;                   // no items on list yet
      last = null;
      }
```

LISTING 5.8   Continued

```
// --------------------------------------------------------------
   public boolean isEmpty()          // true if no links
      { return first==null; }
// --------------------------------------------------------------
   public void insertFirst(long dd)  // insert at front of list
      {
      Link newLink = new Link(dd);   // make new link

      if( isEmpty() )                // if empty list,
         last = newLink;             // newLink <-- last
      else
         first.previous = newLink;   // newLink <-- old first
      newLink.next = first;          // newLink --> old first
      first = newLink;               // first --> newLink
      }
// --------------------------------------------------------------
   public void insertLast(long dd)   // insert at end of list
      {
      Link newLink = new Link(dd);   // make new link
      if( isEmpty() )                // if empty list,
         first = newLink;            // first --> newLink
      else
         {
         last.next = newLink;        // old last --> newLink
         newLink.previous = last;    // old last <-- newLink
         }
      last = newLink;                // newLink <-- last
      }
// --------------------------------------------------------------
   public Link deleteFirst()         // delete first link
      {                              // (assumes non-empty list)
      Link temp = first;
      if(first.next == null)         // if only one item
         last = null;                // null <-- last
      else
         first.next.previous = null; // null <-- old next
      first = first.next;            // first --> old next
      return temp;
      }
// --------------------------------------------------------------
   public Link deleteLast()          // delete last link
```

LISTING 5.8   Continued

```
    {                              // (assumes non-empty list)
    Link temp = last;
    if(first.next == null)         // if only one item
       first = null;               // first --> null
    else
       last.previous.next = null;  // old previous --> null
    last = last.previous;          // old previous <-- last
    return temp;
    }
// ------------------------------------------------------------
                                   // insert dd just after key
    public boolean insertAfter(long key, long dd)
    {                              // (assumes non-empty list)
    Link current = first;          // start at beginning
    while(current.dData != key)    // until match is found,
       {
       current = current.next;     // move to next link
       if(current == null)
          return false;            // didn't find it
       }
    Link newLink = new Link(dd);   // make new link

    if(current==last)              // if last link,
       {
       newLink.next = null;        // newLink --> null
       last = newLink;             // newLink <-- last
       }
    else                           // not last link,
       {
       newLink.next = current.next; // newLink --> old next
                                    // newLink <-- old next
       current.next.previous = newLink;
       }
    newLink.previous = current;    // old current <-- newLink
    current.next = newLink;        // old current --> newLink
    return true;                   // found it, did insertion
    }
// ------------------------------------------------------------
    public Link deleteKey(long key) // delete item w/ given key
    {                              // (assumes non-empty list)
    Link current = first;          // start at beginning
```

LISTING 5.8   Continued

```
      while(current.dData != key)     // until match is found,
         {
         current = current.next;      // move to next link
         if(current == null)
            return null;              // didn't find it
         }
      if(current==first)              // found it; first item?
         first = current.next;        // first --> old next
      else                            // not first
                                      // old previous --> old next
         current.previous.next = current.next;

      if(current==last)               // last item?
         last = current.previous;     // old previous <-- last
      else                            // not last
                                      // old previous <-- old next
         current.next.previous = current.previous;
      return current;                 // return value
      }
// ------------------------------------------------------------
   public void displayForward()
      {
      System.out.print("List (first-->last): ");
      Link current = first;           // start at beginning
      while(current != null)          // until end of list,
         {
         current.displayLink();       // display data
         current = current.next;      // move to next link
         }
      System.out.println("");
      }
// ------------------------------------------------------------
   public void displayBackward()
      {
      System.out.print("List (last-->first): ");
      Link current = last;            // start at end
      while(current != null)          // until start of list,
         {
         current.displayLink();       // display data
         current = current.previous;  // move to previous link
         }
```

LISTING 5.8   Continued

```
      System.out.println("");
      }
// ------------------------------------------------------------
   }  // end class DoublyLinkedList
/////////////////////////////////////////////////////////////////
class DoublyLinkedApp
   {
   public static void main(String[] args)
      {                              // make a new list
      DoublyLinkedList theList = new DoublyLinkedList();

      theList.insertFirst(22);     // insert at front
      theList.insertFirst(44);
      theList.insertFirst(66);

      theList.insertLast(11);      // insert at rear
      theList.insertLast(33);
      theList.insertLast(55);

      theList.displayForward();    // display list forward
      theList.displayBackward();   // display list backward

      theList.deleteFirst();       // delete first item
      theList.deleteLast();        // delete last item
      theList.deleteKey(11);       // delete item with key 11

      theList.displayForward();    // display list forward

      theList.insertAfter(22, 77); // insert 77 after 22
      theList.insertAfter(33, 88); // insert 88 after 33

      theList.displayForward();    // display list forward
      }  // end main()
   }  // end class DoublyLinkedApp
/////////////////////////////////////////////////////////////////
```

In main() we insert some items at the beginning of the list and at the end, display
the items going both forward and backward, delete the first and last items and the
item with key 11, display the list again (forward only), insert two items using the
insertAfter() method, and display the list again. Here's the output:

```
List (first-->last): 66 44 22 11 33 55
List (last-->first): 55 33 11 22 44 66
List (first-->last): 44 22 33
List (first-->last): 44 22 77 33 88
```

The deletion methods and the insertAfter() method assume that the list isn't empty. Although for simplicity we don't show it in main(), isEmpty() should be used to verify that there's something in the list before attempting such insertions and deletions.

Doubly Linked List as Basis for Deques

A doubly linked list can be used as the basis for a deque, mentioned in the preceding chapter. In a deque you can insert and delete at either end, and the doubly linked list provides this capability.

Iterators

We've seen how the user of a list can find a link with a given key using a find() method. The method starts at the beginning of the list and examines each link until it finds one matching the search key. Other operations we've looked at, such as deleting a specified link or inserting before or after a specified link, also involve searching through the list to find the specified link. However, these methods don't give the user any control over the traversal to the specified item.

Suppose you wanted to traverse a list, performing some operation on certain links. For example, imagine a personnel file stored as a linked list. You might want to increase the wages of all employees who were being paid minimum wage, without affecting employees already above the minimum. Or suppose that in a list of mail-order customers, you decided to delete all customers who had not ordered anything in six months.

In an array, such operations are easy because you can use an array index to keep track of your position. You can operate on one item, then increment the index to point to the next item, and see if that item is a suitable candidate for the operation. However, in a linked list, the links don't have fixed index numbers. How can we provide a list's user with something analogous to an array index? You could repeatedly use find() to look for appropriate items in a list, but that approach requires many comparisons to find each link. It's far more efficient to step from link to link, checking whether each one meets certain criteria and performing the appropriate operation if it does.

A Reference in the List Itself?

As users of a list class, what we need is access to a reference that can point to any arbitrary link. This way, we can examine or modify the link. We should be able to increment the reference so we can traverse along the list, looking at each link in turn, and we should be able to access the link pointed to by the reference.

Assuming we create such a reference, where will it be installed? One possibility is to use a field in the list itself, called current or something similar. You could access a link using current and increment current to move to the next link.

One problem with this approach is that you might need more than one such reference, just as you often use several array indices at the same time. How many would be appropriate? There's no way to know how many the user might need. Thus, it seems easier to allow the user to create as many such references as necessary. To make this possible in an object-oriented language, it's natural to embed each reference in a class object. This object can't be the same as the list class because there's only one list object, so it is normally implemented as a separate class.

An Iterator Class

Objects containing references to items in data structures, used to traverse these structures, are commonly called *iterators* (or sometimes, as in certain Java classes, *enumerators*). Here's a preliminary idea of how they look:

```
class ListIterator()
   {
   private Link current;
   ...
   }
```

The current field contains a reference to the link the iterator currently points to. (The term *points* as used here doesn't refer to pointers in C++; we're using it in its generic sense to mean "refers to.")

To use such an iterator, the user might create a list and then create an iterator object associated with the list. Actually, as it turns out, letting the list create the iterator is easier, so it can pass the iterator certain information, such as a reference to its first field. Thus, we add a getIterator() method to the list class; this method returns a suitable iterator object to the user. Here's some abbreviated code in main() that shows how the class user would invoke an iterator:

```
public static void main(...)
   {
   LinkList theList = new LinkList();          // make list
   ListIterator iter1 = theList.getIterator(); // make iter
```

```
    Link aLink = iter1.getCurrent();    // access link at iterator
    iter1.nextLink();                    // move iter to next link
    }
```

After we've made the iterator object, we can use it to access the link it points to or increment it so it points to the next link, as shown in the second two statements. We call the iterator object iter1 to emphasize that you could make more iterators (iter2 and so on) the same way.

The iterator always points to some link in the list. It's associated with the list, but it's not the same as the list or the same as a link. Figure 5.17 shows two iterators pointing to links in a list.

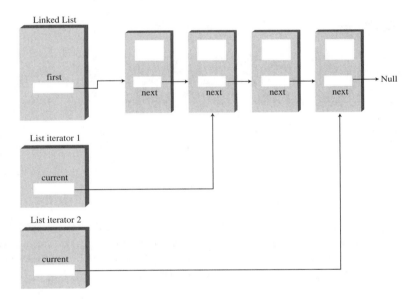

FIGURE 5.17 List iterators.

Additional Iterator Features

We've seen several programs in which the use of a previous field made performing certain operations simpler, such as deleting a link from an arbitrary location. Such a field is also useful in an iterator.

Also, it may be that the iterator will need to change the value of the list's first field—for instance, if an item is inserted or deleted at the beginning of the list. If the iterator is an object of a separate class, how can it access a private field, such as first, in the list? One solution is for the list to pass a reference from itself to the iterator when it creates the iterator. This reference is stored in a field in the iterator.

The list must then provide public methods that allow the iterator to change first. These LinkList methods are getFirst() and setFirst(). (The weakness of this approach is that these methods allow anyone to change first, which introduces an element of risk.)

Here's a revised (although still incomplete) iterator class that incorporates these additional fields, along with reset() and nextLink() methods:

```
class ListIterator()
   {
   private Link current;        // reference to current link
   private Link previous;       // reference to previous link
   private LinkList ourList;    // reference to "parent" list

   public void reset()          // set to start of list
      {
      current = ourList.getFirst();  // current --> first
      previous = null;               // previous --> null
      }
   public void nextLink()       // go to next link
      {
      previous = current;       // set previous to this
      current = current.next;   // set this to next
      }
   ...
   }
```

We might note, for you old-time C++ programmers, that in C++ the connection between the iterator and the list is typically provided by making the iterator class a *friend* of the list class. However, Java has no friend classes, which are controversial in any case because they are a chink in the armor of data hiding.

Iterator Methods

Additional methods can make the iterator a flexible and powerful class. All operations previously performed by the class that involve iterating through the list, such as insertAfter(), are more naturally performed by the iterator. In our example the iterator includes the following methods:

- reset()—Sets the iterator to the start of the list
- nextLink()—Moves the iterator to the next link
- getCurrent()—Returns the link at the iterator
- atEnd()—Returns true if the iterator is at the end of the list

- `insertAfter()`—Inserts a new link after the iterator

- `insertBefore()`—Inserts a new link before the iterator

- `deleteCurrent()`—Deletes the link at the iterator

The user can position the iterator using `reset()` and `nextLink()`, check whether it's at the end of the list with `atEnd()`, and perform the other operations shown.

Deciding which tasks should be carried out by an iterator and which by the list itself is not always easy. An `insertBefore()` method works best in the iterator, but an `insertFirst()` routine that always inserts at the beginning of the list might be more appropriate in the list class. We've kept a `displayList()` routine in the list, but this operation could also be handled with `getCurrent()` and `nextLink()` calls to the iterator.

The `interIterator.java` Program

The `interIterator.java` program includes an interactive interface that permits the user to control the iterator directly. After you've started the program, you can perform the following actions by typing the appropriate letter:

- s—Show the list contents

- r—Reset the iterator to the start of the list

- n—Go to the next link

- g—Get the contents of the current link

- b—Insert before the current link

- a—Insert a new link after the current link

- d—Delete the current link

Listing 5.9 shows the complete `interIterator.java` program.

LISTING 5.9   The `interIterator.java` Program

```
// interIterator.java
// demonstrates iterators on a linked listListIterator
// to run this program: C>java InterIterApp
import java.io.*;               // for I/O
//////////////////////////////////////////////////////////////
class Link
   {
   public long dData;            // data item
```

LISTING 5.9   Continued

```java
   public Link next;              // next link in list
// -------------------------------------------------------------
   public Link(long dd)           // constructor
      { dData = dd; }
// -------------------------------------------------------------
   public void displayLink()      // display ourself
      { System.out.print(dData + " "); }
   }  // end class Link
////////////////////////////////////////////////////////////////
class LinkList
   {
   private Link first;            // ref to first item on list

// -------------------------------------------------------------
   public LinkList()              // constructor
      { first = null; }           // no items on list yet
// -------------------------------------------------------------
   public Link getFirst()         // get value of first
      { return first; }
// -------------------------------------------------------------
   public void setFirst(Link f)   // set first to new link
      { first = f; }
// -------------------------------------------------------------
   public boolean isEmpty()       // true if list is empty
      { return first==null; }
// -------------------------------------------------------------
   public ListIterator getIterator()  // return iterator
      {
      return new ListIterator(this);  // initialized with
      }                               //   this list
// -------------------------------------------------------------
   public void displayList()
      {
      Link current = first;       // start at beginning of list
      while(current != null)      // until end of list,
         {
         current.displayLink();   // print data
         current = current.next;  // move to next link
         }
      System.out.println("");
      }
```

LISTING 5.9   Continued

```
// ------------------------------------------------------------
   } // end class LinkList
/////////////////////////////////////////////////////////////
class ListIterator
   {
   private Link current;          // current link
   private Link previous;         // previous link
   private LinkList ourList;      // our linked list
//-------------------------------------------------------------
   public ListIterator(LinkList list) // constructor
      {
      ourList = list;
      reset();
      }
//-------------------------------------------------------------
   public void reset()            // start at 'first'
      {
      current = ourList.getFirst();
      previous = null;
      }
//-------------------------------------------------------------
   public boolean atEnd()         // true if last link
      { return (current.next==null); }
//-------------------------------------------------------------
   public void nextLink()         // go to next link
      {
      previous = current;
      current = current.next;
      }
//-------------------------------------------------------------
   public Link getCurrent()       // get current link
      { return current; }
//-------------------------------------------------------------
   public void insertAfter(long dd)   // insert after
      {                              // current link
      Link newLink = new Link(dd);

      if( ourList.isEmpty() )     // empty list
         {
         ourList.setFirst(newLink);
         current = newLink;
```

LISTING 5.9   Continued

```
      }
    else                        // not empty
      {
      newLink.next = current.next;
      current.next = newLink;
      nextLink();               // point to new link
      }
    }
//----------------------------------------------------------------
  public void insertBefore(long dd)    // insert before
    {                                  // current link
    Link newLink = new Link(dd);

    if(previous == null)       // beginning of list
      {                        // (or empty list)
      newLink.next = ourList.getFirst();
      ourList.setFirst(newLink);
      reset();
      }
    else                        // not beginning
      {
      newLink.next = previous.next;
      previous.next = newLink;
      current = newLink;
      }
    }
//----------------------------------------------------------------
  public long deleteCurrent()     // delete item at current
    {
    long value = current.dData;
    if(previous == null)       // beginning of list
      {
      ourList.setFirst(current.next);
      reset();
      }
    else                        // not beginning
      {
      previous.next = current.next;
      if( atEnd() )
        reset();
      else
```

LISTING 5.9   Continued

```
                    current = current.next;
            }
        return value;
        }
//----------------------------------------------------------------
    }  // end class ListIterator
////////////////////////////////////////////////////////////////
class InterIterApp
    {
    public static void main(String[] args) throws IOException
        {
        LinkList theList = new LinkList();            // new list
        ListIterator iter1 = theList.getIterator();  // new iter
        long value;

        iter1.insertAfter(20);              // insert items
        iter1.insertAfter(40);
        iter1.insertAfter(80);
        iter1.insertBefore(60);

        while(true)
            {
            System.out.print("Enter first letter of show, reset, ");
            System.out.print("next, get, before, after, delete: ");
            System.out.flush();
            int choice = getChar();         // get user's option
            switch(choice)
                {
                case 's':                   // show list
                    if( !theList.isEmpty() )
                        theList.displayList();
                    else
                        System.out.println("List is empty");
                    break;
                case 'r':                   // reset (to first)
                    iter1.reset();
                    break;
                case 'n':                   // advance to next item
                    if( !theList.isEmpty() && !iter1.atEnd() )
                        iter1.nextLink();
                    else
```

LISTING 5.9   Continued

```
                            System.out.println("Can't go to next link");
                  break;
               case 'g':                      // get current item
                  if( !theList.isEmpty() )
                     {
                     value = iter1.getCurrent().dData;
                     System.out.println("Returned " + value);
                     }
                  else
                     System.out.println("List is empty");
                  break;
               case 'b':                      // insert before current
                  System.out.print("Enter value to insert: ");
                  System.out.flush();
                  value = getInt();
                  iter1.insertBefore(value);
                  break;
               case 'a':                      // insert after current
                  System.out.print("Enter value to insert: ");
                  System.out.flush();
                  value = getInt();
                  iter1.insertAfter(value);
                  break;
               case 'd':                      // delete current item
                  if( !theList.isEmpty() )
                     {
                     value = iter1.deleteCurrent();
                     System.out.println("Deleted " + value);
                     }
                  else
                     System.out.println("Can't delete");
                  break;
               default:
                  System.out.println("Invalid entry");
               }  // end switch
            }  // end while
         }  // end main()
//--------------------------------------------------------------
public static String getString() throws IOException
      {
      InputStreamReader isr = new InputStreamReader(System.in);
```

LISTING 5.9   Continued

```
    BufferedReader br = new BufferedReader(isr);
    String s = br.readLine();
    return s;
    }
//-------------------------------------------------------------
  public static char getChar() throws IOException
    {
    String s = getString();
    return s.charAt(0);
    }
//-------------------------------------------------------------
  public static int getInt() throws IOException
    {
    String s = getString();
    return Integer.parseInt(s);
    }
//-------------------------------------------------------------
  }  // end class InterIterApp
/////////////////////////////////////////////////////////////////
```

The main() routine inserts four items into the list, using an iterator and its
insertAfter() method. Then it waits for the user to interact with it. In the following
sample interaction, the user displays the list, resets the iterator to the beginning,
goes forward two links, gets the current link's key value (which is 60), inserts 100
before this, inserts 7 after the 100, and displays the list again:

```
Enter first letter of
   show, reset, next, get, before, after, delete: s
20 40 60 80
Enter first letter of
   show, reset, next, get, before, after, delete: r
Enter first letter of
   show, reset, next, get, before, after, delete: n
Enter first letter of
   show, reset, next, get, before, after, delete: n
Enter first letter of
   show, reset, next, get, before, after, delete: g
Returned 60
Enter first letter of
   show, reset, next, get, before, after, delete: b
Enter value to insert: 100
```

```
Enter first letter of
    show, reset, next, get, before, after, delete: a
Enter value to insert: 7
Enter first letter of
    show, reset, next, get, before, after, delete: s
20 40 100 7 60 80
```

Experimenting with the `interIterator.java` program will give you a feeling for how the iterator moves along the links and how it can insert and delete links anywhere in the list.

Where Does the Iterator Point?

One of the design issues in an iterator class is deciding where the iterator should point following various operations.

When you delete an item with `deleteCurrent()`, should the iterator end up pointing to the next item, to the previous item, or back at the beginning of the list? Keeping the iterator in the vicinity of the deleted item is convenient because the chances are the class user will be carrying out other operations there. However, you can't move it to the previous item because there's no way to reset the list's `previous` field to the previous item. (You would need a doubly linked list for that task.) Our solution is to move the iterator to the link following the deleted link. If we've just deleted the item at the end of the list, the iterator is set to the beginning of the list.

Following calls to `insertBefore()` and `insertAfter()`, we return with `current` pointing to the newly inserted item.

The `atEnd()` Method

There's another question about the `atEnd()` method. It could return `true` when the iterator points to the last valid link in the list, or it could return `true` when the iterator points *past* the last link (and is thus not pointing to a valid link).

With the first approach, a loop condition used to iterate through the list becomes awkward because you need to perform an operation on the last link before checking whether it is the last link (and terminating the loop if it is).

However, the second approach doesn't allow you to find out you're at the end of the list until it's too late to do anything with the last link. (You couldn't look for the last link and then delete it, for example.) This is because when `atEnd()` became `true`, the iterator would no longer point to the last link (or indeed any valid link), and you can't "back up" the iterator in a singly linked list.

We take the first approach. This way, the iterator always points to a valid link, although you must be careful when writing a loop that iterates through the list, as we'll see next.

Iterative Operations

As we noted, an iterator allows you to traverse the list, performing operations on certain data items. Here's a code fragment that displays the list contents, using an iterator instead of the list's displayList() method:

```
iter1.reset();                          // start at first
long value = iter1.getCurrent().dData;  // display link
System.out.println(value + " ");
while( !iter1.atEnd() )                  // until end,
   {
   iter1.nextLink();                     // go to next link,
   long value = iter1.getCurrent().dData;  // display it
   System.out.println(value + " ");
   }
```

Although we don't do so here, you should check with isEmpty() to be sure the list is not empty before calling getCurrent().

The following code shows how you could delete all items with keys that are multiples of 3. We show only the revised main() routine; everything else is the same as in interIterator.java (Listing 5.9).

```
class InterIterApp
   {
   public static void main(String[] args) throws IOException
      {
      LinkList theList = new LinkList();          // new list
      ListIterator iter1 = theList.getIterator();  // new iter

      iter1.insertAfter(21);                // insert links
      iter1.insertAfter(40);
      iter1.insertAfter(30);
      iter1.insertAfter(7);
      iter1.insertAfter(45);

      theList.displayList();               // display list

      iter1.reset();                       // start at first link
      Link aLink = iter1.getCurrent();     // get it
      if(aLink.dData % 3 == 0)             // if divisible by 3,
         iter1.deleteCurrent();           // delete it
      while( !iter1.atEnd() )              // until end of list,
         {
```

```
        iter1.nextLink();              // go to next link

        aLink = iter1.getCurrent();    // get link
        if(aLink.dData % 3 == 0)       // if divisible by 3,
           iter1.deleteCurrent();      // delete it
        }
     theList.displayList();            // display list
     }  // end main()
  }  // end class InterIterApp
```

We insert five links and display the list. Then we iterate through the list, deleting those links with keys divisible by 3, and display the list again. Here's the output:

```
21 40 30 7 45
40 7
```

Again, although we don't show it here, it's important to check whether the list is empty before calling deleteCurrent().

Other Methods

You could create other useful methods for the ListIterator class. For example, a find() method would return an item with a specified key value, as we've seen when find() is a list method. A replace() method could replace items that had certain key values with other items.

Because it's a singly linked list, you can iterate along it only in the forward direction. If a doubly linked list were used, you could go either way, allowing operations such as deletion from the end of the list, just as with non-iterators. This capability would probably be a convenience in some applications.

Summary

- A linked list consists of one linkedList object and a number of Link objects.

- The linkedList object contains a reference, often called first, to the first link in the list.

- Each Link object contains data and a reference, often called next, to the next link in the list.

- A next value of null signals the end of the list.

- Inserting an item at the beginning of a linked list involves changing the new link's next field to point to the old first link and changing first to point to the new item.

- Deleting an item at the beginning of a list involves setting `first` to point to `first.next`.

- To traverse a linked list, you start at `first` and then go from link to link, using each link's `next` field to find the next link.

- A link with a specified key value can be found by traversing the list. Once found, an item can be displayed, deleted, or operated on in other ways.

- A new link can be inserted before or after a link with a specified key value, following a traversal to find this link.

- A double-ended list maintains a pointer to the last link in the list, often called last, as well as to the first.

- A double-ended list allows insertion at the end of the list.

- An Abstract Data Type (ADT) is a data storage class considered without reference to its implementation.

- Stacks and queues are ADTs. They can be implemented using either arrays or linked lists.

- In a sorted linked list, the links are arranged in order of ascending (or sometimes descending) key value.

- Insertion in a sorted list takes O(N) time because the correct insertion point must be found. Deletion of the smallest link takes O(1) time.

- In a doubly linked list, each link contains a reference to the previous link as well as the next link.

- A doubly linked list permits backward traversal and deletion from the end of the list.

- An iterator is a reference, encapsulated in a class object, that points to a link in an associated list.

- Iterator methods allow the user to move the iterator along the list and access the link currently pointed to.

- An iterator can be used to traverse through a list, performing some operation on selected links (or all links).

Questions

These questions are intended as a self-test for readers. Answers may be found in Appendix C.

1. Which of the following is *not* true? A reference to a class object

    a. can be used to access public methods in the object.

    b. has a size dependant on its class.

    c. has the data type of the class.

    d. does not hold the object itself.

2. Access to the links in a linked list is usually through the _____ link.

3. When you create a reference to a link in a linked list, it

    a. must refer to the first link.

    b. must refer to the link pointed to by `current`.

    c. must refer to the link pointed to by `next`.

    d. can refer to any link you want.

4. How many references must you change to insert a link in the middle of a singly linked list?

5. How many references must you change to insert a link at the end of a singly linked list?

6. In the `insertFirst()` method in the `linkList.java` program (Listing 5.1), the statement `newLink.next=first;` means that

    a. the next new link to be inserted will refer to `first`.

    b. `first` will refer to the new link.

    c. the `next` field of the new link will refer to the old first link.

    d. `newLink.next` will refer to the new first link in the list.

7. Assuming `current` points to the next-to-last link in a singly linked list, what statement will delete the last link from the list?

8. When all references to a link are changed to refer to something else, what happens to the link?

9. A double-ended list

    a. can be accessed from either end.

    b. is a different name for a doubly linked list.

    c. has pointers running both forward and backward between links.

    d. has its first link connected to its last link.

10. A special case often occurs for insertion and deletion routines when a list is _____.

11. Assuming a copy takes longer than a comparison, is it faster to delete an item with a certain key from a linked list or from an unsorted array?

12. How many times would you need to traverse a singly linked list to delete the item with the largest key?

13. Of the lists discussed in this chapter, which one would be best for implementing a queue?

14. Which of the following is *not* true? Iterators would be useful if you wanted to

   a. do an insertion sort on a linked list.

   b. insert a new link at the beginning of a list.

   c. swap two links at arbitrary locations.

   d. delete all links with a certain key value.

15. Which do you think would be a better choice to implement a stack: a singly linked list or an array?

Experiments

Carrying out these experiments will help to provide insights into the topics covered in the chapter. No programming is involved.

1. Use the LinkList Workshop applet to execute insert, find, and delete operations on both sorted and unsorted lists. For the operations demonstrated by this applet, is there any advantage to the sorted list?

2. Modify main() in the linkList.java program (Listing 5.1) so that it continuously inserts links into the list until memory is exhausted. After each 1,000 items, have it display the number of items inserted so far. This way, you can learn approximately how many links a list can hold in your particular machine. (Of course, the number will vary depending on what other programs are in memory and many other factors.) Don't try this experiment if it will crash your institution's network.

Programming Projects

Writing programs that solve the Programming Projects helps to solidify your understanding of the material and demonstrates how the chapter's concepts are applied.

(As noted in the Introduction, qualified instructors may obtain completed solutions to the Programming Projects on the publisher's Web site.)

5.1 Implement a priority queue based on a sorted linked list. The remove operation on the priority queue should remove the item with the smallest key.

5.2 Implement a deque based on a doubly linked list. (See Programming Project 4.2 in the preceding chapter.) The user should be able to carry out the normal operations on the deque.

5.3 A circular list is a linked list in which the last link points back to the first link. There are many ways to design a circular list. Sometimes there is a pointer to the "start" of the list. However, this makes the list less like a real circle and more like an ordinary list that has its end attached to its beginning. Make a class for a singly linked circular list that has no end and no beginning. The only access to the list is a single reference, current, that can point to any link on the list. This reference can move around the list as needed. (See Programming Project 5.5 for a situation in which such a circular list is ideally suited.) Your list should handle insertion, searching, and deletion. You may find it convenient if these operations take place one link downstream of the link pointed to by current. (Because the upstream link is singly linked, you can't get at it without going all the way around the circle.) You should also be able to display the list (although you'll need to break the circle at some arbitrary point to print it on the screen). A step() method that moves current along to the next link might come in handy too.

5.4 Implement a stack class based on the circular list of Programming Project 5.3. This exercise is not too difficult. (However, implementing a queue can be harder, unless you make the circular list doubly linked.)

5.5 The Josephus Problem is a famous mathematical puzzle that goes back to ancient times. There are many stories to go with the puzzle. One is that Josephus was one of a group of Jews who were about to be captured by the Romans. Rather than be enslaved, they chose to commit suicide. They arranged themselves in a circle and, starting at a certain person, started counting off around the circle. Every n^{th} person had to leave the circle and commit suicide. Josephus decided he didn't want to die, so he arranged the rules so he would be the last person left. If there were (say) 20 people, and he was the seventh person from the start of the circle, what number should he tell them to use for counting off? The problem is made much more complicated because the circle shrinks as the counting continues.

Create an application that uses a circular linked list (like that in Programming Project 5.3) to model this problem. Inputs are the number of people in the circle, the number used for counting off, and the number of the person where counting starts (usually 1). The output is the list of persons being eliminated. When a person drops out of the circle, counting starts again from the person who was on his left (assuming you go around clockwise). Here's an example. There are seven people numbered 1 through 7, and you start at 1 and count off by threes. People will be eliminated in the order 4, 1, 6, 5, 7, 3. Number 2 will be left.

5.6 Let's try something a little different: a two-dimensional linked list, which we'll call a matrix. This is the list analogue of a two-dimensional array. It might be useful in applications such as spreadsheet programs. If a spreadsheet is based on an array, and you insert a new row near the top, you must move every cell in the lower rows N*M cells, which is potentially a slow process. If the spreadsheet is implemented by a matrix, you need only change N pointers.

For simplicity, we'll assume a singly linked approach (although a double-linked approach would probably be more appropriate for a spreadsheet). Each link (except those on the top row and left side) is pointed to by the link directly above it and by the link on its left. You can start at the upper-left link and navigate to, say, the link on the third row and fifth column by following the pointers down two rows and right four columns. Assume your matrix is created with specified dimensions (7 by 10, for example). You should be able to insert values in specified links and display the contents of the matrix.

6

Recursion

Recursion is a programming technique in which a method (function) calls itself. This may sound like a strange thing to do, or even a catastrophic mistake. Recursion is, however, one of the most interesting, and one of the most surprisingly effective, techniques in programming. Like pulling yourself up by your bootstraps (you do have bootstraps, don't you?), recursion seems incredible when you first encounter it. However, it not only works, it also provides a unique conceptual framework for solving many problems.

In this chapter we'll examine numerous examples to show the wide variety of situations to which recursion can be applied. We will calculate triangular numbers and factorials, generate anagrams, perform a recursive binary search, solve the Towers of Hanoi puzzle, and investigate a sorting technique called mergesort. Workshop applets are provided to demonstrate the Towers of Hanoi and mergesort.

We'll also discuss the strengths and weaknesses of recursion, and show how a recursive approach can be transformed into a stack-based approach.

Triangular Numbers

It's said that the Pythagorians, a band of mathematicians in ancient Greece who worked under Pythagoras (of Pythagorian theorem fame), felt a mystical connection with the series of numbers 1, 3, 6, 10, 15, 21, ... (where the ... means the series continues indefinitely). Can you find the next member of this series?

The nth term in the series is obtained by adding n to the previous term. Thus, the second term is found by adding 2 to the first term (which is 1), giving 3. The third term is 3 added to the second term (which is 3) giving 6, and so on.

The numbers in this series are called *triangular numbers* because they can be visualized as a triangular arrangement of objects, shown as little squares in Figure 6.1.

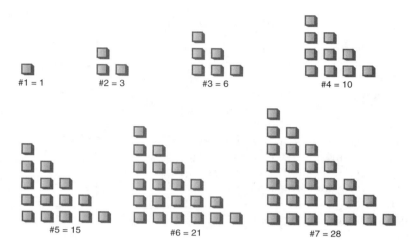

FIGURE 6.1 The triangular numbers.

Finding the nth Term Using a Loop

Suppose you wanted to find the value of some arbitrary nth term in the series—say the fourth term (whose value is 10). How would you calculate it? Looking at Figure 6.2, you might decide that the value of any term can be obtained by adding up all the vertical columns of squares.

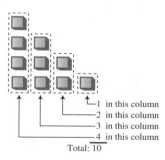

FIGURE 6.2 Triangular number as columns.

In the fourth term, the first column has four little squares, the second column has three, and so on. Adding 4+3+2+1 gives 10.

The following `triangle()` method uses this column-based technique to find a triangular number. It sums all the columns, from a height of n to a height of 1:

```
int triangle(int n)
   {
   int total = 0;

   while(n > 0)              // until n is 1
      {
      total = total + n;   // add n (column height) to total
      --n;                 // decrement column height
      }
   return total;
   }
```

The method cycles around the loop n times, adding n to total the first time, n-1 the second time, and so on down to 1, quitting the loop when n becomes 0.

Finding the nth Term Using Recursion

The loop approach may seem straightforward, but there's another way to look at this problem. The value of the nth term can be thought of as the sum of only two things, instead of a whole series. They are

1. The first (tallest) column, which has the value n.

2. The sum of all the remaining columns.

This is shown in Figure 6.3.

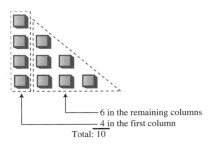

6 in the remaining columns
4 in the first column
Total: 10

FIGURE 6.3 Triangular number as column plus triangle.

Finding the Remaining Columns

If we knew about a method that found the sum of all the remaining columns, we could write our `triangle()` method, which returns the value of the nth triangular number, like this:

```
int triangle(int n)
    {
    return( n + sumRemainingColumns(n) );   // (incomplete version)
    }
```

But what have we gained here? It looks like writing the `sumRemainingColumns()` method is just as hard as writing the `triangle()` method in the first place.

Notice in Figure 6.3, however, that the sum of all the remaining columns for term n is the same as the sum of *all* the columns for term n-1. Thus, if we knew about a method that summed all the columns for term n, we could call it with an argument of n-1 to find the sum of all the remaining columns for term n:

```
int triangle(int n)
    {
    return( n + sumAllColumns(n-1) );   // (incomplete version)
    }
```

But when you think about it, the `sumAllColumns()` method is doing exactly the same thing the `triangle()` method is: summing all the columns for some number n passed as an argument. So why not use the `triangle()` method itself, instead of some other method? That would look like this:

```
int triangle(int n)
    {
    return( n + triangle(n-1) );   // (incomplete version)
    }
```

You may be amazed that a method can call itself, but why shouldn't it be able to? A method call is (among other things) a transfer of control to the start of the method. This transfer of control can take place from within the method as well as from outside.

Passing the Buck

All these approaches may seem like passing the buck. Someone tells me to find the 9th triangular number. I know this is 9 plus the 8th triangular number, so I call Harry and ask him to find the 8th triangular number. When I hear back from him, I'll add 9 to whatever he tells me, and that will be the answer.

Harry knows the 8th triangular number is 8 plus the 7th triangular number, so he calls Sally and asks her to find the 7th triangular number. This process continues with each person passing the buck to another one.

Where does this buck-passing end? Someone at some point must be able to figure out an answer that doesn't involve asking another person to help. If this didn't happen, there would be an infinite chain of people asking other people questions—a sort of arithmetic Ponzi scheme that would never end. In the case of triangle(), this would mean the method calling itself over and over in an infinite series that would eventually crash the program.

The Buck Stops Here

To prevent an infinite regress, the person who is asked to find the first triangular number of the series, when n is 1, must know, without asking anyone else, that the answer is 1. There are no smaller numbers to ask anyone about, there's nothing left to add to anything else, so the buck stops there. We can express this by adding a condition to the triangle() method:

```
int triangle(int n)
   {
   if(n==1)
      return 1;
   else
      return( n + triangle(n-1) );
   }
```

The condition that leads to a recursive method returning without making another recursive call is referred to as the *base case*. It's critical that every recursive method have a base case to prevent infinite recursion and the consequent demise of the program.

The triangle.java Program

Does recursion actually work? If you run the triangle.java program, you'll see that it does. Enter a value for the term number, n, and the program will display the value of the corresponding triangular number. Listing 6.1 shows the triangle.java program.

LISTING 6.1   The triangle.java Program

```
// triangle.java
// evaluates triangular numbers
// to run this program: C>java TriangleApp
import java.io.*;                    // for I/O
//////////////////////////////////////////////////////////////
```

LISTING 6.1   Continued

```
class TriangleApp
    {
    static int theNumber;

    public static void main(String[] args) throws IOException
        {
        System.out.print("Enter a number: ");
        theNumber = getInt();
        int theAnswer = triangle(theNumber);
        System.out.println("Triangle="+theAnswer);
        }  // end main()
//-------------------------------------------------------------
    public static int triangle(int n)
        {
        if(n==1)
            return 1;
        else
            return( n + triangle(n-1) );
        }
//-------------------------------------------------------------
    public static String getString() throws IOException
        {
        InputStreamReader isr = new InputStreamReader(System.in);
        BufferedReader br = new BufferedReader(isr);
        String s = br.readLine();
        return s;
        }
//-------------------------------------------------------------
    public static int getInt() throws IOException
        {
        String s = getString();
        return Integer.parseInt(s);
        }
//-------------------------------------------------------------
    }  // end class TriangleApp
/////////////////////////////////////////////////////////////////
```

The main() routine prompts the user for a value for n, calls triangle(), and displays the return value. The triangle() method calls itself repeatedly to do all the work.

Here's some sample output:

```
Enter a number: 1000
Triangle = 500500
```

Incidentally, if you're skeptical of the results returned from `triangle()`, you can check them by using the following formula:

nth triangular number = $(n^2+n)/2$

What's Really Happening?

Let's modify the `triangle()` method to provide an insight into what's happening when it executes. We'll insert some output statements to keep track of the arguments and return values:

```
public static int triangle(int n)
   {
   System.out.println("Entering: n=" + n);
   if(n==1)
      {
      System.out.println("Returning 1");
      return 1;
      }
   else
      {
      int temp = n + triangle(n-1);
      System.out.println("Returning " + temp);
      return temp;
      }
   }
```

Here's the interaction when this method is substituted for the earlier `triangle()` method and the user enters 5:

```
Enter a number: 5

Entering: n=5
Entering: n=4
Entering: n=3
Entering: n=2
Entering: n=1
Returning 1
Returning 3
Returning 6
```

```
Returning 10
Returning 15

Triangle = 15
```

Each time the `triangle()` method calls itself, its argument, which starts at 5, is reduced by 1. The method plunges down into itself again and again until its argument is reduced to 1. Then it returns. This triggers an entire series of returns. The method rises back up, phoenix-like, out of the discarded versions of itself. Each time it returns, it adds the value of n it was called with to the return value from the method it called.

The return values recapitulate the series of triangular numbers, until the answer is returned to `main()`. Figure 6.4 shows how each invocation of the `triangle()` method can be imagined as being "inside" the previous one.

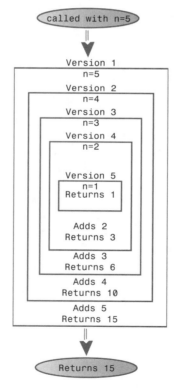

FIGURE 6.4   The recursive `triangle()` method.

Notice that, just before the innermost version returns a 1, there are actually five different incarnations of triangle() in existence at the same time. The outer one was passed the argument 5; the inner one was passed the argument 1.

Characteristics of Recursive Methods

Although it's short, the triangle() method possesses the key features common to all recursive routines:

- It calls itself.

- When it calls itself, it does so to solve a smaller problem.

- There's some version of the problem that is simple enough that the routine can solve it, and return, without calling itself.

In each successive call of a recursive method to itself, the argument becomes smaller (or perhaps a range described by multiple arguments becomes smaller), reflecting the fact that the problem has become "smaller" or easier. When the argument or range reaches a certain minimum size, a condition is triggered and the method returns without calling itself.

Is Recursion Efficient?

Calling a method involves certain overhead. Control must be transferred from the location of the call to the beginning of the method. In addition, the arguments to the method and the address to which the method should return must be pushed onto an internal stack so that the method can access the argument values and know where to return.

In the case of the triangle() method, it's probable that, as a result of this overhead, the while loop approach executes more quickly than the recursive approach. The penalty may not be significant, but if there are a large number of method calls as a result of a recursive method, it might be desirable to eliminate the recursion. We'll talk about this issue more at the end of this chapter.

Another inefficiency is that memory is used to store all the intermediate arguments and return values on the system's internal stack. This may cause problems if there is a large amount of data, leading to stack overflow.

Recursion is usually used because it simplifies a problem conceptually, not because it's inherently more efficient.

Mathematical Induction

Recursion is the programming equivalent of mathematical induction. Mathematical induction is a way of defining something in terms of itself. (The term is also used to describe a related approach to proving theorems.) Using induction, we could define the triangular numbers mathematically by saying

tri(n) = 1   if n = 1

tri(n) = n + tri(n–1)   if n > 1

Defining something in terms of itself may seem circular, but in fact it's perfectly valid (provided there's a base case).

Factorials

Factorials are similar in concept to triangular numbers, except that multiplication is used instead of addition. The triangular number corresponding to n is found by adding n to the triangular number of n-1, while the factorial of n is found by multiplying n by the factorial of n-1. That is, the fifth triangular number is 5+4+3+2+1, while the factorial of 5 is 5*4*3*2*1, which equals 120. Table 6.1 shows the factorials of the first 10 numbers.

TABLE 6.1   Factorials

Number	Calculation	Factorial
0	by definition	1
1	1 * 1	1
2	2 * 1	2
3	3 * 2	6
4	4 * 6	24
5	5 * 24	120
6	6 * 120	720
7	7 * 720	5,040
8	8 * 5,040	40,320
9	9 * 40,320	362,880

The factorial of 0 is defined to be 1. Factorial numbers grow large very rapidly, as you can see.

A recursive method similar to `triangle()` can be used to calculate factorials. It looks like this:

```
int factorial(int n)
   {
   if(n==0)
      return 1;
   else
      return (n * factorial(n-1) );
   }
```

There are only two differences between `factorial()` and `triangle()`. First, `factorial()` uses a * instead of a + in the expression

```
n * factorial(n-1)
```

Second, the base condition occurs when n is 0, not 1. Here's some sample interaction when this method is used in a program similar to `triangle.java`:

```
Enter a number: 6
Factorial =720
```

Figure 6.5 shows how the various incarnations of `factorial()` call themselves when initially entered with n=4.

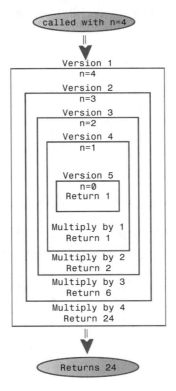

FIGURE 6.5   The recursive `factorial()` method.

Calculating factorials is the classic demonstration of recursion, although factorials aren't as easy to visualize as triangular numbers.

Various other numerological entities lend themselves to calculation using recursion in a similar way, such as finding the greatest common denominator of two numbers (which is used to reduce a fraction to lowest terms), raising a number to a power, and so on. Again, while these calculations are interesting for demonstrating recursion, they probably wouldn't be used in practice because a loop-based approach is more efficient.

Anagrams

Here's a different kind of situation in which recursion provides a neat solution to a problem. A permutation is an arrangement of things in a definite order. Suppose you want to list all the anagrams of a specified word—that is, all possible permutations (whether they make a real English word or not) that can be made from the letters of the original word. We'll call this *anagramming* a word. Anagramming cat, for example, would produce

- cat
- cta
- atc
- act
- tca
- tac

Try anagramming some words yourself. You'll find that the number of possibilities is the factorial of the number of letters. For 3 letters there are 6 possible words; for 4 letters there are 24 words; for 5 letters, 120; and so on. (This assumes that all letters are distinct; if there are multiple instances of the same letter, there will be fewer possible words.)

How would you write a program to anagram a word? Here's one approach. Assume the word has n letters.

1. Anagram the rightmost n-1 letters.

2. Rotate all n letters.

3. Repeat these steps n times.

To *rotate* the word means to shift all the letters one position left, except for the leftmost letter, which "rotates" back to the right, as shown in Figure 6.6.

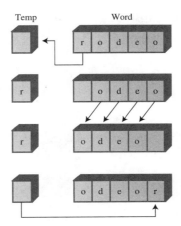

FIGURE 6.6   Rotating a word.

Rotating the word n times gives each letter a chance to begin the word. While the selected letter occupies this first position, all the other letters are then anagrammed (arranged in every possible position). For cat, which has only three letters, rotating the remaining two letters simply switches them. The sequence is shown in Table 6.2.

TABLE 6.2   Anagramming the Word cat

Word	Display Word?	First Letter	Remaining Letters	Action
cat	Yes	c	at	Rotate at
cta	Yes	c	ta	Rotate ta
cat	No	c	at	Rotate cat
atc	Yes	a	tc	Rotate tc
act	Yes	a	ct	Rotate ct
atc	No	a	tc	Rotate atc
tca	Yes	t	ca	Rotate ca
tac	Yes	t	ac	Rotate ac
tca	No	t	ca	Rotate tca
cat	No	c	at	Done

Notice that we must rotate back to the starting point with two letters before performing a three-letter rotation. This leads to sequences like cat, cta, cat. The redundant sequences aren't displayed.

How do we anagram the rightmost n-1 letters? By calling ourselves. The recursive doAnagram() method takes the size of the word to be anagrammed as its only parameter. This word is understood to be the rightmost n letters of the complete word. Each time `doAnagram()` calls itself, it does so with a word one letter smaller than before, as shown in Figure 6.7.

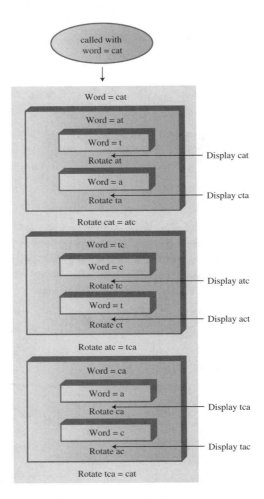

FIGURE 6.7   The recursive `doAnagram()` method.

The base case occurs when the size of the word to be anagrammed is only one letter. There's no way to rearrange one letter, so the method returns immediately. Otherwise, it anagrams all but the first letter of the word it was given and then rotates the entire word. These two actions are performed n times, where n is the size of the word. Here's the recursive routine `doAnagram()`:

```java
public static void doAnagram(int newSize)
   {
   if(newSize == 1)                 // if too small,
      return;                       // go no further
   for(int j=0; j<newSize; j++)     // for each position,
      {
      doAnagram(newSize-1);         // anagram remaining
      if(newSize==2)                // if innermost,
         displayWord();             // display it
      rotate(newSize);              // rotate word
      }
   }
```

Each time the doAnagram() method calls itself, the size of the word is one letter
smaller, and the starting position is one cell further to the right, as shown in
Figure 6.8.

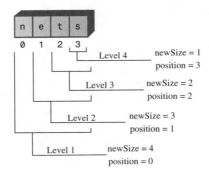

FIGURE 6.8   Smaller and smaller words.

Listing 6.2 shows the complete anagram.java program. The main() routine gets a word
from the user, inserts it into a character array so it can be dealt with conveniently,
and then calls doAnagram().

LISTING 6.2   The anagram.java Program

```java
// anagram.java
// creates anagrams
// to run this program: C>java AnagramApp
import java.io.*;
////////////////////////////////////////////////////////////////
class AnagramApp
   {
```

LISTING 6.2   Continued

```
static int size;
static int count;
static char[] arrChar = new char[100];

public static void main(String[] args) throws IOException
   {
   System.out.print("Enter a word: ");      // get word
   String input = getString();
   size = input.length();                   // find its size
   count = 0;
   for(int j=0; j<size; j++)                 // put it in array
      arrChar[j] = input.charAt(j);
   doAnagram(size);                          // anagram it
   }  // end main()
//-----------------------------------------------------------
public static void doAnagram(int newSize)
   {
   if(newSize == 1)                          // if too small,
      return;                                // go no further
   for(int j=0; j<newSize; j++)              // for each position,
      {
      doAnagram(newSize-1);                  // anagram remaining
      if(newSize==2)                         // if innermost,
         displayWord();                      // display it
      rotate(newSize);                       // rotate word
      }
   }
//-----------------------------------------------------------
// rotate left all chars from position to end
public static void rotate(int newSize)
   {
   int j;
   int position = size - newSize;
   char temp = arrChar[position];            // save first letter
   for(j=position+1; j<size; j++)            // shift others left
      arrChar[j-1] = arrChar[j];
   arrChar[j-1] = temp;                      // put first on right
   }
//-----------------------------------------------------------
public static void displayWord()
   {
```

LISTING 6.2   Continued

```
      if(count < 99)
         System.out.print(" ");
      if(count < 9)
         System.out.print(" ");
      System.out.print(++count + " ");
      for(int j=0; j<size; j++)
         System.out.print( arrChar[j] );
      System.out.print("   ");
      System.out.flush();
      if(count%6 == 0)
         System.out.println("");
      }
   //------------------------------------------------------------
   public static String getString() throws IOException
      {
      InputStreamReader isr = new InputStreamReader(System.in);
      BufferedReader br = new BufferedReader(isr);
      String s = br.readLine();
      return s;
      }
   //------------------------------------------------------------
   } // end class AnagramApp
///////////////////////////////////////////////////////////////
```

The rotate() method rotates the word one position left as described earlier. The displayWord() method displays the entire word and adds a count to make it easy to see how many words have been displayed. Here's some sample interaction with the program:

```
Enter a word: cats
   1 cats      2 cast      3 ctsa      4 ctas      5 csat      6 csta
   7 atsc      8 atcs      9 asct     10 astc     11 acts     12 acst
  13 tsca     14 tsac     15 tcas     16 tcsa     17 tasc     18 tacs
  19 scat     20 scta     21 satc     22 sact     23 stca     24 stac
```

(Is it only coincidence that *scat* is an anagram of *cats*?) You can use the program to anagram five-letter or even six-letter words. However, because the factorial of 6 is 720, anagramming such long sequences may generate more words than you want to know about.

A Recursive Binary Search

Remember the binary search we discussed in Chapter 2, "Arrays"? We wanted to find a given cell in an ordered array using the fewest number of comparisons. The solution was to divide the array in half, see which half the desired cell lay in, divide that half in half again, and so on. Here's what the original find() method looked like:

```
//-----------------------------------------------------------
public int find(long searchKey)
   {
   int lowerBound = 0;
   int upperBound = nElems-1;
   int curIn;

   while(true)
      {
      curIn = (lowerBound + upperBound ) / 2;
      if(a[curIn]==searchKey)
         return curIn;              // found it
      else if(lowerBound > upperBound)
         return nElems;             // can't find it
      else                          // divide range
         {
         if(a[curIn] < searchKey)
            lowerBound = curIn + 1; // it's in upper half
         else
            upperBound = curIn - 1; // it's in lower half
         } // end else divide range
      } // end while
   } // end find()
//-----------------------------------------------------------
```

You might want to reread the section on binary searches in ordered arrays in Chapter 2, which describes how this method works. Also, run the Ordered Workshop applet from that chapter if you want to see a binary search in action.

We can transform this loop-based method into a recursive method quite easily. In the loop-based method, we change lowerBound or upperBound to specify a new range and then cycle through the loop again. Each time through the loop we divide the range (roughly) in half.

Recursion Replaces the Loop

In the recursive approach, instead of changing lowerBound or upperBound, we call find() again with the new values of lowerBound or upperBound as arguments. The loop disappears, and its place is taken by the recursive calls. Here's how that looks:

```
private int recFind(long searchKey, int lowerBound,
                                    int upperBound)
   {
   int curIn;

   curIn = (lowerBound + upperBound ) / 2;
   if(a[curIn]==searchKey)
      return curIn;              // found it
   else if(lowerBound > upperBound)
      return nElems;             // can't find it
   else                         // divide range
      {
      if(a[curIn] < searchKey)   // it's in upper half
         return recFind(searchKey, curIn+1, upperBound);
      else                       // it's in lower half
         return recFind(searchKey, lowerBound, curIn-1);
      } // end else divide range
   } // end recFind()
```

The class user, represented by main(), may not know how many items are in the array when it calls find(), and in any case shouldn't be burdened with having to know what values of upperBound and lowerBound to set initially. Therefore, we supply an intermediate public method, find(), which main() calls with only one argument, the value of the search key. The find() method supplies the proper initial values of lowerBound and upperBound (0 and nElems-1) and then calls the private, recursive method recFind(). The find() method looks like this:

```
public int find(long searchKey)
   {
   return recFind(searchKey, 0, nElems-1);
   }
```

Listing 6.3 shows the complete listing for the binarySearch.java program.

LISTING 6.3   The binarySearch.java Program

```
// binarySearch.java
// demonstrates recursive binary search
// to run this program: C>java BinarySearchApp
////////////////////////////////////////////////////////////
class ordArray
   {
   private long[] a;                    // ref to array a
```

LISTING 6.3   Continued

```
private int nElems;              // number of data items
//--------------------------------------------------------------
public ordArray(int max)         // constructor
   {
   a = new long[max];            // create array
   nElems = 0;
   }
//--------------------------------------------------------------
public int size()
   { return nElems; }
//--------------------------------------------------------------
public int find(long searchKey)
   {
   return recFind(searchKey, 0, nElems-1);
   }
//--------------------------------------------------------------
private int recFind(long searchKey, int lowerBound,
                                    int upperBound)
   {
   int curIn;

   curIn = (lowerBound + upperBound ) / 2;
   if(a[curIn]==searchKey)
      return curIn;              // found it
   else if(lowerBound > upperBound)
      return nElems;             // can't find it
   else                          // divide range
      {
      if(a[curIn] < searchKey)   // it's in upper half
         return recFind(searchKey, curIn+1, upperBound);
      else                       // it's in lower half
         return recFind(searchKey, lowerBound, curIn-1);
      }  // end else divide range
   }  // end recFind()
//--------------------------------------------------------------
public void insert(long value)   // put element into array
   {
   int j;
   for(j=0; j<nElems; j++)        // find where it goes
      if(a[j] > value)            // (linear search)
         break;
```

LISTING 6.3 Continued

```
    for(int k=nElems; k>j; k--)    // move bigger ones up
        a[k] = a[k-1];
    a[j] = value;                  // insert it
    nElems++;                      // increment size
    }  // end insert()
//-----------------------------------------------------------
public void display()              // displays array contents
    {
    for(int j=0; j<nElems; j++)        // for each element,
        System.out.print(a[j] + " ");  // display it
    System.out.println("");
    }
//-----------------------------------------------------------
}  // end class ordArray
////////////////////////////////////////////////////////////////
class BinarySearchApp
    {
    public static void main(String[] args)
        {
        int maxSize = 100;          // array size
        ordArray arr;               // reference to array
        arr = new ordArray(maxSize); // create the array

        arr.insert(72);             // insert items
        arr.insert(90);
        arr.insert(45);
        arr.insert(126);
        arr.insert(54);
        arr.insert(99);
        arr.insert(144);
        arr.insert(27);
        arr.insert(135);
        arr.insert(81);
        arr.insert(18);
        arr.insert(108);
        arr.insert(9);
        arr.insert(117);
        arr.insert(63);
        arr.insert(36);
```

LISTING 6.3   Continued

```
    arr.display();                 // display array

    int searchKey = 27;            // search for item
    if( arr.find(searchKey) != arr.size() )
       System.out.println("Found " + searchKey);
    else
       System.out.println("Can't find " + searchKey);
    }  // end main()
  }  // end class BinarySearchApp
////////////////////////////////////////////////////////////////
```

In main() we insert 16 items into the array. The insert() method arranges them in sorted order; they're then displayed. Finally, we use find() to try to find the item with a key value of 27. Here's some sample output:

```
9 18 27 36 45 54 63 72 81 90 99 108 117 126 135 144
Found 27
```

In binarySearch.java there are 16 items in an array. Figure 6.9 shows how the recFind() method in this program calls itself over and over, each time with a smaller range than before. When the innermost version of the method finds the desired item, which has the key value 27, it returns with the index value of the item, which is 2 (as can be seen in the display of ordered data). This value is then returned from each version of recFind() in turn; finally, find() returns it to the class user.

The recursive binary search has the same big O efficiency as the non-recursive version: O(logN). It is somewhat more elegant, but may be slightly slower.

Divide-and-Conquer Algorithms

The recursive binary search is an example of the *divide-and-conquer* approach. You divide the big problem into two smaller problems and solve each one separately. The solution to each smaller problem is the same: You divide it into two even smaller problems and solve them. The process continues until you get to the base case, which can be solved easily, with no further division into halves.

The divide-and-conquer approach is commonly used with recursion, although, as we saw in the binary search in Chapter 2, you can also use a non-recursive approach.

A divide-and-conquer approach usually involves a method that contains two recursive calls to itself, one for each half of the problem. In the binary search, there are two such calls, but only one of them is actually executed. (Which one depends on the value of the key.) The mergesort, which we'll encounter later in this chapter, actually executes both recursive calls (to sort two halves of an array).

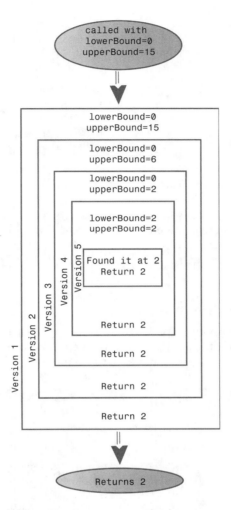

FIGURE 6.9   The recursive binarySearch() method.

The Towers of Hanoi

The Towers of Hanoi is an ancient puzzle consisting of a number of disks placed on three columns, as shown in Figure 6.10.

The disks all have different diameters and holes in the middle so they will fit over the columns. All the disks start out on column A. The object of the puzzle is to transfer all the disks from column A to column C. Only one disk can be moved at a time, and no disk can be placed on a disk that's smaller than itself.

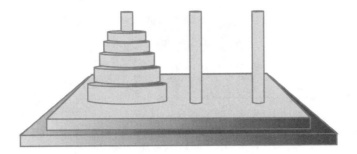

FIGURE 6.10   The Towers of Hanoi.

There's an ancient myth that somewhere in India, in a remote temple, monks labor day and night to transfer 64 golden disks from one of three diamond-studded towers to another. When they are finished, the world will end. Any alarm you may feel, however, will be dispelled when you see how long it takes to solve the puzzle for far fewer than 64 disks.

The Towers Workshop Applet

Start up the Towers Workshop applet. You can attempt to solve the puzzle yourself by using the mouse to drag the topmost disk to another tower. Figure 6.11 shows how the towers look after several moves have been made.

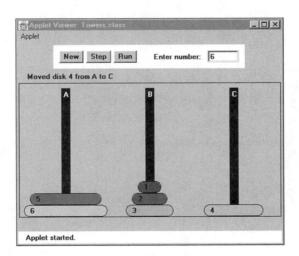

FIGURE 6.11   The Towers Workshop applet.

There are three ways to use the Workshop applet:

- You can attempt to solve the puzzle manually, by dragging the disks from tower to tower.

- You can repeatedly press the Step button to watch the algorithm solve the puzzle. At each step in the solution, a message is displayed, telling you what the algorithm is doing.

- You can press the Run button and watch the algorithm solve the puzzle with no intervention on your part; the disks zip back and forth between the posts.

To restart the puzzle, type in the number of disks you want to use, from 1 to 10, and press New twice. (After the first time, you're asked to verify that restarting is what you want to do.) The specified number of disks will be arranged on tower A. Once you drag a disk with the mouse, you can't use Step or Run; you must start over with New. However, you can switch to manual in the middle of stepping or running, and you can switch to Step when you're running, and Run when you're stepping.

Try solving the puzzle manually with a small number of disks, say three or four. Work up to higher numbers. The applet gives you the opportunity to learn intuitively how the problem is solved.

Moving Subtrees

Let's call the initial tree-shaped (or pyramid-shaped) arrangement of disks on tower A a *tree*. (This kind of tree has nothing to do with the trees that are data storage structures, described elsewhere in this book.) As you experiment with the applet, you'll begin to notice that smaller tree-shaped stacks of disks are generated as part of the solution process. Let's call these smaller trees, containing fewer than the total number of disks, *subtrees*. For example, if you're trying to transfer four disks, you'll find that one of the intermediate steps involves a subtree of three disks on tower B, as shown in Figure 6.12.

FIGURE 6.12   A subtree on tower B.

These subtrees form many times in the solution of the puzzle. This happens because the creation of a subtree is the only way to transfer a larger disk from one tower to another: All the smaller disks must be placed on an intermediate tower, where they naturally form a subtree.

Here's a rule of thumb that may help when you try to solve the puzzle manually. If the subtree you're trying to move has an odd number of disks, start by moving the topmost disk directly to the tower where you want the subtree to go. If you're trying to move a subtree with an even number of disks, start by moving the topmost disk to the intermediate tower.

The Recursive Algorithm

The solution to the Towers of Hanoi puzzle can be expressed recursively using the notion of subtrees. Suppose you want to move all the disks from a source tower (call it S) to a destination tower (call it D). You have an intermediate tower available (call it I). Assume there are n disks on tower S. Here's the algorithm:

1. Move the subtree consisting of the top n-1 disks from S to I.

2. Move the remaining (largest) disk from S to D.

3. Move the subtree from I to D.

When you begin, the source tower is A, the intermediate tower is B, and the destination tower is C. Figure 6.13 shows the three steps for this situation.

First, the subtree consisting of disks 1, 2, and 3 is moved to the intermediate tower B. Then the largest disk, 4, is moved to tower C. Then the subtree is moved from B to C.

Of course, this solution doesn't solve the problem of how to move the subtree consisting of disks 1, 2, and 3 to tower B, because you can't move a subtree all at once; you must move it one disk at a time. Moving the three-disk subtree is not so easy. However, it's easier than moving four disks.

As it turns out, moving three disks from A to the destination tower B can be done with the same three steps as moving four disks. That is, move the subtree consisting of the top two disks from tower A to intermediate tower C; then move disk 3 from A to B. Then move the subtree back from C to B.

How do you move a subtree of two disks from A to C? Move the subtree consisting of only one disk (1) from A to B. This is the base case: When you're moving only one disk, you just move it; there's nothing else to do. Then move the larger disk (2) from A to C, and replace the subtree (disk 1) on it.

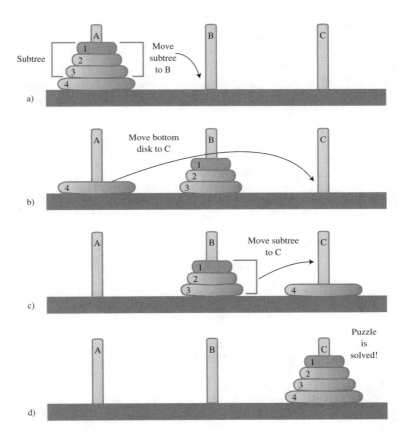

FIGURE 6.13   Recursive solution to towers puzzle.

The towers.java Program

The towers.java program solves the Towers of Hanoi puzzle using this recursive approach. It communicates the moves by displaying them; this approach requires much less code than displaying the towers. It's up to the human reading the list to actually carry out the moves.

The code is simplicity itself. The main() routine makes a single call to the recursive method doTowers(). This method then calls itself recursively until the puzzle is solved. In this version, shown in Listing 6.4, there are initially only three disks, but you can recompile the program with any number.

LISTING 6.4   The towers.java Program

```
// towers.java
// solves the towers of Hanoi puzzle
// to run this program: C>java TowersApp
////////////////////////////////////////////////////////////
class TowersApp
   {
   static int nDisks = 3;

   public static void main(String[] args)
      {
      doTowers(nDisks, 'A', 'B', 'C');
      }
   //-----------------------------------------------------------
   public static void doTowers(int topN,
                               char from, char inter, char to)
      {
      if(topN==1)
         System.out.println("Disk 1 from " + from + " to "+ to);
      else
         {
         doTowers(topN-1, from, to, inter);   // from-->inter

         System.out.println("Disk " + topN +
                            " from " + from + " to "+ to);
         doTowers(topN-1, inter, from, to);   // inter-->to
         }
      }
   //-----------------------------------------------------------
   }  // end class TowersApp
////////////////////////////////////////////////////////////
```

Remember that three disks are moved from A to C. Here's the output from the
program:

```
Disk 1 from A to C
Disk 2 from A to B
Disk 1 from C to B
Disk 3 from A to C
Disk 1 from B to A
Disk 2 from B to C
Disk 1 from A to C
```

The arguments to doTowers() are the number of disks to be moved, and the source (from), intermediate (inter), and destination (to) towers to be used. The number of disks decreases by 1 each time the method calls itself. The source, intermediate, and destination towers also change.

Here is the output with additional notations that show when the method is entered and when it returns, its arguments, and whether a disk is moved because it's the base case (a subtree consisting of only one disk) or because it's the remaining bottom disk after a subtree has been moved:

```
Enter (3 disks): s=A, i=B, d=C
   Enter (2 disks): s=A, i=C, d=B
      Enter (1 disk): s=A, i=B, d=C
         Base case: move disk 1 from A to C
      Return (1 disk)
      Move bottom disk 2 from A to B
      Enter (1 disk): s=C, i=A, d=B
         Base case: move disk 1 from C to B
      Return (1 disk)
   Return (2 disks)
   Move bottom disk 3 from A to C
   Enter (2 disks): s=B, i=A, d=C
      Enter (1 disk): s=B, i=C, d=A
         Base case: move disk 1 from B to A
      Return (1 disk)
      Move bottom disk 2 from B to C
      Enter (1 disk): s=A, i=B, d=C
         Base case: move disk 1 from A to C
      Return (1 disk)
   Return (2 disks)
Return (3 disks)
```

If you study this output along with the source code for doTower(), it should become clear exactly how the method works. It's amazing that such a small amount of code can solve such a seemingly complicated problem.

mergesort

Our final example of recursion is the mergesort. This is a much more efficient sorting technique than those we saw in Chapter 3, "Simple Sorting," at least in terms of speed. While the bubble, insertion, and selection sorts take $O(N^2)$ time, the mergesort is $O(N*logN)$. The graph in Figure 2.9 (in Chapter 2) shows how much faster this is. For example, if N (the number of items to be sorted) is 10,000, then N^2 is

100,000,000, while N*logN is only 40,000. If sorting this many items required 40 seconds with the mergesort, it would take almost 28 hours for the insertion sort.

The mergesort is also fairly easy to implement. It's conceptually easier than quicksort and the Shell short, which we'll encounter in the next chapter.

The downside of the mergesort is that it requires an additional array in memory, equal in size to the one being sorted. If your original array barely fits in memory, the mergesort won't work. However, if you have enough space, it's a good choice.

Merging Two Sorted Arrays

The heart of the mergesort algorithm is the merging of two already-sorted arrays. Merging two sorted arrays A and B creates a third array, C, that contains all the elements of A and B, also arranged in sorted order. We'll examine the merging process first; later we'll see how it's used in sorting.

Imagine two sorted arrays. They don't need to be the same size. Let's say array A has 4 elements and array B has 6. They will be merged into an array C that starts with 10 empty cells. Figure 6.14 shows these arrays.

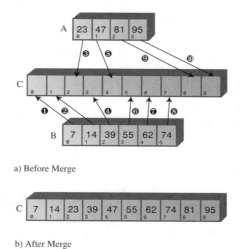

a) Before Merge

b) After Merge

FIGURE 6.14   Merging two arrays.

In the figure, the circled numbers indicate the order in which elements are transferred from A and B to C. Table 6.3 shows the comparisons necessary to determine which element will be copied. The steps in the table correspond to the steps in the figure. Following each comparison, the smaller element is copied to A.

TABLE 6.3 Merging Operations

Step	Comparison (If Any)	Copy
1	Compare 23 and 7	Copy 7 from B to C
2	Compare 23 and 14	Copy 14 from B to C
3	Compare 23 and 39	Copy 23 from A to C
4	Compare 39 and 47	Copy 39 from B to C
5	Compare 55 and 47	Copy 47 from A to C
6	Compare 55 and 81	Copy 55 from B to C
7	Compare 62 and 81	Copy 62 from B to C
8	Compare 74 and 81	Copy 74 from B to C
9		Copy 81 from A to C
10		Copy 95 from A to C

Notice that, because B is empty following step 8, no more comparisons are necessary; all the remaining elements are simply copied from A into C.

Listing 6.5 shows a Java program that carries out the merge shown in Figure 6.14 and Table 6.3. This is not a recursive program; it is a prelude to understanding mergesort.

LISTING 6.5 The `merge.java` Program

```
// merge.java
// demonstrates merging two arrays into a third
// to run this program: C>java MergeApp
////////////////////////////////////////////////////////////////
class MergeApp
   {
   public static void main(String[] args)
      {
      int[] arrayA = {23, 47, 81, 95};
      int[] arrayB = {7, 14, 39, 55, 62, 74};
      int[] arrayC = new int[10];

      merge(arrayA, 4, arrayB, 6, arrayC);
      display(arrayC, 10);
      }  // end main()
//--------------------------------------------------------------
                                       // merge A and B into C
   public static void merge( int[] arrayA, int sizeA,
                             int[] arrayB, int sizeB,
                             int[] arrayC )
      {
```

LISTING 6.5   Continued

```
      int aDex=0, bDex=0, cDex=0;

      while(aDex < sizeA && bDex < sizeB)  // neither array empty
         if( arrayA[aDex] < arrayB[bDex] )
            arrayC[cDex++] = arrayA[aDex++];
         else
            arrayC[cDex++] = arrayB[bDex++];

      while(aDex < sizeA)                  // arrayB is empty,
         arrayC[cDex++] = arrayA[aDex++];  // but arrayA isn't

      while(bDex < sizeB)                  // arrayA is empty,
         arrayC[cDex++] = arrayB[bDex++];  // but arrayB isn't
      }  // end merge()
//------------------------------------------------------------
                                   // display array
   public static void display(int[] theArray, int size)
      {
      for(int j=0; j<size; j++)
         System.out.print(theArray[j] + " ");
      System.out.println("");
      }
//------------------------------------------------------------
   }  // end class MergeApp
////////////////////////////////////////////////////////////////
```

In `main()` the arrays `arrayA`, `arrayB`, and `arrayC` are created; then the `merge()` method is called to merge `arrayA` and `arrayB` into `arrayC`, and the resulting contents of `arrayC` are displayed. Here's the output:

7 14 23 39 47 55 62 74 81 95

The `merge()` method has three `while` loops. The first steps along both `arrayA` and `arrayB`, comparing elements and copying the smaller of the two into `arrayC`.

The second `while` loop deals with the situation when all the elements have been transferred out of `arrayB`, but `arrayA` still has remaining elements. (This is what happens in the example, where 81 and 95 remain in `arrayA`.) The loop simply copies the remaining elements from `arrayA` into `arrayC`.

The third loop handles the similar situation when all the elements have been transferred out of `arrayA`, but `arrayB` still has remaining elements; they are copied to `arrayC`.

Sorting by Merging

The idea in the mergesort is to divide an array in half, sort each half, and then use the merge() method to merge the two halves into a single sorted array. How do you sort each half? This chapter is about recursion, so you probably already know the answer: You divide the half into two quarters, sort each of the quarters, and merge them to make a sorted half.

Similarly, each pair of 8ths is merged to make a sorted quarter, each pair of 16ths is merged to make a sorted 8th, and so on. You divide the array again and again until you reach a subarray with only one element. This is the base case; it's assumed an array with one element is already sorted.

We've seen that generally something is reduced in size each time a recursive method calls itself, and built back up again each time the method returns. In mergeSort() the range is divided in half each time this method calls itself, and each time it returns it merges two smaller ranges into a larger one.

As mergeSort() returns from finding two arrays of one element each, it merges them into a sorted array of two elements. Each pair of resulting 2-element arrays is then merged into a 4-element array. This process continues with larger and larger arrays until the entire array is sorted. This is easiest to see when the original array size is a power of 2, as shown in Figure 6.15.

First, in the bottom half of the array, range 0-0 and range 1-1 are merged into range 0-1. Of course, 0-0 and 1-1 aren't really ranges; they're only one element, so they are base cases. Similarly, 2-2 and 3-3 are merged into 2-3. Then ranges 0-1 and 2-3 are merged into 0-3.

In the top half of the array, 4-4 and 5-5 are merged into 4-5, 6-6 and 7-7 are merged into 6-7, and 4-5 and 6-7 are merged into 4-7. Finally, the top half, 0-3, and the bottom half, 4-7, are merged into the complete array, 0-7, which is now sorted.

When the array size is not a power of 2, arrays of different sizes must be merged. For example, Figure 6.16 shows the situation when the array size is 12. Here an array of size 2 must be merged with an array of size 1 to form an array of size 3.

First, the 1-element ranges 0-0 and 1-1 are merged into the 2-element range 0-1. Then range 0-1 is merged with the 1-element range 2-2. This creates a 3-element range 0-2. It's merged with the 3-element range 3-5. The process continues until the array is sorted.

Notice that in mergesort we don't merge two separate arrays into a third one, as we demonstrated in the merge.java program. Instead, we merge parts of a single array into itself.

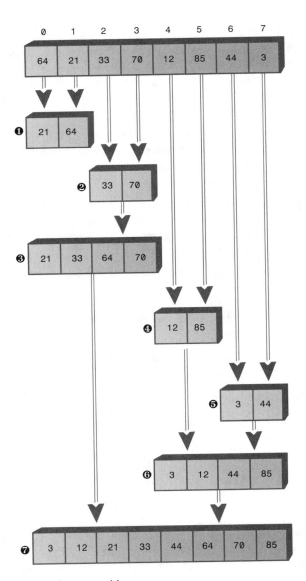

FIGURE 6.15   Merging larger and larger arrays.

You may wonder where all these subarrays are located in memory. In the algorithm, a workspace array of the same size as the original array is created. The subarrays are stored in sections of the workspace array. This means that subarrays in the original array are copied to appropriate places in the workspace array. After each merge, the workspace array is copied back into the original array.

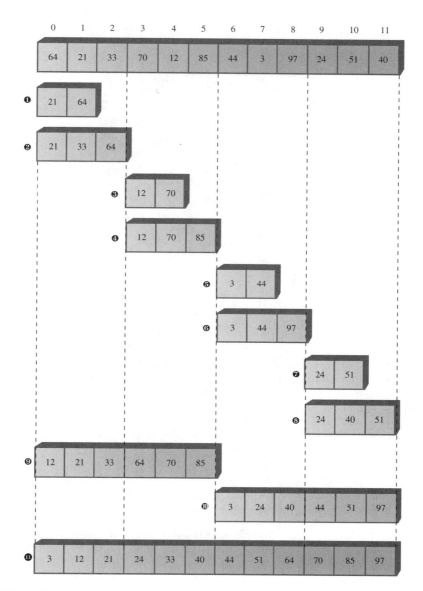

FIGURE 6.16   Array size not a power of 2.

The MergeSort Workshop Applet

This sorting process is easier to appreciate when you see it happening before your very eyes. Start up the MergeSort Workshop applet. Repeatedly pressing the Step button will execute mergesort step by step. Figure 6.17 shows what it looks like after the first three presses.

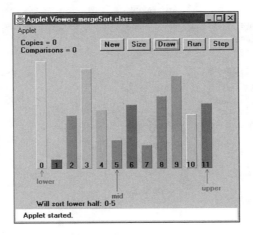

FIGURE 6.17   The MergeSort Workshop applet.

The Lower and Upper arrows show the range currently being considered by the algorithm, and the Mid arrow shows the middle part of the range. The range starts as the entire array and then is halved each time the mergeSort() method calls itself. When the range is one element, mergeSort() returns immediately; that's the base case. Otherwise, the two subarrays are merged. The applet provides messages, such as Entering mergeSort: 0-5, to tell you what it's doing and the range it's operating on.

Many steps involve the mergeSort() method calling itself or returning. Comparisons and copies are performed only during the merge process, when you'll see messages like Merged 0-0 and 1-1 into workspace. You can't see the merge happening because the workspace isn't shown. However, you can see the result when the appropriate section of the workspace is copied back into the original (visible) array: The bars in the specified range will appear in sorted order.

First, the first two bars will be sorted, then the first three bars, then the two bars in the range 3-4, then the three bars in the range 3-5, then the six bars in the range 0-5, and so on, corresponding to the sequence shown in Figure 6.16. Eventually, all the bars will be sorted.

You can cause the algorithm to run continuously by pressing the Run button. You can stop this process at any time by pressing Step, single-step as many times as you want, and resume running by pressing Run again.

As in the other sorting Workshop applets, pressing New resets the array with a new group of unsorted bars, and toggles between random and inverse arrangements. The Size button toggles between 12 bars and 100 bars.

Watching the algorithm run with 100 inversely sorted bars is especially instructive. The resulting patterns show clearly how each range is sorted individually and merged with its other half, and how the ranges grow larger and larger.

The mergeSort.java **Program**

In a moment we'll look at the entire mergeSort.java program. First, let's focus on the method that carries out the mergesort. Here it is:

```
private void recMergeSort(long[] workSpace, int lowerBound,
                                           int upperBound)
   {
   if(lowerBound == upperBound)            // if range is 1,
      return;                              // no use sorting
   else
      {                                    // find midpoint
      int mid = (lowerBound+upperBound) / 2;
                                           // sort low half
      recMergeSort(workSpace, lowerBound, mid);
                                           // sort high half
      recMergeSort(workSpace, mid+1, upperBound);
                                           // merge them
      merge(workSpace, lowerBound, mid+1, upperBound);
      }  // end else
   }  // end recMergeSort
```

As you can see, besides the base case, there are only four statements in this method. One computes the midpoint, there are two recursive calls to recMergeSort() (one for each half of the array), and finally a call to merge() to merge the two sorted halves. The base case occurs when the range contains only one element (lowerBound==upperBound) and results in an immediate return.

In the mergeSort.java program, the mergeSort() method is the one actually seen by the class user. It creates the array workSpace[] and then calls the recursive routine recMergeSort() to carry out the sort. The creation of the workspace array is handled in mergeSort() because doing it in recMergeSort() would cause the array to be created anew with each recursive call, an inefficiency.

The merge() method in the previous merge.java program (Listing 6.5) operated on three separate arrays: two source arrays and a destination array. The merge() routine in the mergeSort.java program operates on a single array: the theArray member of the DArray class. The arguments to this merge() method are the starting point of the low-half subarray, the starting point of the high-half subarray, and the upper bound of the high-half subarray. The method calculates the sizes of the subarrays based on this information.

Listing 6.6 shows the complete `mergeSort.java` program, which uses a variant of the array classes from Chapter 2, adding the `mergeSort()` and `recMergeSort()` methods to the `DArray` class. The `main()` routine creates an array, inserts 12 items, displays the array, sorts the items with `mergeSort()`, and displays the array again.

LISTING 6.6   The `mergeSort.java` Program

```
// mergeSort.java
// demonstrates recursive merge sort
// to run this program: C>java MergeSortApp
//////////////////////////////////////////////////////////////////
class DArray
   {
   private long[] theArray;          // ref to array theArray
   private int nElems;               // number of data items
   //--------------------------------------------------------------
   public DArray(int max)            // constructor
      {
      theArray = new long[max];      // create array
      nElems = 0;
      }
   //--------------------------------------------------------------
   public void insert(long value)    // put element into array
      {
      theArray[nElems] = value;      // insert it
      nElems++;                      // increment size
      }
   //--------------------------------------------------------------
   public void display()             // displays array contents
      {
      for(int j=0; j<nElems; j++)    // for each element,
         System.out.print(theArray[j] + " ");  // display it
      System.out.println("");
      }
   //--------------------------------------------------------------
   public void mergeSort()           // called by main()
      {                              // provides workspace
      long[] workSpace = new long[nElems];
      recMergeSort(workSpace, 0, nElems-1);
      }
   //--------------------------------------------------------------
   private void recMergeSort(long[] workSpace, int lowerBound,
                                               int upperBound)
      {
```

LISTING 6.6   Continued

```
      if(lowerBound == upperBound)            // if range is 1,
         return;                              // no use sorting
      else
         {                                    // find midpoint
         int mid = (lowerBound+upperBound) / 2;
                                              // sort low half
         recMergeSort(workSpace, lowerBound, mid);
                                              // sort high half
         recMergeSort(workSpace, mid+1, upperBound);
                                              // merge them
         merge(workSpace, lowerBound, mid+1, upperBound);
         }  // end else
      }  // end recMergeSort()
   //-----------------------------------------------------------
   private void merge(long[] workSpace, int lowPtr,
                          int highPtr, int upperBound)
      {
      int j = 0;                              // workspace index
      int lowerBound = lowPtr;
      int mid = highPtr-1;
      int n = upperBound-lowerBound+1;        // # of items

      while(lowPtr <= mid && highPtr <= upperBound)
         if( theArray[lowPtr] < theArray[highPtr] )
            workSpace[j++] = theArray[lowPtr++];
         else
            workSpace[j++] = theArray[highPtr++];

      while(lowPtr <= mid)
         workSpace[j++] = theArray[lowPtr++];

      while(highPtr <= upperBound)
         workSpace[j++] = theArray[highPtr++];

      for(j=0; j<n; j++)
         theArray[lowerBound+j] = workSpace[j];
      }  // end merge()
   //-----------------------------------------------------------
   }  // end class DArray
////////////////////////////////////////////////////////////////
class MergeSortApp
   {
```

LISTING 6.6   Continued

```
public static void main(String[] args)
    {
    int maxSize = 100;           // array size
    DArray arr;                  // reference to array
    arr = new DArray(maxSize);   // create the array

    arr.insert(64);              // insert items
    arr.insert(21);
    arr.insert(33);
    arr.insert(70);
    arr.insert(12);
    arr.insert(85);
    arr.insert(44);
    arr.insert(3);
    arr.insert(99);
    arr.insert(0);
    arr.insert(108);
    arr.insert(36);

    arr.display();               // display items

    arr.mergeSort();             // merge sort the array

    arr.display();               // display items again
    } // end main()
    } // end class MergeSortApp
///////////////////////////////////////////////////////////////
```

The output from the program is simply the display of the unsorted and sorted arrays:

```
64 21 33 70 12 85 44 3 99 0 108 36
0 3 12 21 33 36 44 64 70 85 99 108
```

If we put additional statements in the recMergeSort() method, we could generate a running commentary on what the program does during a sort. The following output shows how this might look for the four-item array {64, 21, 33, 70}. (You can think of this as the lower half of the array in Figure 6.15.)

```
Entering 0-3
    Will sort low half of 0-3
    Entering 0-1
        Will sort low half of 0-1
```

```
        Entering 0-0
        Base-Case Return 0-0
      Will sort high half of 0-1
        Entering 1-1
        Base-Case Return 1-1
      Will merge halves into 0-1
    Return 0-1                        theArray=21 64 33 70
  Will sort high half of 0-3
  Entering 2-3
    Will sort low half of 2-3
        Entering 2-2
        Base-Case Return 2-2
      Will sort high half of 2-3
        Entering 3-3
        Base-Case Return 3-3
      Will merge halves into 2-3
    Return 2-3                        theArray=21 64 33 70
  Will merge halves into 0-3
Return 0-3                            theArray=21 33 64 70
```

This is roughly the same content as would be generated by the MergeSort Workshop applet if it could sort four items. Study of this output, and comparison with the code for recMergeSort() and Figure 6.15, will reveal the details of the sorting process.

Efficiency of the mergesort

As we noted, the mergesort runs in O(N*logN) time. How do we know this? Let's see how we can figure out the number of times a data item must be copied and the number times it must be compared with another data item during the course of the algorithm. We assume that copying and comparing are the most time-consuming operations; that the recursive calls and returns don't add much overhead.

Number of Copies
Consider Figure 6.15. Each cell below the top line represents an element copied from the array into the workspace.

Adding up all the cells in Figure 6.15 (the seven numbered steps) shows there are 24 copies necessary to sort 8 items. $\log_2 8$ is 3, so $8*\log_2 8$ equals 24. This shows that, for the case of 8 items, the number of copies is proportional to $N*\log_2 N$.

Another way to look at this calculation is that, to sort 8 items requires 3 *levels*, each of which involves 8 copies. A level means all copies into the same size subarray. In the first level, there are four 2-element subarrays; in the second level, there are two 4-element subarrays; and in the third level, there is one 8-element subarray. Each level has 8 elements, so again there are 3*8 or 24 copies.

In Figure 6.15, by considering only half the graph, you can see that 8 copies are necessary for an array of 4 items (steps 1, 2, and 3), and 2 copies are necessary for 2 items. Similar calculations provide the number of copies necessary for larger arrays. Table 6.4 summarizes this information.

TABLE 6.4   Number of Operations When N Is a Power of 2

N	$\log_2 N$	Number of Copies into Workspace ($N*\log_2 N$)	Total Copies	Comparisons Max (Min)
2	1	2	4	1 (1)
4	2	8	16	5 (4)
8	3	24	48	17 (12)
16	4	64	128	49 (32)
32	5	160	320	129 (80)
64	6	384	768	321 (192)
128	7	896	1792	769 (448)

Actually, the items are not only copied into the workspace, they're also copied back into the original array. This doubles the number of copies, as shown in the Total Copies column. The final column of Table 6.4 shows comparisons, which we'll return to in a moment.

It's harder to calculate the number of copies and comparisons when N is not a multiple of 2, but these numbers fall between those that are a power of 2. For 12 items, there are 88 total copies, and for 100 items, 1,344 total copies.

Number of Comparisons

In the mergesort algorithm, the number of comparisons is always somewhat less than the number of copies. How much less? Assuming the number of items is a power of 2, for each individual merging operation, the maximum number of comparisons is always one less than the number of items being merged, and the minimum is half the number of items being merged. You can see why this is true in Figure 6.18, which shows two possibilities when trying to merge two arrays of four items each.

In the first case, the items interleave, and seven comparisons must be made to merge them. In the second case, all the items in one array are smaller than all the items in the other, so only four comparisons need be made.

There are many merges for each sort, so we must add the comparisons for each one. Referring back to Figure 6.15, you can see that seven merge operations are required to sort eight items. The number of items being merged and the resulting number of comparisons are shown in Table 6.5.

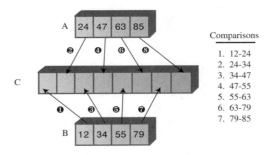

a) Worst-case Scenario

Comparisons

1. 12-24
2. 24-34
3. 34-47
4. 47-55
5. 55-63
6. 63-79
7. 79-85

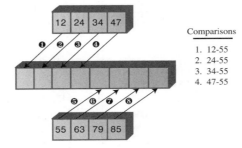

b) Best-case Scenario

Comparisons

1. 12-55
2. 24-55
3. 34-55
4. 47-55

FIGURE 6.18 Maximum and minimum comparisons.

TABLE 6.5 Comparisons Involved in Sorting 8 Items

Step Number	1	2	3	4	5	6	7	Totals
Number of items being merged (N)	2	2	4	2	2	4	8	24
Maximum comparisons (N-1)	1	1	3	1	1	3	7	17
Minimum comparisons (N/2)	1	1	2	1	1	2	4	12

For each merge, the maximum number of comparisons is one less than the number of items. Adding these figures for all the merges gives us a total of 17.

The minimum number of comparisons is always half the number of items being merged, and adding these figures for all the merges results in 12 comparisons. Similar arithmetic results in the Comparisons columns for Table 6.4. The actual number of comparisons to sort a specific array depends on how the data is arranged, but it will be somewhere between the maximum and minimum values.

Eliminating Recursion

Some algorithms lend themselves to a recursive approach, some don't. As we've seen, the recursive `triangle()` and `factorial()` methods can be implemented more efficiently using a simple loop. However, various divide-and-conquer algorithms, such as mergesort, work very well as recursive routines.

Often an algorithm is easy to conceptualize as a recursive method, but in practice the recursive approach proves to be inefficient. In such cases, it's useful to transform the recursive approach into a non-recursive approach. Such a transformation can often make use of a stack.

Recursion and Stacks

There is a close relationship between recursion and stacks. In fact, most compilers implement recursion by using stacks. As we noted, when a method is called, the compiler pushes the arguments to the method and the return address (where control will go when the method returns) on the stack, and then transfers control to the method. When the method returns, it pops these values off the stack. The arguments disappear, and control returns to the return address.

Simulating a Recursive Method

In this section we'll demonstrate how any recursive solution can be transformed into a stack-based solution. Remember the recursive `triangle()` method from the first section in this chapter? Here it is again:

```
int triangle(int n)
   {
   if(n==1)
      return 1;
   else
      return( n + triangle(n-1) );
   }
```

We're going to break this algorithm down into its individual operations, making each operation one `case` in a `switch` statement. (You can perform a similar decomposition using `goto` statements in C++ and some other languages, but Java doesn't support `goto`.)

The `switch` statement is enclosed in a method called `step()`. Each call to `step()` causes one `case` section within the `switch` to be executed. Calling `step()` repeatedly will eventually execute all the code in the algorithm.

The `triangle()` method we just saw performs two kinds of operations. First, it carries out the arithmetic necessary to compute triangular numbers. This involves checking if n is 1, and adding n to the results of previous recursive calls. However, `triangle()` also performs the operations necessary to manage the method itself, including transfer of control, argument access, and the return address. These operations are not visible by looking at the code; they're built into all methods. Here, roughly speaking, is what happens during a call to a method:

- When a method is called, its arguments and the return address are pushed onto a stack.

- A method can access its arguments by peeking at the top of the stack.

- When a method is about to return, it peeks at the stack to obtain the return address, and then pops both this address and its arguments off the stack and discards them.

The `stackTriangle.java` program contains three classes: `Params`, `StackX`, and `StackTriangleApp`. The `Params` class encapsulates the return address and the method's argument, n; objects of this class are pushed onto the stack. The `StackX` class is similar to those in other chapters, except that it holds objects of class `Params`. The `StackTriangleApp` class contains four methods: `main()`, `recTriangle()`, `step()`, and the usual `getInt()` method for numerical input.

The `main()` routine asks the user for a number, calls the `recTriangle()` method to calculate the triangular number corresponding to n, and displays the result.

The `recTriangle()` method creates a `StackX` object and initializes `codePart` to 1. It then settles into a `while` loop, where it repeatedly calls `step()`. It won't exit from the loop until `step()` returns true by reaching case 6, its exit point. The `step()` method is basically a large `switch` statement in which each `case` corresponds to a section of code in the original `triangle()` method. Listing 6.7 shows the `stackTriangle.java` program.

LISTING 6.7 The `stackTriangle.java` Program

```
// stackTriangle.java
// evaluates triangular numbers, stack replaces recursion
// to run this program: C>java StackTriangleApp
import java.io.*;               // for I/O
//////////////////////////////////////////////////////////////
class Params      // parameters to save on stack
   {
   public int n;
   public int returnAddress;
```

LISTING 6.7   Continued

```
   public Params(int nn, int ra)
      {
      n=nn;
      returnAddress=ra;
      }
   } // end class Params
//////////////////////////////////////////////////////////////////
class StackX
   {
   private int maxSize;         // size of StackX array
   private Params[] stackArray;
   private int top;             // top of stack
//--------------------------------------------------------------
   public StackX(int s)         // constructor
      {
      maxSize = s;             // set array size
      stackArray = new Params[maxSize];  // create array
      top = -1;                // no items yet
      }
//--------------------------------------------------------------
   public void push(Params p)   // put item on top of stack
      {
      stackArray[++top] = p;    // increment top, insert item
      }
//--------------------------------------------------------------
   public Params pop()          // take item from top of stack
      {
      return stackArray[top--]; // access item, decrement top
      }
//--------------------------------------------------------------
   public Params peek()         // peek at top of stack
      {
      return stackArray[top];
      }
//--------------------------------------------------------------
   } // end class StackX
//////////////////////////////////////////////////////////////////
class StackTriangleApp
   {
   static int theNumber;
   static int theAnswer;
```

LISTING 6.7   Continued

```
   static StackX theStack;
   static int codePart;
   static Params theseParams;
//------------------------------------------------------------
   public static void main(String[] args) throws IOException
      {
      System.out.print("Enter a number: ");
      theNumber = getInt();
      recTriangle();
      System.out.println("Triangle="+theAnswer);
      } // end main()
//------------------------------------------------------------
   public static void recTriangle()
      {
      theStack = new StackX(10000);
      codePart = 1;
      while( step() == false)  // call step() until it's true
         ;                     // null statement
      }
//------------------------------------------------------------
   public static boolean step()
      {
      switch(codePart)
         {
         case 1:                           // initial call
            theseParams = new Params(theNumber, 6);
            theStack.push(theseParams);
            codePart = 2;
            break;
         case 2:                           // method entry
            theseParams = theStack.peek();
            if(theseParams.n == 1)         // test
               {
               theAnswer = 1;
               codePart = 5;   // exit
               }
            else
               codePart = 3;   // recursive call
            break;
         case 3:                           // method call
            Params newParams = new Params(theseParams.n - 1, 4);
```

LISTING 6.7   Continued

```
                    theStack.push(newParams);
                    codePart = 2;   // go enter method
                    break;
                case 4:                              // calculation
                    theseParams = theStack.peek();
                    theAnswer = theAnswer + theseParams.n;
                    codePart = 5;
                    break;
                case 5:                              // method exit
                    theseParams = theStack.peek();
                    codePart = theseParams.returnAddress; // (4 or 6)
                    theStack.pop();
                    break;
                case 6:                              // return point
                    return true;
                }  // end switch
            return false;
            }  // end triangle
//--------------------------------------------------------------
    public static String getString() throws IOException
        {
        InputStreamReader isr = new InputStreamReader(System.in);
        BufferedReader br = new BufferedReader(isr);
        String s = br.readLine();
        return s;
        }
//--------------------------------------------------------------
    public static int getInt() throws IOException
        {
        String s = getString();
        return Integer.parseInt(s);
        }
//--------------------------------------------------------------
    }  // end class StackTriangleApp
////////////////////////////////////////////////////////////////
```

This program calculates triangular numbers, just as the triangle.java program (Listing 6.1) at the beginning of the chapter did. Here's some sample output:

```
Enter a number: 100
Triangle=5050
```

Figure 6.19 shows how the sections of code in each case relate to the various parts of the algorithm.

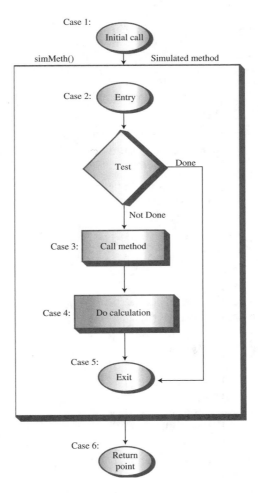

FIGURE 6.19   The cases and the step() method.

The program simulates a method, but it has no name in the listing because it isn't a real Java method. Let's call this simulated method simMeth(). The initial call to simMeth() (at case 1) pushes the value entered by the user and a return value of 6 onto the stack and moves to the entry point of simMeth() (case 2).

At its entry (case 2), simMeth() tests whether its argument is 1. It accesses the argument by peeking at the top of the stack. If the argument is 1, this is the base case and control goes to simMeth()'s exit (case 5). If not, it calls itself recursively (case 3). This recursive call consists of pushing n-1 and a return address of 4 onto the stack, and going to the method entry at case 2.

On the return from the recursive call, simMeth() adds its argument n to the value returned from the call. Finally, it exits (case 5). When it exits, it pops the last Params object off the stack; this information is no longer needed.

The return address given in the initial call was 6, so case 6 is the place where control goes when the method returns. This code returns true to let the while loop in recTriangle() know that the loop is over.

Note that in this description of simMeth()'s operation we use terms like *argument*, *recursive call*, and *return address* to mean simulations of these features, not the normal Java versions.

If you inserted some output statements in each case to see what simMeth() was doing, you could arrange for output like this:

```
Enter a number: 4
case 1. theAnswer=0  Stack:
case 2. theAnswer=0  Stack: (4, 6)
case 3. theAnswer=0  Stack: (4, 6)
case 2. theAnswer=0  Stack: (4, 6) (3, 4)
case 3. theAnswer=0  Stack: (4, 6) (3, 4)
case 2. theAnswer=0  Stack: (4, 6) (3, 4) (2, 4)
case 3. theAnswer=0  Stack: (4, 6) (3, 4) (2, 4)
case 2. theAnswer=0  Stack: (4, 6) (3, 4) (2, 4) (1, 4)
case 5. theAnswer=1  Stack: (4, 6) (3, 4) (2, 4) (1, 4)
case 4. theAnswer=1  Stack: (4, 6) (3, 4) (2, 4)
case 5. theAnswer=3  Stack: (4, 6) (3, 4) (2, 4)
case 4. theAnswer=3  Stack: (4, 6) (3, 4)
case 5. theAnswer=6  Stack: (4, 6) (3, 4)
case 4. theAnswer=6  Stack: (4, 6)
case 5. theAnswer=10 Stack: (4, 6)
case 6. theAnswer=10 Stack:
Triangle=10
```

The case number shows what section of code is being executed. The contents of the stack (consisting of Params objects containing n followed by a return address) are also shown. The simMeth() method is entered four times (case 2) and returns four times (case 5). Only when it starts returning does theAnswer begin to accumulate the results of the calculations.

What Does This Prove?

In stackTriangle.java (Listing 6.7) we have a program that more or less systematically transforms a program that uses recursion into a program that uses a stack. This suggests that such a transformation is possible for any program that uses recursion, and in fact this is the case.

With some additional work, you can systematically refine the code we show here, simplifying it and even eliminating the switch statement entirely to make the code more efficient.

In practice, however, it's usually more practical to rethink the algorithm from the beginning, using a stack-based approach instead of a recursive approach. Listing 6.8 shows what happens when we do that with the triangle() method.

LISTING 6.8   The stackTriangle2.java Program

```
// stackTriangle2.java
// evaluates triangular numbers, stack replaces recursion
// to run this program: C>java StackTriangle2App
import java.io.*;                  // for I/O
////////////////////////////////////////////////////////////////
class StackX
   {
   private int maxSize;        // size of stack array
   private int[] stackArray;
   private int top;            // top of stack
//-------------------------------------------------------------
   public StackX(int s)          // constructor
      {
      maxSize = s;
      stackArray = new int[maxSize];
      top = -1;
      }
//-------------------------------------------------------------
   public void push(int p)     // put item on top of stack
      { stackArray[++top] = p; }
//-------------------------------------------------------------
   public int pop()            // take item from top of stack
      { return stackArray[top--]; }
//-------------------------------------------------------------
   public int peek()           // peek at top of stack
      { return stackArray[top]; }
//-------------------------------------------------------------
```

LISTING 6.8   Continued

```java
    public boolean isEmpty()      // true if stack is empty
       { return (top == -1); }
//--------------------------------------------------------------
    }  // end class StackX
////////////////////////////////////////////////////////////////
class StackTriangle2App
    {
    static int theNumber;
    static int theAnswer;
    static StackX theStack;

    public static void main(String[] args) throws IOException
       {
       System.out.print("Enter a number: ");
       theNumber = getInt();
       stackTriangle();
       System.out.println("Triangle="+theAnswer);
       }  // end main()
//--------------------------------------------------------------
    public static void stackTriangle()
       {
       theStack = new StackX(10000);     // make a stack

       theAnswer = 0;                    // initialize answer

       while(theNumber > 0)              // until n is 1,
          {
          theStack.push(theNumber);   // push value
          --theNumber;                // decrement value
          }
       while( !theStack.isEmpty() )      // until stack empty,
          {
          int newN = theStack.pop();  // pop value,
          theAnswer += newN;          // add to answer
          }
       }
//--------------------------------------------------------------
    public static String getString() throws IOException
       {
       InputStreamReader isr = new InputStreamReader(System.in);
       BufferedReader br = new BufferedReader(isr);
```

LISTING 6.8 Continued

```
    String s = br.readLine();
    return s;
    }
//-------------------------------------------------------------
  public static int getInt() throws IOException
    {
    String s = getString();
    return Integer.parseInt(s);
    }
//-------------------------------------------------------------
  }  // end class StackTriangle2App
```

Here two short while loops in the stackTriangle() method substitute for the entire step() method of the stackTriangle.java program. Of course, in this program you can see by inspection that you can eliminate the stack entirely and use a simple loop. However, in more complicated algorithms the stack must remain.

Often you'll need to experiment to see whether a recursive method, a stack-based approach, or a simple loop is the most efficient (or practical) way to handle a particular situation.

Some Interesting Recursive Applications

Let's look briefly at some other situations in which recursion is useful. You will see from the diversity of these examples that recursion can pop up in unexpected places. We'll examine three problems: raising a number to a power, fitting items into a knapsack, and choosing members of a mountain-climbing team. We'll explain the concepts and leave the implementations as exercises.

Raising a Number to a Power

The more sophisticated pocket calculators allow you to raise a number to an arbitrary power. They usually have a key labeled something like x^y, where the circumflex indicates that x is raised to the y power. How would you do this calculation if your calculator lacked this key? You might assume you would need to multiply x by itself y times. That is, if x was 2 and y was 8 (2^8), you would carry out the arithmetic for 2*2*2*2*2*2*2. However, for large values of y, this approach might prove tedious. Is there a quicker way?

One solution is to rearrange the problem so you multiply by multiples of 2 whenever possible, instead of by 2. Take 2^8 as an example. Eventually, we must involve eight 2s in the multiplication process. Let's say we start with 2*2=4. We've used up two of the

2s, but there are still six to go. However, we now have a new number to work with: 4. So we try 4*4=16. This uses four 2s (because each 4 is two 2s multiplied together). We need to use up four more 2s, but now we have 16 to work with, and 16*16=256 uses exactly eight 2s (because each 16 has four 2s).

So we've found the answer to 2^8 with only three multiplications instead of seven. That's O(log N) time instead O(N).

Can we make this process into an algorithm that a computer can execute? The scheme is based on the mathematical equality $x^y = (x^2)^{y/2}$. In our example, $2^8 = (2^2)^{8/2}$, or $2^8 = (2^2)^4$. This is true because raising a power to another power is the same as multiplying the powers.

However, we're assuming our computer can't raise a number to a power, so we can't handle $(2^2)^4$. Let's see if we can transform this into an expression that involves only multiplication. The trick is to start by substituting a new variable for 2^2.

Let's say that $2^2=a$. Then 2^8 equals $(2^2)^4$, which is a^4. However, according to the original equality, a^4 can be written $(a^2)^2$, so $2^8 = (a^2)^2$.

Again we substitute a new variable for a^2, say $a^2=c$, then $(c)^2$ can be written $(c^2)^1$, which also equals 2^8.

Now we have a problem we can handle with simple multiplication: c times c.

You can imbed this scheme in a recursive method—let's call it power()—for calculating powers. The arguments are x and y, and the method returns x^y. We don't need to worry about variables like a and c anymore because x and y get new values each time the method calls itself. Its arguments are x*x and y/2. For the x=2 and y=8, the sequence of arguments and return values would be

```
x=2, y=8
x=4, y=4
x=16, y=2
x=256, y=1
Returning 256, x=256, y=1
Returning 256, x=16, y=2
Returning 256, x=4, y=4
Returning 256, x=2, y=8
```

When y is 1, we return. The answer, 256, is passed unchanged back up the sequence of methods.

We've shown an example in which y is an even number throughout the entire sequence of divisions. This will not usually be the case. Here's how to revise the algorithm to deal with the situation where y is odd. Use integer division on the way down and don't worry about a remainder when dividing y by 2. However, during the return process, whenever y is an odd number, do an additional multiplication by x. Here's the sequence for 3^{18}:

```
x=3, y=18
x=9, y=9
x=81, y=4
x=6561, y=2
x=43046721, y=1
Returning 43046721, x=43046721, y=1
Returning 43046721, x=6561, y=2
Returning 43046721, x=81, y=4
Returning 387420489, x=9, y=9  // y is odd; so multiply by x
Returning 387420489, x=3, y=18
```

The Knapsack Problem

The Knapsack Problem is a classic in computer science. In its simplest form it involves trying to fit items of different weights into a knapsack so that the knapsack ends up with a specified total weight. You don't need to fit in all the items.

For example, suppose you want your knapsack to weigh exactly 20 pounds, and you have five items, with weights of 11, 8, 7, 6, and 5 pounds. For small numbers of items, humans are pretty good at solving this problem by inspection. So you can probably figure out that only the 8, 7, and 5 combination of items adds up to 20.

If we want a computer to solve this problem, we'll need to give it more detailed instructions. Here's the algorithm:

1. If at any point in this process the sum of the items you selected adds up to the target, you're done.

2. Start by selecting the first item. The remaining items must add up to the knapsack's target weight minus the first item; this is a new target weight.

3. Try, one by one, each of the possible combinations of the remaining items. Notice, however, that you don't really need to try all the combinations, because whenever the sum of the items is more than the target weight, you can stop adding items.

4. If none of the combinations work, discard the first item, and start the whole process again with the second item.

5. Continue this with the third item and so on until you've tried all the combinations, at which point you know there is no solution.

In the example just described, start with 11. Now we want the remaining items to add up to 9 (20 minus 11). Of these, we start with 8, which is too small. Now we want the remaining items to add up to 1 (9 minus 8). We start with 7, but that's bigger than 1, so we try 6 and then 5, which are also too big. We've run out of items, so we know that any combination that includes 8 won't add up to 9. Next we try 7, so now we're looking for a target of 2 (9 minus 7). We continue in the same way, as summarized here:

```
Items: 11, 8, 7, 6, 5
=========================================
11         // Target = 20, 11 is too small
11, 8      // Target = 9, 8 is too small
11, 8, 7   // Target = 1, 7 is too big
11, 8, 6   // Target = 1, 6 is too big
11, 8, 5   // Target = 1, 5 is too big. No more items
11, 7      // Target = 9, 7 is too small
11, 7, 6   // Target = 2, 6 is too big
11, 7, 5   // Target = 2, 5 is too big. No more items
11, 6      // Target = 9, 6 is too small
11, 6, 5   // Target = 3, 5 is too big. No more items
11, 5      // Target = 9, 5 is too small. No more items
8,         // Target = 20, 8 is too small
8, 7       // Target = 12, 7 is too small
8, 7, 6    // Target = 5, 6 is too big
8, 7, 5    // Target = 5, 5 is just right. Success!
```

As you may recognize, a recursive routine can pick the first item, and, if the item is smaller than the target, the routine can call itself with a new target to investigate the sums of all the remaining items.

Combinations: Picking a Team

In mathematics, a *combination* is a selection of things in which their order doesn't matter. For example, suppose there is a group of five mountain climbers named A, B, C, D, and E. From this group you want to select a team of three to scale steep and icy Mount Anaconda. However, you're worried about how the team members will get along, so you decide to list all the possible teams; that is, all the possible combinations of three climbers. But then you think it would be nice to have a computer program print out all the combinations for you. Such a program would show you the 10 possible combinations:

ABC, ABD, ABE, ACD, ACE, ADE, BCD, BCE, BDE, CDE

How would you write such a program? It turns out there's an elegant recursive solution. It involves dividing these combinations into two groups: those that begin with A and those that don't. Suppose we abbreviate the idea of 3 people selected from a group of 5 as (5,3). Let's say n is the size of the group and k is the size of a team. A theorem says that

$(n, k) = (n - 1, k - 1) + (n - 1, k)$

For our example of 3 people selected from a group of 5, we have

$(5, 3) = (4, 2) + (4, 3)$

We've broken a large problem into two smaller ones. Instead of selecting from a group of 5, we're selecting twice from a group of 4: First, all the ways to select 2 people from a group of 4, then all the ways to select 3 people from a group of 4.

There are 6 ways to select 2 people from a group of 4. In the (4, 2) term—which we'll call the left term—these 6 combinations are

    BC, BD, BE, CD, CE, DE

A is the missing group member, so to make three-person teams we precede these combinations with A:

    ABC, ABD, ABE, ACD, ACE, ADE

There are four ways to select 3 people from a group of 4. In the (4, 3) term—the right term—we have

    BCD, BCE, BDE, CDE

When these 4 combinations from the right term are added to the 6 from the left term, we get the 10 combinations for (5, 3).

You can apply the same decomposition process to each of the groups of 4. For example, (4, 2) is (3, 1) added to (3, 2). As you can see, this is a natural place to apply recursion.

You can think of this problem as a tree with (5,3) on the top row, (4,3) and (4,2) on the next row, and so on, where the nodes in the tree correspond to recursive function calls. Figure 6.20 shows what this looks like for the (5,3) example.

The base cases are combinations that make no sense: those with a 0 for either number and those where the team size is greater than the group size. The combination (1,1) is valid but there's no point trying to break it down further. In the figure, dotted lines show the base cases; you return rather than following them.

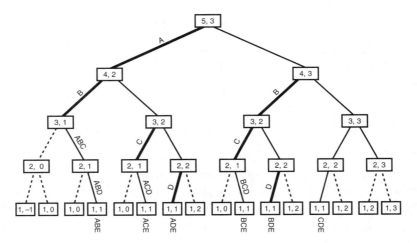

FIGURE 6.20 Picking a team of 3 from a group of 5.

The recursion depth corresponds to the group members: The node on the top row represents group member A, the two nodes on the next row represent group member B, and so on. If there are 5 group members, you'll have 5 levels.

As you descend the tree you need to remember the sequence of members you visit. Here's how to do that: Whenever you make a call to a left term, you record the node you're leaving by adding its letter to a sequence. These left calls and the letters to add to the sequence are shown by the darker lines in the figure. You'll need to role the sequence back up as you return.

To record all the combinations, you can display them as you go along. You don't display anything when making left calls. However, when you make calls to the right, you check the sequence; if you're at a valid node, and adding one member will complete the team, then add the node to the sequence and display the complete team.

Summary

- A recursive method calls itself repeatedly, with different argument values each time.

- Some value of its arguments causes a recursive method to return without calling itself. This is called the base case.

- When the innermost instance of a recursive method returns, the process "unwinds" by completing pending instances of the method, going from the latest back to the original call.

- A triangular number is the sum of itself and all numbers smaller than itself. (*Number* means *integer* in this context.) For example, the triangular number of 4 is 10, because 4+3+2+1 = 10.

- The factorial of a number is the product of itself and all numbers smaller than itself. For example, the factorial of 4 is 4*3*2*1 = 24.

- Both triangular numbers and factorials can be calculated using either a recursive method or a simple loop.

- The anagram of a word (all possible combinations of its n letters) can be found recursively by repeatedly rotating all its letters and anagramming the rightmost n-1 of them.

- A binary search can be carried out recursively by checking which half of a sorted range the search key is in, and then doing the same thing with that half.

- The Towers of Hanoi puzzle consists of three towers and an arbitrary number of rings.

- The Towers of Hanoi puzzle can be solved recursively by moving all but the bottom disk of a subtree to an intermediate tower, moving the bottom disk to the destination tower, and finally moving the subtree to the destination.

- Merging two sorted arrays means to create a third array that contains all the elements from both arrays in sorted order.

- In mergesort, 1-element subarrays of a larger array are merged into 2-element subarrays, 2-element subarrays are merged into 4-element subarrays, and so on until the entire array is sorted.

- mergesort requires O(N*logN) time.

- mergesort requires a workspace equal in size to the original array.

- For triangular numbers, factorials, anagrams, and the binary search, the recursive method contains only one call to itself. (There are two shown in the code for the binary search, but only one is used on any given pass through the method's code.)

- For the Towers of Hanoi and mergesort, the recursive method contains two calls to itself.

- Any operation that can be carried out with recursion can be carried out with a stack.

- A recursive approach may be inefficient. If so, it can sometimes be replaced with a simple loop or a stack-based approach.

Questions

These questions are intended as a self-test for readers. Answers may be found in
Appendix C.

1. If the user enters 10 in the triangle.java program (Listing 6.1), what is the
   maximum number of "copies" of the triangle() method (actually just copies of
   its argument) that exist at any one time?

2. Where are the copies of the argument, mentioned in question 1, stored?

   a. in a variable in the triangle() method

   b. in a field of the TriangleApp class

   c. in a variable of the getString() method

   d. on a stack

3. Assume the user enters 10 as in question 1. What is the value of n when the
   triangle() method first returns a value other than 1?

4. Assume the same situation as in question 1. What is the value of n when the
   triangle() method is about to return to main()?

5. True or false: In the triangle() method, the return values are stored on the
   stack.

6. In the anagram.java program (Listing 6.2), at a certain depth of recursion, a
   version of the doAnagram() method is working with the string "led". When this
   method calls a new version of itself, what letters will the new version be
   working with?

7. We've seen that recursion can take the place of a loop, as in the loop-oriented
   orderedArray.java program (Listing 2.4) and the recursive binarySearch.java
   program (Listing 6.3). Which of the following is *not* true?

   a. Both programs divide the range repeatedly in half.

   b. If the key is not found, the loop version returns because the range
      bounds cross, but the recursive version occurs because it reaches the
      bottom recursion level.

   c. If the key is found, the loop version returns from the entire method,
      whereas the recursive version returns from only one level of recursion.

   d. In the recursive version the range to be searched must be specified in the
      arguments, while in the loop version it need not be.

8. In the `recFind()` method in the `binarySearch.java` program (Listing 6.3), what takes the place of the loop in the non-recursive version?

   a. the `recFind()` method

   b. arguments to `recFind()`

   c. recursive calls to `recFind()`

   d. the call from `main()` to `recFind()`

9. The `binarySearch.java` program is an example of the _____ approach to solving a problem.

10. What gets smaller as you make repeated recursive calls in the `redFind()` method?

11. What becomes smaller with repeated recursive calls in the `towers.java` program (Listing 6.4)?

12. The algorithm in the towers.java program involves

    a. "trees" that are data storage devices.

    b. secretly putting small disks under large disks.

    c. changing which columns are the source and destination.

    d. moving one small disk and then a stack of larger disks.

13. Which is *not* true about the `merge()` method in the `merge.java` program (Listing 6.5)?

    a. Its algorithm can handle arrays of different sizes.

    b. It must search the target array to find where to put the next item.

    c. It is not recursive.

    d. It continuously takes the smallest item irrespective of what array it's in.

14. The disadvantage of mergesort is that

    a. it is not recursive.

    b. it uses more memory.

    c. although faster than the insertion sort, it is much slower than quicksort.

    d. it is complicated to implement.

15. Besides a loop, a _____ can often be used instead of recursion.

Experiments

Carrying out these experiments will help to provide insights into the topics covered in the chapter. No programming is involved.

1. In the triangle.java program (Listing 6.1), remove the code for the base case (the if(n==1), the return 1;, and the else). Then run the program and see what happens.

2. Use the Towers Workshop applet in manual mode to solve the puzzle with seven or more disks.

3. Rewrite the main() part of mergeSort.java (Listing 6.6) so you can fill the array with hundreds of thousands of random numbers. Run the program to sort these numbers and compare its speed with the sorts in Chapter 3, "Simple Sorting."

Programming Projects

Writing programs that solve the Programming Projects helps to solidify your understanding of the material and demonstrates how the chapter's concepts are applied. (As noted in the Introduction, qualified instructors may obtain completed solutions to the Programming Projects on the publisher's Web site.)

6.1 Suppose you buy a budget-priced pocket PC and discover that the chip inside can't do multiplication, only addition. You program your way out of this quandary by writing a recursive method, mult(), that performs multiplication of x and y by adding x to itself y times. Its arguments are x and y and its return value is the product of x and y. Write such a method and a main() program to call it. Does the addition take place when the method calls itself or when it returns?

6.2 In Chapter 8, "Binary Trees," we'll look at binary trees, where every branch has (potentially) exactly two sub-branches. If we draw a binary tree on the screen using characters, we might have 1 branch on the top row, 2 on the next row, then 4, 8, 16, and so on. Here's what that looks like for a tree 16 characters wide:

```
--------X-------
----X-------X---
--X---X---X---X-
-X-X-X-X-X-X-X-X
XXXXXXXXXXXXXXXX
```

(Note that the bottom line should be shifted a half character-width right, but there's nothing we can do about that with character-mode graphics.) You can draw this tree using a recursive makeBranches() method with arguments left and right, which are the endpoints of a horizontal range. When you first enter the routine, left is 0 and right is the number of characters (including dashes) in all the lines, minus 1. You draw an X in the center of this range. Then the method calls itself twice: once for the left half of the range and once for the right half. Return when the range gets too small. You will probably want to put all the dashes and Xs into an array and display the array all at once, perhaps with a display() method. Write a main() program to draw the tree by calling makeBranches() and display(). Allow main() to determine the line length of the display (32, 64, or whatever). Ensure that the array that holds the characters for display is no larger than it needs to be. What is the relationship of the number of lines (five in the picture here) to the line width?

6.3 Implement the recursive approach to raising a number to a power, as described in the "Raising a Number to a Power" section near the end of this chapter. Write the recursive power() function and a main() routine to test it.

6.4 Write a program that solves the knapsack problem for an arbitrary knapsack capacity and series of weights. Assume the weights are stored in an array. Hint: The arguments to the recursive knapsack() function are the target weight and the array index where the remaining items start.

6.5 Implement a recursive approach to showing all the teams that can be created from a group (n things taken k at a time). Write the recursive showTeams() method and a main() method to prompt the user for the group size and the team size to provide arguments for showTeam(), which then displays all the possible combinations.

Advanced Sorting

We discussed simple sorting in the aptly titled Chapter 3, "Simple Sorting." The sorts described there—the bubble, selection, and insertion sorts—are easy to implement but are rather slow. In Chapter 6, "Recursion," we described the mergesort. It runs much faster than the simple sorts but requires twice as much space as the original array; this is often a serious drawback.

This chapter covers two advanced approaches to sorting: Shellsort and quicksort. These sorts both operate much faster than the simple sorts: the Shellsort in about $O(N*(logN)^2)$ time, and quicksort in $O(N*logN)$ time. Neither of these sorts requires a large amount of extra space, as mergesort does. The Shellsort is almost as easy to implement as mergesort, while quicksort is the fastest of all the general-purpose sorts. We'll conclude the chapter with a brief mention of the radix sort, an unusual and interesting approach to sorting.

We'll examine the Shellsort first. Quicksort is based on the idea of partitioning, so we'll then examine partitioning separately, before examining quicksort itself.

Shellsort

The Shellsort is named for Donald L. Shell, the computer scientist who discovered it in 1959. It's based on the insertion sort, but adds a new feature that dramatically improves the insertion sort's performance.

The Shellsort is good for medium-sized arrays, perhaps up to a few thousand items, depending on the particular implementation. It's not quite as fast as quicksort and other $O(N*logN)$ sorts, so it's not optimum for very large files. However, it's much faster than the $O(N^2)$ sorts like the selection sort and the insertion sort, and it's very easy to implement: The code is short and simple.

The worst-case performance is not significantly worse than the average performance. (We'll see later in this chapter that the worst-case performance for quicksort can be much worse unless precautions are taken.) Some experts (see Sedgewick in Appendix B, "Further Reading") recommend starting with a Shellsort for almost any sorting project and changing to a more advanced sort, like quicksort, only if Shellsort proves too slow in practice.

Insertion Sort: Too Many Copies

Because Shellsort is based on the insertion sort, you might want to review the section titled "Insertion Sort" in Chapter 3. Recall that partway through the insertion sort the items to the left of a marker are internally sorted (sorted among themselves) and items to the right are not. The algorithm removes the item at the marker and stores it in a temporary variable. Then, beginning with the item to the left of the newly vacated cell, it shifts the sorted items right one cell at a time, until the item in the temporary variable can be reinserted in sorted order.

Here's the problem with the insertion sort. Suppose a small item is on the far right, where the large items should be. To move this small item to its proper place on the left, all the intervening items (between the place where it is and where it should be) must be shifted one space right. This step takes close to N copies, just for one item. Not all the items must be moved a full N spaces, but the average item must be moved N/2 spaces, which takes N times N/2 shifts for a total of $N^2/2$ copies. Thus, the performance of insertion sort is $O(N^2)$.

This performance could be improved if we could somehow move a smaller item many spaces to the left without shifting all the intermediate items individually.

N-Sorting

The Shellsort achieves these large shifts by insertion-sorting widely spaced elements. After they are sorted, it sorts somewhat less widely spaced elements, and so on. The spacing between elements for these sorts is called the *increment* and is traditionally represented by the letter h. Figure 7.1 shows the first step in the process of sorting a 10-element array with an increment of 4. Here the elements 0, 4, and 8 are sorted.

After 0, 4, and 8 are sorted, the algorithm shifts over one cell and sorts 1, 5, and 9. This process continues until all the elements have been *4-sorted*, which means that all items spaced four cells apart are sorted among themselves. The process is shown (using a more compact visual metaphor) in Figure 7.2.

After the complete 4-sort, the array can be thought of as comprising four subarrays: (0,4,8), (1,5,9), (2,6), and (3,7), each of which is completely sorted. These subarrays are interleaved but otherwise independent.

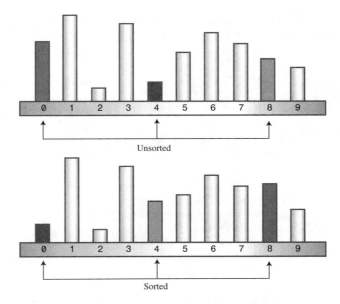

FIGURE 7.1   4-sorting 0, 4, and 8.

Notice that, in this particular example, at the end of the 4-sort no item is more than two cells from where it would be if the array were completely sorted. This is what is meant by an array being "almost" sorted and is the secret of the Shellsort. By creating interleaved, internally sorted sets of items, we minimize the amount of work that must be done to complete the sort.

Now, as we noted in Chapter 3, the insertion sort is very efficient when operating on an array that's almost sorted. If it needs to move items only one or two cells to sort the file, it can operate in almost O(N) time. Thus, after the array has been 4-sorted, we can 1-sort it using the ordinary insertion sort. The combination of the 4-sort and the 1-sort is much faster than simply applying the ordinary insertion sort without the preliminary 4-sort.

Diminishing Gaps

We've shown an initial interval—or gap—of 4 cells for sorting a 10-cell array. For larger arrays the interval should start out much larger. The interval is then repeatedly reduced until it becomes 1.

For instance, an array of 1,000 items might be 364-sorted, then 121-sorted, then 40-sorted, then 13-sorted, then 4-sorted, and finally 1-sorted. The sequence of numbers used to generate the intervals (in this example, 364, 121, 40, 13, 4, 1) is called the *interval sequence* or *gap sequence*. The particular interval sequence shown here, attributed to Knuth (see Appendix B), is a popular one. In reversed form, starting from 1, it's generated by the recursive expression

h = 3*h + 1

where the initial value of h is 1. The first two columns of Table 7.1 show how this formula generates the sequence.

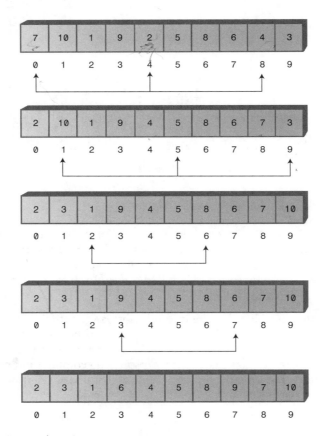

FIGURE 7.2   A complete 4-sort.

TABLE 7.1   Knuth's Interval Sequence

h	3*h + 1	(h–1) / 3
1	4	
4	13	1
13	40	4
40	121	13

TABLE 7.1   Continued

h	3*h + 1	(h–1) / 3
121	364	40
364	1093	121
1093	3280	364

There are other approaches to generating the interval sequence; we'll return to this issue later. First, we'll explore how the Shellsort works using Knuth's sequence.

In the sorting algorithm, the sequence-generating formula is first used in a short loop to figure out the initial gap. A value of 1 is used for the first value of h, and the h=h*3+1 formula is applied to generate the sequence 1, 4, 13, 40, 121, 364, and so on. This process ends when the gap is larger than the array. For a 1,000-element array, the seventh number in the sequence, 1,093, is too large. Thus, we begin the sorting process with the sixth-largest number, creating a 364-sort. Then, each time through the outer loop of the sorting routine, we reduce the interval using the inverse of the formula previously given:

h = (h–1) / 3

This is shown in the third column of Table 7.1. This inverse formula generates the reverse sequence 364, 121, 40, 13, 4, 1. Starting with 364, each of these numbers is used to n-sort the array. When the array has been 1-sorted, the algorithm is done.

The Shellsort Workshop Applet

You can use the Shellsort Workshop applet to see how this sort works. Figure 7.3 shows the applet after all the bars have been 4-sorted, just as the 1-sort begins.

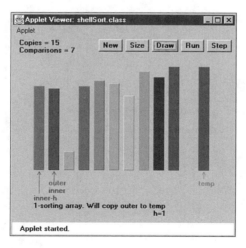

FIGURE 7.3   The Shellsort Workshop applet.

As you single-step through the algorithm, you'll notice that the explanation we gave in the preceding discussion is slightly simplified. The sequence for the 4-sort is not actually (0,4,8), (1,5,9), (2,6), and (3,7). Instead, the first two elements of each group of three are sorted first, then the first two elements of the second group, and so on. Once the first two elements of all the groups are sorted, the algorithm returns and sorts three-element groups. The actual sequence is (0,4), (1,5), (2,6), (3,7), (0,4,8), (1,5,9).

It might seem more obvious for the algorithm to 4-sort each complete subarray first—(0,4), (0,4,8), (1,5), (1,5,9), (2,6), (3,7)—but the algorithm handles the array indices more efficiently using the first scheme.

The Shellsort is actually not very efficient with only 10 items, making almost as many swaps and comparisons as the insertion sort. However, with 100 bars the improvement becomes significant.

It's instructive to run the Workshop applet starting with 100 inversely sorted bars. (Remember that, as in Chapter 3, the first press of New creates a random sequence of bars, while the second press creates an inversely sorted sequence.) Figure 7.4 shows how the bars look after the first pass, when the array has been completely 40-sorted. Figure 7.5 shows the situation after the next pass, when it is 13-sorted. With each new value of h, the array becomes more nearly sorted.

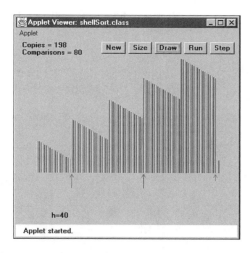

FIGURE 7.4   After the 40-sort.

Why is the Shellsort so much faster than the insertion sort, on which it's based? When h is large, the number of items per pass is small, and items move long distances. This is very efficient. As h grows smaller, the number of items per pass increases, but the items are already closer to their final sorted positions, which is more efficient for the insertion sort. It's the combination of these trends that makes the Shellsort so effective.

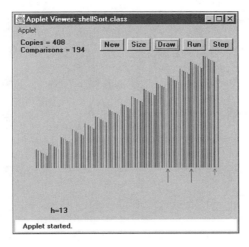

FIGURE 7.5   After the 13-sort.

Notice that later sorts (small values of h) don't undo the work of earlier sorts (large values of h). An array that has been 40-sorted remains 40-sorted after a 13-sort, for example. If this wasn't so, the Shellsort couldn't work.

Java Code for the Shellsort

The Java code for the Shellsort is scarcely more complicated than for the insertion sort. Starting with the insertion sort, you substitute h for 1 in appropriate places and add the formula to generate the interval sequence. We've made shellSort() a method in the ArraySh class, a version of the array classes from Chapter 2, "Arrays." Listing 7.1 shows the complete shellSort.java program.

LISTING 7.1   The shellSort.java Program

```
// shellSort.java
// demonstrates shell sort
// to run this program: C>java ShellSortApp
//--------------------------------------------------------------
class ArraySh
   {
   private long[] theArray;        // ref to array theArray
   private int nElems;             // number of data items
//--------------------------------------------------------------
```

LISTING 7.1   Continued

```
    public ArraySh(int max)          // constructor
       {
       theArray = new long[max];     // create the array
       nElems = 0;                   // no items yet
       }
//------------------------------------------------------------
    public void insert(long value)   // put element into array
       {
       theArray[nElems] = value;     // insert it
       nElems++;                     // increment size
       }
//------------------------------------------------------------
    public void display()            // displays array contents
       {
       System.out.print("A=");
       for(int j=0; j<nElems; j++)    // for each element,
          System.out.print(theArray[j] + " ");   // display it
       System.out.println("");
       }
//------------------------------------------------------------
    public void shellSort()
       {
       int inner, outer;
       long temp;

       int h = 1;                    // find initial value of h
       while(h <= nElems/3)
          h = h*3 + 1;               // (1, 4, 13, 40, 121, ...)

       while(h>0)                    // decreasing h, until h=1
          {
                                     // h-sort the file
          for(outer=h; outer<nElems; outer++)
             {
             temp = theArray[outer];
             inner = outer;
                                     // one subpass (eg 0, 4, 8)
             while(inner > h-1 && theArray[inner-h] >=  temp)
                {
                theArray[inner] = theArray[inner-h];
                inner -= h;
```

LISTING 7.1   Continued

```
                    }
                theArray[inner] = temp;
                }  // end for
            h = (h-1) / 3;                   // decrease h
            }  // end while(h>0)
        }  // end shellSort()
//-------------------------------------------------------------
    }  // end class ArraySh
/////////////////////////////////////////////////////////////////
class ShellSortApp
    {
    public static void main(String[] args)
        {
        int maxSize = 10;              // array size
        ArraySh arr;
        arr = new ArraySh(maxSize);    // create the array

        for(int j=0; j<maxSize; j++)   // fill array with
            {                          // random numbers
            long n = (int)(java.lang.Math.random()*99);
            arr.insert(n);
            }
        arr.display();                 // display unsorted array
        arr.shellSort();               // shell sort the array
        arr.display();                 // display sorted array
        }  // end main()
    }  // end class ShellSortApp
```

In `main()` we create an object of type `ArraySh`, able to hold 10 items, fill it with random data, display it, Shellsort it, and display it again. Here's some sample output:

```
A=20 89 6 42 55 59 41 69 75 66
A=6 20 41 42 55 59 66 69 75 89
```

You can change `maxSize` to higher numbers, but don't go too high; 10,000 items take a fraction of a minute to sort.

The Shellsort algorithm, although it's implemented in just a few lines, is not simple to follow. To see the details of its operation, step through a 10-item sort with the Workshop applet, comparing the messages generated by the applet with the code in the `shellSort()` method.

Other Interval Sequences

Picking an interval sequence is a bit of a black art. Our discussion so far used the formula h=h*3+1 to generate the interval sequence, but other interval sequences have been used with varying degrees of success. The only absolute requirement is that the diminishing sequence ends with 1, so the last pass is a normal insertion sort.

In Shell's original paper, he suggested an initial gap of N/2, which was simply divided in half for each pass. Thus, the descending sequence for N=100 is 50, 25, 12, 6, 3, 1. This approach has the advantage that you don't need to calculate the sequence before the sort begins to find the initial gap; you just divide N by 2. However, this turns out not to be the best sequence. Although it's still better than the insertion sort for most data, it sometimes degenerates to $O(N^2)$ running time, which is no better than the insertion sort.

A variation of this approach is to divide each interval by 2.2 instead of 2. For n=100 this leads to 45, 20, 9, 4, 1. This is considerably better than dividing by 2, as it avoids some worst-case circumstances that lead to $O(N^2)$ behavior. Some extra code is needed to ensure that the last value in the sequence is 1, no matter what N is. This gives results comparable to Knuth's sequence shown in the listing.

Another possibility for a descending sequence (from Flamig; see Appendix B) is

```
if(h < 5)
    h = 1;
else
    h = (5*h-1) / 11;
```

It's generally considered important that the numbers in the interval sequence are relatively prime; that is, they have no common divisors except 1. This constraint makes it more likely that each pass will intermingle all the items sorted on the previous pass. The inefficiency of Shell's original N/2 sequence is due to its failure to adhere to this rule.

You may be able to invent a gap sequence of your own that does just as well (or possibly even better) than those shown. Whatever it is, it should be quick to calculate so as not to slow down the algorithm.

Efficiency of the Shellsort

No one so far has been able to analyze the Shellsort's efficiency theoretically, except in special cases. Based on experiments, there are various estimates, which range from $O(N^{3/2})$ down to $O(N^{7/6})$.

Table 7.2 shows some of these estimated O() values, compared with the slower insertion sort and the faster quicksort. The theoretical times corresponding to various values of N are shown. Note that $N^{x/y}$ means the yth root of N raised to the x power. Thus, if N is 100, $N^{3/2}$ is the square root of 100^3, which is 1,000. Also, $(\log N)^2$ means the log of N, squared. This is often written $\log^2 N$, but that's easy to confuse with $\log_2 N$, the logarithm to the base 2 of N.

TABLE 7.2 Estimates of Shellsort Running Time

O() Value	Type of Sort	10 Items	100 Items	1,000 Items	10,000 Items
N^2	Insertion, etc.	100	10,000	1,000,000	100,000,000
$N^{3/2}$	Shellsort	32	1,000	32,000	1,000,000
$N*(\log N)^2$	Shellsort	10	400	9,000	160,000
$N^{5/4}$	Shellsort	18	316	5,600	100,000
$N^{7/6}$	Shellsort	14	215	3,200	46,000
$N*\log N$	Quicksort, etc.	10	200	3,000	40,000

For most data, the higher estimates, such as $N^{3/2}$, are probably more realistic.

Partitioning

Partitioning is the underlying mechanism of quicksort, which we'll explore next, but it's also a useful operation on its own, so we'll cover it here in its own section.

To *partition* data is to divide it into two groups, so that all the items with a key value higher than a specified amount are in one group, and all the items with a lower key value are in another.

You can easily imagine situations in which you would want to partition data. Maybe you want to divide your personnel records into two groups: employees who live within 15 miles of the office and those who live farther away. Or a school administrator might want to divide students into those with grade point averages higher and lower than 3.5, so as to know who deserves to be on the Dean's list.

The Partition Workshop Applet

Our Partition Workshop applet demonstrates the partitioning process. Figure 7.6 shows 12 bars before partitioning, and Figure 7.7 shows them again after partitioning.

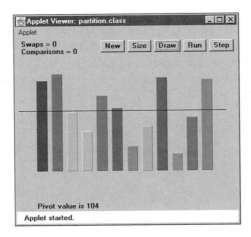

FIGURE 7.6   Twelve bars before partitioning.

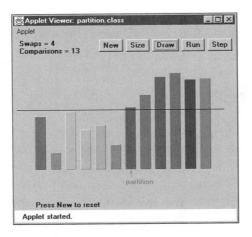

FIGURE 7.7   Twelve bars after partitioning.

The horizontal line represents the *pivot value*, which is the value used to determine into which of the two groups an item is placed. Items with a key value less than the pivot value go in the left part of the array, and those with a greater (or equal) key go in the right part. (In the section on quicksort, we'll see that the pivot value can be the key value of an actual data item, called the pivot. For now, it's just a number.)

The arrow labeled *partition* points to the leftmost item in the right (higher) subarray. This value is returned from the partitioning method, so it can be used by other methods that need to know where the division is.

For a more vivid display of the partitioning process, set the Partition Workshop applet to 100 bars and press the Run button. The leftScan and rightScan pointers will zip toward each other, swapping bars as they go. When they meet, the partition is complete.

You can choose any value you want for the pivot value, depending on why you're doing the partition (such as choosing a grade point average of 3.5). For variety, the Workshop applet chooses a random number for the pivot value (the horizontal black line) each time New or Size is pressed, but the value is never too far from the average bar height.

After being partitioned, the data is by no means sorted; it has simply been divided into two groups. However, it's more sorted than it was before. As we'll see in the next section, it doesn't take much more trouble to sort it completely.

Notice that partitioning is not *stable*. That is, each group is not in the same order it was originally. In fact, partitioning tends to reverse the order of some of the data in each group.

The partition.java **Program**

How is the partitioning process carried out? Let's look at some example code. Listing 7.2 shows the partition.java program, which includes the partitionIt() method for partitioning an array.

LISTING 7.2   The partition.java Program

```java
// partition.java
// demonstrates partitioning an array
// to run this program: C>java PartitionApp
////////////////////////////////////////////////////////////////
class ArrayPar
   {
   private long[] theArray;        // ref to array theArray
   private int nElems;             // number of data items
//--------------------------------------------------------------
   public ArrayPar(int max)        // constructor
      {
      theArray = new long[max];    // create the array
      nElems = 0;                  // no items yet
      }
//--------------------------------------------------------------
   public void insert(long value)  // put element into array
      {
      theArray[nElems] = value;    // insert it
```

LISTING 7.2   Continued

```
      nElems++;                      // increment size
      }
//-------------------------------------------------------------
   public int size()               // return number of items
      { return nElems; }
//-------------------------------------------------------------
   public void display()           // displays array contents
      {
      System.out.print("A=");
      for(int j=0; j<nElems; j++)   // for each element,
         System.out.print(theArray[j] + " ");  // display it
      System.out.println("");
      }
//-------------------------------------------------------------
    public int partitionIt(int left, int right, long pivot)
      {
      int leftPtr = left - 1;          // right of first elem
      int rightPtr = right + 1;        // left of pivot
      while(true)
         {
         while(leftPtr < right &&      // find bigger item
               theArray[++leftPtr] < pivot)
            ;  // (nop)

         while(rightPtr > left &&      // find smaller item
               theArray[--rightPtr] > pivot)
            ;  // (nop)
         if(leftPtr >= rightPtr)       // if pointers cross,
            break;                     //    partition done
         else                          // not crossed, so
            swap(leftPtr, rightPtr);   //    swap elements
         }  // end while(true)
      return leftPtr;                  // return partition
      }  // end partitionIt()
//-------------------------------------------------------------
   public void swap(int dex1, int dex2)  // swap two elements
      {
      long temp;
      temp = theArray[dex1];           // A into temp
      theArray[dex1] = theArray[dex2]; // B into A
      theArray[dex2] = temp;           // temp into B
```

LISTING 7.2   Continued

```
      } // end swap()
//-------------------------------------------------------------
  } // end class ArrayPar
/////////////////////////////////////////////////////////////////
class PartitionApp
  {
  public static void main(String[] args)
    {
    int maxSize = 16;              // array size
    ArrayPar arr;                  // reference to array
    arr = new ArrayPar(maxSize);   // create the array

    for(int j=0; j<maxSize; j++)   // fill array with
      {                            // random numbers
      long n = (int)(java.lang.Math.random()*199);
      arr.insert(n);
      }
    arr.display();                 // display unsorted array

    long pivot = 99;               // pivot value
    System.out.print("Pivot is " + pivot);
    int size = arr.size();
                                   // partition array
    int partDex = arr.partitionIt(0, size-1, pivot);

    System.out.println(", Partition is at index " + partDex);
    arr.display();                 // display partitioned array
    } // end main()
```

The main() routine creates an ArrayPar object that holds 16 items of type long. The pivot value is fixed at 99. The routine inserts 16 random values into ArrayPar, displays them, partitions them by calling the partitionIt() method, and displays them again. Here's some sample output:

```
A=149 192 47 152 159 195 61 66 17 167 118 64 27 80 30 105
Pivot is 99, partition is at index 8
A=30 80 47 27 64 17 61 66 195 167 118 159 152 192 149 105
```

You can see that the partition is successful: The first eight numbers are all smaller than the pivot value of 99; the last eight are all larger.

Notice that the partitioning process doesn't necessarily divide the array in half as it does in this example; that depends on the pivot value and key values of the data. There may be many more items in one group than in the other.

The Partition Algorithm

The partitioning algorithm works by starting with two pointers, one at each end of the array. (We use the term *pointers* to mean indices that point to array elements, not C++ pointers.) The pointer on the left, leftPtr, moves toward the right, and the one on the right, rightPtr, moves toward the left. Notice that leftPtr and rightPtr in the partition.java program correspond to leftScan and rightScan in the Partition Workshop applet.

Actually, leftPtr is initialized to one position to the left of the first cell, and rightPtr to one position to the right of the last cell, because they will be incremented and decremented, respectively, before they're used.

Stopping and Swapping

When leftPtr encounters a data item smaller than the pivot value, it keeps going because that item is already on the correct side of the array. However, when it encounters an item larger than the pivot value, it stops. Similarly, when rightPtr encounters an item larger than the pivot, it keeps going, but when it finds a smaller item, it also stops. Two inner while loops, the first for leftPtr and the second for rightPtr, control the scanning process. A pointer stops because its while loop exits. Here's a simplified version of the code that scans for out-of-place items:

```
while( theArray[++leftPtr] < pivot )    // find bigger item
    ;  // (nop)
while( theArray[--rightPtr] > pivot )   // find smaller item
    ;  // (nop)
swap(leftPtr, rightPtr);                // swap elements
```

The first while loop exits when an item larger than pivot is found; the second loop exits when an item smaller than pivot is found. When both these loops exit, both leftPtr and rightPtr point to items that are in the wrong sides of the array, so these items are swapped.

After the swap, the two pointers continue on, again stopping at items that are in the wrong side of the array and swapping them. All this activity is nested in an outer while loop, as can be seen in the partitionIt() method in Listing 7.2. When the two pointers eventually meet, the partitioning process is complete and this outer while loop exits.

You can watch the pointers in action when you run the Partition Workshop applet with 100 bars. These pointers, represented by blue arrows, start at opposite ends of the array and move toward each other, stopping and swapping as they go. The bars between them are unpartitioned; those they've already passed over are partitioned. When they meet, the entire array is partitioned.

Handling Unusual Data

If we were sure that there was a data item at the right end of the array that was smaller than the pivot value, and an item at the left end that was larger, the simplified while loops previously shown would work fine. Unfortunately, the algorithm may be called upon to partition data that isn't so well organized.

If all the data is smaller than the pivot value, for example, the leftPtr variable will go all the way across the array, looking in vain for a larger item, and fall off the right end, creating an *array index out of bounds* exception. A similar fate will befall rightPtr if all the data is larger than the pivot value.

To avoid these problems, extra tests must be placed in the while loops to check for the ends of the array: leftPtr<right in the first loop and rightPtr>left in the second. You can see these tests in context in Listing 7.2.

In the section on quicksort, we'll see that a clever pivot-selection process can eliminate these end-of-array tests. Eliminating code from inner loops is always a good idea if you want to make a program run faster.

Delicate Code

The code in the while loops is rather delicate. For example, you might be tempted to remove the increment operators from the inner while loops and use them to replace the nop statements. (Nop refers to a statement consisting only of a semicolon, and means *no operation*). For example, you might try to change this:

```
while(leftPtr < right && theArray[++leftPtr] < pivot)
   ;  // (nop)
```

to this:

```
while(leftPtr < right && theArray[leftPtr] < pivot)
   ++leftPtr;
```

and similarly for the other inner while loop. These changes would make it possible for the initial values of the pointers to be left and right, which is somewhat clearer than left-1 and right+1.

However, these changes result in the pointers being incremented only when the condition is satisfied. The pointers must move in any case, so two extra statements within the outer while loop would be required to bump the pointers. The nop version is the most efficient solution.

Equal Keys

Here's another subtle change you might be tempted to make in the `partitionIt()` code. If you run the `partitionIt()` method on items that are all equal to the pivot value, you will find that every comparison leads to a swap. Swapping items with equal keys seems like a waste of time. The < and > operators that compare `pivot` with the array elements in the `while` loops cause the extra swapping. However, suppose you try to fix this by replacing them with <= and >= operators. This indeed prevents the swapping of equal elements, but it also causes `leftPtr` and `rightPtr` to end up at the ends of the array when the algorithm has finished. As we'll see in the section on quicksort, it's good for the pointers to end up in the middle of the array, and very bad for them to end up at the ends. So if `partitionIt()` is going to be used for quicksort, the < and > operators are the right way to go, even if they cause some unnecessary swapping.

Efficiency of the Partition Algorithm

The partition algorithm runs in O(N) time. It's easy to see why this is so when running the Partition Workshop applet: The two pointers start at opposite ends of the array and move toward each other at a more or less constant rate, stopping and swapping as they go. When they meet, the partition is complete. If there were twice as many items to partition, the pointers would move at the same rate, but they would have twice as many items to compare and swap, so the process would take twice as long. Thus, the running time is proportional to N.

More specifically, for each partition there will be N+1 or N+2 comparisons. Every item will be encountered and used in a comparison by one or the other of the pointers, leading to N comparisons, but the pointers overshoot each other before they find out they've "crossed" or gone beyond each other, so there are one or two extra comparisons before the partition is complete. The number of comparisons is independent of how the data is arranged (except for the uncertainty between one or two extra comparisons at the end of the scan).

The number of swaps, however, does depend on how the data is arranged. If it's inversely ordered, and the pivot value divides the items in half, then every pair of values must be swapped, which is N/2 swaps. (Remember in the Partition Workshop applet that the pivot value is selected randomly, so that the number of swaps for inversely sorted bars won't always be exactly N/2.)

For random data, there will be fewer than N/2 swaps in a partition, even if the pivot value is such that half the bars are shorter and half are taller. This is because some bars will already be in the right place (short bars on the left, tall bars on the right). If the pivot value is higher (or lower) than most of the bars, there will be even fewer swaps because only those few bars that are higher (or lower) than the pivot will need to be swapped. On average, for random data, about half the maximum number of swaps take place.

Although there are fewer swaps than comparisons, they are both proportional to N. Thus, the partitioning process runs in O(N) time. Running the Workshop applet, you can see that for 12 random bars there are about 3 swaps and 14 comparisons, and for 100 random bars there are about 25 swaps and 102 comparisons.

Quicksort

Quicksort is undoubtedly the most popular sorting algorithm, and for good reason: In the majority of situations, it's the fastest, operating in O(N*logN) time. (This is only true for *internal* or in-memory sorting; for sorting data in disk files, other algorithms may be better.) Quicksort was discovered by C.A.R. Hoare in 1962.

To understand quicksort, you should be familiar with the partitioning algorithm described in the preceding section. Basically, the quicksort algorithm operates by partitioning an array into two subarrays and then calling itself recursively to quicksort each of these subarrays. However, there are some embellishments we can make to this basic scheme. They have to do with the selection of the pivot and the sorting of small partitions. We'll examine these refinements after we've looked at a simple version of the main algorithm.

It's difficult to understand *what* quicksort is doing before you understand *how* it does it, so we'll reverse our usual presentation and show the Java code for quicksort before presenting the QuickSort1 Workshop applet.

The Quicksort Algorithm

The code for a basic recursive quicksort method is fairly simple. Here's an example:

```java
public void recQuickSort(int left, int right)
   {
   if(right-left <= 0)           // if size is 1,
      return;                    //    it's already sorted
   else                          // size is 2 or larger
      {
                                          // partition range
      int partition = partitionIt(left, right);
      recQuickSort(left, partition-1);   // sort left side
      recQuickSort(partition+1, right);  // sort right side
      }
   }
```

As you can see, there are three basic steps:

1. Partition the array or subarray into left (smaller keys) and right (larger keys) groups.

2. Call ourselves to sort the left group.

3. Call ourselves again to sort the right group.

After a partition, all the items in the left subarray are smaller than all those on the right. If we then sort the left subarray and sort the right subarray, the entire array will be sorted. How do we sort these subarrays? By calling ourself recursively.

The arguments to the recQuickSort() method determine the left and right ends of the array (or subarray) it's supposed to sort. The method first checks if this array consists of only one element. If so, the array is by definition already sorted, and the method returns immediately. This is the base case in the recursion process.

If the array has two or more cells, the algorithm calls the partitionIt() method, described in the preceding section, to partition it. This method returns the index number of the *partition*: the left element in the right (larger keys) subarray. The partition marks the boundary between the subarrays. This situation is shown in Figure 7.8.

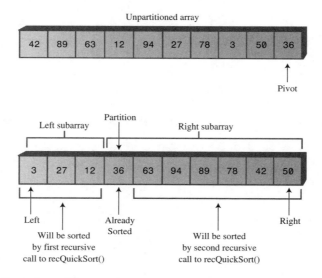

FIGURE 7.8   Recursive calls sort subarrays.

After the array is partitioned, recQuickSort() calls itself recursively, once for the left part of its array, from left to partition-1, and once for the right, from partition+1 to right. Note that the data item at the index partition is not included in either of the recursive calls. Why not? Doesn't it need to be sorted? The explanation lies in how the pivot value is chosen.

Choosing a Pivot Value

What pivot value should the partitionIt() method use? Here are some relevant ideas:

- The pivot value should be the key value of an actual data item; this item is called the *pivot*.

- You can pick a data item to be the pivot more or less at random. For simplicity, let's say we always pick the item on the right end of the subarray being partitioned.

- After the partition, if the pivot is inserted at the boundary between the left and right subarrays, it will be in its final sorted position.

This last point may sound unlikely, but remember that, because the pivot's key value is used to partition the array, following the partition the left subarray holds items smaller than the pivot, and the right subarray holds items larger. The pivot starts out on the right, but if it could somehow be placed between these two subarrays, it would be in the correct place—that is, in its final sorted position. Figure 7.9 shows how this looks with a pivot whose key value is 36.

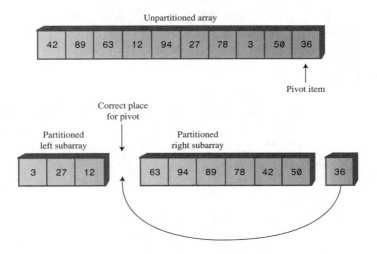

FIGURE 7.9   The pivot and the subarrays.

This figure is somewhat fanciful because you can't actually take an array apart as we've shown. So how do we move the pivot to its proper place?

We could shift all the items in the right subarray to the right one cell to make room for the pivot. However, this is inefficient and unnecessary. Remember that all the items in the right subarray, although they are larger than the pivot, are not yet sorted, so they can be moved around, within the right subarray, without affecting anything. Therefore, to simplify inserting the pivot in its proper place, we can simply swap the pivot (36) and the left item in the right subarray, which is 63. This swap places the pivot in its proper position between the left and right groups. The 63 is switched to the right end, but because it remains in the right (larger) group, the partitioning is undisturbed. This situation is shown in Figure 7.10.

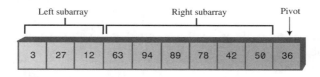

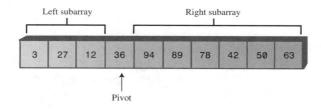

FIGURE 7.10   Swapping the pivot.

When it's swapped into the partition's location, the pivot is in its final resting place. All subsequent activity will take place on one side of it or on the other, but the pivot itself won't be moved (or indeed even accessed) again.

To incorporate the pivot selection process into our recQuickSort() method, let's make it an overt statement, and send the pivot value to partitionIt() as an argument. Here's how that looks:

```
public void recQuickSort(int left, int right)
    {
    if(right-left <= 0)        // if size <= 1,
        return;                //    already sorted
    else                       // size is 2 or larger
        {
        long pivot = theArray[right];     // rightmost item
                                          // partition range
        int partition = partitionIt(left, right, pivot);
```

```
         recQuickSort(left, partition-1);    // sort left side
         recQuickSort(partition+1, right);   // sort right side
         }
      }  // end recQuickSort()
```

When we use this scheme of choosing the rightmost item in the array as the pivot, we'll need to modify the partitionIt() method to exclude this rightmost item from the partitioning process; after all, we already know where it should go after the partitioning process is complete: at the partition, between the two groups. Also, after the partitioning process is completed, we need to swap the pivot from the right end into the partition's location. Listing 7.3 shows the quickSort1.java program, which incorporates these features.

LISTING 7.3   The quickSort1.java Program

```
// quickSort1.java
// demonstrates simple version of quick sort
// to run this program: C>java QuickSort1App
//////////////////////////////////////////////////////////////////
class ArrayIns
   {
   private long[] theArray;         // ref to array theArray
   private int nElems;              // number of data items
//--------------------------------------------------------------
   public ArrayIns(int max)         // constructor
      {
      theArray = new long[max];     // create the array
      nElems = 0;                   // no items yet
      }
//--------------------------------------------------------------
   public void insert(long value)   // put element into array
      {
      theArray[nElems] = value;     // insert it
      nElems++;                     // increment size
      }
//--------------------------------------------------------------
   public void display()            // displays array contents
      {
      System.out.print("A=");
      for(int j=0; j<nElems; j++)    // for each element,
         System.out.print(theArray[j] + " ");  // display it
      System.out.println("");
      }
```

LISTING 7.3   Continued

```
//--------------------------------------------------------------
   public void quickSort()
      {
      recQuickSort(0, nElems-1);
      }
//--------------------------------------------------------------
   public void recQuickSort(int left, int right)
      {
      if(right-left <= 0)               // if size <= 1,
          return;                       //    already sorted
      else                              // size is 2 or larger
         {
         long pivot = theArray[right];      // rightmost item
                                            // partition range
         int partition = partitionIt(left, right, pivot);
         recQuickSort(left, partition-1);   // sort left side
         recQuickSort(partition+1, right);  // sort right side
         }
      }  // end recQuickSort()
//--------------------------------------------------------------
   public int partitionIt(int left, int right, long pivot)
      {
      int leftPtr = left-1;             // left    (after ++)
      int rightPtr = right;             // right-1 (after --)
      while(true)
         {                              // find bigger item
         while( theArray[++leftPtr] < pivot )
           ;  // (nop)
                                        // find smaller item
         while(rightPtr > 0 && theArray[--rightPtr] > pivot)
           ;  // (nop)

         if(leftPtr >= rightPtr)        // if pointers cross,
            break;                      //    partition done
         else                           // not crossed, so
            swap(leftPtr, rightPtr);    //    swap elements
         }  // end while(true)
      swap(leftPtr, right);             // restore pivot
      return leftPtr;                   // return pivot location
      }  // end partitionIt()
//--------------------------------------------------------------
```

LISTING 7.3   Continued

```
   public void swap(int dex1, int dex2)   // swap two elements
      {
      long temp = theArray[dex1];          // A into temp
      theArray[dex1] = theArray[dex2];     // B into A
      theArray[dex2] = temp;               // temp into B
      }  // end swap(
//----------------------------------------------------------------
   }  // end class ArrayIns
/////////////////////////////////////////////////////////////////
class QuickSort1App
   {
   public static void main(String[] args)
      {
      int maxSize = 16;              // array size
      ArrayIns arr;
      arr = new ArrayIns(maxSize);   // create array

      for(int j=0; j<maxSize; j++)   // fill array with
         {                           // random numbers
         long n = (int)(java.lang.Math.random()*99);
         arr.insert(n);
         }
      arr.display();                 // display items
      arr.quickSort();               // quicksort them
      arr.display();                 // display them again
      }  // end main()
   }  // end class QuickSort1App
```

The `main()` routine creates an object of type `ArrayIns`, inserts 16 random data items of type `long` in it, displays it, sorts it with the `quickSort()` method, and displays the results. Here's some typical output:

```
A=69 0 70 6 38 38 24 56 44 26 73 77 30 45 97 65
A=0 6 24 26 30 38 38 44 45 56 65 69 70 73 77 97
```

An interesting aspect of the code in the `partitionIt()` method is that we've been able to remove the test for the end of the array in the first inner `while` loop. This test, seen in the earlier `partitionIt()` method in the `partition.java` program in Listing 7.2, was

```
leftPtr < right
```

It prevented `leftPtr` running off the right end of the array if no item there was larger than `pivot`. Why can we eliminate the test? Because we selected the rightmost item as the pivot, so `leftPtr` will always stop there. However, the test is still necessary for `rightPtr` in the second `while` loop. (Later we'll see how this test can be eliminated as well.)

Choosing the rightmost item as the pivot is thus not an entirely arbitrary choice; it speeds up the code by removing an unnecessary test. Picking the pivot from some other location would not provide this advantage.

The QuickSort1 Workshop Applet

At this point you know enough about the quicksort algorithm to understand the nuances of the QuickSort1 Workshop applet.

The Big Picture

For the big picture, use the Size button to set the applet to sort 100 random bars, and press the Run button. Following the sorting process, the display will look something like Figure 7.11.

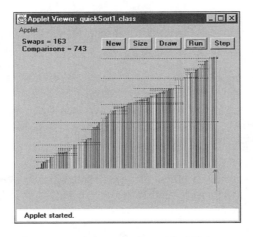

FIGURE 7.11   The QuickSort1 Workshop applet with 100 bars.

Watch how the algorithm partitions the array into two parts, then sorts each of these parts by partitioning it into two parts, and so on, creating smaller and smaller subarrays.

When the sorting process is complete, each dotted line provides a visual record of one of the sorted subarrays. The horizontal range of the line shows which bars were part of the subarray, and its vertical position is the pivot value (the height of the pivot). The total length of all these lines on the display is a measure of how much work the algorithm has done to sort the array; we'll return to this topic later.

Each dotted line (except the shortest ones) should have a line below it (probably separated by other, shorter lines) and a line above it that together add up to the same length as the original line (less one bar). These are the two partitions into which each subarray is divided.

The Details

For a more detailed examination of quicksort's operation, switch to the 12-bar display in the QuickSort1 Workshop applet and step through the sorting process. You'll see how the pivot value corresponds to the height of the pivot on the right side of the array and how the algorithm partitions the array, swaps the pivot into the space between the two sorted groups, sorts the shorter group (using many recursive calls), and then sorts the larger group.

Figure 7.12 shows all the steps involved in sorting 12 bars. The horizontal brackets under the arrays show which subarray is being partitioned at each step, and the circled numbers show the order in which these partitions are created. A pivot being swapped into place is shown with a dotted arrow. The final position of the pivot is shown as a dotted cell to emphasize that this cell contains a sorted item that will not be changed thereafter. Horizontal brackets under single cells (steps 5, 6, 7, 11, and 12) are base case calls to recQuickSort(); they return immediately.

Sometimes, as in steps 4 and 10, the pivot ends up in its original position on the right side of the array being sorted. In this situation, there is only one subarray remaining to be sorted: the one to the left of the pivot. There is no second subarray to its right.

The different steps in Figure 7.12 occur at different levels of recursion, as shown in Table 7.3. The initial call from main() to recQuickSort() is the first level, recQuickSort() calling two new instances of itself is the second level, these two instances calling four more instances is the third level, and so on.

TABLE 7.3 Recursion Levels for Figure 7.12

Step	Recursion Level
1	1
2, 8	2
3, 7, 9, 12	3
4, 10	4
5, 6, 11	5

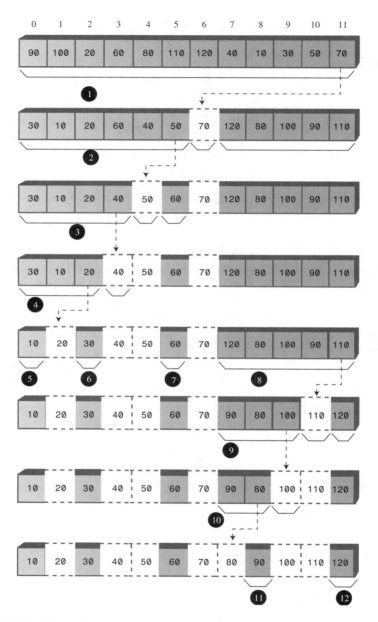

FIGURE 7.12   The quicksort process.

The order in which the partitions are created, corresponding to the step numbers, does not correspond with depth. It's not the case that all the first-level partitions are done first, then all the second level ones, and so on. Instead, the left group at every level is handled before any of the right groups.

In theory there should be 8 steps in the fourth level and 16 in the fifth level, but in this small array we run out of items before these steps are necessary.

The number of levels in the table shows that with 12 data items, the machine stack needs enough space for 5 sets of arguments and return values; one for each recursion level. This is, as we'll see later, somewhat greater than the logarithm to the base 2 of the number of items: $\log_2 N$. The size of the machine stack is determined by your particular system. Sorting very large numbers of data items using recursive procedures may cause this stack to overflow, leading to memory errors.

Things to Notice
Here are some details you may notice as you run the QuickSort1 Workshop applet.

You might think that a powerful algorithm like quicksort would not be able to handle subarrays as small as two or three items. However, this version of the quicksort algorithm is quite capable of sorting such small subarrays; leftScan and rightScan just don't go very far before they meet. For this reason we don't need to use a different sorting scheme for small subarrays. (Although, as we'll see later, handling small subarrays differently may have advantages.)

At the end of each scan, the leftScan variable ends up pointing to the partition— that is, the left element of the right subarray. The pivot is then swapped with the partition to put the pivot in its proper place, as we've seen. As we noted, in steps 3 and 9 of Figure 7.12, leftScan ends up pointing to the pivot itself, so the swap has no effect. This may seem like a wasted swap; you might decide that leftScan should stop one bar sooner. However, it's important that leftScan scan all the way to the pivot; otherwise, a swap would unsort the pivot and the partition.

Be aware that leftScan and rightScan start at left-1 and right. This may look peculiar on the display, especially if left is 0; then leftScan will start at –1. Similarly, rightScan initially points to the pivot, which is not included in the partitioning process. These pointers start outside the subarray being partitioned because they will be incremented and decremented, respectively, before they're used the first time.

The applet shows ranges as numbers in parentheses; for example, (2-5) means the subarray from index 2 to index 5. The range given in some of the messages may be negative: from a higher number to a lower one, such as *Array partitioned; left (7-6), right (8-8)*. The (8-8) range means a single cell (8), but what does (7-6) mean? This range isn't real; it simply reflects the values that left and right, the arguments to recQuickSort(), have when this method is called. Here's the code in question:

```
int partition = partitionIt(left, right, pivot);
recQuickSort(left, partition-1);   // sort left side
recQuickSort(partition+1, right);  // sort right side
```

If `partitionIt()` is called with `left = 7` and `right = 8`, for example, and happens to return 7 as the partition, then the range supplied in the first call to `recQuickSort()` will be (7-6) and the range to the second will be (8-8). This is normal. The base case in `recQuickSort()` is activated by array sizes less than 1 as well as by 1, so it will return immediately for negative ranges. Negative ranges are not shown in Figure 7.12, although they do cause (brief) calls to `recQuickSort()`.

Degenerates to O(N²) Performance

If you use the QuickSort1 Workshop applet to sort 100 inversely sorted bars, you'll see that the algorithm runs much more slowly and that many more dotted horizontal lines are generated, indicating more and larger subarrays are being partitioned. What's happening here?

The problem is in the selection of the pivot. Ideally, the pivot should be the median of the items being sorted. That is, half the items should be larger than the pivot, and half smaller. This would result in the array being partitioned into two subarrays of equal size. Having two equal subarrays is the optimum situation for the quicksort algorithm. If it has to sort one large and one small array, it's less efficient because the larger subarray has to be subdivided more times.

The worst situation results when a subarray with N elements is divided into one subarray with 1 element and the other with N-1 elements. (This division into 1 cell and N-1 cells can also be seen in steps 3 and 9 in Figure 7.12.) If this 1 and N-1 division happens with every partition, then every element requires a separate partition step. This is in fact what takes place with inversely sorted data: In all the subarrays, the pivot is the smallest item, so every partition results in N-1 elements in one subarray and only the pivot in the other.

To see this unfortunate process in action, step through the QuickSort1 Workshop applet with 12 inversely sorted bars. Notice how many more steps are necessary than with random data. In this situation the advantage gained by the partitioning process is lost and the performance of the algorithm degenerates to O(N²).

Besides being slow, there's another potential problem when quicksort operates in O(N²) time. When the number of partitions increases, the number of recursive function calls also increases. Every function call takes up room on the machine stack. If there are too many calls, the machine stack may overflow and paralyze the system.

To summarize: In the QuickSort1 applet, we select the rightmost element as the pivot. If the data is truly random, this isn't too bad a choice because usually the pivot won't be too close to either end of the array. However, when the data is sorted or inversely sorted, choosing the pivot from one end or the other is a bad idea. Can we improve on our approach to selecting the pivot?

Median-of-Three Partitioning

Many schemes have been devised for picking a better pivot. The method should be simple but have a good chance of avoiding the largest or smallest value. Picking an element at random is simple but—as we've seen—doesn't always result in a good selection. However, we could examine all the elements and actually calculate which one was the median. This would be the ideal pivot choice, but the process isn't practical, as it would take more time than the sort itself.

A compromise solution is to find the median of the first, last, and middle elements of the array, and use this for the pivot. Picking the median of the first, last, and middle elements is called the *median-of-three* approach and is shown in Figure 7.13.

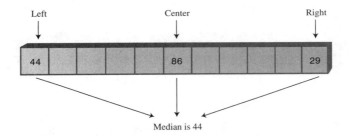

FIGURE 7.13   The median of three.

Finding the median of three items is obviously much faster than finding the median of all the items, and yet it successfully avoids picking the largest or smallest item in cases where the data is already sorted or inversely sorted. There are probably some pathological arrangements of data where the median-of-three scheme works poorly, but normally it's a fast and effective technique for finding the pivot.

Besides picking the pivot more effectively, the median-of-three approach has an additional benefit: We can dispense with the rightPtr>left test in the second inside while loop, leading to a small increase in the algorithm's speed. How is this possible?

The test can be eliminated because we can use the median-of-three approach to not only select the pivot, but also to sort the three elements used in the selection process. Figure 7.14 shows this operation.

When these three elements are sorted, and the median item is selected as the pivot, we are guaranteed that the element at the left end of the subarray is less than (or equal to) the pivot, and the element at the right end is greater than (or equal to) the pivot. This means that the `leftPtr` and `rightPtr` indices can't step beyond the right or left ends of the array, respectively, even if we remove the `leftPtr>right` and `rightPtr<left` tests. (The pointer will stop, thinking it needs to swap the item, only to find that it has crossed the other pointer and the partition is complete.) The values at `left` and `right` act as *sentinels* to keep `leftPtr` and `rightPtr` confined to valid array values.

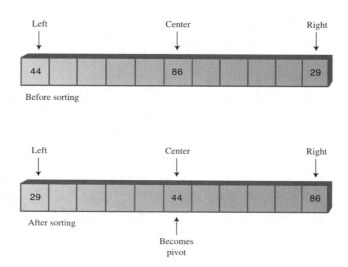

FIGURE 7.14   Sorting the left, center, and right elements.

Another small benefit to median-of-three partitioning is that after the left, center, and right elements are sorted, the partition process doesn't need to examine these elements again. The partition can begin at `left+1` and `right-1` because `left` and `right` have in effect already been partitioned. We know that `left` is in the correct partition because it's on the left and it's less than the pivot, and `right` is in the correct place because it's on the right and it's greater than the pivot.

Thus, median-of-three partitioning not only avoids $O(N^2)$ performance for already-sorted data, it also allows us to speed up the inner loops of the partitioning algorithm and reduce slightly the number of items that must be partitioned.

The `quickSort2.java` Program

Listing 7.4 shows the `quickSort2.java` program, which incorporates median-of-three partitioning. We use a separate method, `medianOf3()`, to sort the left, center, and right elements of a subarray. This method returns the value of the pivot, which is then sent to the `partitionIt()` method.

LISTING 7.4   The quickSort2.java Program

```java
// quickSort2.java
// demonstrates quick sort with median-of-three partitioning
// to run this program: C>java QuickSort2App
////////////////////////////////////////////////////////////////
class ArrayIns
   {
   private long[] theArray;          // ref to array theArray
   private int nElems;               // number of data items
//--------------------------------------------------------------
   public ArrayIns(int max)          // constructor
      {
      theArray = new long[max];      // create the array
      nElems = 0;                    // no items yet
      }
//--------------------------------------------------------------
   public void insert(long value)    // put element into array
      {
      theArray[nElems] = value;      // insert it
      nElems++;                      // increment size
      }
//--------------------------------------------------------------
   public void display()             // displays array contents
      {
      System.out.print("A=");
      for(int j=0; j<nElems; j++)     // for each element,
         System.out.print(theArray[j] + " ");  // display it
      System.out.println("");
      }
//--------------------------------------------------------------
   public void quickSort()
      {
      recQuickSort(0, nElems-1);
      }
//--------------------------------------------------------------
   public void recQuickSort(int left, int right)
      {
      int size = right-left+1;
      if(size <= 3)                  // manual sort if small
         manualSort(left, right);
      else                           // quicksort if large
         {
```

LISTING 7.4   Continued

```
        long median = medianOf3(left, right);
        int partition = partitionIt(left, right, median);
        recQuickSort(left, partition-1);
        recQuickSort(partition+1, right);
        }
      }  // end recQuickSort()
//-------------------------------------------------------------
   public long medianOf3(int left, int right)
      {
      int center = (left+right)/2;
                                          // order left & center
      if( theArray[left] > theArray[center] )
         swap(left, center);
                                          // order left & right
      if( theArray[left] > theArray[right] )
         swap(left, right);
                                          // order center & right
      if( theArray[center] > theArray[right] )
         swap(center, right);

      swap(center, right-1);              // put pivot on right
      return theArray[right-1];           // return median value
      }  // end medianOf3()
//-------------------------------------------------------------
   public void swap(int dex1, int dex2)  // swap two elements
      {
      long temp = theArray[dex1];         // A into temp
      theArray[dex1] = theArray[dex2];    // B into A
      theArray[dex2] = temp;              // temp into B
      }  // end swap(
//-------------------------------------------------------------
    public int partitionIt(int left, int right, long pivot)
      {
      int leftPtr = left;                 // right of first elem
      int rightPtr = right - 1;           // left of pivot

      while(true)
        {
        while( theArray[++leftPtr] < pivot )  // find bigger
           ;                              //     (nop)
        while( theArray[--rightPtr] > pivot ) // find smaller
```

LISTING 7.4   Continued

```
                      ;                          //      (nop)
            if(leftPtr >= rightPtr)      // if pointers cross,
               break;                    //    partition done
            else                         // not crossed, so
               swap(leftPtr, rightPtr);  // swap elements
            }  // end while(true)
         swap(leftPtr, right-1);         // restore pivot
         return leftPtr;                 // return pivot location
         }  // end partitionIt()
//--------------------------------------------------------------
   public void manualSort(int left, int right)
      {
      int size = right-left+1;
      if(size <= 1)
         return;             // no sort necessary
      if(size == 2)
         {                   // 2-sort left and right
         if( theArray[left] > theArray[right] )
            swap(left, right);
         return;
         }
      else                   // size is 3
         {                   // 3-sort left, center, & right
         if( theArray[left] > theArray[right-1] )
            swap(left, right-1);               // left, center
         if( theArray[left] > theArray[right] )
            swap(left, right);                 // left, right
         if( theArray[right-1] > theArray[right] )
            swap(right-1, right);              // center, right
         }
      }  // end manualSort()
//--------------------------------------------------------------
   }  // end class ArrayIns
////////////////////////////////////////////////////////////////
class QuickSort2App
   {
   public static void main(String[] args)
      {
      int maxSize = 16;              // array size
      ArrayIns arr;                  // reference to array
      arr = new ArrayIns(maxSize);   // create the array
```

LISTING 7.4   Continued

```
        for(int j=0; j<maxSize; j++)   // fill array with
           {                           // random numbers
           long n = (int)(java.lang.Math.random()*99);
           arr.insert(n);
           }
        arr.display();                 // display items
        arr.quickSort();               // quicksort them
        arr.display();                 // display them again
        }  // end main()
      }  // end class QuickSort2App
```

This program uses another new method, manualSort(), to sort subarrays of three or fewer elements. It returns immediately if the subarray is one cell (or less), swaps the cells if necessary if the range is 2, and sorts three cells if the range is 3. The recQuickSort() routine can't be used to sort ranges of 2 or 3 because median partitioning requires at least four cells.

The main() routine and the output of quickSort2.java are similar to those of quickSort1.java.

The QuickSort2 Workshop Applet

The Quicksort2 Workshop applet demonstrates the quicksort algorithm using median-of-three partitioning. This applet is similar to the QuickSort1 Workshop applet, but starts off sorting the first, center, and left elements of each subarray and selecting the median of these as the pivot value. At least, it does this if the array size is greater than 3. If the subarray is two or three units, the applet simply sorts it "by hand" without partitioning or recursive calls.

Notice the dramatic improvement in performance when the applet is used to sort 100 inversely ordered bars. No longer is every subarray partitioned into 1 cell and N-1 cells; instead, the subarrays are partitioned roughly in half.

Other than this improvement for ordered data, the QuickSort2 Workshop applet produces results similar to QuickSort1. It is no faster when sorting random data; it's advantages become evident only when sorting ordered data.

Handling Small Partitions

If you use the median-of-three partitioning method, it follows that the quicksort algorithm won't work for partitions of three or fewer items. The number 3 in this case is called a *cutoff* point. In the examples above we sorted subarrays of two or three items by hand. Is this the best way?

Using an Insertion Sort for Small Partitions

Another option for dealing with small partitions is to use the insertion sort. When
you do this, you aren't restricted to a cutoff of 3. You can set the cutoff to 10, 20, or
any other number. It's interesting to experiment with different values of the cutoff to
see where the best performance lies. Knuth (see Appendix B) recommends a cutoff of
9. However, the optimum number depends on your computer, operating system,
compiler (or interpreter), and so on.

The quickSort3.java program, shown in Listing 7.5, uses an insertion sort to handle
subarrays of fewer than 10 cells.

LISTING 7.5   The quickSort3.java Program

```java
// quickSort3.java
// demonstrates quick sort; uses insertion sort for cleanup
// to run this program: C>java QuickSort3App
/////////////////////////////////////////////////////////////////
class ArrayIns
   {
   private long[] theArray;          // ref to array theArray
   private int nElems;               // number of data items
//--------------------------------------------------------------
   public ArrayIns(int max)          // constructor
      {
      theArray = new long[max];      // create the array
      nElems = 0;                    // no items yet
      }
//--------------------------------------------------------------
   public void insert(long value)    // put element into array
      {
      theArray[nElems] = value;      // insert it
      nElems++;                      // increment size
      }
//--------------------------------------------------------------
   public void display()             // displays array contents
      {
      System.out.print("A=");
      for(int j=0; j<nElems; j++)    // for each element,
         System.out.print(theArray[j] + " ");  // display it
      System.out.println("");
      }
//--------------------------------------------------------------
   public void quickSort()
```

LISTING 7.5   Continued

```
       {
    recQuickSort(0, nElems-1);
    // insertionSort(0, nElems-1); // the other option
       }
//-----------------------------------------------------------
   public void recQuickSort(int left, int right)
       {
       int size = right-left+1;
       if(size < 10)                    // insertion sort if small
          insertionSort(left, right);
       else                             // quicksort if large
          {
          long median = medianOf3(left, right);
          int partition = partitionIt(left, right, median);
          recQuickSort(left, partition-1);
          recQuickSort(partition+1, right);
          }
       }  // end recQuickSort()
//-----------------------------------------------------------
   public long medianOf3(int left, int right)
       {
       int center = (left+right)/2;
                                        // order left & center
       if( theArray[left] > theArray[center] )
          swap(left, center);
                                        // order left & right
       if( theArray[left] > theArray[right] )
          swap(left, right);
                                        // order center & right
       if( theArray[center] > theArray[right] )
          swap(center, right);

       swap(center, right-1);           // put pivot on right
       return theArray[right-1];        // return median value
       }  // end medianOf3()
//-----------------------------------------------------------
   public void swap(int dex1, int dex2)  // swap two elements
       {
       long temp = theArray[dex1];       // A into temp
       theArray[dex1] = theArray[dex2];  // B into A
       theArray[dex2] = temp;            // temp into B
```

LISTING 7.5 Continued

```
         } // end swap(
//------------------------------------------------------------
    public int partitionIt(int left, int right, long pivot)
        {
        int leftPtr = left;            // right of first elem
        int rightPtr = right - 1;      // left of pivot
        while(true)
            {
            while( theArray[++leftPtr] < pivot )  // find bigger
                ;                                 // (nop)
            while( theArray[--rightPtr] > pivot ) // find smaller
                ;                                 // (nop)
            if(leftPtr >= rightPtr)    // if pointers cross,
                break;                 //     partition done
            else                       // not crossed, so
                swap(leftPtr, rightPtr); // swap elements
            } // end while(true)
        swap(leftPtr, right-1);        // restore pivot
        return leftPtr;                // return pivot location
        } // end partitionIt()
//------------------------------------------------------------
                                       // insertion sort
   public void insertionSort(int left, int right)
        {
        int in, out;
                                       //  sorted on left of out
        for(out=left+1; out<=right; out++)
            {
            long temp = theArray[out];  // remove marked item
            in = out;                   // start shifts at out
                                        // until one is smaller,
            while(in>left && theArray[in-1] >= temp)
                {
                theArray[in] = theArray[in-1]; // shift item to right
                --in;                   // go left one position
                }
            theArray[in] = temp;        // insert marked item
            } // end for
        } // end insertionSort()
//------------------------------------------------------------
    } // end class ArrayIns
```

LISTING 7.5   Continued

```
///////////////////////////////////////////////////////////
class QuickSort3App
    {
    public static void main(String[] args)
        {
        int maxSize = 16;                // array size
        ArrayIns arr;                    // reference to array
        arr = new ArrayIns(maxSize);     // create the array

        for(int j=0; j<maxSize; j++)     // fill array with
            {                            // random numbers
            long n = (int)(java.lang.Math.random()*99);
            arr.insert(n);
            }
        arr.display();                   // display items
        arr.quickSort();                 // quicksort them
        arr.display();                   // display them again
        }  // end main()
```

Using the insertion sort for small subarrays turns out to be the fastest approach on
our particular installation, but it is not much faster than sorting subarrays of three or
fewer cells by hand, as in quickSort2.java. The numbers of comparisons and copies
are reduced substantially in the quicksort phase, but are increased by an almost
equal amount in the insertion sort, so the time savings are not dramatic. However,
this approach is probably worthwhile if you are trying to squeeze the last ounce of
performance out of quicksort.

Insertion Sort Following Quicksort

Another option is to completely quicksort the array without bothering to sort partitions smaller than the cutoff. This is shown with a commented-out line in the
quickSort() method. (If this call is used, the call to insertionSort() should be
removed from recQuickSort().) When quicksort is finished, the array will be almost
sorted. You then apply the insertion sort to the entire array. The insertion sort is
supposed to operate efficiently on almost-sorted arrays, and this approach is recommended by some experts, but on our installation it runs very slowly. The insertion
sort appears to be happier doing a lot of small sorts than one big one.

Removing Recursion

Another embellishment recommended by many writers is removing recursion from
the quicksort algorithm. This involves rewriting the algorithm to store deferred

subarray bounds (left and right) on a stack, and using a loop instead of recursion to oversee the partitioning of smaller and smaller subarrays. The idea in doing this is to speed up the program by removing method calls. However, this idea arose with older compilers and computer architectures, which imposed a large time penalty for each method call. It's not clear that removing recursion is much of an improvement for modern systems, which handle method calls more efficiently.

Efficiency of Quicksort

We've said that quicksort operates in O(N*logN) time. As we saw in the discussion of mergesort in Chapter 6, this is generally true of the divide-and-conquer algorithms, in which a recursive method divides a range of items into two groups and then calls itself to handle each group. In this situation the logarithm actually has a base of 2: The running time is proportional to $N*\log_2 N$.

You can get an idea of the validity of this $N*\log_2 N$ running time for quicksort by running one of the quickSort Workshop applets with 100 random bars and examining the resulting dotted horizontal lines.

Each dotted line represents an array or subarray being partitioned: the pointers leftScan and rightScan moving toward each other, comparing each data item and swapping when appropriate. We saw in the "Partitioning" section that a single partition runs in O(N) time. This tells us that the total length of all the dotted lines is proportional to the running time of quicksort. But how long are all the lines? Measuring them with a ruler on the screen would be tedious, but we can visualize them a different way.

There is always 1 line that runs the entire width of the graph, spanning N bars. This results from the first partition. There will also be 2 lines (one below and one above the first line) that have an average length of N/2 bars; together they are again N bars long. Then there will be 4 lines with an average length of N/4 that again total N bars, then 8 lines, 16 lines, and so on. Figure 7.15 shows how this looks for 1, 2, 4, and 8 lines.

In this figure solid horizontal lines represent the dotted horizontal lines in the quicksort applets, and captions like *N/4 cells long* indicate average, not actual, line lengths. The circled numbers on the left show the order in which the lines are created.

Each series of lines (the eight N/8 lines, for example) corresponds to a level of recursion. The initial call to recQuickSort() is the first level and makes the first line; the two calls from within the first call—the second level of recursion—make the next two lines; and so on. If we assume we start with 100 cells, the results are shown in Table 7.4.

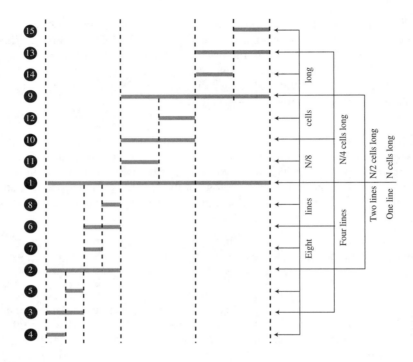

FIGURE 7.15   Lines correspond to partitions.

TABLE 7.4   Line Lengths and Recursion

Recursion Level	Step Numbers in Figure 7.15	Average Line Length (Cells)	Number of Lines	Total Length (Cells)
1	1	100	1	100
2	2, 9	50	2	100
3	3, 6, 10, 13	25	4	100
4	4, 5, 7, 8, 11, 12, 14, 15	12	8	96
5	Not shown	6	16	96
6	Not shown	3	32	96
7	Not shown	1	64	64
				Total = 652

Where does this division process stop? If we keep dividing 100 by 2, and count how many times we do this, we get the series 100, 50, 25, 12, 6, 3, 1, which is about seven levels of recursion. This looks about right on the workshop applets: If you pick some point on the graph and count all the dotted lines directly above and below it, there will be an average of approximately seven. (In Figure 7.15, because not all levels of recursion are shown, only four lines intersect any vertical slice of the graph.)

Table 7.4 shows a total of 652 cells. This is only an approximation because of round-off errors, but it's close to 100 times the logarithm to the base 2 of 100, which is 6.65. Thus, this informal analysis suggests the validity of the $N*\log_2 N$ running time for quicksort.

More specifically, in the section on partitioning, we found that there should be N+2 comparisons and fewer than N/2 swaps. Multiplying these quantities by $\log_2 N$ for various values of N gives the results shown in Table 7.5.

TABLE 7.5   Swaps and Comparisons in Quicksort

N	8	12	16	64	100	128
$\log_2 N$	3	3.59	4	6	6.65	7
$N*\log_2 N$	24	43	64	384	665	896
Comparisons: $(N+2)*\log_2 N$	30	50	72	396	678	910
Swaps: fewer than $N/2*\log_2 N$	12	21	32	192	332	448

The $\log_2 N$ quantity used in Table 7.5 is actually true only in the best-case scenario, where each subarray is partitioned exactly in half. For random data the figure is slightly greater. Nevertheless, the QuickSort1 and QuickSort2 Workshop applets approximate these results for 12 and 100 bars, as you can see by running them and observing the Swaps and Comparisons fields.

Because they have different cutoff points and handle the resulting small partitions differently, QuickSort1 performs fewer swaps but more comparisons than QuickSort2. The number of swaps shown in Table 7.5 is the maximum (which assumes the data is inversely sorted). For random data the actual number of swaps turns out to be one-half to two-thirds of the figures shown.

Radix Sort

We'll close this chapter by briefly mentioning a sort that uses a different approach. The sorts we've looked at so far treat the key as a simple numerical value that is compared with other values to sort the data. The radix sort disassembles the key into digits and arranges the data items according to the value of the digits. Amazingly, no comparisons are necessary.

Algorithm for the Radix Sort

We'll discuss the radix sort in terms of normal base-10 arithmetic, which is easier to visualize. However, an efficient implementation of the radix sort would use base-2 arithmetic to take advantage of the computer's speed in bit manipulation. We'll look at the *radix sort* rather than the similar but somewhat more complex *radix-exchange sort*. The word *radix* means the base of a system of numbers. Ten is the radix of the decimal system and 2 is the radix of the binary system. The sort involves examining each digit of the key separately, starting with the 1s (least significant) digit.

1. All the data items are divided into 10 groups, according to the value of their 1s digit.

2. These 10 groups are then reassembled: All the keys ending with 0 go first, followed by all the keys ending in 1, and so on up to 9. We'll call these steps a sub-sort.

3. In the second sub-sort, all data is divided into 10 groups again, but this time according to the value of their 10s digit. This must be done without changing the order of the previous sort. That is, within each of the 10 groups, the ordering of the items remains the same as it was after step 2; the sub-sorts must be stable.

4. Again the 10 groups are recombined, those with a 10s digit of 0 first, then those with a 10s digit of 1, and so on up to 9.

5. This process is repeated for the remaining digits. If some keys have fewer digits than others, their higher-order digits are considered to be 0.

Here's an example, using seven data items, each with three digits. Leading zeros are shown for clarity.

```
421 240 035 532 305 430 124              // unsorted array
(240 430) (421) (532) (124) (035 305)    // sorted on 1s digit
(305) (421 124) (430 532 035) (240)      // sorted on 10s digit
(035) (124) (240) (305) (421 430) (532)  // sorted on 100s digit
035 124 240 305 421 430 532              // sorted array
```

The parentheses delineate the groups. Within each group the digits in the appropriate position are the same. To convince yourself that this approach really works, try it on a piece of paper with some numbers you make up.

Designing a Program

In practice the original data probably starts out in an ordinary array. Where should the 10 groups go? There's a problem with using another array or an array of 10 arrays. It's not likely there will be exactly the same number of 0s, 1s, 2s, and so on in every digit position, so it's hard to know how big to make the arrays. One way to solve this problem is to use 10 linked lists instead of 10 arrays. Linked lists expand and contract as needed. We'll use this approach.

An outer loop looks at each digit of the keys in turn. There are two inner loops: The first takes the data from the array and puts it on the lists; the second copies it from the lists back to the array. You need to use the right kind of linked list. To keep the sub-sorts stable, you need the data to come out of each list in the same order it went in. Which kind of linked list makes this easy? We'll leave the coding details as an exercise.

Efficiency of the Radix Sort

At first glance the efficiency of the radix sort seems too good to be true. All you do is copy the original data from the array to the lists and back again. If there are 10 data items, this is 20 copies. You repeat this procedure once for each digit. If you assume, say, 5-digit numbers, then you'll have 20*5 equals 100 copies. If you have 100 data items, there are 200*5 equals 1,000 copies. The number of copies is proportional to the number of data items, which is O(N), the most efficient sorting algorithm we've seen.

Unfortunately, it's generally true that if you have more data items, you'll need longer keys. If you have 10 times as much data, you may need to add another digit to the key. The number of copies is proportional to the number of data items times the number of digits in the key. The number of digits is the log of the key values, so in most situations we're back to O(N*logN) efficiency, the same as quicksort.

There are no comparisons, although it takes time to extract each digit from the number. This must be done once for every two copies. It may be, however, that a given computer can do the digit-extraction in binary more quickly than it can do a comparison. Of course, like mergesort, the radix sort uses about twice as much memory as quicksort.

Summary

- The Shellsort applies the insertion sort to widely spaced elements, then less widely spaced elements, and so on.

- The expression *n-sorting* means sorting every nth element.

- A sequence of numbers, called the *interval sequence*, or *gap sequence*, is used to determine the sorting intervals in the Shellsort.

- A widely used interval sequence is generated by the recursive expression h=3*h+1, where the initial value of h is 1.

- If an array holds 1,000 items, it could be 364-sorted, 121-sorted, 40-sorted, 13-sorted, 4-sorted, and finally 1-sorted.

- The Shellsort is hard to analyze, but runs in approximately $O(N*(logN)^2)$ time. This is much faster than the $O(N^2)$ algorithms like insertion sort, but slower than the $O(N*logN)$ algorithms like quicksort.

- To *partition* an array is to divide it into two subarrays, one of which holds items with key values less than a specified value, while the other holds items with keys greater than or equal to this value.

- The *pivot value* is the value that determines into which group an item will go during partitioning. Items smaller than the pivot value go in the left group; larger items go in the right group.

- In the partitioning algorithm, two array indices, each in its own `while` loop, start at opposite ends of the array and step toward each other, looking for items that need to be swapped.

- When an index finds an item that needs to be swapped, its `while` loop exits.

- When both `while` loops exit, the items are swapped.

- When both `while` loops exit, and the indices have met or passed each other, the partition is complete.

- Partitioning operates in linear $O(N)$ time, making N plus 1 or 2 comparisons and fewer than N/2 swaps.

- The partitioning algorithm may require extra tests in its inner `while` loops to prevent the indices running off the ends of the array.

- Quicksort partitions an array and then calls itself twice recursively to sort the two resulting subarrays.

- Subarrays of one element are already sorted; this can be a base case for quicksort.

- The pivot value for a partition in quicksort is the key value of a specific item, called the pivot.

- In a simple version of quicksort, the pivot can always be the item at the right end of the subarray.

- During the partition the pivot is placed out of the way on the right, and is not involved in the partitioning process.

- Later the pivot is swapped again, into the space between the two partitions. This is its final sorted position.

- In the simple version of quicksort, performance is only $O(N^2)$ for already-sorted (or inversely sorted) data.

- In a more advanced version of quicksort, the pivot can be the median of the first, last, and center items in the subarray. This is called *median-of-three* partitioning.

- Median-of-three partitioning effectively eliminates the problem of $O(N^2)$ performance for already-sorted data.

- In median-of-three partitioning, the left, center, and right items are sorted at the same time the median is determined.

- This sort eliminates the need for the end-of-array tests in the inner while loops in the partitioning algorithm.

- Quicksort operates in $O(N*\log_2 N)$ time (except when the simpler version is applied to already-sorted data).

- Subarrays smaller than a certain size (the *cutoff*) can be sorted by a method other than quicksort.

- The insertion sort is commonly used to sort subarrays smaller than the cutoff.

- The insertion sort can also be applied to the entire array, after it has been sorted down to a cutoff point by quicksort.

- The radix sort is about as fast as quicksort but uses twice as much memory.

Questions

These questions are intended as a self-test for readers. Answers may be found in Appendix C.

1. The Shellsort works by

    a. partitioning the array.

    b. swapping adjacent elements.

    c. dealing with widely separated elements.

    d. starting with the normal insertion sort.

2. If an array has 100 elements, then Knuth's algorithm would start with an interval of _____.

3. To transform the insertion sort into the Shellsort, which of the following do you *not* do?

   a. Substitute h for 1.

   b. Insert an algorithm for creating gaps of decreasing width.

   c. Enclose the normal insertion sort in a loop.

   d. Change the direction of the indices in the inner loop.

4. True or false: A good interval sequence for the Shellsort is created by repeatedly dividing the array size in half.

5. Fill in the big O values: The speed of the Shellsort is more than _____ but less than _____.

6. Partitioning is

   a. putting all elements larger than a certain value on one end of the array.

   b. dividing an array in half.

   c. partially sorting parts of an array.

   d. sorting each half of an array separately.

7. When partitioning, each array element is compared to the _____.

8. In partitioning, if an array element is equal to the answer to question 7,

   a. it is passed over.

   b. it is passed over or not, depending on the other array element.

   c. it is placed in the pivot position.

   d. it is swapped.

9. True or false: In quicksort, the pivot can be an arbitrary element of the array.

10. Assuming larger keys on the right, the partition is

    a. the element between the left and right subarrays.

    b. the key value of the element between the left and right subarrays.

    c. the left element in the right subarray.

    d. the key value of the left element in the right subarray.

11. Quicksort involves partitioning the original array and then _____.

12. After a partition in a simple version of quicksort, the pivot may be

    a. used to find the median of the array.

    b. exchanged with an element of the right subarray.

    c. used as the starting point of the next partition.

    d. discarded.

13. Median-of-three partitioning is a way of choosing the _____ .

14. In quicksort, for an array of N elements, the partitionIt() method will examine each element approximately _____ times.

15. True or false: You can speed up quicksort if you stop partitioning when the partition size is 5 and finish by using a different sort.

Experiments

Carrying out these experiments will help to provide insights into the topics covered in the chapter. No programming is involved.

1. Find out what happens when you use the Partition Workshop applet on 100 inversely sorted bars. Is the result almost sorted?

2. Modify the shellSort.java program (Listing 7.1) so it prints the entire contents of the array after completing each n-sort. The array should be small enough so its contents fit on one line. Analyze these intermediate steps to see if the algorithm is operating the way you think should.

3. Modify the shellSort.java (Listing 7.1) and the quickSort3.java (Listing 7.5) programs to sort appropriately large arrays, and compare their speeds. Also, compare these speeds with those of the sorts in Chapter 3.

Programming Projects

Writing programs that solve the Programming Projects helps to solidify your understanding of the material and demonstrates how the chapter's concepts are applied. (As noted in the Introduction, qualified instructors may obtain completed solutions to the Programming Projects on the publisher's Web site.)

7.1 Modify the partition.java program (Listing 7.2) so that the partitionIt() method always uses the highest-index (right) element as the pivot, rather than an arbitrary number. (This is similar to what happens in the quickSort1.java program in Listing 7.3.) Make sure your routine will work for arrays of three or fewer elements. To do so, you may need a few extra statements.

7.2 Modify the `quickSort2.java` program (Listing 7.4) to count the number of copies and comparisons it makes during a sort and then display the totals. This program should duplicate the performance of the QuickSort2 Workshop applet, so the copies and comparisons for inversely sorted data should agree. (Remember that a swap is three copies.)

7.3 In Exercise 3.2 in Chapter 3, we suggested that you could find the median of a set of data by sorting the data and picking the middle element. You might think using quicksort and picking the middle element would be the fastest way to find the median, but there's an even faster way. It uses the partition algorithm to find the median without completely sorting the data.

To see how this works, imagine that you partition the data, and, by chance, the pivot happens to end up at the middle element. You're done! All the items to the right of the pivot are larger (or equal), and all the items to the left are smaller (or equal), so if the pivot falls in the exact center of the array, then it's the median. The pivot won't end up in the center very often, but we can fix that by repartitioning the partition that contains the middle element.

Suppose your array has seven elements numbered from 0 to 6. The middle is element 3. If you partition this array and the pivot ends up at 4, then you need to partition again from 0 to 4 (the partition that contains 3), not 5 to 6. If the pivot ends up at 2, you need to partition from 2 to 6, not 0 to 1. You continue partitioning the appropriate partitions recursively, always checking if the pivot falls on the middle element. Eventually, it will, and you're done. Because you need fewer partitions than in quicksort, this algorithm is faster.

Extend Programming Project 7.1 to find the median of an array. You'll make recursive calls somewhat like those in quicksort, but they will only partition each subarray, not completely sort it. The process stops when the median is found, not when the array is sorted.

7.4 Selection means finding the kth largest or kth smallest element from an array. For example, you might want to select the 7th largest element. Finding the median (as in Programming Project 7.2) is a special case of selection. The same partitioning process can be used, but you look for an element with a specified index number rather than the middle element. Modify the program from Programming Project 7.2 to allow the selection of an arbitrary element. How small an array can your program handle?

7.5 Implement a radix sort as described in the last section of this chapter. It should handle variable amounts of data and variable numbers of digits in the key. You could make the number-base variable as well (so it can be something other than 10), but it will be hard to see what's happening unless you develop a routine to print values in different bases.

8

Binary Trees

In this chapter we switch from algorithms, the focus of Chapter 7, "Advanced Sorting," to data structures. Binary trees are one of the fundamental data storage structures used in programming. They provide advantages that the data structures we've seen so far cannot. In this chapter we'll learn why you would want to use trees, how they work, and how to go about creating them.

Why Use Binary Trees?

Why might you want to use a tree? Usually, because it combines the advantages of two other structures: an ordered array and a linked list. You can search a tree quickly, as you can an ordered array, and you can also insert and delete items quickly, as you can with a linked list. Let's explore these topics a bit before delving into the details of trees.

Slow Insertion in an Ordered Array

Imagine an array in which all the elements are arranged in order—that is, an ordered array, such as we saw in Chapter 2, "Arrays." As we learned, you can quickly search such an array for a particular value, using a binary search. You check in the center of the array; if the object you're looking for is greater than what you find there, you narrow your search to the top half of the array; if it's less, you narrow your search to the bottom half. Applying this process repeatedly finds the object in O(logN) time. You can also quickly iterate through an ordered array, visiting each object in sorted order.

On the other hand, if you want to insert a new object into an ordered array, you first need to find where the object will go, and then move all the objects with greater keys up one space in the array to make room for it. These multiple moves are time-consuming, requiring, on the average, moving half the items (N/2 moves). Deletion involves the same multimove operation and is thus equally slow.

If you're going to be doing a lot of insertions and deletions, an ordered array is a bad choice.

Slow Searching in a Linked List

On the other hand, as we saw in Chapter 5, "Linked Lists," insertions and deletions are quick to perform on a linked list. They are accomplished simply by changing a few references. These operations require O(1) time (the fastest Big O time).

Unfortunately, however, *finding* a specified element in a linked list is not so easy. You must start at the beginning of the list and visit each element until you find the one you're looking for. Thus, you will need to visit an average of N/2 objects, comparing each one's key with the desired value. This process is slow, requiring O(N) time. (Notice that times considered fast for a sort are slow for data structure operations.)

You might think you could speed things up by using an ordered linked list, in which the elements were arranged in order, but this doesn't help. You still must start at the beginning and visit the elements in order, because there's no way to access a given element without following the chain of references to it. (Of course, in an ordered list it's much quicker to visit the nodes in order than it is in a non-ordered list, but that doesn't help to find an arbitrary object.)

Trees to the Rescue

It would be nice if there were a data structure with the quick insertion and deletion of a linked list, and also the quick searching of an ordered array. Trees provide both these characteristics, and are also one of the most interesting data structures.

What Is a Tree?

We'll be mostly interested in a particular kind of tree called a *binary tree*, but let's start by discussing trees in general before moving on to the specifics of binary trees.

A tree consists of *nodes* connected by *edges*. Figure 8.1 shows a tree. In such a picture of a tree (or in our Workshop applet) the nodes are represented as circles, and the edges as lines connecting the circles.

Trees have been studied extensively as abstract mathematical entities, so there's a large amount of theoretical knowledge about them. A tree is actually an instance of a more general category called a *graph*, but we don't need to worry about that here. We'll discuss graphs in Chapter 13, "Graphs," and Chapter 14, "Weighted Graphs."

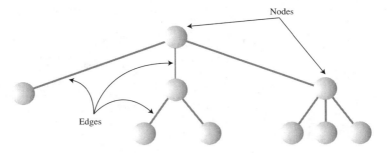

FIGURE 8.1   A general (non-binary) tree.

In computer programs, nodes often represent such entities as people, car parts, airline reservations, and so on—in other words, the typical items we store in any kind of data structure. In an OOP language like Java these real-world entities are represented by objects.

The lines (edges) between the nodes represent the way the nodes are related. Roughly speaking, the lines represent convenience: It's easy (and fast) for a program to get from one node to another if there is a line connecting them. In fact, the *only* way to get from node to node is to follow a path along the lines. Generally, you are restricted to going in one direction along edges: from the root downward.

Edges are likely to be represented in a program by references, if the program is written in Java (or by pointers if the program is written in C or C++).

Typically, there is one node in the top row of a tree, with lines connecting to more nodes on the second row, even more on the third, and so on. Thus, trees are small on the top and large on the bottom. This may seem upside-down compared with real trees, but generally a program starts an operation at the small end of the tree, and it's (arguably) more natural to think about going from top to bottom, as in reading text.

There are different kinds of trees. The tree shown in Figure 8.1 has more than two children per node. (We'll see what "children" means in a moment.) However, in this chapter we'll be discussing a specialized form of tree called a *binary tree*. Each node in a binary tree has a maximum of two children. More general trees, in which nodes can have more than two children, are called multiway trees. We'll see an example in Chapter 10, "2-3-4 Trees and External Storage."

Tree Terminology

Many terms are used to describe particular aspects of trees. You need to know a few of them so our discussion will be comprehensible. Fortunately, most of these terms are related to real-world trees or to family relationships (as in parents and children), so they're not hard to remember. Figure 8.2 shows many of these terms applied to a binary tree.

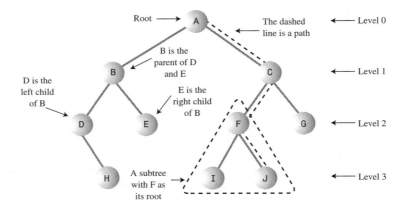

FIGURE 8.2   Tree terms.

Path

Think of someone walking from node to node along the edges that connect them. The resulting sequence of nodes is called a *path*.

Root

The node at the top of the tree is called the *root*. There is only one root in a tree. For a collection of nodes and edges to be defined as a tree, there must be one (and only one!) path from the root to any other node. Figure 8.3 shows a non-tree. You can see that it violates this rule.

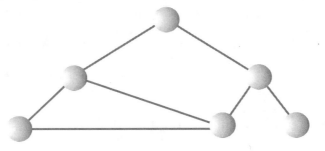

FIGURE 8.3   A non-tree.

Parent

Any node (except the root) has exactly one edge running upward to another node. The node above it is called the *parent* of the node.

Child

Any node may have one or more lines running downward to other nodes. These nodes below a given node are called its *children*.

Leaf

A node that has no children is called a *leaf node* or simply a *leaf*. There can be only one root in a tree, but there can be many leaves.

Subtree

Any node may be considered to be the root of a *subtree*, which consists of its children, and its children's children, and so on. If you think in terms of families, a node's subtree contains all its descendants.

Visiting

A node is *visited* when program control arrives at the node, usually for the purpose of carrying out some operation on the node, such as checking the value of one of its data fields or displaying it. Merely passing over a node on the path from one node to another is not considered to be visiting the node.

Traversing

To *traverse* a tree means to visit all the nodes in some specified order. For example, you might visit all the nodes in order of ascending key value. There are other ways to traverse a tree, as we'll see later.

Levels

The *level* of a particular node refers to how many generations the node is from the root. If we assume the root is Level 0, then its children will be Level 1, its grandchildren will be Level 2, and so on.

Keys

We've seen that one data field in an object is usually designated a *key value*. This value is used to search for the item or perform other operations on it. In tree diagrams, when a circle represents a node holding a data item, the key value of the item is typically shown in the circle. (We'll see many figures later on that show nodes containing keys.)

Binary Trees

If every node in a tree can have at most two children, the tree is called a *binary tree*. In this chapter we'll focus on binary trees because they are the simplest and the most common.

The two children of each node in a binary tree are called the *left child* and the *right child*, corresponding to their positions when you draw a picture of a tree, as shown in Figure 8.2. A node in a binary tree doesn't necessarily have the maximum of two children; it may have only a left child, or only a right child, or it can have no children at all (in which case it's a leaf).

The kind of binary tree we'll be dealing with in this discussion is technically called a *binary search tree*. Figure 8.4 shows a binary search tree.

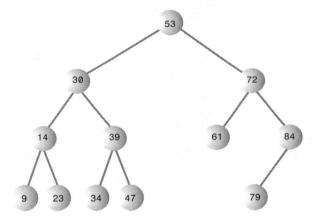

FIGURE 8.4   A binary search tree.

> **NOTE**
>
> The defining characteristic of a binary search tree is this: A node's left child must have a key less than its parent, and a node's right child must have a key greater than or equal to its parent.

An Analogy

One commonly encountered tree is the hierarchical file structure in a computer system. The root directory of a given device (designated with the backslash, as in C:\, on many systems) is the tree's root. The directories one level below the root directory are its children. There may be many levels of subdirectories. Files represent leaves; they have no children of their own.

Clearly, a hierarchical file structure is not a binary tree, because a directory may have many children. A complete pathname, such as C:\SALES\EAST\NOVEMBER\SMITH.DAT, corresponds to the path from the root to the SMITH.DAT leaf. Terms used for the file structure, such as *root* and *path*, were borrowed from tree theory.

A hierarchical file structure differs in a significant way from the trees we'll be discussing here. In the file structure, subdirectories contain no data; they contain only references to other subdirectories or to files. Only files contain data. In a tree, every node contains data (a personnel record, car-part specifications, or whatever). In addition to the data, all nodes except leaves contain references to other nodes.

How Do Binary Search Trees Work?

Let's see how to carry out the common binary tree operations of finding a node with a given key, inserting a new node, traversing the tree, and deleting a node. For each of these operations we'll first show how to use the Binary Tree Workshop applet to carry it out; then we'll look at the corresponding Java code.

The Binary Tree Workshop Applet

Start up the Binary Tree Workshop applet. You'll see a screen something like that shown in Figure 8.5. However, because the tree in the Workshop applet is randomly generated, it won't look exactly the same as the tree in the figure.

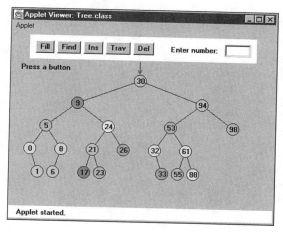

FIGURE 8.5   The Binary Tree Workshop applet.

Using the Workshop Applet

The key values shown in the nodes range from 0 to 99. Of course, in a real tree, there would probably be a larger range of key values. For example, if employees' Social Security numbers were used for key values, they would range up to 999,999,999.

Another difference between the Workshop applet and a real tree is that the Workshop applet is limited to a depth of five; that is, there can be no more than five levels from the root to the bottom. This restriction ensures that all the nodes in the tree will be visible on the screen. In a real tree the number of levels is unlimited (until you run out of memory).

Using the Workshop applet, you can create a new tree whenever you want. To do this, click the Fill button. A prompt will ask you to enter the number of nodes in the tree. This can vary from 1 to 31, but 15 will give you a representative tree. After typing in the number, press Fill twice more to generate the new tree. You can experiment by creating trees with different numbers of nodes.

Unbalanced Trees

Notice that some of the trees you generate are *unbalanced*; that is, they have most of their nodes on one side of the root or the other, as shown in Figure 8.6. Individual subtrees may also be unbalanced.

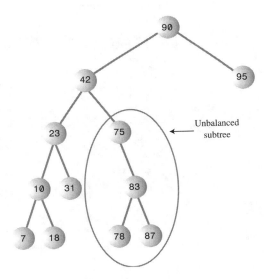

FIGURE 8.6   An unbalanced tree (with an unbalanced subtree).

Trees become unbalanced because of the order in which the data items are inserted. If these key values are inserted randomly, the tree will be more or less balanced. However, if an ascending sequence (like 11, 18, 33, 42, 65, and so on) or a descending sequence is generated, all the values will be right children (if ascending) or left children (if descending) and the tree will be unbalanced. The key values in the Workshop applet are generated randomly, but of course some short ascending or descending sequences will be created anyway, which will lead to local imbalances. When you learn how to insert items into the tree in the Workshop applet, you can try building up a tree by inserting such an ordered sequence of items and see what happens.

If you ask for a large number of nodes when you use Fill to create a tree, you may not get as many nodes as you requested. Depending on how unbalanced the tree becomes, some branches may not be able to hold a full number of nodes. This is because the depth of the applet's tree is limited to five; the problem would not arise in a real tree.

If a tree is created by data items whose key values arrive in random order, the problem of unbalanced trees may not be too much of a problem for larger trees because the chances of a long run of numbers in sequence is small. But key values can arrive in strict sequence; for example, when a data-entry person arranges a stack of personnel files into order of ascending employee number before entering the data. When this happens, tree efficiency can be seriously degraded. We'll discuss unbalanced trees and what to do about them in Chapter 9, "Red-Black Trees."

Representing the Tree in Java Code

Let's see how we might implement a binary tree in Java. As with other data structures, there are several approaches to representing a tree in the computer's memory. The most common is to store the nodes at unrelated locations in memory, and connect them using references in each node that point to its children.

You can also represent a tree in memory as an array, with nodes in specific positions stored in corresponding positions in the array. We'll return to this possibility at the end of this chapter. For our sample Java code we'll use the approach of connecting the nodes using references.

NOTE

As we discuss individual operations, we'll show code fragments pertaining to that operation. The complete program from which these fragments are extracted can be seen toward the end of this chapter in Listing 8.1.

The Node **Class**

First, we need a class of node objects. These objects contain the data representing the objects being stored (employees in an employee database, for example) and also references to each of the node's two children. Here's how that looks:

```
class Node
    {
    int iData;                  // data used as key value
    double fData;               // other data
    node leftChild;             // this node's left child
    node rightChild;            // this node's right child

    public void displayNode()
        {
        // (see Listing 8.1 for method body)
        }
    }
```

Some programmers also include a reference to the node's parent. This simplifies some operations but complicates others, so we don't include it. We do include a method called displayNode() to display the node's data, but its code isn't relevant here.

There are other approaches to designing class Node. Instead of placing the data items directly into the node, you could use a reference to an object representing the data item:

```
class Node
    {
    person p1;                  // reference to person object
    node leftChild;             // this node's left child
    node rightChild;            // this node's right child
    }

class person
    {
    int iData;
    double fData;
    }
```

This approach makes it conceptually clearer that the node and the data item it holds aren't the same thing, but it results in somewhat more complicated code, so we'll stick to the first approach.

The Tree **Class**

We'll also need a class from which to instantiate the tree itself: the object that holds all the nodes. We'll call this class Tree. It has only one field: a Node variable that holds the root. It doesn't need fields for the other nodes because they are all accessed from the root.

The Tree class has a number of methods. They are used for finding, inserting, and deleting nodes; for different kinds of traverses; and for displaying the tree. Here's a skeleton version:

```
class Tree
   {
   private Node root;            // the only data field in Tree

   public void find(int key)
      {
      }
   public void insert(int id, double dd)
      {
      }
   public void delete(int id)
      {
      }
   // various other methods
   } // end class Tree
```

The TreeApp **Class**

Finally, we need a way to perform operations on the tree. Here's how you might write a class with a main() routine to create a tree, insert three nodes into it, and then search for one of them. We'll call this class TreeApp:

```
class TreeApp
   {
   public static void main(String[] args)
      {
      Tree theTree = new Tree;      // make a tree

      theTree.insert(50, 1.5);      // insert 3 nodes
      theTree.insert(25, 1.7);
      theTree.insert(75, 1.9);

      node found = theTree.find(25); // find node with key 25
      if(found != null)
```

```
        System.out.println("Found the node with key 25");
     else
        System.out.println("Could not find node with key 25");
     }  // end main()
  }  // end class TreeApp
```

TIP

In Listing 8.1 the main() routine also provides a primitive user interface so you can decide
from the keyboard whether you want to insert, find, delete, or perform other operations.

Next we'll look at individual tree operations: finding a node, inserting a node,
traversing the tree, and deleting a node.

Finding a Node

Finding a node with a specific key is the simplest of the major tree operations, so
let's start with that.

Remember that the nodes in a binary search tree correspond to objects containing
information. They could be *person objects*, with an employee number as the key and
also perhaps name, address, telephone number, salary, and other fields. Or they
could represent car parts, with a part number as the key value and fields for quantity
on hand, price, and so on. However, the only characteristics of each node that we
can see in the Workshop applet are a number and a color. A node is created with
these two characteristics, and keeps them throughout its life.

Using the Workshop Applet to Find a Node

Look at the Workshop applet, and pick a node, preferably one near the bottom of
the tree (as far from the root as possible). The number shown in this node is its *key
value*. We're going to demonstrate how the Workshop applet finds the node, given
the key value.

For purposes of this discussion we'll assume you've decided to find the node representing the item with key value 57, as shown in Figure 8.7. Of course, when you run
the Workshop applet, you'll get a different tree and will need to pick a different key
value.

Click the Find button. The prompt will ask for the value of the node to find. Enter 57
(or whatever the number is on the node you chose). Click Find twice more.

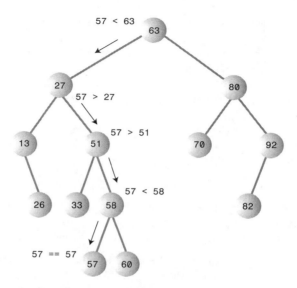

FIGURE 8.7 Finding node 57.

As the Workshop applet looks for the specified node, the prompt will display either `Going to left child` or `Going to right child`, and the red arrow will move down one level to the right or left.

In Figure 8.7 the arrow starts at the root. The program compares the key value 57 with the value at the root, which is 63. The key is less, so the program knows the desired node must be on the left side of the tree—either the root's left child or one of this child's descendants. The left child of the root has the value 27, so the comparison of 57 and 27 will show that the desired node is in the right subtree of 27. The arrow will go to 51, the root of this subtree. Here, 57 is again greater than the 51 node, so we go to the right, to 58, and then to the left, to 57. This time the comparison shows 57 equals the node's key value, so we've found the node we want.

The Workshop applet doesn't do anything with the node after finding it, except to display a message saying it has been found. A serious program would perform some operation on the found node, such as displaying its contents or changing one of its fields.

Java Code for Finding a Node

Here's the code for the `find()` routine, which is a method of the `Tree` class:

```
public Node find(int key)          // find node with given key
   {                               // (assumes non-empty tree)
   Node current = root;                 // start at root
```

```
    while(current.iData != key)       // while no match,
      {
      if(key < current.iData)         // go left?
         current = current.leftChild;
      else
         current = current.rightChild; // or go right?
      if(current == null)             // if no child,
         return null;                 // didn't find it
      }
   return current;                    // found it
   }
```

This routine uses a variable current to hold the node it is currently examining. The argument key is the value to be found. The routine starts at the root. (It has to; this is the only node it can access directly.) That is, it sets current to the root.

Then, in the while loop, it compares the value to be found, key, with the value of the iData field (the key field) in the current node. If key is less than this field, current is set to the node's left child. If key is greater than (or equal) to the node's iData field, current is set to the node's right child.

Can't Find the Node
If current becomes equal to null, we couldn't find the next child node in the sequence; we've reached the end of the line without finding the node we were looking for, so it can't exist. We return null to indicate this fact.

Found the Node
If the condition of the while loop is not satisfied, so that we exit from the bottom of the loop, the iData field of current is equal to key; that is, we've found the node we want. We return the node so that the routine that called find() can access any of the node's data.

Tree Efficiency
As you can see, the time required to find a node depends on how many levels down it is situated. In the Workshop applet there can be up to 31 nodes, but no more than five levels—so you can find any node using a maximum of only five comparisons. This is O(logN) time, or more specifically O($\log_2$N) time, the logarithm to the base 2. We'll discuss this further toward the end of this chapter.

Inserting a Node

To insert a node, we must first find the place to insert it. This is much the same process as trying to find a node that turns out not to exist, as described in the "Can't

Find the Node" section. We follow the path from the root to the appropriate node, which will be the parent of the new node. When this parent is found, the new node is connected as its left or right child, depending on whether the new node's key is less or greater than that of the parent.

Using the Workshop Applet to Insert a Node

To insert a new node with the Workshop applet, press the Ins button. You'll be asked to type the key value of the node to be inserted. Let's assume we're going to insert a new node with the value 45. Type this number into the text field.

The first step for the program in inserting a node is to find where it should be inserted. Figure 8.8a shows how this step looks.

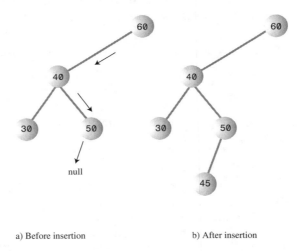

a) Before insertion   b) After insertion

FIGURE 8.8   Inserting a node.

The value 45 is less than 60 but greater than 40, so we arrive at node 50. Now we want to go left because 45 is less than 50, but 50 has no left child; its leftChild field is null. When it sees this null, the insertion routine has found the place to attach the new node. The Workshop applet does this by creating a new node with the value 45 (and a randomly generated color) and connecting it as the left child of 50, as shown in Figure 8.8b.

Java Code for Inserting a Node

The insert() function starts by creating the new node, using the data supplied as arguments.

Next, `insert()` must determine where to insert the new node. This is done using roughly the same code as finding a node, described in the section "Java Code for Finding a Node." The difference is that when you're simply trying to *find* a node and you encounter a `null` (non-existent) node, you know the node you're looking for doesn't exist so you return immediately. When you're trying to *insert* a node, you insert it (creating it first, if necessary) before returning.

The value to be searched for is the data item passed in the argument id. The `while` loop uses `true` as its condition because it doesn't care if it encounters a node with the same value as id; it treats another node with the same key value as if it were simply greater than the key value. (We'll return to the subject of duplicate nodes later in this chapter.)

A place to insert a new node will always be found (unless you run out of memory); when it is, and the new node is attached, the `while` loop exits with a `return` statement.

Here's the code for the `insert()` function:

```
public void insert(int id, double dd)
   {
   Node newNode = new Node();      // make new node
   newNode.iData = id;             // insert data
   newNode.dData = dd;
   if(root==null)                  // no node in root
      root = newNode;
   else                            // root occupied
      {
      Node current = root;         // start at root
      Node parent;
      while(true)                  // (exits internally)
         {
         parent = current;
         if(id < current.iData)    // go left?
            {
            current = current.leftChild;
            if(current == null)    // if end of the line,
               {                   // insert on left
               parent.leftChild = newNode;
               return;
               }
            }  // end if go left
         else                      // or go right?
            {
```

```
            current = current.rightChild;
            if(current == null)  // if end of the line
               {                 // insert on right
               parent.rightChild = newNode;
               return;
               }
            }  // end else go right
         }  // end while
      }  // end else not root
   }  // end insert()
// -----------------------------------------------------------
```

We use a new variable, parent (the parent of current), to remember the last non-null node we encountered (50 in Figure 8.8). This is necessary because current is set to null in the process of discovering that its previous value did not have an appropriate child. If we didn't save parent, we would lose track of where we were.

To insert the new node, change the appropriate child pointer in parent (the last non-null node you encountered) to point to the new node. If you were looking unsuccessfully for parent's left child, you attach the new node as parent's left child; if you were looking for its right child, you attach the new node as its right child. In Figure 8.8, 45 is attached as the left child of 50.

Traversing the Tree

Traversing a tree means visiting each node in a specified order. This process is not as commonly used as finding, inserting, and deleting nodes. One reason for this is that traversal is not particularly fast. But traversing a tree is useful in some circumstances, and it's theoretically interesting. (It's also simpler than deletion, the discussion of which we want to defer as long as possible.)

There are three simple ways to traverse a tree. They're called *preorder*, *inorder*, and *postorder*. The order most commonly used for binary search trees is inorder, so let's look at that first and then return briefly to the other two.

Inorder Traversal

An inorder traversal of a binary search tree will cause all the nodes to be visited in *ascending order*, based on their key values. If you want to create a sorted list of the data in a binary tree, this is one way to do it.

The simplest way to carry out a traversal is the use of recursion (discussed in Chapter 6, "Recursion"). A recursive method to traverse the entire tree is called with a node as an argument. Initially, this node is the root. The method needs to do only three things:

1. Call itself to traverse the node's left subtree.

2. Visit the node.

3. Call itself to traverse the node's right subtree.

Remember that *visiting* a node means doing something to it: displaying it, writing it to a file, or whatever.

Traversals work with any binary tree, not just with binary search trees. The traversal mechanism doesn't pay any attention to the key values of the nodes; it only concerns itself with whether a node has children.

Java Code for Traversing

The actual code for inorder traversal is so simple we show it before seeing how traversal looks in the Workshop applet. The routine, inOrder(), performs the three steps already described. The visit to the node consists of displaying the contents of the node. Like any recursive function, it must have a base case—the condition that causes the routine to return immediately, without calling itself. In inOrder() this happens when the node passed as an argument is null. Here's the code for the inOrder() method:

```
private void inOrder(node localRoot)
    {
    if(localRoot != null)
        {
        inOrder(localRoot.leftChild);

        System.out.print(localRoot.iData + " ");
        inOrder(localRoot.rightChild);
        }
    }
```

This method is initially called with the root as an argument:

```
inOrder(root);
```

After that, it's on its own, calling itself recursively until there are no more nodes to visit.

Traversing a Three-Node Tree

Let's look at a simple example to get an idea of how this recursive traversal routine works. Imagine traversing a tree with only three nodes: a root (A), with a left child (B), and a right child (C), as shown in Figure 8.9.

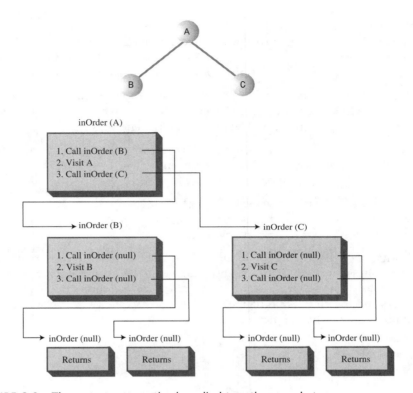

FIGURE 8.9   The `inOrder()` method applied to a three-node tree.

We start by calling `inOrder()` with the root A as an argument. This incarnation of `inOrder()` we'll call `inOrder(A)`. `inOrder(A)` first calls `inOrder()` with its left child, B, as an argument. This second incarnation of `inOrder()` we'll call `inOrder(B)`.

`inOrder(B)` now calls itself with its left child as an argument. However, it has no left child, so this argument is `null`. This creates an invocation of `inorder()` we could call `inOrder(null)`. There are now three instances of `inOrder()` in existence: `inOrder(A)`, `inOrder(B)`, and `inOrder(null)`. However, `inOrder(null)` returns immediately when it finds its argument is `null`. (We all have days like that.)

Now `inOrder(B)` goes on to visit B; we'll assume this means to display it. Then `inOrder(B)` calls `inOrder()` again, with its right child as an argument. Again this argument is `null`, so the second `inorder(null)` returns immediately. Now `inOrder(B)` has carried out steps 1, 2, and 3, so it returns (and thereby ceases to exist).

Now we're back to inOrder(A), just returning from traversing A's left child. We visit A and then call inOrder() again with C as an argument, creating inOrder(C). Like inOrder(B), inOrder(C) has no children, so step 1 returns with no action, step 2 visits C, and step 3 returns with no action. inOrder(B) now returns to inOrder(A).

However, inOrder(A) is now done, so it returns and the entire traversal is complete. The order in which the nodes were visited is A, B, C; they have been visited *inorder*. In a binary search tree this would be the order of ascending keys.

More complex trees are handled similarly. The inOrder() function calls itself for each node, until it has worked its way through the entire tree.

Traversing with the Workshop Applet

To see what a traversal looks like with the Workshop applet, repeatedly press the Trav button. (You don't need to type in any numbers.)

Here's what happens when you use the Tree Workshop applet to traverse inorder the tree shown in Figure 8.10. This is slightly more complex than the three-node tree seen previously. The red arrow starts at the root. Table 8.1 shows the sequence of node keys and the corresponding messages. The key sequence is displayed at the bottom of the Workshop applet screen.

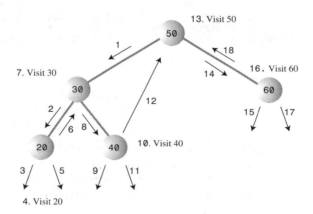

FIGURE 8.10   Traversing a tree inorder.

TABLE 8.1   Workshop Applet Traversal

Step Number	Red Arrow on Node	Message	List of Nodes Visited
1	50 (root)	Will check left child	
2	30	Will check left child	
3	20	Will check left child	
4	20	Will visit this node	
5	20	Will check right child	20
6	20	Will go to root of previous subtree	20
7	30	Will visit this node	20
8	30	Will check right child	20 30
9	40	Will check left child	20 30
10	40	Will visit this node	20 30
11	40	Will check right child	20 30 40
12	40	Will go to root of previous subtree	20 30 40
13	50	Will visit this node	20 30 40
14	50	Will check right child	20 30 40 50
15	60	Will check left child	20 30 40 50
16	60	Will visit this node	20 30 40 50
17	60	Will check right child	20 30 40 50 60
18	60	Will go to root of previous subtree	20 30 40 50 60
19	50	Done traversal	20 30 40 50 60

It may not be obvious, but for each node, the routine traverses the node's left subtree, visits the node, and traverses the right subtree. For example, for node 30 this happens in steps 2, 7, and 8.

The traversal algorithm isn't as complicated as it looks. The best way to get a feel for what's happening is to traverse a variety of different trees with the Workshop applet.

Preorder and Postorder Traversals

You can traverse the tree in two ways besides inorder; they're called preorder and postorder. It's fairly clear why you might want to traverse a tree inorder, but the motivation for preorder and postorder traversals is more obscure. However, these traversals are indeed useful if you're writing programs that *parse* or analyze algebraic expressions. Let's see why that should be true.

A binary tree (not a binary search tree) can be used to represent an algebraic expression that involves the binary arithmetic operators +, –, /, and *. The root node holds an operator, and the other nodes hold either a variable name (like A, B, or C), or another operator. Each subtree is a valid algebraic expression.

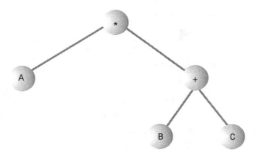

```
Infix: A*(B+C)
Prefix: *A+BC
Postfix: ABC+*
```

FIGURE 8.11   Tree representing an algebraic expression.

For example, the binary tree shown in Figure 8.11 represents the algebraic expression

A*(B+C)

This is called *infix* notation; it's the notation normally used in algebra. (For more on infix and postfix, see the section "Parsing Arithmetic Expressions" in Chapter 4, "Stacks and Queues.") Traversing the tree inorder will generate the correct inorder sequence A*B+C, but you'll need to insert the parentheses yourself.

What does all this have to do with preorder and postorder traversals? Let's see what's involved. For these other traversals the same three steps are used as for inorder, but in a different sequence. Here's the sequence for a preorder() method:

1. Visit the node.

2. Call itself to traverse the node's left subtree.

3. Call itself to traverse the node's right subtree.

Traversing the tree shown in Figure 8.11 using preorder would generate the expression

*A+BC

This is called *prefix* notation. One of the nice things about it is that parentheses are never required; the expression is unambiguous without them. Starting on the left, each operator is applied to the next two things in the expression. For the first operator, *, these two things are A and +BC. For the second operator, +, the two things are B and C, so this last expression is B+C in inorder notation. Inserting that into the original expression *A+BC (preorder) gives us A*(B+C) in inorder. By using different traversals of the tree, we can transform one form of the algebraic expression into another.

The third kind of traversal, postorder, contains the three steps arranged in yet another way:

1. Call itself to traverse the node's left subtree.

2. Call itself to traverse the node's right subtree.

3. Visit the node.

For the tree in Figure 8.11, visiting the nodes with a postorder traversal would generate the expression

ABC+*

This is called *postfix* notation. It means "apply the last operator in the expression, *, to the first and second things." The first thing is A, and the second thing is BC+.

BC+ means "apply the last operator in the expression, +, to the first and second things." The first thing is B and the second thing is C, so this gives us (B+C) in infix. Inserting this in the original expression ABC+* (postfix) gives us A*(B+C) postfix.

> **NOTE**
>
> The code in Listing 8.1 contains methods for preorder and postorder traversals, as well as for inorder.

We won't show the details here, but you can fairly easily construct a tree like that in Figure 8.11 using a postfix expression as input. The approach is analogous to that of evaluating a postfix expression, which we saw in the postfix.java program (Listing 4.8 in Chapter 4). However, instead of storing operands on the stack, we store entire subtrees. We read along the postfix string as we did in postfix.java. Here are the steps when we encounter an operand:

1. Make a tree with one node that holds the operand.

2. Push this tree onto the stack.

Here are the steps when we encounter an operator:

1. Pop two operand trees B and C off the stack.

2. Create a new tree A with the operator in its root.

3. Attach B as the right child of A.

4. Attach C as the left child of A.

5. Push the resulting tree back on the stack.

When you're done evaluating the postfix string, you pop the one remaining item off the stack. Somewhat amazingly, this item is a complete tree depicting the algebraic expression. You can see the prefix and infix representations of the original postfix (and recover the postfix expression) by traversing the tree as we described. We'll leave an implementation of this process as an exercise.

Finding Maximum and Minimum Values

Incidentally, we should note how easy it is to find the maximum and minimum values in a binary search tree. In fact, this process is so easy we don't include it as an option in the Workshop applet, nor show code for it in Listing 8.1. Still, understanding how it works is important.

For the minimum, go to the left child of the root; then go to the left child of that child, and so on, until you come to a node that has no left child. This node is the minimum, as shown in Figure 8.12.

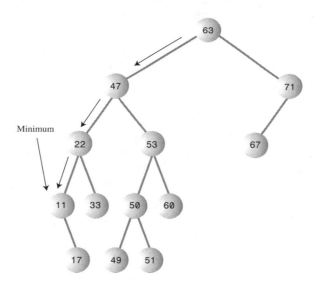

FIGURE 8.12   Minimum value of a tree.

Here's some code that returns the node with the minimum key value:

```
public Node minimum()      // returns node with minimum key value
   {
   Node current, last;
   current = root;                  // start at root
   while(current != null)           // until the bottom,
```

```
        {
    last = current;              // remember node
    current = current.leftChild;  // go to left child
        }
    return last;
    }
```

We'll need to know about finding the minimum value when we set about deleting a node.

For the *maximum* value in the tree, follow the same procedure, but go from right child to right child until you find a node with no right child. This node is the maximum. The code is the same except that the last statement in the loop is

```
current = current.rightChild;  // go to right child
```

Deleting a Node

Deleting a node is the most complicated common operation required for binary search trees. However, deletion is important in many tree applications, and studying the details builds character.

You start by finding the node you want to delete, using the same approach we saw in find() and insert(). When you've found the node, there are three cases to consider:

1. The node to be deleted is a leaf (has no children).

2. The node to be deleted has one child.

3. The node to be deleted has two children.

We'll look at these three cases in turn. The first is easy; the second, almost as easy; and the third, quite complicated.

Case 1: The Node to Be Deleted Has No Children

To delete a leaf node, you simply change the appropriate child field in the node's parent to point to null, instead of to the node. The node will still exist, but it will no longer be part of the tree. This is shown in Figure 8.13.

Because of Java's garbage collection feature, we don't need to worry about explicitly deleting the node itself. When Java realizes that nothing in the program refers to the node, it will be removed from memory. (In C and C++ you would need to execute free() or delete() to remove the node from memory.)

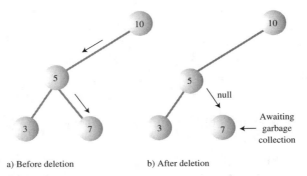

a) Before deletion   b) After deletion

FIGURE 8.13   Deleting a node with no children.

Using the Workshop Applet to Delete a Node with No Children

Assume you're going to delete node 7 in Figure 8.13. Press the Del button and enter 7 when prompted. Again, the node must be found before it can be deleted. Repeatedly pressing Del will take you from 10 to 5 to 7. When the node is found, it's deleted without incident.

Java Code to Delete a Node with No Children

The first part of the delete() routine is similar to find() and insert(). It involves finding the node to be deleted. As with insert(), we need to remember the parent of the node to be deleted so we can modify its child fields. If we find the node, we drop out of the while loop with parent containing the node to be deleted. If we can't find it, we return from delete() with a value of false.

```
public boolean delete(int key) // delete node with given key
   {                           // (assumes non-empty list)
   Node current = root;
   Node parent = root;
   boolean isLeftChild = true;

   while(current.iData != key)      // search for node
      {
      parent = current;
      if(key < current.iData)       // go left?
         {
         isLeftChild = true;
         current = current.leftChild;
         }
      else                          // or go right?
         {
```

```
        isLeftChild = false;
        current = current.rightChild;
        }
    if(current == null)           // end of the line,
        return false;             // didn't find it
    } // end while
// found node to delete
// continues...
    }
```

After we've found the node, we check first to verify that it has no children. When this is true, we check the special case of the root. If that's the node to be deleted, we simply set it to null; this empties the tree. Otherwise, we set the parent's leftChild or rightChild field to null to disconnect the parent from the node.

```
// delete() continued...
// if no children, simply delete it
if(current.leftChild==null &&
                         current.rightChild==null)
    {
    if(current == root)           // if root,
        root = null;              // tree is empty
    else if(isLeftChild)
        parent.leftChild = null;  // disconnect
    else                          // from parent
        parent.rightChild = null;
    }
// continues...
```

Case 2: The Node to Be Deleted Has One Child

This second case isn't so bad either. The node has only two connections: to its parent and to its only child. You want to "snip" the node out of this sequence by connecting its parent directly to its child. This process involves changing the appropriate reference in the parent (leftChild or rightChild) to point to the deleted node's child. This situation is shown in Figure 8.14.

Using the Workshop Applet to Delete a Node with One Child

Let's assume we're using the Workshop applet on the tree in Figure 8.14 and deleting node 71, which has a left child but no right child. Press Del and enter 71 when prompted. Keep pressing Del until the arrow rests on 71. Node 71 has only one child, 63. It doesn't matter whether 63 has children of its own; in this case it has one: 67.

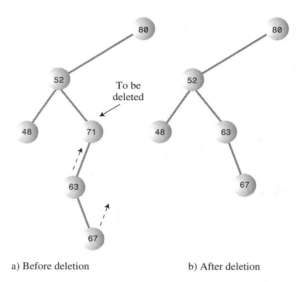

a) Before deletion   b) After deletion

FIGURE 8.14   Deleting a node with one child.

Pressing Del once more causes 71 to be deleted. Its place is taken by its left child, 63. In fact, the entire subtree of which 63 is the root is moved up and plugged in as the new right child of 52.

Use the Workshop applet to generate new trees with one-child nodes, and see what happens when you delete them. Look for the subtree whose root is the deleted node's child. No matter how complicated this subtree is, it's simply moved up and plugged in as the new child of the deleted node's parent.

Java Code to Delete a Node with One Child

The following code shows how to deal with the one-child situation. There are four variations: The child of the node to be deleted may be either a left or right child, and for each of these cases the node to be deleted may be either the left or right child of its parent.

There is also a specialized situation: the node to be deleted may be the root, in which case it has no parent and is simply replaced by the appropriate subtree. Here's the code (which continues from the end of the no-child code fragment shown earlier):

```
// delete() continued...
// if no right child, replace with left subtree
else if(current.rightChild==null)
   if(current == root)
```

```
            root = current.leftChild;
      else if(isLeftChild)           // left child of parent
         parent.leftChild = current.leftChild;
      else                           // right child of parent
         parent.rightChild = current.leftChild;

// if no left child, replace with right subtree
else if(current.leftChild==null)
   if(current == root)
      root = current.rightChild;
   else if(isLeftChild)              // left child of parent
      parent.leftChild = current.rightChild;
   else                              // right child of parent
      parent.rightChild = current.rightChild;
// continued...
```

Notice that working with references makes it easy to move an entire subtree. You do this by simply disconnecting the old reference to the subtree and creating a new reference to it somewhere else. Although there may be lots of nodes in the subtree, you don't need to worry about moving them individually. In fact, they "move" only in the sense of being conceptually in different positions relative to the other nodes. As far as the program is concerned, only the reference to the root of the subtree has changed.

Case 3: The Node to Be Deleted Has Two Children

Now the fun begins. If the deleted node has two children, you can't just replace it with one of these children, at least if the child has its own children. Why not? Examine Figure 8.15, and imagine deleting node 25 and replacing it with its right subtree, whose root is 35. Which left child would 35 have? The deleted node's left child, 15, or the new node's left child, 30? In either case 30 would be in the wrong place, but we can't just throw it away.

We need another approach. The good news is that there's a trick. The bad news is that, even with the trick, there are a lot of special cases to consider. Remember that in a binary search tree the nodes are arranged in order of ascending keys. For each node, the node with the next-highest key is called its *inorder successor*, or simply its successor. In Figure 8.15a, node 30 is the successor of node 25.

Here's the trick: To delete a node with two children, *replace the node with its inorder successor*. Figure 8.16 shows a deleted node being replaced by its successor. Notice that the nodes are still in order. (There's more to it if the successor itself has children; we'll look at that possibility in a moment.)

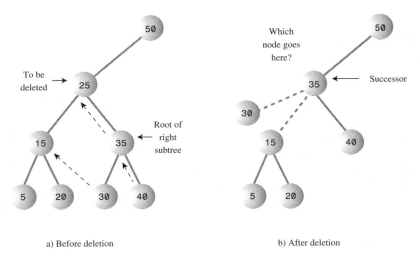

a) Before deletion   b) After deletion

FIGURE 8.15   Cannot replace with subtree.

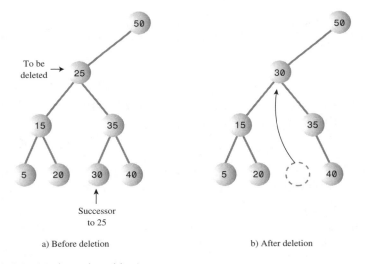

a) Before deletion   b) After deletion

FIGURE 8.16   Node replaced by its successor.

Finding the Successor

How do you find the successor of a node? As a human being, you can do this quickly (for small trees, anyway). Just take a quick glance at the tree and find the next-largest number following the key of the node to be deleted. In Figure 8.16 it doesn't take long to see that the successor of 25 is 30. There's just no other number that is greater than 25 and also smaller than 35. However, the computer can't do things "at a glance"; it needs an algorithm. Here it is:

First, the program goes to the original node's right child, which must have a key larger than the node. Then it goes to this right child's left child (if it has one), and to this left child's left child, and so on, following down the path of left children. The last left child in this path is the successor of the original node, as shown in Figure 8.17.

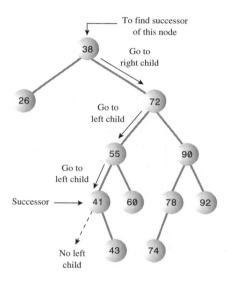

FIGURE 8.17   Finding the successor.

Why does this algorithm work? What we're really looking for is *the smallest of the set of nodes that are larger than the original node.* When you go to the original node's right child, all the nodes in the resulting subtree are greater than the original node because this is how a binary search tree is defined. Now we want the smallest value in this subtree. As we learned, you can find the minimum value in a subtree by following the path down all the left children. Thus, this algorithm finds the minimum value that is greater than the original node; this is what we mean by its successor.

If the right child of the original node has no left children, this right child is itself the successor, as shown in Figure 8.18.

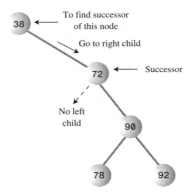

FIGURE 8.18   The right child is the successor.

Using the Workshop Applet to Delete a Node with Two Children
Generate a tree with the Workshop applet, and pick a node with two children. Now
mentally figure out which node is its successor, by going to its right child and then
following down the line of this right child's left children (if it has any). You may
want to make sure the successor has no children of its own. If it does, the situation
gets more complicated because entire subtrees are moved around, rather than a
single node.

After you've chosen a node to delete, click the Del button. You'll be asked for the key
value of the node to delete. When you've specified it, repeated presses of the Del
button will show the red arrow searching down the tree to the designated node.
When the node is deleted, it's replaced by its successor.

Let's assume you use the Workshop applet to delete the node with key 30 from the
example shown earlier in Figure 8.15. The red arrow will go from the root at 50 to
25; then 25 will be replaced by 30.

Java Code to Find the Successor
Here's some code for a method getSuccessor(), which returns the successor of the
node specified as its delNode argument. (This routine assumes that delNode does
indeed have a right child, but we know this is true because we've already determined
that the node to be deleted has two children.)

```
// returns node with next-highest value after delNode
// goes to right child, then right child's left descendants

private node getSuccessor(node delNode)
   {
   Node successorParent = delNode;
```

```
        Node successor = delNode;
        Node current = delNode.rightChild;    // go to right child
        while(current != null)                // until no more
            {                                 // left children,
            successorParent = successor;
            successor = current;
            current = current.leftChild;      // go to left child
            }
                                              // if successor not
        if(successor != delNode.rightChild)   // right child,
            {                                 // make connections
            successorParent.leftChild = successor.rightChild;
            successor.rightChild = delNode.rightChild;
            }
        return successor;
        }
```

The routine first goes to delNode's right child and then, in the while loop, follows down the path of all this right child's left children. When the while loop exits, successor contains delNode's successor.

When we've found the successor, we may need to access its parent, so within the while loop we also keep track of the parent of the current node.

The getSuccessor() routine carries out two additional operations in addition to finding the successor. However, to understand them, we need to step back and consider the big picture.

As we've seen, the successor node can occupy one of two possible positions relative to current, the node to be deleted. The successor can be current's right child, or it can be one of this right child's left descendants. We'll look at these two situations in turn.

Successor Is Right Child of delNode

If successor is the right child of current, things are simplified somewhat because we can simply move the subtree of which successor is the root and plug it in where the deleted node was. This operation requires only two steps:

1. Unplug current from the rightChild field of its parent (or leftChild field if appropriate), and set this field to point to successor.

2. Unplug current's left child from current, and plug it into the leftChild field of successor.

Here are the code statements that carry out these steps, excerpted from `delete()`:

1. `parent.rightChild = successor;`

2. `successor.leftChild = current.leftChild;`

This situation is summarized in Figure 8.19, which shows the connections affected by these two steps.

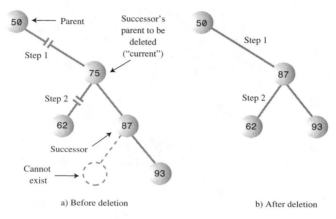

a) Before deletion   b) After deletion

FIGURE 8.19   Deletion when the successor is the right child.

Here's the code in context (a continuation of the `else-if` ladder shown earlier):

```
// delete() continued
else  // two children, so replace with inorder successor
   {
   // get successor of node to delete (current)
   Node successor = getSuccessor(current);

   // connect parent of current to successor instead
   if(current == root)
      root = successor;
   else if(isLeftChild)
      parent.leftChild = successor;
   else
      parent.rightChild = successor;
   // connect successor to current's left child
   successor.leftChild = current.leftChild;
   }  // end else two children
// (successor cannot have a left child)
```

```
return true;
}   // end delete()
```

Notice that this is—finally—the end of the delete() routine. Let's review the code for these two steps:

- Step 1: If the node to be deleted, current, is the root, it has no parent so we merely set the root to the successor. Otherwise, the node to be deleted can be either a left or right child (Figure 8.19 shows it as a right child), so we set the appropriate field in its parent to point to successor. When delete() returns and current goes out of scope, the node referred to by current will have no references to it, so it will be discarded during Java's next garbage collection.

- Step 2: We set the left child of successor to point to current's left child.

What happens if the successor has children of its own? First of all, *a successor node is guaranteed not to have a left child*. This is true whether the successor is the right child of the node to be deleted or one of this right child's left children. How do we know this?

Well, remember that the algorithm we use to determine the successor goes to the right child first and then to any left children of that right child. It stops when it gets to a node with no left child, so the algorithm itself determines that the successor can't have any left children. If it did, that left child would be the successor instead.

You can check this out on the Workshop applet. No matter how many trees you make, you'll never find a situation in which a node's successor has a left child (assuming the original node has two children, which is the situation that leads to all this trouble in the first place).

On the other hand, the successor may very well have a right child. This isn't much of a problem when the successor is the right child of the node to be deleted. When we move the successor, its right subtree simply follows along with it. There's no conflict with the right child of the node being deleted because the successor *is* this right child.

In the next section we'll see that a successor's right child needs more attention if the successor is not the right child of the node to be deleted.

Successor Is Left Descendant of Right Child of delNode

If successor is a left descendant of the right child of the node to be deleted, four steps are required to perform the deletion:

1. Plug the right child of successor into the leftChild field of the successor's parent.

2. Plug the right child of the node to be deleted into the rightChild field of successor.

3. Unplug `current` from the `rightChild` field of its parent, and set this field to point to `successor`.

4. Unplug `current`'s left child from `current`, and plug it into the `leftChild` field of `successor`.

Steps 1 and 2 are handled in the `getSuccessor()` routine, while 3 and 4 are carried out in `delete()`. Figure 8.20 shows the connections affected by these four steps.

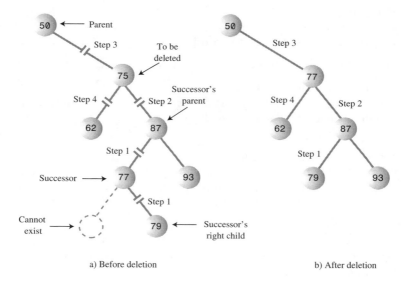

a) Before deletion   b) After deletion

FIGURE 8.20   Deletion when the successor is the left child.

Here's the code for these four steps:

1. `successorParent.leftChild = successor.rightChild;`

2. `successor.rightChild = delNode.rightChild;`

3. `parent.rightChild = successor;`

4. `successor.leftChild = current.leftChild;`

(Step 3 could also refer to the left child of its parent.) The numbers in Figure 8.20 show the connections affected by the four steps. Step 1 in effect *replaces the successor with its right subtree*. Step 2 keeps the right child of the deleted node in its proper place (this happens automatically when the successor is the right child of the deleted node). Steps 1 and 2 are carried out in the `if` statement that ends the `getSuccessor()` method shown earlier. Here's that statement again:

```
                                        // if successor not
if(successor != delNode.rightChild)     // right child,
    {                                   // make connections
    successorParent.leftChild = successor.rightChild;
    successor.rightChild = delNode.rightChild;
    }
```

These steps are more convenient to perform here than in delete(), because in getSuccessor() we can easily figure out where the successor's parent is while we're descending the tree to find the successor.

Steps 3 and 4 we've seen already; they're the same as steps 1 and 2 in the case where the successor is the right child of the node to be deleted, and the code is in the if statement at the end of delete().

Is Deletion Necessary?

If you've come this far, you can see that deletion is fairly involved. In fact, it's so complicated that some programmers try to sidestep it altogether. They add a new Boolean field to the node class, called something like isDeleted. To delete a node, they simply set this field to true. Then other operations, like find(), check this field to be sure the node isn't marked as deleted before working with it. This way, deleting a node doesn't change the structure of the tree. Of course, it also means that memory can fill up with "deleted" nodes.

This approach is a bit of a cop-out, but it may be appropriate where there won't be many deletions in a tree. (If ex-employees remain in the personnel file forever, for example.)

The Efficiency of Binary Trees

As you've seen, most operations with trees involve descending the tree from level to level to find a particular node. How long does it take to do this? In a full tree, about half the nodes are on the bottom level. (More accurately, if it's full, there's one more node on the bottom row than in the rest of the tree.) Thus, about half of all searches or insertions or deletions require finding a node on the lowest level. (An additional quarter of these operations require finding the node on the next-to-lowest level, and so on.)

During a search we need to visit one node on each level. So we can get a good idea how long it takes to carry out these operations by knowing how many levels there are. Assuming a full tree, Table 8.2 shows how many levels are necessary to hold a given number of nodes.

TABLE 8.2   Number of Levels for Specified Number of Nodes

Number of Nodes	Number of Levels
1	1
3	2
7	3
15	4
31	5
...	...
1,023	10
...	...
32,767	15
...	...
1,048,575	20
...	...
33,554,432	25
...	...
1,073,741,824	30

This situation is very much like the ordered array discussed in Chapter 2. In that case, the number of comparisons for a binary search was approximately equal to the base 2 logarithm of the number of cells in the array. Here, if we call the number of nodes in the first column N, and the number of levels in the second column L, we can say that N is 1 less than 2 raised to the power L, or

$$N = 2^L - 1$$

Adding 1 to both sides of the equation, we have

$$N + 1 = 2^L$$

This is equivalent to

$$L = \log_2(N + 1)$$

Thus, the time needed to carry out the common tree operations is proportional to the base 2 log of N. In Big O notation we say such operations take O(logN) time.

If the tree isn't full, analysis is difficult. We can say that for a tree with a given number of levels, average search times will be shorter for the non-full tree than the full tree because fewer searches will proceed to lower levels.

Compare the tree to the other data storage structures we've discussed so far. In an unordered array or a linked list containing 1,000,000 items, finding the item you want takes, on the average, 500,000 comparisons. But in a tree of 1,000,000 items, only 20 (or fewer) comparisons are required.

In an ordered array you can find an item equally quickly, but inserting an item requires, on the average, moving 500,000 items. Inserting an item in a tree with 1,000,000 items requires 20 or fewer comparisons, plus a small amount of time to connect the item.

Similarly, deleting an item from a 1,000,000-item array requires moving an average of 500,000 items, while deleting an item from a 1,000,000-node tree requires 20 or fewer comparisons to find the item, plus (possibly) a few more comparisons to find its successor, plus a short time to disconnect the item and connect its successor.

Thus, a tree provides high efficiency for all the common data storage operations.

Traversing is not as fast as the other operations. However, traversals are probably not very commonly carried out in a typical large database. They're more appropriate when a tree is used as an aid to parsing algebraic or similar expressions, which are probably not too long anyway.

Trees Represented as Arrays

Our code examples are based on the idea that a tree's edges are represented by leftChild and rightChild references in each node. However, there's a completely different way to represent a tree: with an array.

In the array approach, the nodes are stored in an array and are not linked by references. The position of the node in the array corresponds to its position in the tree. The node at index 0 is the root, the node at index 1 is the root's left child, and so on, progressing from left to right along each level of the tree. This is shown in Figure 8.21.

Every position in the tree, whether it represents an existing node or not, corresponds to a cell in the array. Adding a node at a given position in the tree means inserting the node into the equivalent cell in the array. Cells representing tree positions with no nodes are filled with 0 or null.

With this scheme, a node's children and parent can be found by applying some simple arithmetic to the node's index number in the array. If a node's index number is index, this node's left child is

```
2*index + 1
```

its right child is

```
2*index + 2
```

and its parent is

```
(index-1) / 2
```

(where the / character indicates integer division with no remainder). You can check this out by looking at Figure 8.21.

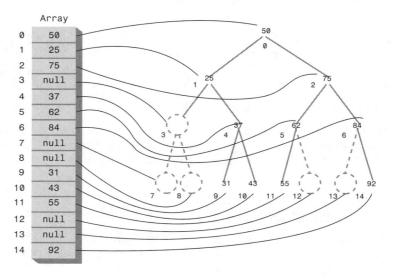

FIGURE 8.21   Tree represented by an array.

In most situations, representing a tree with an array isn't very efficient. Unfilled nodes and deleted nodes leave holes in the array, wasting memory. Even worse, when deletion of a node involves moving subtrees, every node in the subtree must be moved to its new location in the array, which is time-consuming in large trees.

However, if deletions aren't allowed, the array representation may be useful, especially if obtaining memory for each node dynamically is, for some reason, too time-consuming. The array representation may also be useful in special situations. The tree in the Workshop applet, for example, is represented internally as an array to make it easy to map the nodes from the array to fixed locations on the screen display.

Duplicate Keys

As in other data structures, the problem of duplicate keys must be addressed. In the code shown for insert(), and in the Workshop applet, a node with a duplicate key will be inserted as the right child of its twin.

The problem is that the find() routine will find only the first of two (or more) duplicate nodes. The find() routine could be modified to check an additional data item, to distinguish data items even when the keys were the same, but this would be (at least somewhat) time-consuming.

One option is to simply forbid duplicate keys. When duplicate keys are excluded by the nature of the data (employee ID numbers, for example), there's no problem. Otherwise, you need to modify the insert() routine to check for equality during the insertion process, and abort the insertion if a duplicate is found.

The Fill routine in the Workshop applet excludes duplicates when generating the random keys.

The Complete tree.java Program

In this section we'll show the complete program that includes all the methods and code fragments we've looked at so far in this chapter. It also features a primitive user interface. This allows the user to choose an operation (finding, inserting, deleting, traversing, and displaying the tree) by entering characters. The display routine uses character output to generate a picture of the tree. Figure 8.22 shows the display generated by the program.

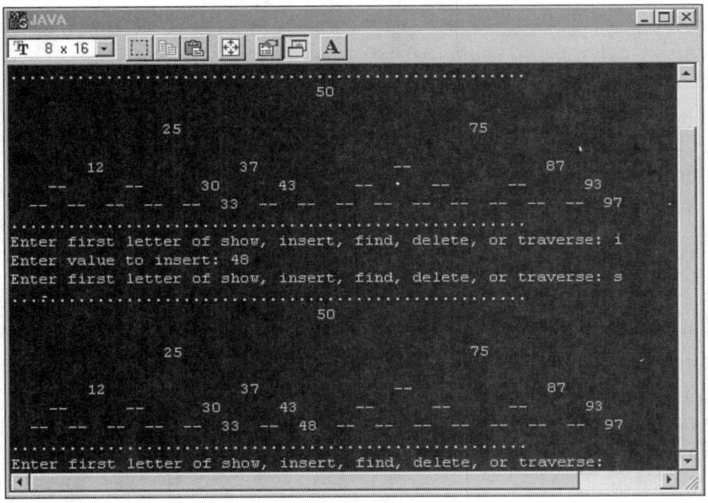

FIGURE 8.22 Output of the tree.java program.

In the figure, the user has typed s to display the tree, then typed i and 48 to insert a node with that value, and then s again to display the tree with the additional node. The 48 appears in the lower display.

The available commands are the characters s, i, f, d, and t, for show, insert, find, delete, and traverse. The i, f, and d options ask for the key value of the node to be operated on. The t option gives you a choice of traversals: 1 for preorder, 2 for inorder, and 3 for postorder. The key values are then displayed in that order.

The display shows the key values arranged in something of a tree shape; however, you'll need to imagine the edges. Two dashes (—) represent a node that doesn't exist at a particular position in the tree. The program initially creates some nodes so the user will have something to see before any insertions are made. You can modify this initialization code to start with any nodes you want, or with no nodes (which is good nodes).

You can experiment with the program in Listing 8.1 as you can with the Workshop applet. Unlike the Workshop applet, however, it doesn't show you the steps involved in carrying out an operation; it does everything at once.

LISTING 8.1   The tree.java Program

```
// tree.java
// demonstrates binary tree
// to run this program: C>java TreeApp
import java.io.*;
import java.util.*;              // for Stack class
////////////////////////////////////////////////////////////
class Node
   {
   public int iData;            // data item (key)
   public double dData;         // data item
   public Node leftChild;       // this node's left child
   public Node rightChild;      // this node's right child

   public void displayNode()    // display ourself
      {
      System.out.print('{');
      System.out.print(iData);
      System.out.print(", ");
      System.out.print(dData);
      System.out.print("} ");
      }
   } // end class Node
////////////////////////////////////////////////////////////
class Tree
   {
   private Node root;           // first node of tree

// -------------------------------------------------------------
   public Tree()                // constructor
      { root = null; }          // no nodes in tree yet
```

LISTING 8.1   Continued

```java
// ------------------------------------------------------------
   public Node find(int key)       // find node with given key
      {                            // (assumes non-empty tree)
      Node current = root;                   // start at root
      while(current.iData != key)            // while no match,
         {
         if(key < current.iData)         // go left?
            current = current.leftChild;
         else                            // or go right?
            current = current.rightChild;
         if(current == null)             // if no child,
            return null;                 // didn't find it
         }
      return current;                    // found it
      }  // end find()
// ------------------------------------------------------------
   public void insert(int id, double dd)
      {
      Node newNode = new Node();       // make new node
      newNode.iData = id;              // insert data
      newNode.dData = dd;
      if(root==null)                   // no node in root
         root = newNode;
      else                             // root occupied
         {
         Node current = root;          // start at root
         Node parent;
         while(true)                   // (exits internally)
            {
            parent = current;
            if(id < current.iData)  // go left?
               {
               current = current.leftChild;
               if(current == null)  // if end of the line,
                  {                 // insert on left
                  parent.leftChild = newNode;
                  return;
                  }
               }  // end if go left
            else                      // or go right?
               {
```

LISTING 8.1   Continued

```
                current = current.rightChild;
                if(current == null)  // if end of the line
                    {                   // insert on right
                    parent.rightChild = newNode;
                    return;
                    }
                }  // end else go right
            }  // end while
        }  // end else not root
    }  // end insert()
// ---------------------------------------------------------------
    public boolean delete(int key) // delete node with given key
        {                          // (assumes non-empty list)
        Node current = root;
        Node parent = root;
        boolean isLeftChild = true;

        while(current.iData != key)      // search for node
            {
            parent = current;
            if(key < current.iData)        // go left?
                {
                isLeftChild = true;
                current = current.leftChild;
                }
            else                          // or go right?
                {
                isLeftChild = false;
                current = current.rightChild;
                }
            if(current == null)           // end of the line,
                return false;             // didn't find it
            }  // end while
        // found node to delete

        // if no children, simply delete it
        if(current.leftChild==null &&
                             current.rightChild==null)
            {
            if(current == root)           // if root,
                root = null;              // tree is empty
```

LISTING 8.1 Continued

```
            else if(isLeftChild)
               parent.leftChild = null;      // disconnect
            else                             // from parent
               parent.rightChild = null;
            }

         // if no right child, replace with left subtree
         else if(current.rightChild==null)
            if(current == root)
               root = current.leftChild;
            else if(isLeftChild)
               parent.leftChild = current.leftChild;
            else
               parent.rightChild = current.leftChild;

         // if no left child, replace with right subtree
         else if(current.leftChild==null)
            if(current == root)
               root = current.rightChild;
            else if(isLeftChild)
               parent.leftChild = current.rightChild;
            else
               parent.rightChild = current.rightChild;

         else   // two children, so replace with inorder successor
            {
            // get successor of node to delete (current)
            Node successor = getSuccessor(current);

            // connect parent of current to successor instead
            if(current == root)
               root = successor;
            else if(isLeftChild)
               parent.leftChild = successor;
            else
               parent.rightChild = successor;

            // connect successor to current's left child
            successor.leftChild = current.leftChild;
            }  // end else two children
         // (successor cannot have a left child)
```

LISTING 8.1   Continued

```
        return true;                            // success
        }  // end delete()
// ------------------------------------------------------------
    // returns node with next-highest value after delNode
    // goes to right child, then right child's left descendents
    private Node getSuccessor(Node delNode)
        {
        Node successorParent = delNode;
        Node successor = delNode;
        Node current = delNode.rightChild;    // go to right child
        while(current != null)                // until no more
            {                                 // left children,
            successorParent = successor;
            successor = current;
            current = current.leftChild;      // go to left child
            }
                                              // if successor not
        if(successor != delNode.rightChild)   // right child,
            {                                 // make connections
            successorParent.leftChild = successor.rightChild;
            successor.rightChild = delNode.rightChild;
            }
        return successor;
        }
// ------------------------------------------------------------
    public void traverse(int traverseType)
        {
        switch(traverseType)
            {
            case 1: System.out.print("\nPreorder traversal: ");
                    preOrder(root);
                    break;
            case 2: System.out.print("\nInorder traversal:  ");
                    inOrder(root);
                    break;
            case 3: System.out.print("\nPostorder traversal: ");
                    postOrder(root);
                    break;
            }
        System.out.println();
        }
```

LISTING 8.1   Continued

```java
// -------------------------------------------------------------
   private void preOrder(Node localRoot)
      {
      if(localRoot != null)
         {
         System.out.print(localRoot.iData + " ");
         preOrder(localRoot.leftChild);
         preOrder(localRoot.rightChild);
         }
      }
// -------------------------------------------------------------
   private void inOrder(Node localRoot)
      {
      if(localRoot != null)
         {
         inOrder(localRoot.leftChild);
         System.out.print(localRoot.iData + " ");
         inOrder(localRoot.rightChild);
         }
      }
// -------------------------------------------------------------
   private void postOrder(Node localRoot)
      {
      if(localRoot != null)
         {
         postOrder(localRoot.leftChild);
         postOrder(localRoot.rightChild);
         System.out.print(localRoot.iData + " ");
         }
      }
// -------------------------------------------------------------
   public void displayTree()
      {
      Stack globalStack = new Stack();
      globalStack.push(root);
      int nBlanks = 32;
      boolean isRowEmpty = false;
      System.out.println(
      "......................................................");
      while(isRowEmpty==false)
         {
```

LISTING 8.1   Continued

```
        Stack localStack = new Stack();
        isRowEmpty = true;

        for(int j=0; j<nBlanks; j++)
           System.out.print(' ');

        while(globalStack.isEmpty()==false)
           {
           Node temp = (Node)globalStack.pop();
           if(temp != null)
              {
              System.out.print(temp.iData);
              localStack.push(temp.leftChild);
              localStack.push(temp.rightChild);

              if(temp.leftChild != null ||
                             temp.rightChild != null)
                 isRowEmpty = false;
              }
           else
              {
              System.out.print("--");
              localStack.push(null);
              localStack.push(null);
              }
           for(int j=0; j<nBlanks*2-2; j++)
              System.out.print(' ');
           }  // end while globalStack not empty
        System.out.println();
        nBlanks /= 2;
        while(localStack.isEmpty()==false)
           globalStack.push( localStack.pop() );
        }  // end while isRowEmpty is false
     System.out.println(
     "...................................................");
     }  // end displayTree()
// -------------------------------------------------------------
   }  // end class Tree
/////////////////////////////////////////////////////////////////
class TreeApp
   {
```

LISTING 8.1   Continued

```java
public static void main(String[] args) throws IOException
   {
   int value;
   Tree theTree = new Tree();

   theTree.insert(50, 1.5);
   theTree.insert(25, 1.2);
   theTree.insert(75, 1.7);
   theTree.insert(12, 1.5);
   theTree.insert(37, 1.2);
   theTree.insert(43, 1.7);
   theTree.insert(30, 1.5);
   theTree.insert(33, 1.2);
   theTree.insert(87, 1.7);
   theTree.insert(93, 1.5);
   theTree.insert(97, 1.5);

   while(true)
      {
      System.out.print("Enter first letter of show, ");
      System.out.print("insert, find, delete, or traverse: ");
      int choice = getChar();
      switch(choice)
         {
         case 's':
            theTree.displayTree();
            break;
         case 'i':
            System.out.print("Enter value to insert: ");
            value = getInt();
            theTree.insert(value, value + 0.9);
            break;
         case 'f':
            System.out.print("Enter value to find: ");
            value = getInt();
            Node found = theTree.find(value);
            if(found != null)
               {
               System.out.print("Found: ");
               found.displayNode();
               System.out.print("\n");
```

LISTING 8.1   Continued

```
                    }
                 else
                    System.out.print("Could not find ");
                    System.out.print(value + '\n');
                 break;
              case 'd':
                 System.out.print("Enter value to delete: ");
                 value = getInt();
                 boolean didDelete = theTree.delete(value);
                 if(didDelete)
                    System.out.print("Deleted " + value + '\n');
                 else
                    System.out.print("Could not delete ");
                    System.out.print(value + '\n');
                 break;
              case 't':
                 System.out.print("Enter type 1, 2 or 3: ");
                 value = getInt();
                 theTree.traverse(value);
                 break;
              default:
                 System.out.print("Invalid entry\n");
              }  // end switch
           }  // end while
        }  // end main()
// -------------------------------------------------------------
   public static String getString() throws IOException
      {
      InputStreamReader isr = new InputStreamReader(System.in);
      BufferedReader br = new BufferedReader(isr);
      String s = br.readLine();
      return s;
      }
// -------------------------------------------------------------
   public static char getChar() throws IOException
      {
      String s = getString();
      return s.charAt(0);
      }
//--------------------------------------------------------------
   public static int getInt() throws IOException
```

LISTING 8.1   Continued

```
      {
      String s = getString();
      return Integer.parseInt(s);
      }
// ------------------------------------------------------------
    } // end class TreeApp
//////////////////////////////////////////////////////////////
```

You can use the [control]+[ ] key combination to exit from this program; in the interest of simplicity there's no single-letter key for this action.

The Huffman Code

You shouldn't get the idea that binary trees are always search trees. Many binary trees are used in other ways. We saw an example in Figure 8.11, where a binary tree represents an algebraic expression.

In this section we'll discuss an algorithm that uses a binary tree in a surprising way to compress data. It's called the Huffman code, after David Huffman who discovered it in 1952. Data compression is important in many situations. An example is sending data over the Internet, where, especially over a dial-up connection, transmission can take a long time. An implementation of this scheme is somewhat lengthy, so we won't show a complete program. Instead, we'll focus on the concepts and leave the implementation as an exercise.

Character Codes

Each character in a normal uncompressed text file is represented in the computer by one byte (for the venerable ASCII code) or by two bytes (for the newer Unicode, which is designed to work for all languages.) In these schemes, every character requires the same number of bits. Table 8.3 shows how some characters are represented in binary using the ASCII code. As you can see, every character takes 8 bits.

TABLE 8.3   Some ASCII Codes

Character	Decimal	Binary
A	65	01000000
B	66	01000001
C	67	01000010
...	...	...
X	88	01011000
Y	89	01011001
Z	90	01011010

There are several approaches to compressing data. For text, the most common approach is to reduce the number of bits that represent the most-used characters. In English, E is often the most common letter, so it seems reasonable to use as few bits as possible to encode it. On the other hand, Z is seldom used, so using a large number of bits is not so bad.

Suppose we use just two bits for E, say 01. We can't encode every letter of the alphabet in two bits because there are only four 2-bit combinations: 00, 01, 10, and 11. Can we use these four combinations for the four most-used characters? Unfortunately not. We must be careful that no character is represented by the same bit combination that appears at the beginning of a longer code used for some other character. For example, if E is 01, and X is 01011000, then anyone decoding 01011000 wouldn't know if the initial 01 represented an E or the beginning of an X. This leads to a rule: *No code can be the prefix of any other code.*

Something else to consider is that in some messages E might not be the most-used character. If the text is a Java source file, for example, the ; (semicolon) character might appear more often than E. Here's the solution to that problem: For each message, we make up a new code tailored to that particular message. Suppose we want to send the message SUSIE SAYS IT IS EASY. The letter S appears a lot, and so does the space character. We might want to make up a table showing how many times each letter appears. This is called a frequency table, as shown in Table 8.4.

TABLE 8.4   Frequency Table

Character	Count
A	2
E	2
I	3
S	6
T	1
U	1
Y	2
Space	4
Linefeed	1

The characters with the highest counts should be coded with a small number of bits. Table 8.5 shows how we might encode the characters in the Susie message.

TABLE 8.5   Huffman Code

Character	Code
A	010
E	1111
I	110

TABLE 8.5   Continued

Character	Code
S	10
T	0110
U	01111
Y	1110
Space	00
Linefeed	01110

We use 10 for S and 00 for the space. We can't use 01 or 11 because they are prefixes for other characters. What about 3-bit combinations? There are eight possibilities: 000, 001, 010, 011, 100, 101, 110, and 111. A is 010 and I is 110. Why aren't any other combinations used? We already know we can't use anything starting with 10 or 00; that eliminates four possibilities. Also, 011 is used at the beginning of U and the linefeed, and 111 is used at the beginning of E and Y. Only two 3-bit codes remain, which we use for A and I. In a similar way we can see why only three 4-bit codes are available.

Thus, the entire message is coded as

10 01111 10 110 1111 00 10 010 1110 10 00 110 0110 00 110 10 00
➡ 1111 010 10 1110 01110

For sanity reasons we show this message broken into the codes for individual characters. Of course, in reality all the bits would run together; there is no space character in a binary message, only 0s and 1s.

Decoding with the Huffman Tree

We'll see later how to create Huffman codes. First, we'll examine the somewhat easier process of decoding. Suppose we received the string of bits shown in the preceding section. How would we transform it back into characters? We can use a kind of binary tree called a *Huffman tree*. Figure 8.23 shows the Huffman tree for the code just discussed.

The characters in the message appear in the tree as leaf nodes. The higher their frequency in the message, the higher up they appear in the tree. The number outside each circle is the frequency. The numbers outside non-leaf nodes are the sums of the frequencies of their children. We'll see later why this is important.

How do we use this tree to decode the message? For each character you start at the root. If you see a 0 bit, you go left to the next node, and if you see a 1 bit, you go right. Try it with the code for A, which is 010. You go left, then right, then left again, and, *mirabile dictu*, you find yourself on the A node. This is shown by the arrows in Figure 8.23.

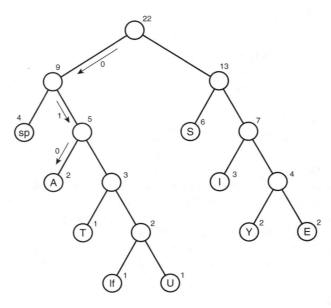

FIGURE 8.23   Huffman tree.

You'll see you can do the same with the other characters. If you have the patience, you can decode the entire bit string this way.

Creating the Huffman Tree

We've seen how to use the Huffman tree for decoding, but how do we create this tree? There are many ways to handle this problem. We'll base our approach on the Node and Tree classes in the tree.java program in Listing 8.1 (although routines that are specific to search trees, like find(), insert(), and delete() are no longer relevant). Here is the algorithm for constructing the tree:

1. Make a Node object (as seen in tree.java) for each character used in the message. For our Susie example that would be nine nodes. Each node has two data items: the character and that character's frequency in the message. Table 8.4 provides this information for the Susie message.

2. Make a tree object for each of these nodes. The node becomes the root of the tree.

3. Insert these trees in a priority queue (as described in Chapter 4). They are ordered by frequency, with the smallest frequency having the highest priority. That is, when you remove a tree, it's always the one with the least-used character.

Now do the following:

1. Remove two trees from the priority queue, and make them into children of a new node. The new node has a frequency that is the sum of the children's frequencies; its character field can be left blank.

2. Insert this new three-node tree back into the priority queue.

3. Keep repeating steps 1 and 2. The trees will get larger and larger, and there will be fewer and fewer of them. When there is only one tree left in the queue, it is the Huffman tree and you're done.

Figures 8.24 and 8.25 show how the Huffman tree is constructed for the Susie message.

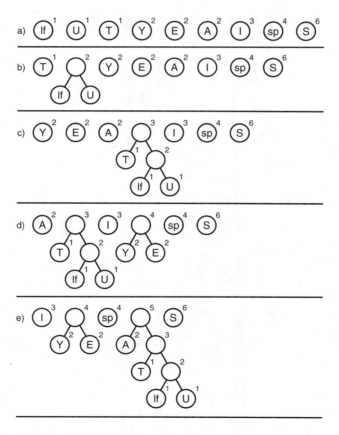

FIGURE 8.24   Growing the Huffman tree, Part 1.

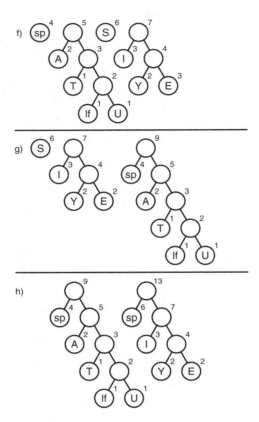

FIGURE 8.25   Growing the Huffman tree, Part 2.

Coding the Message

Now that we have the Huffman tree, how do we code a message? We start by creating a code table, which lists the Huffman code alongside each character. To simplify the discussion, let's assume that, instead of the ASCII code, our computer uses a simplified alphabet that has only uppercase letters with 28 characters. A is 0, B is 1, and so on up to Z, which is 25. A space is 26, and a linefeed is 27. We number these characters so their numerical codes run from 0 to 27. (This is not a compressed code, just a simplification of the ASCII code, the normal way characters are stored in the computer.)

Our code table would be an array of 28 cells. The index of each cell would be the numerical value of the character: 0 for A, 1 for B, and so on. The contents of the cell would be the Huffman code for the corresponding character. Not every cell contains a code; only those that appear in the message. Figure 8.26 shows how this looks for the Susie message.

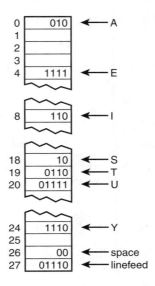

FIGURE 8.26   Code table.

Such a code table makes it easy to generate the coded message: For each character in the original message, we use its code as an index into the code table. We then repeatedly append the Huffman codes to the end of the coded message until it's complete.

Creating the Huffman Code

How do we create the Huffman code to put into the code table? The process is like decoding a message. We start at the root of the Huffman tree and follow every possible path to a leaf node. As we go along the path, we remember the sequence of left and right choices, recording a 0 for a left edge and a 1 for a right edge. When we arrive at the leaf node for a character, the sequence of 0s and 1s is the Huffman code for that character. We put this code into the code table at the appropriate index number.

This process can be handled by calling a method that starts at the root and then calls itself recursively for each child. Eventually, the paths to all the leaf nodes will be explored and the code table will be complete.

CHAPTER 8** Binary Trees

Summary

- Trees consist of nodes (circles) connected by edges (lines).

- The root is the topmost node in a tree; it has no parent.

- In a binary tree, a node has at most two children.

- In a binary search tree, all the nodes that are left descendants of node A have key values less than A; all the nodes that are A's right descendants have key values greater than (or equal to) A.

- Trees perform searches, insertions, and deletions in O(log N) time.

- Nodes represent the data objects being stored in the tree.

- Edges are most commonly represented in a program by references to a node's children (and sometimes to its parent).

- Traversing a tree means visiting all its nodes in some order.

- The simplest traversals are preorder, inorder, and postorder.

- An unbalanced tree is one whose root has many more left descendents than right descendents, or vice versa.

- Searching for a node involves comparing the value to be found with the key value of a node, and going to that node's left child if the key search value is less, or to the node's right child if the search value is greater.

- Insertion involves finding the place to insert the new node and then changing a child field in its new parent to refer to it.

- An inorder traversal visits nodes in order of ascending keys.

- Preorder and postorder traversals are useful for parsing algebraic expressions.

- When a node has no children, it can be deleted by setting the child field in its parent to null.

- When a node has one child, it can be deleted by setting the child field in its parent to point to its child.

- When a node has two children, it can be deleted by replacing it with its successor.

- The successor to a node A can be found by finding the minimum node in the subtree whose root is A's right child.

- In a deletion of a node with two children, different situations arise, depending on whether the successor is the right child of the node to be deleted or one of the right child's left descendants.

- Nodes with duplicate key values may cause trouble in arrays because only the first one can be found in a search.

- Trees can be represented in the computer's memory as an array, although the reference-based approach is more common.

- A Huffman tree is a binary tree (but not a search tree) used in a data-compression algorithm called Huffman Coding.

- In the Huffman code the characters that appear most frequently are coded with the fewest bits, and those that appear rarely are coded with the most bits.

Questions

These questions are intended as a self-test for readers. Answers may be found in Appendix C.

1. Insertion and deletion in a tree require what big O time?

2. A binary tree is a search tree if

    a. every non-leaf node has children whose key values are less than (or equal to) the parent.

    b. every left child has a key less than the parent and every right child has a key greater than (or equal to) the parent.

    c. in the path from the root to every leaf node, the key of each node is greater than (or equal to) the key of its parent.

    d. a node can have a maximum of two children.

3. True or False: Not all trees are binary trees.

4. In a complete binary tree with 20 nodes, and the root considered to be at level 0, how many nodes are there at level 4?

5. A subtree of a binary tree always has

    a. a root that is a child of the main tree's root.

    b. a root unconnected to the main tree's root.

    c. fewer nodes than the main tree.

    d. a sibling with the same number of nodes.

6. In the Java code for a tree, the _____ and the _____ are generally separate classes.

7. Finding a node in a binary search tree involves going from node to node, asking

    a. how big the node's key is in relation to the search key.

    b. how big the node's key is compared to its right or left children.

    c. what leaf node we want to reach.

    d. what level we are on.

8. An unbalanced tree is one

    a. in which most of the keys have values greater than the average.

    b. whose behavior is unpredictable.

    c. in which the root or some other node has many more left children than right children, or vice versa.

    d. that is shaped like an umbrella.

9. Inserting a node starts with the same steps as _____ a node.

10. Suppose a node A has a successor node S. Then S must have a key that is larger than _____ but smaller than or equal to _____.

11. In a binary tree used to represent a mathematical expression, which of the following is not true?

    a. Both children of an operator node must be operands.

    b. Following a postorder traversal, no parentheses need to be added.

    c. Following an inorder traversal, parentheses must be added.

    d. In pre-order traversal a node is visited before either of its children.

12. If a tree is represented by an array, the right child of a node at index n has an index of _____ .

13. True or False: Deleting a node with one child from a binary search tree involves finding that node's successor.

14. A Huffman tree is typically used to _____ text.

15. Which of the following is not true about a Huffman tree?

    a. The most frequently used characters always appear near the top of the tree.

    b. Normally, decoding a message involves repeatedly following a path from the root to a leaf.

c. In coding a character you typically start at a leaf and work upward.

d. The tree can be generated by removal and insertion operations on a priority queue.

Experiments

Carrying out these experiments will help to provide insights into the topics covered in the chapter. No programming is involved.

1. Use the Binary Tree Workshop applet to create 20 trees. What percentage would you say are seriously unbalanced?

2. Create a UML activity diagram (or flowchart, for you old-timers) of the various possibilities when deleting a node from a binary search tree. It should detail the three cases described in the text. Include the variations for left and right children and special cases like deletion of the root. For example, there are two possibilities for case 1 (left and right children). Boxes at the end of each path should describe how to do the deletion in that situation.

3. Use the Binary Tree Workshop applet to delete a node in every possible situation.

Programming Projects

Writing programs to solve the Programming Projects helps to solidify your understanding of the material and demonstrates how the chapter's concepts are applied. (As noted in the Introduction, qualified instructors may obtain completed solutions to the Programming Projects on the publisher's Web site.)

8.1 Start with the tree.java program (Listing 8.1) and modify it to create a binary tree from a string of letters (like A, B, and so on) entered by the user. Each letter will be displayed in its own node. Construct the tree so that all the nodes that contain letters are leaves. Parent nodes can contain some non-letter symbol like +. Make sure that every parent node has exactly two children. Don't worry if the tree is unbalanced. Note that this will not be a search tree; there's no quick way to find a given node. You may end up with something like this:

One way to begin is by making an array of trees. (A group of unconnected trees is called a *forest*.) Take each letter typed by the user and put it in a node. Take each of these nodes and put it in a tree, where it will be the root. Now put all these one-node trees in the array. Start by making a new tree with + at the root and two of the one-node trees as its children. Then keep adding one-node trees from the array to this larger tree. Don't worry if it's an unbalanced tree. You can actually store this intermediate tree in the array by writing over a cell whose contents have already been added to the tree.

The routines find(), insert(), and delete(), which apply only to search trees, can be deleted. Keep the displayTree() method and the traversals because they will work on any binary tree.

8.2   Expand the program in Programming Project 8.1 to create a balanced tree. One way to do this is to make sure that as many leaves as possible appear in the bottom row. You can start by making a three-node tree out of each pair of one-node trees, making a new + node for the root. This results in a forest of three-node trees. Then combine each pair of three-node trees to make a forest of seven-node trees. As the number of nodes per tree grows, the number of trees shrinks, until finally there is only one tree left.

8.3   Again, start with the tree.java program and make a tree from characters typed by the user. This time, make a complete tree—one that is completely full except possibly on the right end of the bottom row. The characters should be ordered from the top down and from left to right along each row, as if writing a letter on a pyramid. (This arrangement does not correspond to any of the three traversals we discussed in this chapter.) Thus, the string ABCDEFGHIJ would be arranged as

```
        A
    B       C
  D   E   F   G
H I   J
```

One way to create this tree is from the top down, rather than the bottom up as in the previous two Programming Projects. Start by creating a node which will be the root of the final tree. If you think of the nodes as being numbered in the same order the letters are arranged, with 1 at the root, then any node numbered n has a left child numbered 2*n and a right child numbered 2*n+1. You might use a recursive routine that makes two children and then calls itself for each child. The nodes don't need to be created in the same order they are arranged on the tree. As in the previous Programming Projects, you can jettison the search-tree routines from the Tree class.

8.4 Write a program that transforms a postfix expression into a tree such as that shown in Figure 8.11 in this chapter. You'll need to modify the Tree class from the tree.java program (Listing 8.1) and the ParsePost class from the postfix.java program (Listing 4.8) in Chapter 4. There are more details in the discussion of Figure 8.11.

After the tree is generated, traversals of the tree will yield the prefix, infix, and postfix equivalents of the algebraic expression. The infix version will need parentheses to avoid generating ambiguous expressions. In the inOrder() method, display an opening parenthesis before the first recursive call and a closing parenthesis after the second recursive call.

8.5 Write a program to implement Huffman coding and decoding. It should do the following:

Accept a text message, possibly of more than one line.

Create a Huffman tree for this message.

Create a code table.

Encode the message into binary.

Decode the message from binary back to text.

If the message is short, the program should be able to display the Huffman tree after creating it. The ideas in Programming Projects 8.1, 8.2, and 8.3 might prove helpful. You can use String variables to store binary numbers as arrangements of the characters 1 and 0. Don't worry about doing actual bit manipulation unless you really want to.

9

Red-Black Trees

As you learned in Chapter 8, "Binary Trees," ordinary binary search trees offer important advantages as data storage devices: You can quickly search for an item with a given key, and you can also quickly insert or delete an item. Other data storage structures, such as arrays, sorted arrays, and linked lists, perform one or the other of these activities slowly. Thus, binary search trees might appear to be the ideal data storage structure.

Unfortunately, ordinary binary search trees suffer from a troublesome problem. They work well if the data is inserted into the tree in random order. However, they become much slower if data is inserted in already-sorted order (17, 21, 28, 36,...) or inversely sorted order (36, 28, 21, 17,...). When the values to be inserted are already ordered, a binary tree becomes unbalanced. With an unbalanced tree, the ability to quickly find (or insert or delete) a given element is lost.

This chapter explores one way to solve the problem of unbalanced trees: the red-black tree, which is a binary search tree with some added features.

There are other ways to ensure that trees are balanced. We'll mention some at the end of this chapter, and examine several, 2-3-4 trees and 2-3 trees, in Chapter 10, "2-3-4 Trees and External Storage." In fact, as we'll see in that chapter, operations on a 2-3-4 tree correspond in a surprising way to operations on a red-black tree.

Our Approach to the Discussion

We'll explain insertion into red-black trees a little differently than we have explained insertion into other data structures. Red-black trees are not trivial to understand. Because of this and also because of a multiplicity of symmetrical cases (for left or right children, inside or outside grandchildren, and so on), the actual code is more lengthy and complex than one might expect. It's therefore hard to learn about the algorithm by examining code. Accordingly, there are no listings in this chapter. You can create similar functionality using a 2-3-4 tree with the code shown in Chapter 10. However, the concepts you learn about here will aid your understanding of 2-3-4 trees and are themselves quite interesting.

Conceptual

For our conceptual understanding of red-black trees, we will be aided by the RBTree Workshop applet. We'll describe how you can work in partnership with the applet to insert new nodes into a tree. Including a human into the insertion routine certainly slows it down but also makes it easier for the human to understand how the process works.

Searching works the same way in a red-black tree as it does in an ordinary binary tree. On the other hand, insertion and deletion, while based on the algorithms in an ordinary tree, are extensively modified. Accordingly, in this chapter we'll be concentrating on the insertion process.

Top-Down Insertion

The approach to insertion that we'll discuss is called *top-down* insertion. This means that some structural changes may be made to the tree as the search routine descends the tree looking for the place to insert the node.

Another approach is *bottom-up* insertion. This involves finding the place to insert the node and then working back up through the tree making structural changes. Bottom-up insertion is less efficient because two passes must be made through the tree.

Balanced and Unbalanced Trees

Before we begin our investigation of red-black trees, let's review how trees become unbalanced. Fire up the Binary Tree Workshop applet from Chapter 8 (not this chapter's RBTree applet). Use the Fill button to create a tree with only one node. Then insert a series of nodes whose keys are in either ascending or descending order. The result will be something like that in Figure 9.1.

The nodes arrange themselves in a line with no branches. Because each node is larger than the previously inserted one, every node is a right child, so all the nodes are on one side of the root. The tree is maximally unbalanced. If you inserted items in descending order, every node would be the left child of its parent, and the tree would be unbalanced on the other side.

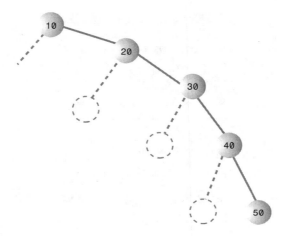

FIGURE 9.1   Items inserted in ascending order.

Degenerates to O(N)

When there are no branches, the tree becomes, in effect, a linked list. The arrangement of data is one-dimensional instead of two-dimensional. Unfortunately, as with a linked list, you must now search through (on the average) half the items to find the one you're looking for. In this situation the speed of searching is reduced to $O(N)$, instead of $O(\log N)$ as it is for a balanced tree. Searching through 10,000 items in such an unbalanced tree would require an average of 5,000 comparisons, whereas for a balanced tree with random insertions it requires only 14. For presorted data you might just as well use a linked list in the first place.

Data that's only partly sorted will generate trees that are only partly unbalanced. If you use the Binary Tree Workshop applet from Chapter 8 to attempt to generate trees with 31 nodes, you'll see that some of them are more unbalanced than others, as shown in Figure 9.2.

Although not as bad as a maximally unbalanced tree, this situation is not optimal for searching times.

In the Binary Tree Workshop applet, trees can become partially unbalanced, even with randomly generated data, because the amount of data is so small that even a short run of ordered numbers will have a big effect on the tree. Also, a very small or very large key value can cause an unbalanced tree by not allowing the insertion of many nodes on one side or the other of its node. A root of 3, for example, allows only two more nodes to be inserted to its left.

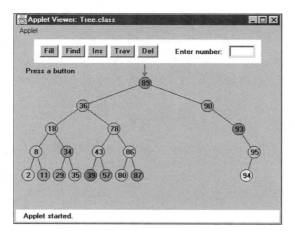

FIGURE 9.2   A partially unbalanced tree.

With a realistic amount of random data, it's not likely a tree would become seriously unbalanced. However, there may be runs of sorted data that will partially unbalance a tree. Searching partially unbalanced trees will take time somewhere between O(N) and O(logN), depending on how badly the tree is unbalanced.

Balance to the Rescue

To guarantee the quick O(log N) search times a tree is capable of, we need to ensure that our tree is always balanced (or at least almost balanced). This means that each node in a tree needs to have roughly the same number of descendents on its left side as it has on its right.

In a red-black tree, balance is achieved during insertion (and also deletion, but we'll ignore that for the moment). As an item is being inserted, the insertion routine checks that certain characteristics of the tree are not violated. If they are, it takes corrective action, restructuring the tree as necessary. By maintaining these characteristics, the tree is kept balanced.

Red-Black Tree Characteristics

What are these mysterious tree characteristics? There are two, one simple and one more complicated:

- The nodes are colored.
- During insertion and deletion, rules are followed that preserve various arrangements of these colors.

Colored Nodes

In a red-black tree, every node is either black or red. These are arbitrary colors; blue and yellow would do just as well. In fact, the whole concept of saying that nodes have "colors" is somewhat arbitrary. Some other analogy could have been used instead: We could say that every node is either heavy or light, or yin or yang. However, colors are convenient labels. A data field, which can be boolean (isRed, for example), is added to the node class to embody this color information.

In the RBTree Workshop applet, the red-black characteristic of a node is shown by its border color. The center color, as it was in the Binary Tree Workshop applet in the preceding chapter, is simply a randomly generated data field of the node.

When we speak of a node's color in this chapter, we'll almost always be referring to its red-black border color. In the figures (except the screenshot of Figure 9.3) we'll show black nodes with a solid black border and red nodes with a white border. (Nodes are sometimes shown with no border to indicate that it doesn't matter whether they're black or red.)

Red-Black Rules

When inserting (or deleting) a new node, certain rules, which we call the *red-black rules*, must be followed. If they're followed, the tree will be balanced. Let's look briefly at these rules:

1. Every node is either red or black.

2. The root is always black.

3. If a node is red, its children must be black (although the converse isn't necessarily true).

4. Every path from the root to a leaf, or to a null child, must contain the same number of black nodes.

The "null child" referred to in Rule 4 is a place where a child could be attached to a non-leaf node. In other words, it's the potential left child of a node with a right child, or the potential right child of a node with a left child. This will make more sense as we go along.

The number of black nodes on a path from root to leaf is called the *black height*. Another way to state Rule 4 is that the black height must be the same for all paths from the root to a leaf.

These rules probably seem completely mysterious. It's not obvious how they will lead to a balanced tree, but they do; some very clever people invented them. Copy them onto a sticky note, and keep it on your computer. You'll need to refer to them often in the course of this chapter.

You can see how the rules work by using the RBTree Workshop applet. We'll do some experiments with the applet in a moment, but first you should understand what actions you can take to fix things if one of the red-black rules is broken.

Duplicate Keys

What happens if there's more than one data item with the same key? This presents a slight problem in red-black trees. It's important that nodes with the same key are distributed on both sides of other nodes with the same key. That is, if keys arrive in the order 50, 50, 50, you want the second 50 to go to the right of the first one, and the third 50 to go to the left of the first one. Otherwise, the tree becomes unbalanced.

Distributing nodes with equal keys could be handled by some kind of randomizing process in the insertion algorithm. However, the search process then becomes more complicated if all items with the same key must be found.

It's simpler to outlaw items with the same key. In this discussion we'll assume duplicates aren't allowed.

Fixing Rule Violations

Suppose you see (or are told by the applet) that the color rules are violated. How can you fix things so your tree is in compliance? There are two, and only two, possible actions you can take:

- You can change the colors of nodes.
- You can perform rotations.

In the applet, changing the color of a node means changing its red-black border color (not the center color). A rotation is a rearrangement of the nodes that, one hopes, leaves the tree more balanced.

At this point such concepts probably seem very abstract, so let's become familiar with the RBTree Workshop applet, which can help to clarify things.

Using the RBTree Workshop Applet

Figure 9.3 shows what the RBTree Workshop applet looks like after some nodes have been inserted. (It may be hard to tell the difference between red and black node borders in the figure, but they should be clear on a color monitor.)

There are quite a few buttons in the RBTree applet. We'll briefly review what they do, although at this point some of the descriptions may be a bit puzzling.

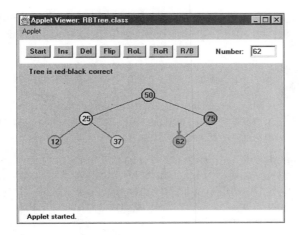

FIGURE 9.3   The RBTree Workshop applet.

Clicking on a Node

The red arrow points to the currently selected node. It's this node whose color is changed or which is the top node in a rotation. You select a node by single-clicking it with the mouse, which moves the red arrow to the node.

The Start Button

When you first start the RBTree Workshop applet, and also when you press the Start button, you'll see that a tree with only one node is created. Because an understanding of red-black trees focuses on using the red-black rules during the insertion process, it's more convenient to begin with the root and build up the tree by inserting additional nodes. To simplify future operations, the initial root node is always given a value of 50. You select your own numbers for subsequent insertions.

The Ins Button

The Ins button causes a new node to be created, with the value that was typed into the Number box, and then inserted into the tree. (At least this is what happens if no color flips are necessary. See the section on the Flip button for more on this possibility.)

Notice that the Ins button does a complete insertion operation with one push; multiple pushes are not required as they were with the Binary Tree Workshop applet in the preceding chapter. It's therefore important to type the key value before pushing the button. The focus in the RBTree applet is not on the process of finding the place to insert the node, which is similar to that in ordinary binary search trees, but on keeping the tree balanced; so the applet doesn't show the individual steps in the insertion.

The Del Button

Pushing the Del button causes the node with the key value typed into the Number box to be deleted. As with the Ins button, this deletion takes place immediately after the first push; multiple pushes are not required.

The Del button and the Ins button use the basic insertion algorithms—the same as those in the Tree Workshop applet. This is how the work is divided between the applet and the user: The applet does the insertion, but it's (mostly) up to the user to make the appropriate changes to the tree to ensure the red-black rules are followed and the tree thereby becomes balanced.

The Flip Button

If there is a black parent with two red children, and you place the red arrow on the parent by clicking on the node with the mouse, then when you press the Flip button, the parent will become red and the children will become black. That is, the colors are flipped between the parent and children. You'll learn later why such a color exchange is desirable.

If you try to flip the root, it will remain black, so as not to violate Rule 2, but its children will change from red to black.

The RoL Button

The RoL button carries out a left rotation. To rotate a group of nodes, first single-click the mouse to position the arrow at the topmost node of the group to be rotated. For a left rotation, the top node must have a right child. Then click the button. We'll examine rotations in detail later.

The RoR Button

The RoR button performs a right rotation. Position the arrow on the top node to be rotated, making sure it has a left child; then click the button.

The R/B Button

The R/B button changes a red node to black, or a black node to red. Single-click the mouse to position the red arrow on the node, and then push the button. (This button changes the color of a single node; don't confuse it with the Flip button, which changes three nodes at once.)

Text Messages

Messages in the text box below the buttons tell you whether the tree is *red-black correct*. The tree is red-black correct if it adheres to rules 1 through 4 listed previously. If it's not correct, you'll see messages advising which rule is being violated. In some cases the red arrow will point to the place where the violation occurred.

Where's the Find Button?

In red-black trees, a search routine operates exactly as it did in the ordinary binary search trees described in the preceding chapter. It starts at the root, and, at each node it encounters (the current node), it decides whether to go to the left or right child by comparing the key of the current node with the search key.

We don't include a Find button in the RBTree applet because you already understand this process and our attention will be on manipulating the red-black aspects of the tree.

Experimenting with the Workshop Applet

Now that you're familiar with the RBTree buttons, let's do some simple experiments to get a feel for what the applet does. The idea here is to learn to manipulate the applet's controls. Later you'll use these skills to balance the tree.

Experiment 1: Inserting Two Red Nodes

Press Start to clear any extra nodes. You'll be left with the root node, which always has the value 50.

Insert a new node with a value smaller than the root, say 25, by typing the number into the Number box and pressing the Ins button. Adding this node doesn't cause any rule violations, so the message continues to say `Tree is red-black correct`.

Insert a second node that's larger than the root, say 75. The tree is still red-black correct. It's also balanced; there are the same number of nodes on the right of the only non-leaf node (the root) as there are on its left. The result is shown in Figure 9.4.

Notice that newly inserted nodes are always colored red (except for the root). This is not an accident. Inserting a red node is less likely to violate the red-black rules than inserting a black one. This is because, if the new red node is attached to a black one, no rule is broken. It doesn't create a situation in which there are two red nodes together (Rule 3), and it doesn't change the black height in any of the paths (Rule 4). Of course, if you attach a new red node to a red node, Rule 3 will be violated. However, with any luck this will happen only half the time. Whereas, if it were possible to add a new black node, it would always change the black height for its path, violating Rule 4.

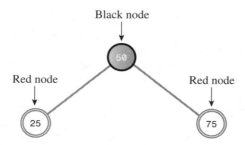

FIGURE 9.4   A balanced tree.

Also, it's easier to fix violations of Rule 3 (parent and child are both red) than Rule 4 (black heights differ), as we'll see later.

Experiment 2: Rotations

Let's try some rotations. Start with the three nodes as shown in Figure 9.4. Position the red arrow on the root (50) by clicking it with the mouse. This node will be the *top node* in the rotation. Now perform a right rotation by pressing the RoR button. The nodes all shift to new positions, as shown in Figure 9.5.

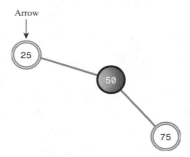

FIGURE 9.5   Following a right rotation.

In this right rotation, the parent or top node moves into the place of its right child, the left child moves up and takes the place of the parent, and the right child moves down to become the grandchild of the new top node.

Notice that the tree is now unbalanced; more nodes appear to the right of the root than to the left. Also, the message indicates that the red-black rules are violated, specifically Rule 2 (the root is always black). Don't worry about this problem yet.

Instead, rotate the other way. Position the red arrow on 25, which is now the root (the arrow should already point to 25 after the previous rotation). Click the RoL button to rotate left. The nodes will return to the position of Figure 9.4.

Experiment 3: Color Flips

Start with the position of Figure 9.4, with nodes 25 and 75 inserted in addition to 50 in the root position. Note that the parent (the root) is black and both its children are red. Now try to insert another node. No matter what value you use, you'll see the message Can't Insert: Needs color flip.

As we mentioned, a color flip is necessary whenever, during the insertion process, a black node with two red children is encountered.

The red arrow should already be positioned on the black parent (the root node), so click the Flip button. The root's two children change from red to black. Ordinarily, the parent would change from black to red, but this is a special case because it's the root: It remains black to avoid violating Rule 2. Now all three nodes are black. The tree is still red-black correct.

Now click the Ins button again to insert the new node. Figure 9.6 shows the result if the newly inserted node has the key value 12.

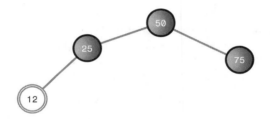

FIGURE 9.6   Colors flipped, new node inserted.

The tree is still red-black correct. The root is black, there's no situation in which a parent and child are both red, and all the paths have the same number of black nodes (two). Adding the new red node didn't change the red-black correctness.

Experiment 4: An Unbalanced Tree

Now let's see what happens when you try to do something that leads to an unbalanced tree. In Figure 9.6 one path has one more node than the other. This isn't very unbalanced, and no red-black rules are violated, so neither we nor the red-black algorithms need to worry about it. However, suppose that one path differs from another by two or more levels (where level is the same as the number of nodes along the path). In this case the red-black rules will always be violated, and we'll need to rebalance the tree.

Insert a 6 into the tree of Figure 9.6. You'll see the message Error: parent and child are both red. Rule 3 has been violated, as shown in Figure 9.7.

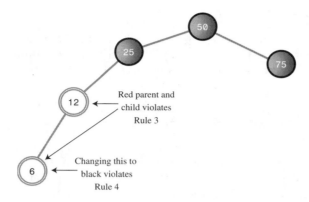

FIGURE 9.7   Parent and child are both red.

How can we fix the tree so Rule 3 isn't violated? An obvious approach is to change one of the offending nodes to black. Let's try changing the child node, 6. Position the red arrow on it and press the R/B button. The node becomes black.

The good news is we fixed the problem of both parent and child being red. The bad news is that now the message says `Error: Black heights differ`. The path from the root to node 6 has three black nodes in it, while the path from the root to node 75 has only two. Thus, Rule 4 is violated. It seems we can't win.

This problem can be fixed with a rotation and some color changes. How to do this will be the topic of later sections.

More Experiments

Experiment with the RBTree Workshop applet on your own. Insert more nodes and see what happens. See if you can use rotations and color changes to achieve a balanced tree. Does keeping the tree red-black correct seem to guarantee an (almost) balanced tree?

Try inserting ascending keys (50, 60, 70, 80, 90) and then restart with the Start button and try descending keys (50, 40, 30, 20, 10). Ignore the messages; we'll see what they mean later. These are the situations that get the ordinary binary search tree into trouble. Can you still balance the tree?

The Red-Black Rules and Balanced Trees

Try to create a tree that is unbalanced by two or more levels but is red-black correct. As it turns out, this is impossible. That's why the red-black rules keep the tree balanced. If one path is more than one node longer than another, it must either have more black nodes, violating Rule 4, or it must have two adjacent red nodes, violating Rule 3. Convince yourself that this is true by experimenting with the applet.

Null Children

Remember that Rule 4 specifies all paths that go from the root to any leaf or to *any null children* must have the same number of black nodes. A null child is a child that a non-leaf node might have, but doesn't. (That is, a missing left child if the node has a right child, or vice versa.) Thus, in Figure 9.8 the path from 50 to 25 to the right child of 25 (its null child) has only one black node, which is not the same as the paths to 6 and 75, which have two. This arrangement violates Rule 4, although both paths to leaf nodes have the same number of black nodes.

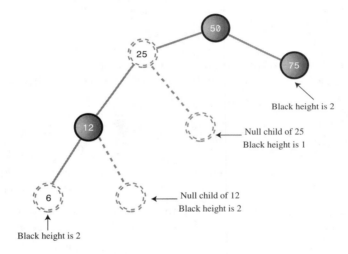

FIGURE 9.8   Path to a null child.

The term *black height* is used to describe the number of black nodes from the root to a given node. In Figure 9.8 the black height of 50 is 1, of 25 is still 1, of 12 is 2, and so on.

Rotations

To balance a tree, you need to physically rearrange the nodes. If all the nodes are on the left of the root, for example, you need to move some of them over to the right side. This is done using *rotations*. In this section we'll learn what rotations are and how to execute them. Rotations must do two things at once:

- Raise some nodes and lower others to help balance the tree.

- Ensure that the characteristics of a binary search tree are not violated.

Recall that in a binary search tree the left children of any node have key values less than the node, while its right children have key values greater than or equal to the node. If the rotation didn't maintain a valid binary search tree, it wouldn't be of much use, because the search algorithm, as we saw in the preceding chapter, relies on the search-tree arrangement.

Note that color rules and node color changes are used only to help decide when to perform a rotation. Fiddling with the colors doesn't accomplish anything by itself; it's the rotation that's the heavy hitter. Color rules are like rules of thumb for building a house (such as "exterior doors open inward"), while rotations are like the hammering and sawing needed to actually build it.

Simple Rotations

In Experiment 2 we tried rotations to the left and right. Those rotations were easy to visualize because they involved only three nodes. Let's clarify some aspects of this process.

What's Rotating?

The term *rotation* can be a little misleading. The nodes themselves aren't rotated; it's the relationship between them that changes. One node is chosen as the "top" of the rotation. If we're doing a right rotation, this "top" node will move down and to the right, into the position of its right child. Its left child will move up to take its place.

Remember that the top node isn't the "center" of the rotation. If we talk about a car tire, the top node doesn't correspond to the axle or the hubcap; it's more like the topmost part of the tire tread.

The rotation we described in Experiment 2 was performed with the root as the top node, but of course any node can be the top node in a rotation, provided it has the appropriate child.

Mind the Children

You must be sure that, if you're doing a right rotation, the top node has a left child. Otherwise, there's nothing to rotate into the top spot. Similarly, if you're doing a left rotation, the top node must have a right child.

The Weird Crossover Node

Rotations can be more complicated than the three-node example we've discussed so far. Click Start, and then, with 50 already at the root, insert nodes with the following values, in this order: 25, 75, 12, 37.

When you try to insert the 12, you'll see the `Can't insert: needs color flip` message. Just click the Flip button. The parent and children change color. Then press Ins again to complete the insertion of the 12. Finally, insert the 37. The resulting arrangement is shown in Figure 9.9a.

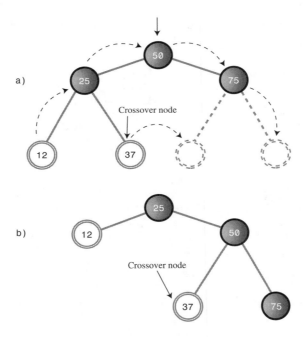

FIGURE 9.9   Rotation with crossover node.

Now we'll try a rotation. Place the arrow on the root (don't forget this!) and press the RoR button. All the nodes move. The 12 follows the 25 up, and the 50 follows the 75 down.

But what's this? The 37 has detached itself from the 25, whose right child it was, and become instead the left child of 50. Some nodes go up, some nodes go down, but the 37 moves *across*. The result is shown in Figure 9.9b. The rotation has caused a violation of Rule 4; we'll see how to fix this problem later.

In the original position of Figure 9.9a, the 37 is called an *inside grandchild* of the top node, 50. (The 12 is an *outside grandchild*.) The inside grandchild, if it's the child of the node that's going up (which is the left child of the top node in a right rotation) is always disconnected from its parent and reconnected to its former grandparent. It's like becoming your own uncle (although it's best not to dwell too long on this analogy).

Subtrees on the Move

We've shown individual nodes changing position during a rotation, but entire subtrees can move as well. To see this, click Start to put 50 at the root, and then insert the following sequence of nodes in order: 25, 75, 12, 37, 62, 87, 6, 18, 31, 43. Click Flip whenever you can't complete an insertion because of the `Can't insert: needs color flip` message. The resulting arrangement is shown in Figure 9.10a.

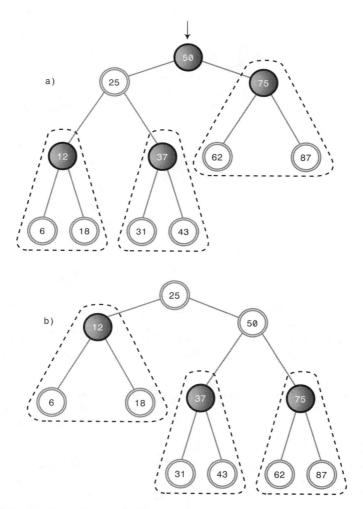

FIGURE 9.10   Subtree motion during rotation.

Position the arrow on the root, 50. Now press RoR. Wow! (Or is it WoW?) A lot of nodes have changed position. The result is shown in Figure 9.10b. Here's what happens:

- The top node (50) goes to its right child.

- The top node's left child (25) goes to the top.

- The entire subtree of which 12 is the root moves up.

- The entire subtree of which 37 is the root moves across to become the left child of 50.

- The entire subtree of which 75 is the root moves down.

You'll see the `Error: root must be black` message, but you can ignore it for the moment. You can flip back and forth by alternately pressing RoR and RoL with the arrow on the top node. Do this and watch what happens to the subtrees, especially the one with 37 as its root.

In Figure 9.10, the subtrees are encircled by dotted triangles. Note that the relations of the nodes within each subtree are unaffected by the rotation. The entire subtree moves as a unit. The subtrees can be larger (have more descendants) than the three nodes we show in this example. No matter how many nodes there are in a subtree, they will all move together during a rotation.

Human Beings Versus Computers

At this point, you've learned pretty much all you need to know about what a rotation does. To cause a rotation, you position the arrow on the top node and then press RoR or RoL. Of course, in a real red-black tree insertion algorithm, rotations happen under program control, without human intervention.

Notice however that, in your capacity as a human being, you could probably balance any tree just by looking at it and performing appropriate rotations. Whenever a node has a lot of left descendants and not too many right ones, you rotate it right, and vice versa.

Unfortunately, computers aren't very good at "just looking" at a pattern. They work better if they can follow a few simple rules. That's what the red-black scheme provides, in the form of color coding and the four color rules.

Inserting a New Node

Now you have enough background to see how a red-black tree's insertion routine uses rotations and the color rules to maintain the tree's balance.

Preview of the Insertion Process

We're going to briefly preview our approach to describing the insertion process. Don't worry if things aren't completely clear in the preview; we'll discuss this process in more detail in a moment.

In the discussion that follows, we'll use X, P, and G to designate a pattern of related nodes. X is a node that has caused a rule violation. (Sometimes X refers to a newly inserted node, and sometimes to the child node when a parent and child have a red-red conflict.)

- X is a particular node.

- P is the parent of X.

- G is the grandparent of X (the parent of P).

On the way down the tree to find the insertion point, you perform a color flip whenever you find a black node with two red children (a violation of Rule 2). Sometimes the flip causes a red-red conflict (a violation of Rule 3). Call the red child X and the red parent P. The conflict can be fixed with a single rotation or a double rotation, depending on whether X is an outside or inside grandchild of G. Following color flips and rotations, you continue down to the insertion point and insert the new node.

After you've inserted the new node X, if P is black, you simply attach the new red node. If P is red, there are two possibilities: X can be an outside or inside grandchild of G. You perform two color changes (we'll see what they are in a moment). If X is an outside grandchild, you perform one rotation, and if it's an inside grandchild, you perform two. This restores the tree to a balanced state.

Now we'll recapitulate this preview in more detail. We'll divide the discussion into three parts, arranged in order of complexity:

1. Color flips on the way down

2. Rotations after the node is inserted

3. Rotations on the way down

If we were discussing these three parts in strict chronological order, we would examine part 3 before part 2. However, it's easier to talk about rotations at the bottom of the tree than in the middle, and operations 1 and 2 are encountered more frequently than operation 3, so we'll discuss 2 before 3.

Color Flips on the Way Down

The insertion routine in a red-black tree starts off doing essentially the same thing it does in an ordinary binary search tree: It follows a path from the root to the place where the node should be inserted, going left or right at each node depending on the relative size of the node's key and the search key.

However, in a red-black tree, getting to the insertion point is complicated by color flips and rotations. We introduced color flips in Experiment 3; now we'll look at them in more detail.

Imagine the insertion routine proceeding down the tree, going left or right at each node, searching for the place to insert a new node. To make sure the color rules aren't broken, it needs to perform color flips when necessary. Here's the rule: Every time the insertion routine encounters a black node that has two red children, it must change the children to black and the parent to red (unless the parent is the root, which always remains black).

How does a color flip affect the red-black rules? For convenience, let's call the node at the top of the triangle, the one that's red before the flip, P for parent. We'll call P's left and right children X1 and X2. This arrangement is shown in Figure 9.11a.

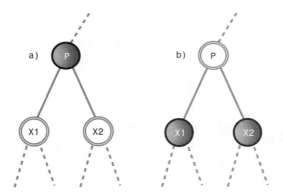

FIGURE 9.11   Color flip.

Black Heights Unchanged

Figure 9.11b shows the nodes after the color flip. The flip leaves unchanged the number of black nodes on the path from the root on down through P to the leaf or null nodes. All such paths go through P, and then through either X1 or X2. Before the flip, only P is black, so the triangle (consisting of P, X1, and X2) adds one black node to each of these paths.

After the flip, P is no longer black, but both X1 and X2 are, so again the triangle contributes one black node to every path that passes through it. So a color flip can't cause Rule 4 to be violated.

Color flips are helpful because they make red leaf nodes into black leaf nodes. This makes it easier to attach new red nodes without violating Rule 3.

Violation of Rule 3

Although Rule 4 is not violated by a color flip, Rule 3 (a node and its parent can't both be red) may be. If the parent of P is black, there's no problem when P is changed from black to red. However, if the parent of P is red, then, after the color change, we'll have two reds in a row.

This problem needs to be fixed before we continue down the path to insert the new node. We can correct the situation with a rotation, as we'll soon see.

The Root Situation

What about the root? Remember that a color flip of the root and its two children leaves the root, as well as its children, black. This color flip avoids violating Rule 2. Does this affect the other red-black rules? Clearly, there are no red-to-red conflicts because we've made more nodes black and none red. Thus, Rule 3 isn't violated. Also, because the root and one or the other of its two children are in every path, the black height of every path is increased the same amount—that is, by 1. Thus, Rule 4 isn't violated either.

Finally, Just Insert It

After you've worked your way down to the appropriate place in the tree, performing color flips (and rotations) if necessary on the way down, you can then insert the new node as described in the preceding chapter for an ordinary binary search tree. However, that's not the end of the story.

Rotations After the Node Is Inserted

The insertion of the new node may cause the red-black rules to be violated. Therefore, following the insertion, we must check for rule violations and take appropriate steps.

Remember that, as described earlier, the newly inserted node, which we'll call X, is always red. X may be located in various positions relative to P and G, as shown in Figure 9.12.

Remember that a node X is an outside grandchild if it's on the same side of its parent P that P is of its parent G. That is, X is an outside grandchild if either it's a left child of P and P is a left child of G (Figure 9.12a), or it's a right child of P and P is a right child of G (Figure 9.12d). Conversely, X is an inside grandchild if it's on the opposite side of its parent P that P is of its parent G (Figures 9.12b and 9.12c).

The multiplicity of what we might call "handed" (left or right) variations shown in Figure 9.12 is one reason the red-black insertion routine is so complex to program.

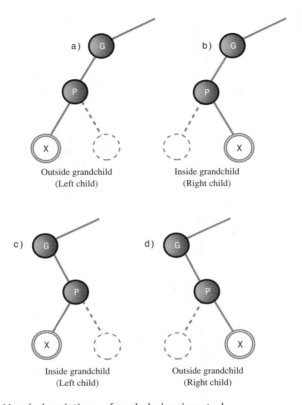

FIGURE 9.12   Handed variations of node being inserted.

The action we take to restore the red-black rules is determined by the colors and configuration of X and its relatives. Perhaps surprisingly, nodes can be arranged in only three major ways (not counting the handed variations already mentioned). Each possibility must be dealt with in a different way to preserve red-black correctness and thereby lead to a balanced tree. We'll list the three possibilities briefly and then discuss each one in detail in its own section. Figure 9.13 shows what they look like. Remember that X is always red.

1. P is black.

2. P is red and X is an outside grandchild of G.

3. P is red and X is an inside grandchild of G.

You might think that this list doesn't cover all the possibilities. We'll return to this question after we've explored these three.

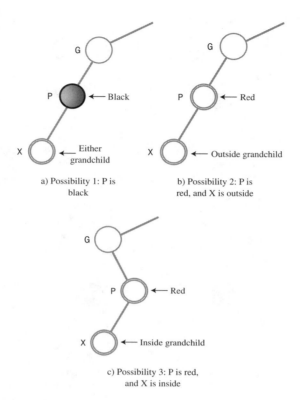

a) Possibility 1: P is
black

b) Possibility 2: P is
red, and X is outside

c) Possibility 3: P is red,
and X is inside

FIGURE 9.13   Three post-insertion possibilities.

Possibility 1: P Is Black

If P is black, we get a free ride. The node we've just inserted is always red. If its
parent is black, there's no red-to-red conflict (Rule 3) and no addition to the number
of black nodes (Rule 4). Thus, no color rules are violated. We don't need to do
anything else. The insertion is complete.

Possibility 2: P Is Red and X Is Outside

If P is red and X is an outside grandchild, we need a single rotation and some color
changes. Let's set this up with the RBTree Workshop applet so we can see what we're
talking about. Start with the usual 50 at the root, and insert 25, 75, and 12. You'll
need to do a color flip before you insert the 12.

Now insert 6, which is X, the new node. Figure 9.14a shows the resulting tree. The
message on the Workshop applet says `Error: parent and child both red`, so we know
we need to take some action.

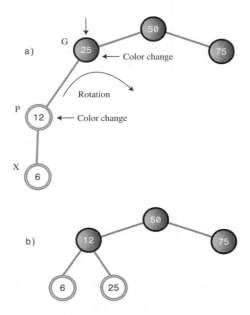

FIGURE 9.14 P is red and X is an outside grandchild.

In this situation, we can take three steps to restore red-black correctness and thereby balance the tree. Here are the steps:

1. Switch the color of X's grandparent G (25 in this example).

2. Switch the color of X's parent P (12).

3. Rotate with X's grandparent G (25) at the top, in the direction that raises X (6). This is a right rotation in the example.

As you've learned, to switch colors, put the arrow on the node and press the R/B button. To rotate right, put the arrow on the top node and press RoR. When you've completed the three steps, the Workshop applet will inform you that the Tree is red/black correct. It's also more balanced than it was, as shown in Figure 9.14b.

In this example, X was an outside grandchild and a left child. There's a symmetrical situation when the X is an outside grandchild but a right child. Try this by creating the tree 50, 25, 75, 87, 93 (with color flips when necessary). Fix it by changing the colors of 75 and 87, and rotating left with 75 at the top. Again, the tree is balanced.

Possibility 3: P Is Red and X Is Inside
If P is red and X is an inside grandchild, we need two rotations and some color changes. To see this one in action, use the RBTree Workshop applet to create the tree

50, 25, 75, 12, 18. (Again, you'll need a color flip before you insert the 12.) The result is shown in Figure 9.15a.

Note that the 18 node is an inside grandchild. It and its parent are both red, so again you see the error message `Error: parent and child both red`.

Fixing this arrangement is slightly more complicated. If we try to rotate right with the grandparent node G (25) at the top, as we did in Possibility 2, the inside grandchild X (18) moves across rather than up, so the tree is no more balanced than before. (Try this, then rotate back, with 12 at the top, to restore it.) A different solution is needed.

The trick when X is an inside grandchild is to perform *two* rotations rather than one. The first changes X from an inside grandchild to an outside grandchild, as shown in Figures 9.15a and 9.15b. Now the situation is similar to Possibility 1, and we can apply the same rotation, with the grandparent at the top, as we did before. The result is shown in Figure 9.15c.

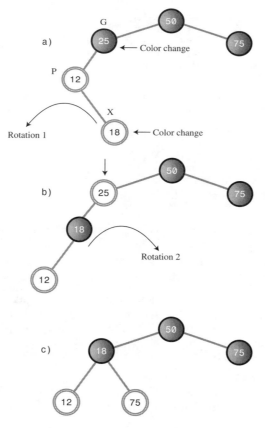

FIGURE 9.15   P is red and X is an inside grandchild.

We must also recolor the nodes. We do this before doing any rotations. (This order doesn't really matter, but if we wait until after the rotations to recolor the nodes, it's hard to know what to call them.) Here are the steps:

1. Switch the color of X's grandparent (25 in this example).

2. Switch the color of X (*not* its parent; X is 18 here).

3. Rotate with X's parent P at the top (*not* the grandparent; the parent is 12), in the direction that raises X (a left rotation in this example).

4. Rotate again with X's grandparent (25) at the top, in the direction that raises X (a right rotation).

The rotations and recoloring restore the tree to red-black correctness and also balance it (as much as possible). As with Possibility 2, there is an analogous case in which P is the right child of G rather than the left.

What About Other Possibilities?

Do the three Post-Insertion Possibilities we just discussed really cover all situations?

Suppose, for example, that X has a sibling S, the other child of P. This scenario might complicate the rotations necessary to insert X. But if P is black, there's no problem inserting X (that's Possibility 1). If P is red, both its children must be black (to avoid violating Rule 3). It can't have a single child S that's black because the black heights would be different for S and the null child. However, we know X is red, so we conclude that it's impossible for X to have a sibling unless P is red.

Another possibility is that G, the grandparent of P, has a child U, the sibling of P and the uncle of X. Again, this scenario would complicate any necessary rotations. However, if P is black, there's no need for rotations when inserting X, as we've seen. So let's assume P is red. Then U must also be red; otherwise, the black height going from G to P would be different from that going from G to U. But a black parent with two red children is flipped on the way down, so this situation can't exist either.

Thus, the three possibilities just discussed are the only ones that can exist (except that, in Possibilities 2 and 3, X can be a right or left child and G can be a right or left child).

What the Color Flips Accomplished

Suppose that performing a rotation and appropriate color changes caused other violations of the red-black rules to appear further up the tree. You can imagine situations in which you would need to work your way all the way back up the tree, performing rotations and color switches, to remove rule violations.

Fortunately, this situation can't arise. Using color flips on the way down has eliminated the situations in which a rotation could introduce any rule violations further up the tree. It ensures that one or two rotations will restore red-black correctness in the entire tree. Actually, proving this is beyond the scope of this book, but such a proof is possible.

The color flips on the way down make insertion in red-black trees more efficient than in other kinds of balanced trees, such as AVL trees. They ensure that you need to pass through the tree only once, on the way down.

Rotations on the Way Down

Now we'll discuss the last of the three operations involved in inserting a node: making rotations on the way down to the insertion point. As we noted, although we're discussing this operation last, it actually takes place before the node is inserted. We've waited until now to discuss it only because it was easier to explain rotations for a just-installed node than for nodes in the middle of the tree.

In the discussion of color flips during the insertion process, we noted that a color flip can cause a violation of Rule 3 (a parent and child can't both be red). We also noted that a rotation can fix this violation.

On the way down there are two possibilities, corresponding to Possibility 2 and Possibility 3 during the insertion phase described earlier. The offending node can be an outside grandchild, or it can be an inside grandchild. (In the situation corresponding to Possibility 1, no action is required.)

Outside Grandchild

First, we'll examine an example in which the offending node is an outside grandchild. By "offending node," we mean the child in the parent-child pair that caused the red-red conflict.

Start a new tree with the 50 node, and insert the following nodes: 25, 75, 12, 37, 6, and 18. You'll need to do color flips when inserting 12 and 6.

Now try to insert a node with the value 3. You'll be told you must do a flip, of 12 and its children 6 and 18. You press the Flip button. The flip is carried out, but now the message says Error: parent and child are both red, referring to 25 and its child 12. The resulting tree is shown in Figure 9.16a.

The procedure used to fix this rule violation is similar to the post-insertion operation with an outside grandchild, described earlier. We must perform two color switches and one rotation. So we can discuss this in the same terms we did when inserting a node, we'll call the node at the top of the triangle that was flipped (which is 12 in this case) X. This looks a little odd, because we're used to thinking of X as the node being inserted, and here it's not even a leaf node. However, these on-the-way-down rotations can take place anywhere within the tree.

The parent of X is P (25 in this case), and the grandparent of X—the parent of P—is G (50 in this case). We follow the same set of rules we did under Possibility 2, discussed earlier:

1. Switch the color of X's grandparent G (50 in this example). Ignore the message that the root must be black.

2. Switch the color of X's parent P (25).

3. Rotate with X's grandparent (50) at the top, in the direction that raises X (here a right rotation).

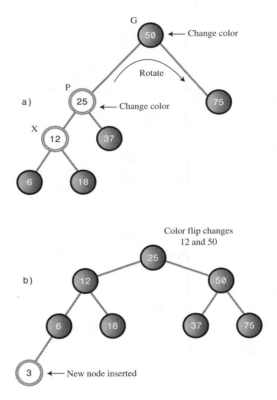

FIGURE 9.16 Outside grandchild on the way down.

Suddenly, the tree is balanced! It has also become pleasantly symmetrical. This appears to be a bit of a miracle, but it's only the result of following the color rules.

Now the node with value 3 can be inserted in the usual way. Because the node it connects to, 6, is black, there's no complexity about the insertion. One color flip (at 50) is necessary. Figure 9.16b shows the tree after 3 is inserted.

Inside Grandchild
If X is an inside grandchild when a red-red conflict occurs on the way down, two rotations are required to set it right. This situation is similar to the inside grandchild in the post-insertion phase, which we called Possibility 3.

Click Start in the RBTree Workshop applet to begin with 50, and insert 25, 75, 12, 37, 31, and 43. You'll need color flips before 12 and 31.

Now try to insert a new node with the value 28. You'll be told you need a color flip (at 37). But when you perform the flip, 37 and 25 are both red, and you get the `Error: parent and child are both red` message. Don't press Ins again.

In this situation G is 50, P is 25, and X is 37, as shown in Figure 9.17a.

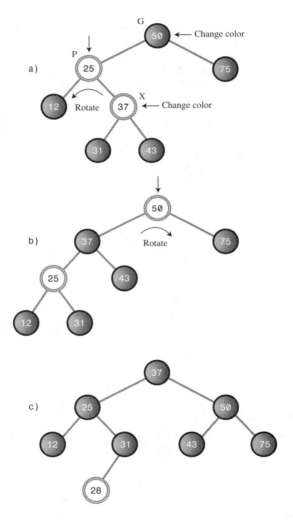

FIGURE 9.17   Inside grandchild on the way down.

To cure the red-red conflict, you must do the same two color changes and two rotations as in Possibility 3:

1. Change the color of G (it's 50; ignore the message that the root must be black).

2. Change the color of X (37).

3. Rotate with P (25) as the top, in the direction that raises X (left in this example). The result is shown in Figure 9.17b.

4. Rotate with G (50) as the top, in the direction that raises X (right in this example).

Now you can insert the 28. A color flip changes 25 and 50 to black as you insert it. The result is shown in Figure 9.17c.

This concludes the description of how a tree is kept red-black correct, and therefore balanced, during the insertion process.

Deletion

As you may recall, coding for deletion in an ordinary binary search tree is considerably harder than for insertion. The same is true in red-black trees, but in addition, the deletion process is, as you might expect, complicated by the need to restore red-black correctness after the node is removed.

In fact, the deletion process is so complicated that many programmers sidestep it in various ways. One approach, as with ordinary binary trees, is to mark a node as deleted without actually deleting it. Any search routine that finds the node knows not to tell anyone about it. This solution works in many situations, especially if deletions are not a common occurrence. In any case, we're going to forgo a discussion of the deletion process. You can refer to Appendix B, "Further Reading," if you want to pursue it.

The Efficiency of Red-Black Trees

Like ordinary binary search trees, a red-black tree allows for searching, insertion, and deletion in $O(\log_2 N)$ time. Search times should be almost the same in the red-black tree as in the ordinary tree because the red-black characteristics of the tree aren't used during searches. The only penalty is that the storage required for each node is increased slightly to accommodate the red-black color (a boolean variable).

More specifically, according to Sedgewick (see Appendix B), in practice a search in a red-black tree takes about $\log_2 N$ comparisons, and it can be shown that the search cannot require more than $2*\log_2 N$ comparisons.

The times for insertion and deletion are increased by a constant factor because of having to perform color flips and rotations on the way down and at the insertion point. On the average, an insertion requires about one rotation. Therefore, insertion still takes $O(\log_2 N)$ time but is slower than insertion in the ordinary binary tree.

Because in most applications there will be more searches than insertions and deletions, there is probably not much overall time penalty for using a red-black tree instead of an ordinary tree. Of course, the advantage is that in a red-black tree sorted data doesn't lead to slow $O(N)$ performance.

Red-Black Tree Implementation

If you're writing an insertion routine for red-black trees, all you need to do (irony intended) is to write code to carry out the operations described in the preceding sections. As we noted, showing and describing such code is beyond the scope of this book. However, here's what you'll need to think about.

You'll need to add a red-black field (which can be type `boolean`) to the `Node` class.

You can adapt the insertion routine from the `tree.java` program (Listing 8.1) in Chapter 8. On the way down to the insertion point, check if the current node is black and its two children are both red. If so, change the color of all three (unless the parent is the root, which must be kept black).

After a color flip, check that there are no violations of Rule 3. If so, perform the appropriate rotations: one for an outside grandchild, two for an inside grandchild.

When you reach a leaf node, insert the new node as in `tree.java`, making sure the node is red. Check again for red-red conflicts, and perform any necessary rotations.

Perhaps surprisingly, your software need not keep track of the black height of different parts of the tree (although you might want to check this during debugging). You need to check only for violations of Rule 3, a red parent with a red child, which can be done locally (unlike checks of black heights, Rule 4, which would require more complex bookkeeping).

If you perform the color flips, color changes, and rotations described earlier, the black heights of the nodes should take care of themselves and the tree should remain balanced. The RBTree Workshop applet reports black-height errors only because the user is not forced to carry out the insertion algorithm correctly.

Other Balanced Trees

The *AVL tree* is the earliest kind of balanced tree. It's named after its inventors: Adelson-Velskii and Landis. In AVL trees each node stores an additional piece of data: the difference between the heights of its left and right subtrees. This difference may not be larger than 1. That is, the height of a node's left subtree may be no more than one level different from the height of its right subtree.

Following insertion, the root of the lowest subtree into which the new node was inserted is checked. If the height of its children differs by more than 1, a single or double rotation is performed to equalize their heights. The algorithm then moves up and checks the node above, equalizing heights if necessary. This check continues all the way back up to the root.

Search times in an AVL tree are O(logN) because the tree is guaranteed to be balanced. However, because two passes through the tree are necessary to insert (or delete) a node, one down to find the insertion point and one up to rebalance the tree, AVL trees are not as efficient as red-black trees and are not used as often.

The other important kind of balanced tree is the *multiway tree*, in which each node can have more than two children. We'll look at one version of multiway trees, the 2-3-4 tree, in the next chapter. One problem with multiway trees is that each node must be larger than for a binary tree because it needs a reference to every one of its children.

Summary

- It's important to keep a binary search tree balanced to ensure that the time necessary to find a given node is kept as short as possible.

- Inserting data that has already been sorted can create a maximally unbalanced tree, which will have search times of O(N).

- In the red-black balancing scheme, each node is given a new characteristic: a color that can be either red or black.

- A set of rules, called red-black rules, specifies permissible ways that nodes of different colors can be arranged.

- These rules are applied while inserting (or deleting) a node.

- A color flip changes a black node with two red children to a red node with two black children.

- In a rotation, one node is designated the top node.

- A right rotation moves the top node into the position of its right child, and the top node's left child into its position.

- A left rotation moves the top node into the position of its left child, and the top node's right child into its position.

- Color flips, and sometimes rotations, are applied while searching down the tree to find where a new node should be inserted. These flips simplify returning the tree to red-black correctness following an insertion.

- After a new node is inserted, red-red conflicts are checked again. If a violation is found, appropriate rotations are carried out to make the tree red-black correct.

- These adjustments result in the tree being balanced, or at least almost balanced.

- Adding red-black balancing to a binary tree has only a small negative effect on average performance, and avoids worst-case performance when the data is already sorted.

Questions

These questions are intended as a self-test for readers. Answers may be found in Appendix C.

1. A balanced binary search tree is desirable because it avoids slow performance when data is inserted _____ .

2. In a balanced tree,

   a. the tree may need to be restructured during searches.

   b. the paths from the root to all the leaf nodes are about the same length.

   c. all left subtrees are the same height as all right subtrees.

   d. the height of all subtrees is closely controlled.

3. True or False: The red-black rules rearrange the nodes in a tree to balance it.

4. A null child is

   a. a child that doesn't exist but will be created next.

   b. a child with no children of its own.

   c. one of the two potential children of a leaf node where an insertion will be made.

   d. a non-existent left child of a node with a right child (or vice versa).

5. Which of the following is *not* a red-black rule?

   a. Every path from a root to a leaf, or to a null child, must contain the same number of black nodes.

   b. If a node is black, its children must be red.

   c. The root is always black.

   d. All three are valid rules.

6. The two possible actions used to balance a tree are _____ and _____ .

7. Newly inserted nodes are always colored _____ .

8. Which of the following is *not* involved in a rotation?

   a. rearranging nodes to restore the characteristics of a binary search tree

   b. changing the color of nodes

   c. ensuring the red-black rules are followed

   d. attempting to balance the tree

9. A "crossover" node or subtree starts as a _____ and becomes a _____ , or vice versa.

10. Which of the following is *not* true? Rotations may need to be made

    a. before a node is inserted.

    b. after a node is inserted.

    c. during a search for the insertion point.

    d. when searching for a node with a given key.

11. A color flip involves changing the color of _____ and _____ .

12. An outside grandchild is

    a. on the opposite side of its parent that its parent is of its sibling.

    b. on the same side of its parent that its parent is of its parent.

    c. one which is the left descendant of a right descendant (or vice versa).

    d. on the opposite side of its parent that its sibling is of their grandparents.

13. True or False: When one rotation immediately follows another, they are in opposite directions.

14. Two rotations are necessary when

    a. the node is an inside grandchild and the parent is red.

    b. the node is an inside grandchild and the parent is black.

    c. the node is an outside grandchild and the parent is red.

    d. the node is an outside grandchild and the parent is black.

15. True or False: Deletion in a red-black tree may require some readjustment of the tree's structure.

Experiments

Carrying out these experiments will help to provide insights into the topics covered in the chapter. No programming is involved.

1. Make an activity diagram or flowchart of all the node arrangements and colors you can encounter when inserting a new node in a red-black tree, and what you should do in each situation to insert a new node.

2. Use the RBTree Workshop applet to set up all the situations depicted in Experiment 1 and follow the instructions for insertion.

3. Do enough insertions to convince yourself that if red-black rules 1, 2, and 3 are followed exactly, Rule 4 will take care of itself.

NOTE

Because no code was shown in this chapter, it hardly seems fair to include any programming projects. If you want a serious challenge, implement a red-black tree. You might start with the `tree.java` program (Listing 8.1) from Chapter 8.

10

2-3-4 Trees and External Storage

In a binary tree, each node has one data item and can have up to two children. If we allow more data items and children per node, the result is a *multiway tree*. 2-3-4 trees, to which we devote the first part of this chapter, are multiway trees that can have up to four children and three data items per node.

2-3-4 trees are interesting for several reasons. First, they're balanced trees like red-black trees. They're slightly less efficient than red-black trees but easier to program. Second, and most important, they serve as an easy-to-understand introduction to B-trees.

A B-tree is another kind of multiway tree that's particularly useful for organizing data in external storage. (*External* means external to main memory; usually this is a disk drive). A node in a B-tree can have dozens or hundreds of children. We'll discuss external storage and B-trees at the end of this chapter.

Introduction to 2-3-4 Trees

In this section we'll look at the characteristics of 2-3-4 trees. Later we'll see how a Workshop applet models a 2-3-4 tree and how we can program a 2-3-4 tree in Java. We'll also look at the surprisingly close relationship between 2-3-4 trees and red-black trees.

Figure 10.1 shows a small 2-3-4 tree. Each lozenge-shaped node can hold one, two, or three data items.

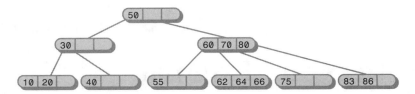

FIGURE 10.1   A 2-3-4 tree.

Here the top three nodes have children, and the six nodes on the bottom row are all leaf nodes, which by definition have no children. In a 2-3-4 tree all the leaf nodes are always on the same level.

What's in a Name?

The 2, 3, and 4 in the name *2-3-4 tree* refer to how many links to child nodes can potentially be contained in a given node. For non-leaf nodes, three arrangements are possible:

- A node with one data item always has two children.
- A node with two data items always has three children.
- A node with three data items always has four children.

In short, a non-leaf node must always have one more child than it has data items. Or, to put it symbolically, if the number of child links is L and the number of data items is D, then

L = D + 1

This critical relationship determines the structure of 2-3-4 trees. A leaf node, by contrast, has no children, but it can nevertheless contain one, two, or three data items. Empty nodes are not allowed.

Because a *2-3-4 tree* can have nodes with up to four children, it's called *a multiway tree of order 4*.

You may wonder why a 2-3-4 tree isn't called a 1-2-3-4 tree. Can't a node have only one child, as nodes in binary trees can? A binary tree (described in Chapter 8, "Binary Trees," and Chapter 9, "Red-Black Trees") can be thought of as a multiway tree of order 2 because each node can have up to two children. However, there's a difference (besides the maximum number of children) between binary trees and 2-3-4 trees. In a binary tree, a node can have *up to* two child links. A single link, to its left or to its right child, is also perfectly permissible. The other link has a null value.

In a 2-3-4 tree, on the other hand, nodes with a single link are not permitted. A node with one data item must always have two links, unless it's a leaf, in which case it has no links.

Figure 10.2 shows the possibilities. A node with two links is called a *2-node*, a node with three links is a *3-node*, and a node with four links is a *4-node*, but there is no such thing as a 1-node.

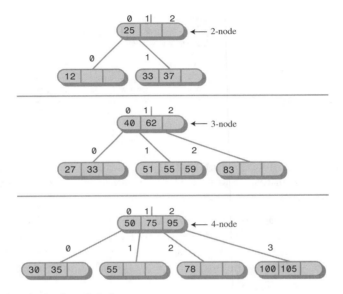

FIGURE 10.2   Nodes in a 2-3-4 tree.

2-3-4 Tree Organization

For convenience we number the data items in a link from 0 to 2, and the child links from 0 to 3, as shown in Figure 10.2. The data items in a node are arranged in ascending key order, by convention from left to right (lower to higher numbers).

An important aspect of any tree's structure is the relationship of its links to the key values of its data items. In a binary tree, all children with keys less than the node's key are in a subtree rooted in the node's left child, and all children with keys larger than or equal to the node's key are rooted in the node's right child. In a 2-3-4 tree the principle is the same, but there's more to it:

- All children in the subtree rooted at child 0 have key values less than key 0.

- All children in the subtree rooted at child 1 have key values greater than key 0 but less than key 1.

- All children in the subtree rooted at child 2 have key values greater than key 1 but less than key 2.

- All children in the subtree rooted at child 3 have key values greater than key 2.

This relationship is shown in Figure 10.3. Duplicate values are not usually permitted in 2-3-4 trees, so we don't need to worry about comparing equal keys.

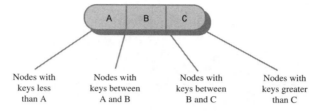

FIGURE 10.3   Keys and children.

Refer back to the tree in Figure 10.1. As in all 2-3-4 trees, the leaves are all on the same level (the bottom row). Upper-level nodes are often not full; that is, they may contain only one or two data items instead of three.

Also, notice that the tree is balanced. It retains its balance even if you insert a sequence of data in ascending (or descending) order. The 2-3-4 tree's self-balancing capability results from the way new data items are inserted, as we'll see in a moment.

Searching a 2-3-4 Tree

Finding a data item with a particular key is similar to the search routine in a binary tree. You start at the root and, unless the search key is found there, select the link that leads to the subtree with the appropriate range of values.

For example, to search for the data item with key 64 in the tree in Figure 10.1, you start at the root. You search the root but don't find the item. Because 64 is larger than 50, you go to child 1, which we will represent as 60/70/80. (Remember that child 1 is on the right because the numbering of children and links starts at 0 on the left.) You don't find the data item in this node either, so you must go to the next child. Here, because 64 is greater than 60 but less than 70, you go again to child 1. This time you find the specified item in the 62/64/66 link.

Insertion

New data items are always inserted in leaves, which are on the bottom row of the tree. If items were inserted in nodes with children, the number of children would need to be changed to maintain the structure of the tree, which stipulates that there should be one more child than data items in a node.

Insertion into a 2-3-4 tree is sometimes quite easy and sometimes rather complicated. In any case the process begins by searching for the appropriate leaf node.

If no full nodes are encountered during the search, insertion is easy. When the
appropriate leaf node is reached, the new data item is simply inserted into it. Figure
10.4 shows a data item with key 18 being inserted into a 2-3-4 tree.

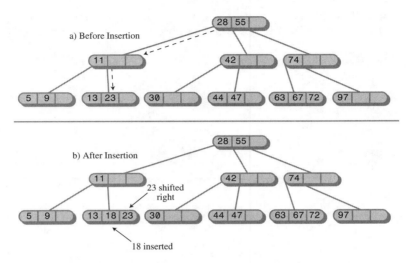

FIGURE 10.4   Insertion with no splits.

Insertion may involve moving one or two other items in a node so the keys will be
in the correct order after the new item is inserted. In this example the 23 had to be
shifted right to make room for the 18.

Node Splits

Insertion becomes more complicated if a full node is encountered on the path down
to the insertion point. When this happens, the node must be *split*. It's this splitting
process that keeps the tree balanced. The kind of 2-3-4 tree we're discussing here is
often called a *top-down* 2-3-4 tree because nodes are split on the way down to the
insertion point.

Let's name the data items in the node that's about to be split A, B, and C. Here's
what happens in a split. (We assume the node being split is not the root; we'll
examine splitting the root later.)

- A new, empty node is created. It's a sibling of the node being split, and is
  placed to its right.

- Data item C is moved into the new node.

- Data item B is moved into the parent of the node being split.

- Data item A remains where it is.

- The rightmost two children are disconnected from the node being split and connected to the new node.

An example of a node split is shown in Figure 10.5. Another way of describing a node split is to say that a 4-node has been transformed into two 2-nodes.

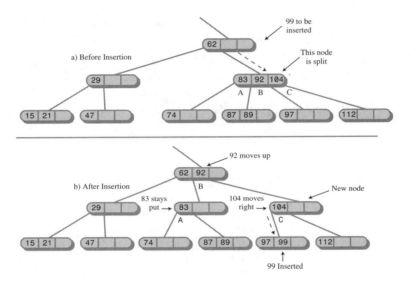

FIGURE 10.5 Splitting a node.

Notice that the effect of the node split is to move data up and to the right. It is this rearrangement that keeps the tree balanced.

Here the insertion required only one node split, but more than one full node may be encountered on the path to the insertion point. When this is the case, there will be multiple splits.

Splitting the Root

When a full root is encountered at the beginning of the search for the insertion point, the resulting split is slightly more complicated:

- A new root is created. It becomes the parent of the node being split.

- A second new node is created. It becomes a sibling of the node being split.

- Data item C is moved into the new sibling.

- Data item B is moved into the new root.

- Data item A remains where it is.

- The two rightmost children of the node being split are disconnected from it and connected to the new right-hand node.

Figure 10.6 shows the root being split. This process creates a new root that's at a higher level than the old one. Thus, the overall height of the tree is increased by one. Another way to describe splitting the root is to say that a 4-node is split into three 2-nodes.

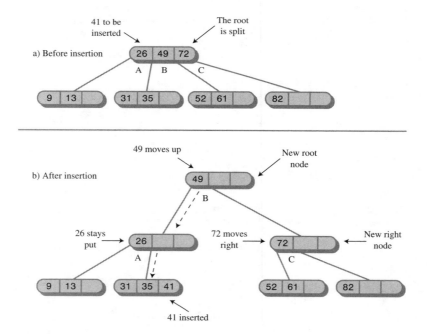

FIGURE 10.6   Splitting the root.

Following a node split, the search for the insertion point continues down the tree. In Figure 10.6, the data item with a key of 41 is inserted into the appropriate leaf.

Splitting on the Way Down

Notice that, because all full nodes are split on the way down, a split can't cause an effect that ripples back up through the tree. The parent of any node that's being split is guaranteed not to be full and can therefore accept data item B without itself needing to be split. Of course, if this parent already had two children when its child was split, it will become full. However, that just means that it will be split when the next search encounters it.

Figure 10.7 shows a series of insertions into an empty tree. There are four node splits: two of the root and two of leaves.

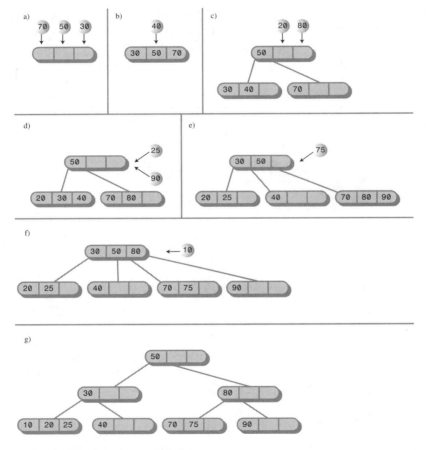

FIGURE 10.7   Insertions into a 2-3-4 tree.

The Tree234 Workshop Applet

Operating the Tree234 Workshop applet provides a quick way to see how 2-3-4 trees work. When you start the applet, you'll see a screen similar to Figure 10.8.

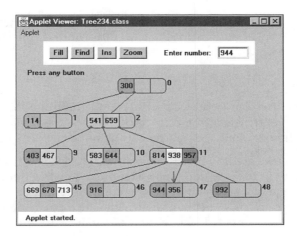

FIGURE 10.8   The Tree234 Workshop applet.

The Fill Button

When it's first started, the Tree234 Workshop applet inserts 7 data items into the tree. You can use the Fill button to create a new tree with a different number of data items from 0 to 45. Click Fill and type the number into the field when prompted. Another click will create the new tree.

The tree may not look very full with 45 nodes, but more nodes require more levels, which won't fit in the display.

The Find Button

You can watch the applet locate a data item with a given key by repeatedly clicking the Find button. When prompted, type in the appropriate key. Then, as you click the button, watch the red arrow move from node to node as it searches for the item.

Messages will say something like Went to child number 1. As we've seen, children are numbered from 0 to 3 from left to right, while data items are numbered from 0 to 2. After a little practice you should be able to predict the path the search will take.

A search involves examining one node on each level. The applet supports a maximum of four levels, so any item can be found by examining a maximum of four nodes. Within each non-leaf node, the algorithm examines each data item, starting on the left, to see if it matches the search key or, if not, which child it should go to next. In a leaf node it examines each data item to see if it matches the search key. If it can't find the specified item in the leaf node, the search fails.

In the Tree234 Workshop applet it's important to complete each operation before attempting a new one. Continue to click the button until the message says Press any button. This is the signal that an operation is complete.

The Ins Button

The Ins button causes a new data item, with a key specified in the text box, to be inserted in the tree. The algorithm first searches for the appropriate node. If it encounters a full node along the way, it splits that node before continuing.

Experiment with the insertion process. Watch what happens when there are no full nodes on the path to the insertion point. This process is straightforward. Then try inserting at the end of a path that includes a full node, either at the root, at the leaf, or somewhere in between. Watch how new nodes are formed and the contents of the node being split are distributed among three different nodes.

The Zoom Button

One of the problems with 2-3-4 trees is that there are a great many nodes and data items just a few levels down. The Tree234 Workshop applet supports only four levels, but there are potentially 64 nodes on the bottom level, each of which can hold up to three data items.

Displaying so many items at once on one row is not practical, so the applet shows only some of them: the children of a selected node. (To see the children of another node, you click on it; we'll discuss that in a moment.) To see a zoomed-out view of the entire tree, click the Zoom button. Figure 10.9 shows what you'll see.

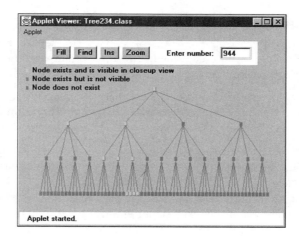

FIGURE 10.9   The zoomed-out view.

In this view nodes are shown as small rectangles; data items are not shown. Nodes that exist and are visible in the zoomed-in view (which you can restore by clicking Zoom again) are shown in green. Nodes that exist but aren't currently visible in the zoomed-out view are shown in magenta, and nodes that don't exist are shown in gray. These colors are hard to distinguish on the figure; you'll need to view the applet on your color monitor to make sense of the display.

Using the Zoom button to toggle back and forth between the zoomed-out and zoomed-in views allows you to see both the big picture and the details and, we hope, put the two together in your mind.

Viewing Different Nodes

In the zoomed-in view you can always see all the nodes in the top two rows: There's only one, the root, in the top row, and only four in the second row. Below the second row things get more complicated because there are too many nodes to fit on the screen: 16 on the third row, 64 on the fourth. However, you can see any node you want by clicking on its parent, or sometimes its grandparent and then its parent.

A blue triangle at the bottom of a node shows where a child is connected to a node. If a node's children are currently visible, the lines to the children can be seen running from the blue triangles to them. If the children aren't currently visible, there are no lines, but the blue triangles indicate that the node nevertheless has children. If you click on the parent node, its children, and the lines to them, will appear. By clicking the appropriate nodes, you can navigate all over the tree.

For convenience, all the nodes are numbered, starting with 0 at the root and continuing up to 85 for the node on the far right of the bottom row. The numbers are displayed to the upper right of each node, as shown in Figure 10.8. Nodes are numbered whether they exist or not, so the numbers on existing nodes probably won't be contiguous.

Figure 10.10 shows a small tree with four nodes in the third row. The user has clicked on node 1, so its two children, numbered 5 and 6, are visible.

If the user clicks on node 2, its children, numbered 9 and 10, will appear, as shown in Figure 10.11.

These figures show how to switch among different nodes in the third row by clicking nodes in the second row. To switch nodes in the fourth row, you'll need to click first on a grandparent in the second row and then on a parent in the third row.

During searches and insertions with the Find and Ins buttons, the view will change automatically to show the node currently being pointed to by the red arrow.

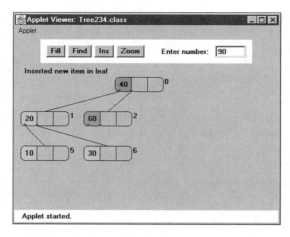

FIGURE 10.10   Selecting the leftmost children.

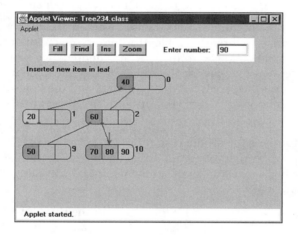

FIGURE 10.11   Selecting the rightmost children.

Experiments

The Tree234 Workshop applet offers a quick way to learn about 2-3-4 trees. Try inserting items into the tree. Watch for node splits. Stop before one is about to happen, and figure out where the three data items from the split node are going to go. Then press Ins again to see whether you're right.

As the tree gets larger you'll need to move around it to see all the nodes. Click on a node to see its children (and their children, and so on). If you lose track of where you are, use the Zoom key to see the big picture.

How many data items can you insert in the tree? There's a limit because only four levels are allowed. Four levels can potentially contain 1 + 4 + 16 + 64 nodes, for a total of 85 nodes (all visible on the zoomed-out display). If there were a full 3 items per node, there would be 255 data items. However, the nodes can't all be full at the same time. Long before they fill up, another root split, leading to five levels, would be necessary, and this is impossible because the applet supports only four levels.

You can insert the most items by deliberately inserting them into nodes that lie on paths with no full nodes, so that no splits are necessary. Of course, this is not a reasonable procedure with real data. For random data you probably can't insert more than about 50 items into the applet. The Fill button allows only 45, to minimize the possibility of overflow.

Java Code for a 2-3-4 Tree

In this section we'll examine a Java program that models a 2-3-4 tree. We'll show the complete `tree234.java` program at the end of the section. This program is relatively complex, and the classes are extensively interrelated, so you'll need to peruse the entire listing to see how it works.

There are four classes: `DataItem`, `Node`, `Tree234`, and `Tree234App`. We'll discuss them in turn.

The `DataItem` Class

Objects of the `DataItem` class represent the data items stored in nodes. In a real-world program each object would contain an entire personnel or inventory record, but here there's only one piece of data, of type `long`, associated with each `DataItem` object.

The only actions that objects of this class can perform are to initialize themselves and display themselves. The display is the data value preceded by a slash: /27. (The display routine in the `Node` class will call this routine to display all the items in a node.)

The `Node` Class

The `Node` class contains two arrays: `childArray` and `itemArray`. The first is four cells long and holds references to whatever children the node might have. The second is three cells long and holds references to objects of type `DataItem` contained in the node.

Note that the data items in `itemArray` comprise an ordered array. New items are added, or existing ones removed, in the same way they would be in any ordered array (as described in Chapter 2, "Arrays"). Items may need to be shifted to make room to insert a new item in order, or to close an empty cell when an item is removed.

We've chosen to store the number of items currently in the node (numItems) and the node's parent (parent) as fields in this class. Neither of these fields is strictly necessary and could be eliminated to make the nodes smaller. However, including them clarifies the programming, and only a small price is paid in increased node size.

Various small utility routines are provided in the Node class to manage the connections to child and parent and to check if the node is full and if it is a leaf. However, the major work is done by the findItem(), insertItem(), and removeItem() routines, which handle individual items within the node. They search through the node for a data item with a particular key; insert a new item into the node, moving existing items if necessary; and remove an item, again moving existing items if necessary. Don't confuse these methods with the find() and insert() routines in the Tree234 class, which we'll look at next.

A display routine displays a node with slashes separating the data items, like /27/56/89/, /14/66/, or /45/.

Don't forget that in Java, references are automatically initialized to null and numbers to 0 when their object is created, so class Node doesn't need a constructor.

The Tree234 Class

An object of the Tree234 class represents the entire tree. The class has only one field: root, of type Node. All operations start at the root, so that's all a tree needs to remember.

Searching
Searching for a data item with a specified key is carried out by the find() routine. It starts at the root and at each node calls that node's findItem() routine to see whether the item is there. If so, it returns the index of the item within the node's item array.

If find() is at a leaf and can't find the item, the search has failed, so it returns –1. If it can't find the item in the current node, and the current node isn't a leaf, find() calls the getNextChild() method, which figures out which of a node's children the routine should go to next.

Inserting
The insert() method starts with code similar to find(), except that it splits a full node if it finds one. Also, it assumes it can't fail; it keeps looking, going to deeper and deeper levels, until it finds a leaf node. At this point the method inserts the new data item into the leaf. (There is always room in the leaf; otherwise, the leaf would have been split.)

Splitting
The split() method is the most complicated in this program. It is passed the node that will be split as an argument. First, the two rightmost data items are removed

from the node and stored. Then the two rightmost children are disconnected; their references are also stored.

A new node, called `newRight`, is created. It will be placed to the right of the node being split. If the node being split is the root, an additional new node is created: a new root.

Next, appropriate connections are made to the parent of the node being split. It may be a pre-existing parent, or if the root is being split, it will be the newly created root node. Assume the three data items in the node being split are called A, B, and C. Item B is inserted in this parent node. If necessary, the parent's existing children are disconnected and reconnected one position to the right to make room for the new data item and new connections. The `newRight` node is connected to this parent. (Refer to Figures 10.5 and 10.6.)

Now the focus shifts to the `newRight` node. Data Item C is inserted in it, and child 2 and child 3, which were previously disconnected from the node being split, are connected to it. The split is now complete, and the `split()` routine returns.

The `Tree234App` Class

In the `Tree234App` class, the `main()` routine inserts a few data items into the tree. It then presents a character-based interface for the user, who can enter s to see the tree, i to insert a new data item, and f to find an existing item. Here's some sample interaction:

```
Enter first letter of show, insert, or find: s
level=0 child=0 /50/
level=1 child=0 /30/40/
level=1 child=1 /60/70/

Enter first letter of show, insert, or find: f
Enter value to find: 40
Found 40

Enter first letter of show, insert, or find: i
Enter value to insert: 20
Enter first letter of show, insert, or find: s
level=0 child=0 /50/
level=1 child=0 /20/30/40/
level=1 child=1 /60/70/

Enter first letter of show, insert, or find: i
Enter value to insert: 10
Enter first letter of show, insert, or find: s
```

```
level=0 child=0 /30/50/
level=1 child=0 /10/20/
level=1 child=1 /40/
level=1 child=2 /60/70/
```

The output is not very intuitive, but there's enough information to draw the tree if you want. The level, starting with 0 at the root, is shown, as well as the child number. The display algorithm is depth-first, so the root is shown first, then its first child and the subtree of which the first child is the root, then the second child and its subtree, and so on.

The output shows two items being inserted: 20 and 10. The second of these caused a node (the root's child 0) to split. Figure 10.12 depicts the tree that results from these insertions, following the final press of the ⌷ key.

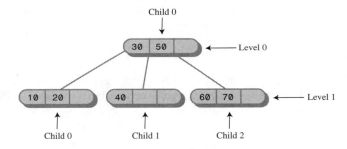

FIGURE 10.12   Sample output of the tree234.java program.

The Complete tree234.java Program

Listing 10.1 shows the complete tree234.java program, including all the classes just discussed. As with most object-oriented programs, it's probably easiest to start by examining the big picture classes first and then work down to the detail-oriented classes. In this program this order is Tree234App, Tree234, Node, DataItem.

LISTING 10.1   The tree234.java Program

```
// tree234.java
// demonstrates 234 tree
// to run this program: C>java Tree234App
import java.io.*;
////////////////////////////////////////////////////////////////
class DataItem
    {
```

LISTING 10.1 Continued

```java
   public long dData;              // one data item
//------------------------------------------------------------
   public DataItem(long dd)      // constructor
      { dData = dd; }
//------------------------------------------------------------
   public void displayItem()   // display item, format "/27"
      { System.out.print("/"+dData); }
//------------------------------------------------------------
   }  // end class DataItem
////////////////////////////////////////////////////////////////
class Node
   {
   private static final int ORDER = 4;
   private int numItems;
   private Node parent;
   private Node childArray[] = new Node[ORDER];
   private DataItem itemArray[] = new DataItem[ORDER-1];
// ------------------------------------------------------------
   // connect child to this node
   public void connectChild(int childNum, Node child)
      {
      childArray[childNum] = child;
      if(child != null)
         child.parent = this;
      }
// ------------------------------------------------------------
   // disconnect child from this node, return it
   public Node disconnectChild(int childNum)
      {
      Node tempNode = childArray[childNum];
      childArray[childNum] = null;
      return tempNode;
      }
// ------------------------------------------------------------
   public Node getChild(int childNum)
      { return childArray[childNum]; }
// ------------------------------------------------------------
   public Node getParent()
      { return parent; }
// ------------------------------------------------------------
   public boolean isLeaf()
```

LISTING 10.1   Continued

```
         { return (childArray[0]==null) ? true : false; }
// --------------------------------------------------------------
   public int getNumItems()
     { return numItems; }
// --------------------------------------------------------------
   public DataItem getItem(int index)    // get DataItem at index
     { return itemArray[index]; }
// --------------------------------------------------------------
   public boolean isFull()
     { return (numItems==ORDER-1) ? true : false; }
// --------------------------------------------------------------
   public int findItem(long key)       // return index of
     {                                 // item (within node)
     for(int j=0; j<ORDER-1; j++)      // if found,
        {                              // otherwise,
        if(itemArray[j] == null)       // return -1
           break;
        else if(itemArray[j].dData == key)
           return j;
        }
     return -1;
     } // end findItem
// --------------------------------------------------------------
   public int insertItem(DataItem newItem)
     {
     // assumes node is not full
     numItems++;                       // will add new item
     long newKey = newItem.dData;      // key of new item

     for(int j=ORDER-2; j>=0; j--)     // start on right,
        {                              //    examine items
        if(itemArray[j] == null)       // if item null,
           continue;                   // go left one cell
        else                           // not null,
           {                           // get its key
           long itsKey = itemArray[j].dData;
           if(newKey < itsKey)                 // if it's bigger
              itemArray[j+1] = itemArray[j]; // shift it right
           else
              {
              itemArray[j+1] = newItem;   // insert new item
```

LISTING 10.1   Continued

```
                  return j+1;                    // return index to
                  }                              //    new item
            }  // end else (not null)
         }  // end for                           // shifted all items,
      itemArray[0] = newItem;                     // insert new item
      return 0;
      }  // end insertItem()
// ---------------------------------------------------------------
   public DataItem removeItem()        // remove largest item
      {
      // assumes node not empty
      DataItem temp = itemArray[numItems-1];   // save item
      itemArray[numItems-1] = null;           // disconnect it
      numItems--;                             // one less item
      return temp;                            // return item
      }
// ---------------------------------------------------------------
   public void displayNode()            // format "/24/56/74/"
      {
      for(int j=0; j<numItems; j++)
         itemArray[j].displayItem();    // "/56"
      System.out.println("/");          // final "/"
      }
// ---------------------------------------------------------------
   }  // end class Node
////////////////////////////////////////////////////////////////
class Tree234
   {
   private Node root = new Node();             // make root node
// ---------------------------------------------------------------
   public int find(long key)
      {
      Node curNode = root;
      int childNumber;
      while(true)
         {
         if(( childNumber=curNode.findItem(key) ) != -1)
            return childNumber;                // found it
         else if( curNode.isLeaf() )
            return -1;                         // can't find it
         else                                  // search deeper
```

LISTING 10.1   Continued

```
                curNode = getNextChild(curNode, key);
            } // end while
        }
// -----------------------------------------------------------
    // insert a DataItem
    public void insert(long dValue)
        {
        Node curNode = root;
        DataItem tempItem = new DataItem(dValue);

        while(true)
            {
            if( curNode.isFull() )              // if node full,
                {
                split(curNode);                 // split it
                curNode = curNode.getParent();  // back up
                                                // search once
                curNode = getNextChild(curNode, dValue);
                }  // end if(node is full)

            else if( curNode.isLeaf() )         // if node is leaf,
                break;                          // go insert
            // node is not full, not a leaf; so go to lower level
            else
                curNode = getNextChild(curNode, dValue);
            }  // end while

        curNode.insertItem(tempItem);       // insert new DataItem
        }  // end insert()
// -----------------------------------------------------------
    public void split(Node thisNode)     // split the node
        {
        // assumes node is full
        DataItem itemB, itemC;
        Node parent, child2, child3;
        int itemIndex;

        itemC = thisNode.removeItem();       // remove items from
        itemB = thisNode.removeItem();       // this node
        child2 = thisNode.disconnectChild(2); // remove children
        child3 = thisNode.disconnectChild(3); // from this node
```

LISTING 10.1 Continued

```
    Node newRight = new Node();      // make new node

    if(thisNode==root)               // if this is the root,
       {
       root = new Node();            // make new root
       parent = root;                // root is our parent
       root.connectChild(0, thisNode); // connect to parent
       }
    else                             // this node not the root
       parent = thisNode.getParent(); // get parent

    // deal with parent
    itemIndex = parent.insertItem(itemB); // item B to parent
    int n = parent.getNumItems();    // total items?

    for(int j=n-1; j>itemIndex; j--)       // move parent's
       {                                   // connections
       Node temp = parent.disconnectChild(j); // one child
       parent.connectChild(j+1, temp);      // to the right
       }
                              // connect newRight to parent
    parent.connectChild(itemIndex+1, newRight);

    // deal with newRight
    newRight.insertItem(itemC);      // item C to newRight
    newRight.connectChild(0, child2); // connect to 0 and 1
    newRight.connectChild(1, child3); // on newRight
    }  // end split()
// -------------------------------------------------------------
    // gets appropriate child of node during search for value
    public Node getNextChild(Node theNode, long theValue)
       {
       int j;
       // assumes node is not empty, not full, not a leaf
       int numItems = theNode.getNumItems();
       for(j=0; j<numItems; j++)            // for each item in node
          {                                 // are we less?
          if( theValue < theNode.getItem(j).dData )
             return theNode.getChild(j);  // return left child
          }  // end for                    // we're greater, so
```

LISTING 10.1   Continued

```
      return theNode.getChild(j);        // return right child
      }
// -----------------------------------------------------------
   public void displayTree()
      {
      recDisplayTree(root, 0, 0);
      }
// -----------------------------------------------------------
   private void recDisplayTree(Node thisNode, int level,
                                              int childNumber)
      {
      System.out.print("level="+level+" child="+childNumber+" ");
      thisNode.displayNode();                // display this node

      // call ourselves for each child of this node
      int numItems = thisNode.getNumItems();
      for(int j=0; j<numItems+1; j++)
         {
         Node nextNode = thisNode.getChild(j);
         if(nextNode != null)
            recDisplayTree(nextNode, level+1, j);
         else
            return;
         }
      }  // end recDisplayTree()
// -------------------------------------------------------------\
   }  // end class Tree234
//////////////////////////////////////////////////////////////////
class Tree234App
   {
   public static void main(String[] args) throws IOException
      {
      long value;
      Tree234 theTree = new Tree234();

      theTree.insert(50);
      theTree.insert(40);
      theTree.insert(60);
      theTree.insert(30);
      theTree.insert(70);
```

LISTING 10.1   Continued

```
      while(true)
         {
         System.out.print("Enter first letter of ");
         System.out.print("show, insert, or find: ");
         char choice = getChar();
         switch(choice)
            {
            case 's':
               theTree.displayTree();
               break;
            case 'i':
               System.out.print("Enter value to insert: ");
               value = getInt();
               theTree.insert(value);
               break;
            case 'f':
               System.out.print("Enter value to find: ");
               value = getInt();
               int found = theTree.find(value);
               if(found != -1)
                  System.out.println("Found "+value);
               else
                  System.out.println("Could not find "+value);
               break;
            default:
               System.out.print("Invalid entry\n");
            }  // end switch
         }  // end while
      }  // end main()
//-------------------------------------------------------------
   public static String getString() throws IOException
      {
      InputStreamReader isr = new InputStreamReader(System.in);
      BufferedReader br = new BufferedReader(isr);
      String s = br.readLine();
      return s;
      }
//-------------------------------------------------------------
   public static char getChar() throws IOException
      {
      String s = getString();
```

LISTING 10.1   Continued

```
        return s.charAt(0);
        }

//--------------------------------------------------------------
    public static int getInt() throws IOException
        {
        String s = getString();
        return Integer.parseInt(s);
        }
//--------------------------------------------------------------
    }  // end class Tree234App
//////////////////////////////////////////////////////////////////
```

2-3-4 Trees and Red-Black Trees

At this point 2-3-4 trees and red-black trees (described in Chapter 9) probably seem like entirely different entities. However, it turns out that in a certain sense they are completely equivalent. One can be transformed into the other by the application of a few simple rules, and even the operations needed to keep them balanced are equivalent. Mathematicians would say they were *isomorphic*.

You probably won't ever need to transform a 2-3-4 tree into a red-black tree, but equivalence of these structures casts additional light on their operation and is useful in analyzing their efficiency.

Historically, the 2-3-4 tree was developed first; later the red-black tree evolved from it.

Transformation from 2-3-4 to Red-Black

A 2-3-4 tree can be transformed into a red-black tree by applying three rules, as shown in Figure 10.13.

- Transform any 2-node in the 2-3-4 tree into a black node in the red-black tree. This is shown in Figure 10.13a.

- Transform any 3-node into a child node and a parent node, as shown in Figure 10.13b. The child node has two children of its own: either W and X or X and Y. The parent has one other child: either Y or W. It doesn't matter which item becomes the child and which the parent. The child is colored red and the parent is colored black.

- Transform any 4-node into a parent and two children, as shown in Figure 10.13c. The first child has its own children W and X; the second child has children Y and Z. As before, the children are colored red and the parent is black.

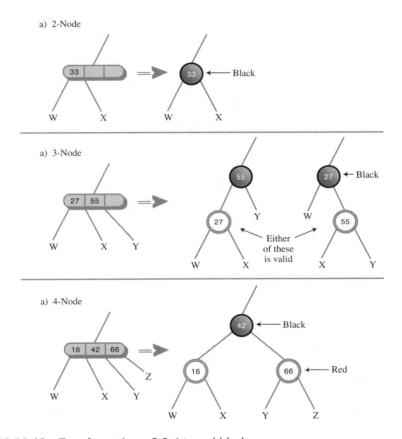

FIGURE 10.13   Transformations: 2-3-4 to red-black.

Figure 10.14 shows a 2-3-4 tree and the corresponding red-black tree obtained by applying these transformations. Dotted lines surround the subtrees that were made from 3-nodes and 4-nodes. The red-black rules are automatically satisfied by the transformation. Check that this is so: Two red nodes are never connected, and there is the same number of black nodes on every path from root to leaf (or null child).

You can say that a 3-node in a 2-3-4 tree is equivalent to a parent with a red child in a red-black tree, and a 4-node is equivalent to a parent with two red children. It follows that a black parent with a black child in a red-black tree does *not* represent a 3-node in a 2-3-4 tree; it simply represents a 2-node with another 2-node child. Similarly, a black parent with two black children does not represent a 4-node.

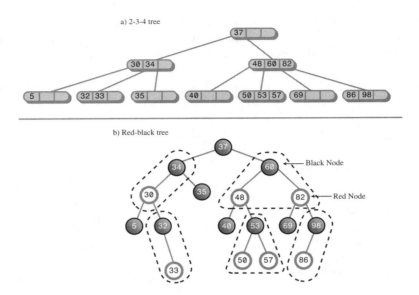

FIGURE 10.14   A 2-3-4 tree and its red-black equivalent.

Operational Equivalence

Not only does the structure of a red-black tree correspond to a 2-3-4 tree, but the operations applied to these two kinds of trees are also equivalent. In a 2-3-4 tree the tree is kept balanced using node splits. In a red-black tree the two balancing methods are color flips and rotations.

4-Node Splits and Color Flips

As you descend a 2-3-4 tree searching for the insertion point for a new node, you split each 4-node into two 2-nodes. In a red-black tree you perform color flips. How are these operations equivalent?

In Figure 10.15a we show a 4-node in a 2-3-4 tree before it is split; Figure 10.15b shows the situation after the split. The 2-node that was the parent of the 4-node becomes a 3-node.

In Figure 10.15c we show the red-black equivalent to the 2-3-4 tree in 10.15a. The dotted line surrounds the equivalent of the 4-node. A color flip results in the red-black tree of Figure 10.15d. Now nodes 40 and 60 are black and 50 is red. Thus, 50 and its parent form the equivalent of a 3-node, as shown by the dotted line. This is the same 3-node formed by the node split in Figure 10.15b.

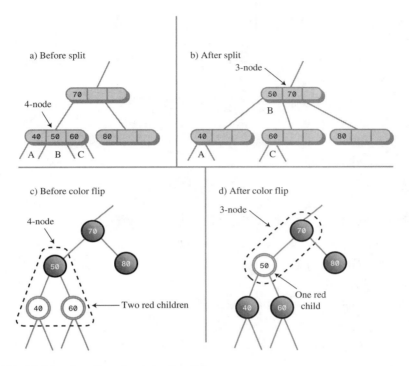

FIGURE 10.15 4-node split and color flip.

Thus, we see that splitting a 4-node during the insertion process in a 2-3-4 tree is equivalent to performing color flips during the insertion process in a red-black tree.

3-Node Splits and Rotations

When a 3-node in a 2-3-4 tree is transformed into its red-black equivalent, two arrangements are possible, as we showed earlier in Figure 10.13b. Either of the two data items can become the parent. Depending on which one is chosen, the child will be either a left child or a right child, and the slant of the line connecting parent and child will be either left or right.

Both arrangements are valid; however, they may not contribute equally to balancing the tree. Let's look at the situation in a slightly larger context.

Figure 10.16a shows a 2-3-4 tree, and 10.16b and 10.16c show two equivalent red-black trees derived from the 2-3-4 tree by applying the transformation rules. The difference between them is the choice of which of the two data items in the 3-node to make the parent: in b) 80 is the parent; in c) it's 70.

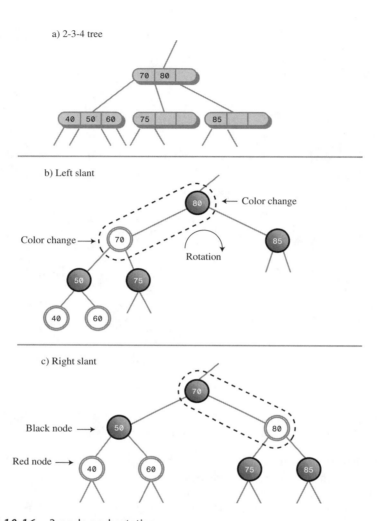

a) 2-3-4 tree

b) Left slant

c) Right slant

FIGURE 10.16   3-node and rotation.

Although these arrangements are equally valid, you can see that the tree in b) is not balanced, while that in c) is. Given the red-black tree in b), we would want to rotate it to the right (and perform two color changes) to balance it. Amazingly, this rotation results in the exact same tree shown in c).

Thus, we see an equivalence between rotations in red-black trees and the choice of which node to make the parent when transforming 2-3-4 trees to red-black trees. Although we don't show it, a similar equivalence can be seen for the double rotation necessary for inside grandchildren.

Efficiency of 2-3-4 Trees

It's harder to analyze the efficiency of a 2-3-4 tree than a red-black tree, but the equivalence of red-black trees and 2-3-4 trees gives us a starting point.

Speed

As we saw in Chapter 8, in a red-black tree one node on each level must be visited during a search, whether to find an existing node or insert a new one. The number of levels in a red-black tree (a balanced binary tree) is about $\log_2(N+1)$, so search times are proportional to this.

One node must be visited at each level in a 2-3-4 tree as well, but the 2-3-4 tree is shorter (has fewer levels) than a red-black tree with the same number of data items. Refer to Figure 10.14, where the 2-3-4 tree has three levels and the red-black tree has five.

More specifically, in 2-3-4 trees there are up to four children per node. If every node were full, the height of the tree would be proportional to $\log_4 N$. Logarithms to the base 2 and to the base 4 differ by a constant factor of 2. Thus, the height of a 2-3-4 tree would be about half that of a red-black tree, provided that all the nodes were full. Since they aren't all full, the height of the 2-3-4 tree is somewhere between $\log_2(N+1)$ and $\log_2(N+1)/2$. The reduced height of the 2-3-4 tree decreases search times slightly compared with red-black trees.

On the other hand, there are more items to examine in each node, which increases the search time. Because the data items in the node are examined using a linear search, this multiplies the search times by an amount proportional to M, the average number of items per node. The result is a search time proportional to $M*\log_4 N$.

Some nodes contain one item, some two, and some three. If we estimate that the average is two, search times will be proportional to $2*\log_4 N$. This is a small constant number that can be ignored in Big O notation.

Thus, for 2-3-4 trees the increased number of items per node tends to cancel out the decreased height of the tree. The search times for a 2-3-4 tree and for a balanced binary tree such as a red-black tree are approximately equal, and are both O(logN).

Storage Requirements

Each node in a 2-3-4 tree contains storage for three references to data items and four references to its children. This space may be in the form of arrays, as shown in tree234.java, or of individual variables. Not all this storage is used. A node with only one data item will waste 2/3 of the space for data and 1/2 the space for children. A node with two data items will waste 1/3 of the space for data and 1/4 of the space for children; or put another way, it will use 5/7 of the available space.

If we take two data items per node as the average utilization, about 2/7 of the available storage is wasted.

You might imagine using linked lists instead of arrays to hold the child and data references, but the overhead of the linked list compared with an array, for only three or four items, would probably not make this a worthwhile approach.

Because they're balanced, red-black trees contain few nodes that have only one child, so almost all the storage for child references is used. Also, every node contains the maximum number of data items, which is one. This makes red-black trees more efficient than 2-3-4 trees in terms of memory usage.

In Java, which stores references to objects instead of the objects themselves, this difference in storage between 2-3-4 trees and red-black trees may not be important, and the programming is certainly simpler for 2-3-4 trees. However, in languages that don't use references this way, the difference in storage efficiency between red-black trees and 2-3-4 trees may be significant.

2-3 Trees

We'll discuss 2-3 trees briefly here because they are historically important and because they are still used in many applications. Also, some of the techniques used with 2-3 trees are applicable to B-trees, which we'll examine in the next section. Finally, it's interesting to see how a small change in the number of children per node can cause a large change in the tree's algorithms.

2-3 trees are similar to 2-3-4 trees except that, as you might have guessed from the name, they hold one less data item and have one less child. They were the first multiway tree, invented by J. E. Hopcroft in 1970. B-trees (of which the 2-3-4 tree is a special case) were not invented until 1972.

In many respects the operation of 2-3 trees is similar to that of 2-3-4 trees. Nodes can hold one or two data items and can have zero, one, two, or three children. Otherwise, the arrangement of the key values of the parent and its children is the same. Inserting a data item into a node is potentially simplified because fewer comparisons and moves are potentially necessary. As in 2-3-4 trees, all insertions are made into leaf nodes, and all leaf nodes are on the bottom level.

Node Splits

Searching for an existing data item is handled just as it is in a 2-3-4 tree except for the number of data items and children. You might guess that insertion is also similar to a 2-3-4 tree, but there is a surprising difference in the way splits are handled.

Here's why the splits are so different. In either kind of tree a node split requires three data items: one to be kept in the node being split, one to move right into the new node, and one to move up to the parent node. A full node in a 2-3-4 tree has three data items, which are moved to these three destinations. However, a full node in a 2-3 tree has only two data items. Where can we get a third item? We must use the new item: the one being inserted in the tree.

In a 2-3-4 tree the new item is inserted after all the splits have taken place. In the 2-3 tree it must participate in the split. It must be inserted in a leaf, so no splits are possible on the way down. If the leaf node where the new item should be inserted is not full, the new item can be inserted immediately, but if the leaf node is full, it must be split. Its two items and the new item are distributed among these three nodes: the existing node, the new node, and the parent node. If the parent is not full, the operation is complete (after connecting the new node). This situation is shown in Figure 10.17.

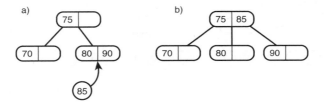

FIGURE 10.17   Insertion with non-full parent.

However, if the parent is full, it too must be split. Its two items and the item passed up from its recently split child must be distributed among the parent, a new sibling of the parent, and the parent's parent. This situation is shown in Figure 10.18.

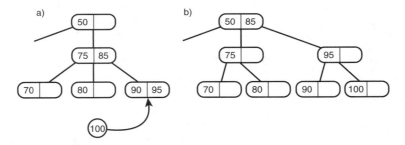

FIGURE 10.18   Insertion with full parent.

If the parent's parent (the grandparent of the leaf node) is full, it too must be split. The splitting process ripples upward until either a non-full parent or the root is encountered. If the root is full, a new root is created that is the parent of the old root, as shown in Figure 10.19.

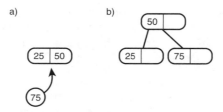

FIGURE 10.19 Splitting the root.

Figure 10.20 shows a node split that ripples up through a tree until it reaches the root.

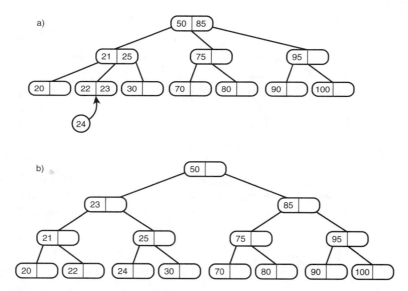

FIGURE 10.20 Splits rippling up a tree.

Implementation

We'll leave a complete Java implementation of a 2-3 tree as an exercise. However, we'll finish with some hints on how to handle splits. This is only one approach (another involves allowing each node to hold a phantom fourth child).

On the way down the insertion routine doesn't care if the nodes it encounters are full or not. It searches down through the tree until it finds the appropriate leaf. If the leaf is not full, it inserts the new value. However, if the leaf is full, it must rearrange the tree to make room. To do this, it calls a split() method. Arguments to this method can be the full leaf node and the new item. It will be the responsibility of split() to make the split and insert the new node in the new leaf.

If split() finds that the leaf's parent is full, it calls itself recursively to split the parent. It keeps calling itself until a non-full leaf or the root is found. The return value of split() is the new right node, which can be used by the previous incarnation of split().

Coding the splitting process is complicated by several factors. In a 2-3-4 tree the three items to be distributed are already sorted, but in the 2-3 tree the new item's key must be compared with the two items in the leaf; the three are then distributed according to the results of the comparison.

Also, splitting a parent creates a second parent, so now we have a left (the original) parent and a new right parent. We need to change the connections from a single parent with three children to two parents with two children each. There are three cases, depending on which child (0, 1, or 2) is being split. This situation is shown in Figure 10.21.

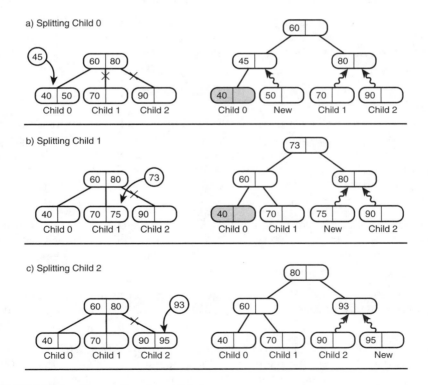

FIGURE 10.21   Connecting the children.

In this figure the new nodes created as the result of a split are shaded, and new connections are shown as wiggly lines.

External Storage

2-3-4 trees are examples of multiway trees, which have more than two children and more than one data item. Another kind of multiway tree, the B-tree, is useful when data resides in external storage. *External storage* typically refers to some kind of disk system, such as the hard disk found in most desktop computers or servers.

In this section we'll begin by describing various aspects of external file handling. We'll talk about a simple approach to organizing external data: sequential ordering. Finally, we'll discuss B-trees and explain why they work so well with disk files. We'll finish with another approach to external storage, indexing, which can be used alone or with a B-tree.

We'll also touch on other aspects of external storage, such as searching techniques. In the next chapter we'll mention a different approach to external storage: hashing.

The details of external storage techniques are dependent on the operating system, language, and even the hardware used in a particular installation. As a consequence, our discussion in this section will be considerably more general than for most topics in this book.

Accessing External Data

The data structures we've discussed so far are all based on the assumption that data is stored entirely in main memory (often called RAM, for Random Access Memory). However, in many situations the amount of data to be processed is too large to fit in main memory all at once. In this case a different kind of storage is necessary. Disk files generally have a much larger capacity than main memory; this is made possible by their lower cost per byte of storage.

Of course, disk files have another advantage: their permanence. When you turn off your computer (or the power fails), the data in main memory is lost. Disk files can retain data indefinitely with the power off. However, it's mostly the size difference that we'll be involved with here.

The disadvantage of external storage is that it's much slower than main memory. This speed difference means that different techniques must be used to handle it efficiently.

As an example of external storage, imagine that you're writing a database program to handle the data found in the phone book for a medium-sized city—perhaps 500,000 entries. Each entry includes a name, address, phone number, and various other data used internally by the phone company. Let's say an entry is stored as a record requiring 512 bytes. The result is a file size of 500,000 × 512, which is 256,000,000 bytes or 256 megabytes. We'll assume that on the target machine this is too large to fit in main memory but small enough to fit on your disk drive.

Thus, you have a large amount of data on your disk drive. How do you structure it to provide the usual desirable characteristics: quick search, insertion, and deletion?

In investigating the answers, you must keep in mind two facts. First, accessing data on a disk drive is much slower than accessing it in main memory. Second, you must access many records at once. Let's explore these points.

Very Slow Access

A computer's main memory works electronically. Any byte can be accessed just as fast as any other byte, in a fraction of a microsecond (a millionth of a second).

Things are more complicated with disk drives. Data is arranged in circular tracks on a spinning disk, something like the tracks on a compact disc (CD) or the grooves in an old-style phonograph record.

To access a particular piece of data on a disk drive, the read-write head must first be moved to the correct track. This is done with a stepping motor or similar device; it's a mechanical activity that requires several milliseconds (thousandths of a second).

Once the correct track is found, the read-write head must wait for the data to rotate into position. On the average, this takes half a revolution. Even if the disk is spinning at 10,000 revolutions per minute, about 3 more milliseconds pass before the data can be read. Once the read-write head is positioned, the actual reading (or writing) process begins; this might take a few more milliseconds.

Thus, disk access times of around 10 milliseconds are common. This is something like 10,000 times slower than main memory.

Technological progress is reducing disk access times every year, but main memory access times are being reduced faster, so the disparity between disk access and main memory access times will grow even larger in the future.

One Block at a Time

When the read-write head is correctly positioned and the reading (or writing) process begins, the drive can transfer a large amount of data to main memory fairly quickly. For this reason, and to simplify the drive control mechanism, data is stored on the disk in chunks called *blocks*, *pages*, *allocation units*, or some other name, depending on the system. We'll call them blocks.

The disk drive always reads or writes a minimum of one block of data at a time. Block size varies, depending on the operating system, the size of the disk drive, and other factors, but it is usually a power of 2. For our phone book example, let's assume a block size of 8,192 bytes (2^{13}). Thus, our phone book database will require 256,000,000 bytes divided by 8,192 bytes per block, which is 31,250 blocks.

Your software is most efficient when it specifies a read or write operation that's a multiple of the block size. If you ask to read 100 bytes, the system will read one block, 8,192 bytes, and throw away all but 100. Or if you ask to read 8,200 bytes, it will read two blocks, or 16,384 bytes, and throw away almost half of them. By organizing your software so that it works with a block of data at a time, you can optimize its performance.

Assuming our phone book record size of 512 bytes, you can store 16 records in a block (8,192 divided by 512), as shown in Figure 10.22. Thus, for maximum efficiency it's important to read 16 records at a time (or multiples of this number).

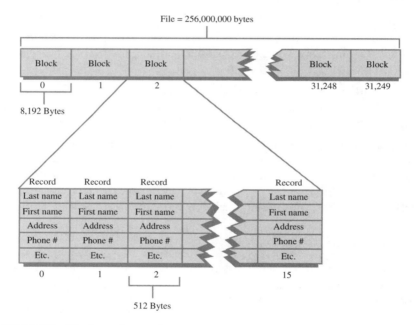

FIGURE 10.22   Blocks and records.

Notice that it's also useful to make your record size a multiple of 2. That way, an integral number of them will always fit in a block.

Of course, the sizes shown in our phone book example for records, blocks, and so on are only illustrative; they will vary widely depending on the number and size of records and other software and hardware constraints. Blocks containing hundreds of records are common, and records may be much larger or smaller than 512 bytes.

Once the read-write head is positioned as described earlier, reading a block is fairly fast, requiring only a few milliseconds. Thus, a disk access to read or write a block is not very dependent on the size of the block. It follows that the larger the block, the more efficiently you can read or write a single record (assuming you use all the records in the block).

Sequential Ordering

One way to arrange the phone book data in the disk file would be to order all the records according to some key, say alphabetically by last name. The record for Joseph Aardvark would come first, and so on. This is shown in Figure 10.23.

FIGURE 10.23   Sequential ordering.

Searching

To search a sequentially ordered file for a particular last name such as Smith, you could use a binary search. You would start by reading a block of records from the middle of the file. The 16 records in the block are all read at once into an 8,192-byte buffer in main memory.

If the keys of these records are too early in the alphabet (Keller, for example), you would go to the 3/4 point in the file (Prince) and read a block there; if the keys were too late, you'd go to the 1/4 point (DeLeon). By continually dividing the range in half, you would eventually find the record you were looking for.

As we saw in Chapter 2, a binary search in main memory takes $\log_2 N$ comparisons, which for 500,000 items would be about 19. If every comparison took, say 10 microseconds, this would be 190 microseconds, or about 2/10,000 of a second, less than an eye blink.

However, we're now dealing with data stored on a disk. Because each disk access is so time-consuming, it's more important to focus on how many disk accesses are necessary than on how many individual records there are. The time to read a block of records will be very much larger than the time to search the 16 records in the block once they're in memory.

Disk accesses are much slower than memory accesses, but on the other hand we access a block at a time, and there are far fewer blocks than records. In our example there are 31,250 blocks. $\log_2$ of this number is about 15, so in theory we'll need about 15 disk accesses to find the record we want.

In practice this number is reduced somewhat because we read 16 records at once. In the beginning stages of a binary search, it doesn't help to have multiple records in memory because the next access will be in a distant part of the file. However, when we get close to the desired record, the next record we want may already be in memory because it's part of the same block of 16. This may reduce the number of comparisons by two or so. Thus, we'll need about 13 disk accesses (15–2), which at 10 milliseconds per access requires about 130 milliseconds, or ½ second. This is much slower than in-memory access, but still not too bad.

Insertion

Unfortunately, the picture is much worse if we want to insert (or delete) an item from a sequentially ordered file. Because the data is ordered, both operations require moving half the records on the average, and therefore about half the blocks.

Moving each block requires two disk accesses: one read and one write. When the insertion point is found, the block containing it is read into a memory buffer. The last record in the block is saved, and the appropriate number of records are shifted up to make room for the new one, which is inserted. Then the buffer contents are written back to the disk file.

Next, the second block is read into the buffer. Its last record is saved, all the other records are shifted up, and the last record from the previous block is inserted at the beginning of the buffer. Then the buffer contents are again written back to disk. This process continues until all the blocks beyond the insertion point have been rewritten.

Assuming there are 31,250 blocks, we must read and write (on the average) 15,625 of them, which at 10 milliseconds per read and write requires more than 5 minutes to insert a single entry. This won't be satisfactory if you have thousands of new names to add to the phone book.

Another problem with the sequential ordering is that it works quickly for only one key. Our file is arranged by last names. But suppose you wanted to search for a particular phone number. You can't use a binary search because the data is ordered by name. You would need to go through the entire file, block by block, using sequential access. This search would require reading an average of half the blocks, which would require about 2.5 minutes, very poor performance for a simple search. It would be nice to have a more efficient way to store disk data.

B-Trees

How can the records of a file be arranged to provide fast search, insertion, and deletion times? We've seen that trees are a good approach to organizing in-memory data. Will trees work with files?

They will, but a different kind of tree must be used for external data than for in-memory data. The appropriate tree is a multiway tree somewhat like a 2-3-4 tree, but with many more data items per node; it's called a *B-tree*. B-trees were first conceived as appropriate structures for external storage by R. Bayer and E. M. McCreight in 1972. (Strictly speaking, 2-3 trees and 2-3-4 trees are B-trees of order 3 and 4, respectively, but the term *B-tree* is often taken to mean many more children per node.)

One Block Per Node

Why do we need so many items per node? We've seen that disk access is most efficient when data is read or written one block at a time. In a tree, the entity containing data is a node. It makes sense then to store an entire block of data in each node of the tree. This way, reading a node accesses a maximum amount of data in the shortest time.

How much data can be put in a node? When we simply stored the 512-byte data records for our phone book example, we could fit 16 into an 8,192-byte block.

In a tree, however, we also need to store the links to other nodes (which means links to other blocks, because a node corresponds to a block). In an in-memory tree, such as those we've discussed in previous chapters, these links are references (or pointers, in languages like C++) to nodes in other parts of memory. For a tree stored in a disk file, the links are block numbers in a file (from 0 to 31,249, in our phone book example). For block numbers we can use a field of type int, a 4-byte type, which can point to more than 2 billion possible blocks, which is probably enough for most files.

Now we can no longer squeeze 16 512-byte records into a block because we need room for the links to child nodes. We could reduce the number of records to 15 to make room for the links, but it's most efficient to have an even number of records per node, so (after appropriate negotiation with management) we reduce the record size to 507 bytes. There will be 17 child links (one more than the number of data items) so the links will require 68 bytes (17×4). This leaves room for 16 507-byte records with 12 bytes left over ($507 \times 16 + 68 = 8,180$). A block in such a tree, and the corresponding node representation, is shown in Figure 10.24.

Within each node the data is ordered sequentially by key, as in a 2-3-4 tree. In fact, the structure of a B-tree is similar to that of a 2-3-4 tree, except that there are more data items per node and more links to children. The order of a B-tree is the number of children each node can potentially have. In our example this is 17, so the tree is an order 17 B-tree.

Searching

A search for a record with a specified key is carried out in much the same way it is in an in-memory 2-3-4 tree. First, the block containing the root is read into memory. The search algorithm then starts examining each of the 15 records (or, if it's not full, as many as the node actually holds), starting at 0. When it finds a record with a greater key, it knows to go to the child whose link lies between this record and the preceding one.

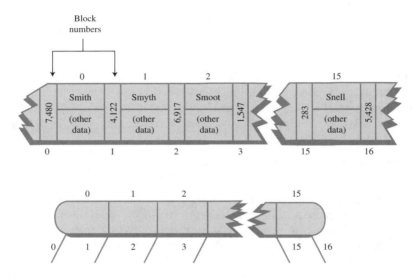

FIGURE 10.24   A node in a B-tree of order 17.

This process continues until the correct node is found. If a leaf is reached without finding the specified key, the search is unsuccessful.

Insertion

The insertion process in a B-tree is more like an insertion in a 2-3 tree than in a 2-3-4 tree. Recall that in a 2-3-4 tree many nodes are not full, and in fact contain only one data item. In particular, a node split always produces two nodes with one item in each. This is not an optimum approach in a B-tree.

In a B-tree it's important to keep the nodes as full as possible so that each disk access, which reads an entire node, can acquire the maximum amount of data. To help achieve this end, the insertion process differs from that of 2-3-4 trees in three ways:

- A node split divides the data items equally: Half go to the newly created node, and half remain in the old one.

- Node splits are performed from the bottom up, as in a 2-3 tree, rather than from the top down.

- Again, as in a 2-3 tree, it's not the middle item in a node that's promoted upward, but the middle item in the sequence formed from the items in the node plus the new item.

We'll demonstrate these features of the insertion process by building a small B-tree, as shown in Figure 10.25. There isn't room to show a realistic number of records per node, so we'll show only four; thus, the tree is an order 5 B-tree.

Figure 10.25a shows a root node that's already full; items with keys 20, 40, 60, and 80 have already been inserted into the tree. A new data item with a key of 70 is inserted, resulting in a node split. Here's how the split is accomplished. Because it's the root that's being split, two new nodes are created (as in a 2-3-4 tree): a new root and a new node to the right of the one being split.

To decide where the data items go, the insertion algorithm arranges their five keys in order, in an internal buffer. Four of these keys are from the node being split, and the fifth is from the new item being inserted. In Figure 10.25, these five-item sequences are shown to the side of the tree. In this first step the sequence 20, 40, 60, 70, 80 is shown.

The center item in this sequence, 60 in this first step, is promoted to the new root node. (In the figure, an arrow indicates that the center item will go upward.) All the items to the left of center remain in the node being split, and all the items to the right go into the new right-hand node. The result is shown in Figure 10.25b. (In our phone book example, eight items would go into each child node, rather than the two shown in the figure.)

In Figure 10.25b we insert two more items, 10 and 30. They fill up the left child, as shown in Figure 10.25c. The next item to be inserted, 15, splits this left child, with the result shown in Figure 10.25d. Here the 20 has been promoted upward into the root.

Next, three items—75, 85, and 90—are inserted into the tree. The first two fill up the third child, and the third splits it, causing the creation of a new node and the promotion of the middle item, 80, to the root. The result is shown in Figure 10.25e.

Again three items—25, 35, and 50—are added to the tree. The first two items fill up the second child, and the third one splits it, causing the creation of a new node and the promotion of the middle item, 35, to the root, as shown in Figure 10.25f.

Now the root is full. However, subsequent insertions don't necessarily cause a node split, because nodes are split only when a new item is inserted into a full node, not when a full node is encountered in the search down the tree. Thus, 22 and 27 are inserted in the second child without causing any splits, as shown in Figure 10.25g.

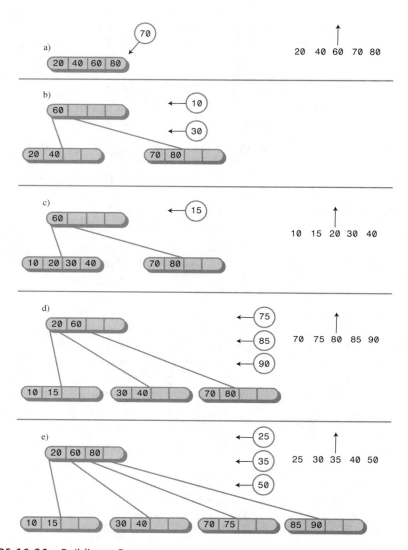

FIGURE 10.25   Building a B-tree.

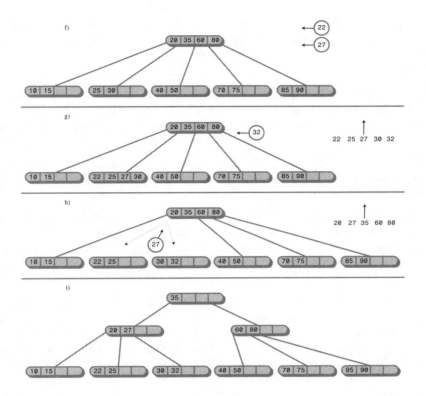

However, the next item to be inserted, 32, does cause a split; in fact it causes two of them. The second node child is full, so it's split, as shown in Figure 10.25h. However, the 27, promoted from this split, has no place to go because the root is full. Therefore, the root must be split as well, resulting in the arrangement of Figure 10.25i.

Notice that throughout the insertion process no node (except the root) is ever less than half full, and many are more than half full. As we noted, this promotes efficiency because a file access that reads a node always acquires a substantial amount of data.

Efficiency of B-Trees

Because there are so many records per node, and so many nodes per level, operations on B-trees are very fast, considering that the data is stored on disk. In our phone book example there are 500,000 records. All the nodes in the B-tree are at least half full, so they contain at least 8 records and 9 links to children. The height of the tree is thus somewhat less than $\log_9 N$ (logarithm to the base 9 of N), where N is 500,000. This is 5.972, so there will be about 6 levels in the tree.

Thus, using a B-tree, only six disk accesses are necessary to find any record in a file of 500,000 records. At 10 milliseconds per access, this takes about 60 milliseconds, or 6/100 of a second. This is dramatically faster than the binary search of a sequentially ordered file.

The more records there are in a node, the fewer levels there are in the tree. We've seen that there are 6 levels in our B-tree, even though the nodes hold only 16 records. In contrast, a binary tree with 500,000 items would have about 19 levels, and a 2-3-4 tree would have 10. If we use blocks with hundreds of records, we can reduce the number of levels in the tree and further improve access times.

Although searching is faster in B-trees than in sequentially ordered disk files, it's for insertion and deletion that B-trees show the greatest advantage.

Let's first consider a B-tree insertion in which no nodes need to be split. This is the most likely scenario, because of the large number of records per node. In our phone book example, as we've seen, only 6 accesses are required to find the insertion point. Then one more access is required to write the block containing the newly inserted record back to the disk, for a total of 7 accesses.

Next let's see how things look if a node must be split. The node being split must be read, have half its records removed, and be written back to disk. The newly created node must be written to the disk, and the parent must be read and, following the insertion of the promoted record, written back to disk. This is 5 accesses in addition to the 6 necessary to find the insertion point, for a total of 12. This is a major improvement over the 500,000 accesses required for insertion in a sequential file.

In some versions of the B-tree, only leaf nodes contain records. Non-leaf nodes contain only keys and block numbers. This may result in faster operation because each block can hold many more block numbers. The resulting higher-order tree will have fewer levels, and access speed will be increased. However, programming may be complicated because there are two kinds of nodes: leaves and non-leaves.

Indexing

A different approach to speeding up file access is to store records in sequential order but use a file *index* along with the data itself. A file index is a list of key/block pairs, arranged with the keys in order. Recall that in our original phone book example we had 500,000 records of 512 bytes each, stored 16 records to a block, in 31,250 blocks. Assuming our search key is the last name, every entry in the index contains two items:

- The key, like Jones.

- The number of the block where the Jones record is located within the file. These numbers run from 0 to 31,249.

Let's say we use a string 28 bytes long for the key (big enough for most last names) and 4 bytes for the block number (a type `int` in Java). Each entry in our index thus requires 32 bytes. This is only 1/16 the amount necessary for each record.

The entries in the index are arranged sequentially by last name. The original records on the disk can be arranged in any convenient order. This usually means that new records are simply appended to the end of the file, so the records are ordered by time of insertion. This arrangement is shown in Figure 10.26.

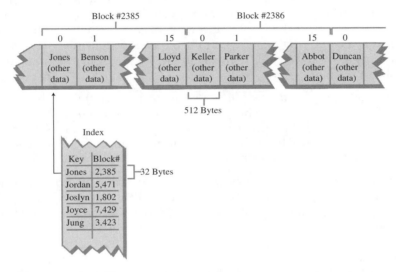

FIGURE 10.26   A file index.

Index File in Memory

The index is much smaller than the file containing actual records. It may even be small enough to fit entirely in main memory. In our example there are 500,000 records. Each one has a 32-byte entry in the index, so the index will be $32 \times 500,000$, or 1,600,000 bytes long (1.6 megabytes). In modern computers there's no problem fitting this in memory. The index can be stored on the disk but read into memory whenever the database program is started up. From then on, operations on the index can take place in memory. At the end of each day (or perhaps more frequently), the index can be written back to disk for permanent storage.

Searching

The index-in-memory approach allows much faster operations on the phone book file than are possible with a file in which the records themselves are arranged sequentially. For example, a binary search requires 19 index accesses. At 20 microseconds per access, that's only about 4/10,000 of a second. Then there's (inevitably) the time to read the actual record from the file, once its block number has been found in the index. However, this is only one disk access of (say) 10 milliseconds.

Insertion

To insert a new item in an indexed file, two steps are necessary. We first insert the item's full record into the main file; then we insert an entry, consisting of the key and the block number where the new record is stored, into the index.

Because the index is in sequential order, to insert a new item, we need to move half the index entries, on the average. Figuring 2 microseconds to move a byte in memory, we have 250,000 times 32 times 2, or about 16 seconds to insert a new entry. This compares with 5 minutes for the unindexed sequential file. (Note that we don't need to move any records in the main file; we simply append the new record at the end of the file.)

Of course, you can use a more sophisticated approach to storing the index in memory. You could store it as a binary tree, 2-3-4 tree, or red-black tree, for example. Any of these would significantly reduce insertion and deletion times. In any case the index-in-memory approach is much faster than the sequential-file approach. In some cases it will also be faster than a B-tree.

The only actual disk accesses necessary for an insertion into an indexed file involve the new record itself. Usually, the last block in the file is read into memory, the new record is appended, and the block is written back out. This process involves only two file accesses.

Multiple Indexes

An advantage of the indexed approach is that multiple indexes, each with a different key, can be created for the same file. In one index the keys can be last names; in another, telephone numbers; in another, addresses. Because the indexes are small compared with the file, this doesn't increase the total data storage very much. Of course, it does present more of a challenge when items are deleted from the file because entries must be deleted from all the indexes, but we won't get into that here.

Index Too Large for Memory

If the index is too large to fit in memory, it must be broken into blocks and stored on the disk. For large files storing the index itself as a B-tree may then be profitable. In the main file the records are stored in any convenient order.

This arrangement can be very efficient. Appending records to the end of the main file is a fast operation, and inserting the index entry for the new record is also quick because the index is a tree. The result is very fast searching and insertion for large files.

Note that when an index is arranged as a B-tree, each node contains n child pointers and n-1 data items. The child pointers are the block numbers of other nodes in the index. The data items consist of a key value and a pointer to a block in the main file. Don't confuse these two kinds of block pointers.

Complex Search Criteria

In complex searches the only practical approach may be to read every block in a file sequentially. Suppose in our phone book example we wanted a list of all entries in the phone book with first name Frank, who lived in Springfield, and who had a phone number with three 7 digits in it. (These were perhaps clues found scrawled on a scrap of paper clutched in the hand of a victim of foul play.)

A file organized by last names would be no help at all. Even if there were index files ordered by first names and cities, there would be no convenient way to find which files contained both Frank and Springfield. In such cases (which are quite common in many kinds of databases), the fastest approach is probably to read the file sequentially, block by block, checking each record to see whether it meets the criteria.

Sorting External Files

Mergesort is the preferred algorithm for sorting external data. This is because, more so than most sorting techniques, disk accesses tend to occur in adjacent records rather than random parts of the file.

Recall from Chapter 6, "Recursion," that mergesort works recursively by calling itself to sort smaller and smaller sequences. Once two of the smallest sequences (one byte each in the internal-memory version) have been sorted, they are then merged into a sorted sequence twice as long. Larger and larger sequences are merged, until eventually the entire file is sorted.

The approach for external storage is similar. However, the smallest sequence that can be read from the disk is a block of records. Thus, a two-stage process is necessary.

In the first phase, a block is read, its records are sorted internally, and the resulting sorted block is written back to disk. The next block is similarly sorted and written back to disk. This process continues until all the blocks are internally sorted.

In the second phase, two sorted blocks are read, merged into a two-block sequence, and written back to disk. This process continues until all pairs of blocks have been merged. Next, each pair of two-block sequences is merged into a four-block sequence. Each time, the size of the sorted sequences doubles, until the entire file is sorted.

Figure 10.27 shows the mergesort process on an external file. The file consists of four blocks of four records each, for a total of 16 records. Only three blocks can fit in internal memory. (Of course, all these sizes would be much larger in a real situation.) Figure 10.27 shows the file before sorting; the number in each record is its key value.

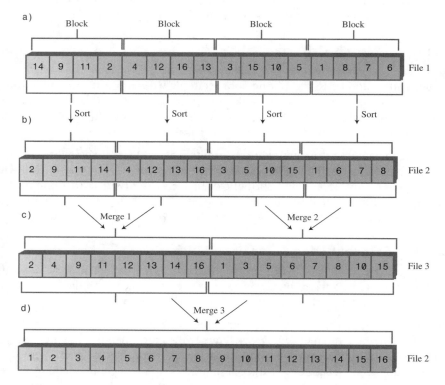

FIGURE 10.27   Mergesort on an external file.

Internal Sort of Blocks

In the first phase all the blocks in the file are sorted internally. This is done by reading the block into memory and sorting it with any appropriate internal sorting algorithm, such as quicksort (or for smaller numbers of records Shellsort or insertion sort). The result of sorting the blocks internally is shown in Figure 10.27b.

A second file may be used to hold the sorted blocks, and we assume that availability of external storage is not a problem. It's often desirable to avoid modifying the original file.

Merging

In the second phase we want to merge the sorted blocks. In the first pass we merge every pair of blocks into a sorted two-block sequence. Thus, the two blocks 2-9-11-14 and 4-12-13-16 are merged into 2-4-9-11-12-13-14-16. Also, 3-5-10-15 and 1-6-7-8 are merged into 1-3-5-6-7-8-10-15. The result is shown in Figure 10.27c. A third file is necessary to hold the result of this merge step.

In the second pass, the two 8-record sequences are merged into a 16-record sequence, which can be written back to File 2, as shown in Figure 10.27d. Now the sort is complete. Of course, more merge steps would be required to sort larger files; the number of such steps is proportional to $\log_2 N$. The merge steps can alternate between two files (File 2 and File 3 in Figure 10.22).

Internal Arrays

Because the computer's internal memory has room for only three blocks, the merging process must take place in stages. Let's say there are three arrays, called arr1, arr2, and arr3, each of which can hold a block.

In the first merge, block 2-9-11-14 is read into arr1, and 4-12-13-16 is read into arr2. These two arrays are then mergesorted into arr3. However, because arr3 holds only one block, it becomes full before the sort is completed. When it becomes full, its contents are written to disk. The sort then continues, filling up arr3 again. This completes the sort, and arr3 is again written to disk. The following lists show the details of each of the three mergesorts.

Mergesort 1:

  1. Read 2-9-11-14 into arr1.

  2. Read 4-12-13-16 into arr2.

  3. Merge 2, 4, 9, 11 into arr3; write to disk.

  4. Merge 12, 13, 14, 16 into arr3; write to disk.

Mergesort 2:

  1. Read 3-5-10-15 into arr1.

  2. Read 1-6-7-8 into arr2.

  3. Merge 1, 3, 5, 6 into arr3; write to disk.

  4. Merge 7, 8, 10, 15 into arr3, write to disk.

Mergesort 3:

  1. Read 2-4-9-11 into arr1.

  2. Read 1-3-5-6 into arr2.

  3. Merge 1, 2, 3, 4 into arr3; write to disk.

4. Merge 5, 6 into arr3 (arr2 is now empty).

5. Read 7-8-10-15 into arr2.

6. Merge 7, 8 into arr3; write to disk.

7. Merge 9, 10, 11 into arr3 (arr1 is now empty).

8. Read 12-13-14-16 into arr1.

9. Merge 12 into arr3; write to disk.

10. Merge 13, 14, 15, 16 into arr3; write to disk.

This last sequence of 10 steps is rather lengthy, so it may be helpful to examine the details of the array contents as the steps are completed. Figure 10.28 shows how these arrays look at various stages of mergesort 3.

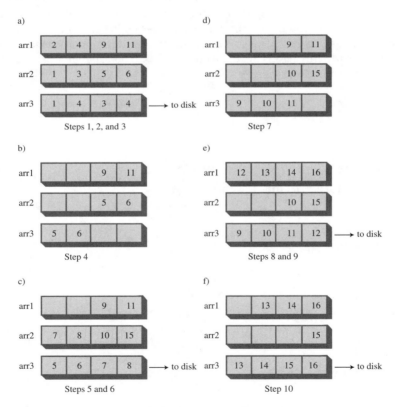

FIGURE 10.28   Array contents during mergesort 3.

Summary

- A multiway tree has more keys and children than a binary tree.

- A 2-3-4 tree is a multiway tree with up to three keys and four children per node.

- In a multiway tree, the keys in a node are arranged in ascending order.

- In a 2-3-4 tree, all insertions are made in leaf nodes, and all leaf nodes are on the same level.

- Three kinds of nodes are possible in a 2-3-4 tree: A 2-node has one key and two children, a 3-node has two keys and three children, and a 4-node has three keys and four children.

- There is no 1-node in a 2-3-4 tree.

- In a search in a 2-3-4 tree, at each node the keys are examined. If the search key is not found, the next node will be child 0 if the search key is less than key 0; child 1 if the search key is between key 0 and key 1; child 2 if the search key is between key 1 and key 2; and child 3 if the search key is greater than key 2.

- Insertion into a 2-3-4 tree requires that any full node be split on the way down the tree, during the search for the insertion point.

- Splitting the root creates two new nodes; splitting any other node creates one new node.

- The height of a 2-3-4 tree can increase only when the root is split.

- There is a one-to-one correspondence between a 2-3-4 tree and a red-black tree.

- To transform a 2-3-4 tree into a red-black tree, make each 2-node into a black node, make each 3-node into a black parent with a red child, and make each 4-node into a black parent with two red children.

- When a 3-node is transformed into a parent and child, either node can become the parent.

- Splitting a node in a 2-3-4 tree is the same as performing a color flip in a red-black tree.

- A rotation in a red-black tree corresponds to changing between the two possible orientations (slants) when transforming a 3-node.

- The height of a 2-3-4 tree is less than $\log_2 N$.

- Search times are proportional to the height.

- The 2-3-4 tree wastes space because many nodes are not even half full.

- A 2-3 tree is similar to a 2-3-4 tree, except that it can have only one or two data items and one, two, or three children.

- Insertion in a 2-3 tree involves finding the appropriate leaf and then performing splits from the leaf upward, until a non-full node is found.

- External storage means storing data outside of main memory, usually on a disk.

- External storage is larger, cheaper (per byte), and slower than main memory.

- Data in external storage is typically transferred to and from main memory a block at a time.

- Data can be arranged in external storage in sequential key order. This gives fast search times but slow insertion (and deletion) times.

- A B-tree is a multiway tree in which each node may have dozens or hundreds of keys and children.

- There is always one more child than there are keys in a B-tree node.

- For the best performance, a B-tree is typically organized so that a node holds one block of data.

- If the search criteria involve many keys, a sequential search of all the records in a file may be the most practical approach.

Questions

These questions are intended as a self-test for readers. Answers may be found in Appendix C.

1. A 2-3-4 tree is so named because a node can have

    a. three children and four data items.

    b. two, three, or four children.

    c. two parents, three children, and four items.

    d. two parents, three items, and four children.

2. A 2-3-4 tree is superior to a binary search tree in that it is _____ .

3. Imagine a parent node with data items 25, 50, and 75. If one of its child nodes had items with values 60 and 70, it would be the child numbered _____ .

4. True or False: Data items are located exclusively in leaf nodes.

5. Which of the following is *not* true each time a node is split?

   a. Exactly one new node is created.

   b. Exactly one new data item is added to the tree.

   c. One data item moves from the split node to its parent.

   d. One data item moves from the split node to its new sibling.

6. A 2-3-4 tree increases its number of levels when _____.

7. Searching a 2-3-4 tree does *not* involve

   a. splitting nodes on the way down if necessary.

   b. picking the appropriate child to go to, based on data items in a node.

   c. ending up at a leaf node if the search key is not found.

   d. examining at least one data item in any node visited.

8. After a non-root node of a 2-3-4 tree is split, does its new right child contain the item previously numbered 0, 1, or 2?

9. A 4-node split in a 2-3-4 tree is equivalent to a _____ in a red-black tree.

10. Which of the following statements about a node-splitting operation in a 2-3 tree (not a 2-3-4 tree) is *not* true?

    a. The parent of a split node must also be split if it is full.

    b. The smallest item in the node being split always stays in that node.

    c. When the parent is split, child 2 must always be disconnected from its old parent and connected to the new parent.

    d. The splitting process starts at a leaf and works upward.

11. What is the big O efficiency of a 2-3 tree?

12. In accessing data on a disk drive,

    a. inserting data is slow but finding the place to write data is fast.

    b. moving data to make room for more data is fast because so many items can be accessed at once.

    c. deleting data is unusually fast.

    d. finding the place to write data is comparatively slow but a lot of data can be written quickly.

13. In a B-tree each node contains _____ data items.

14. True or False: Node splits in a B-tree have similarities to node splits in a 2-3 tree.

15. In external storage, indexing means keeping a file of

    a. keys and their corresponding blocks.

    b. records and their corresponding blocks.

    c. keys and their corresponding records.

    d. last names and their corresponding keys.

Experiments

Carrying out these experiments will help to provide insights into the topics covered in the chapter. No programming is involved.

1. Draw by hand what a 2-3-4 tree looks like after each of the following insertions: 10, 20, 30, 40, 50, 60, 70, 80, and 90. Don't use the Tree234 Workshop applet.

2. Draw by hand what a 2-3 tree looks like after inserting the same sequence of values as in Experiment 1.

3. Think about how you would remove a node from a 2-3-4 tree.

Programming Projects

Writing programs to solve the Programming Projects helps to solidify your understanding of the material and demonstrates how the chapter's concepts are applied. (As noted in the Introduction, qualified instructors may obtain completed solutions to the Programming Projects on the publisher's Web site.)

10.1 This project should be easy. Write a method that returns the minimum value in a 2-3-4 tree.

10.2 Write a method that does an inorder traverse of a 2-3-4 tree. It should display all the items in order.

10.3 A 2-3-4 tree can be used as a sorting machine. Write a sort() method that's passed an array of key values from main() and writes them back to the array in sorted order.

10.4 Modify the tree234.java program (Listing 10.1) so that it creates and works with 2-3 trees instead. It should display the tree and allow searches. It should also allow items to be inserted, but only if the parent of the leaf node (which is being split) does not also need to be split. This implies that the split() routine need not be recursive. In writing insert(), remember that no splits happen until the appropriate leaf has been located. Then the leaf will be split if it's full. You'll need to be able to split the root too, but only when it's a leaf. With this limited routine you can insert fewer than nine items before the program crashes.

10.5 Extend the program in Programming Project 10.4 so that the split() routine is recursive and can handle situations with a full parent of a full child. This will allow insertion of an unlimited number of items. Note that in the revised split() routine you'll need to split the parent before you can decide where the items go and where to attach the children.

11

Hash Tables

A *hash table* is a data structure that offers very fast insertion and searching. When you first hear about them, hash tables sound almost too good to be true. No matter how many data items there are, insertion and searching (and sometimes deletion) can take close to constant time: O(1) in big O notation. In practice this is just a few machine instructions.

For a human user of a hash table, this is essentially instantaneous. It's so fast that computer programs typically use hash tables when they need to look up tens of thousands of items in less than a second (as in spelling checkers). Hash tables are significantly faster than trees, which, as we learned in the preceding chapters, operate in relatively fast O(logN) time. Not only are they fast, hash tables are relatively easy to program.

Hash tables do have several disadvantages. They're based on arrays, and arrays are difficult to expand after they've been created. For some kinds of hash tables, performance may degrade catastrophically when a table becomes too full, so the programmer needs to have a fairly accurate idea of how many data items will need to be stored (or be prepared to periodically transfer data to a larger hash table, a time-consuming process).

Also, there's no convenient way to visit the items in a hash table in any kind of order (such as from smallest to largest). If you need this capability, you'll need to look elsewhere.

However, if you don't need to visit items in order, and you can predict in advance the size of your database, hash tables are unparalleled in speed and convenience.

Introduction to Hashing

In this section we'll introduce hash tables and hashing. One important concept is how a range of key values is transformed into a range of array index values. In a hash table this is accomplished with a hash function. However, for certain kinds of keys, no hash function is necessary; the key values can be used directly as array indices. We'll look at this simpler situation first and then go on to show how hash functions can be used when keys aren't distributed in such an orderly fashion.

Employee Numbers as Keys

Suppose you're writing a program to access employee records for a small company with, say, 1,000 employees. Each employee record requires 1,000 bytes of storage. Thus, you can store the entire database in only 1 megabyte, which will easily fit in your computer's memory.

The company's personnel director has specified that she wants the fastest possible access to any individual record. Also, every employee has been given a number from 1 (for the founder) to 1,000 (for the most recently hired worker). These employee numbers can be used as keys to access the records; in fact access by other keys is deemed unnecessary. Employees are seldom laid off, but even when they are, their records remain in the database for reference (concerning retirement benefits and so on). What sort of data structure should you use in this situation?

Index Numbers As Keys

One possibility is a simple array. Each employee record occupies one cell of the array, and the index number of the cell is the employee number for that record. This type of array is shown in Figure 11.1.

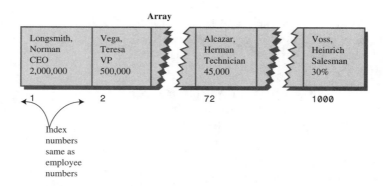

FIGURE 11.1   Employee numbers as array indices.

As you know, accessing a specified array element is very fast if you know its index number. The clerk looking up Herman Alcazar knows that he is employee number 72, so he enters that number, and the program goes instantly to index number 72 in the array. A single program statement is all that's necessary:

```
empRecord rec = databaseArray[72];
```

Adding a new item is also very quick: You insert it just past the last occupied element. The next new record—for Jim Chan, the newly hired employee number 1,001—would go in cell 1,001. Again, a single statement inserts the new record:

```
databaseArray[totalEmployees++] = newRecord;
```

Presumably, the array is made somewhat larger than the current number of employees, to allow room for expansion, but not much expansion is anticipated.

Not Always So Orderly

The speed and simplicity of data access using this array-based database make it very attractive. However, our example works only because the keys are unusually well organized. They run sequentially from 1 to a known maximum, and this maximum is a reasonable size for an array. There are no deletions, so memory-wasting gaps don't develop in the sequence. New items can be added sequentially at the end of the array, and the array doesn't need to be very much larger than the current number of items.

A Dictionary

In many situations the keys are not so well behaved as in the employee database just described. The classic example is a dictionary. If you want to put every word of an English-language dictionary, from *a* to *zyzzyva* (yes, it's a word), into your computer's memory so they can be accessed quickly, a hash table is a good choice.

A similar widely used application for hash tables is in computer-language compilers, which maintain a *symbol table* in a hash table. The symbol table holds all the variable and function names made up by the programmer, along with the address where they can be found in memory. The program needs to access these names very quickly, so a hash table is the preferred data structure.

Let's say we want to store a 50,000-word English-language dictionary in main memory. You would like every word to occupy its own cell in a 50,000-cell array, so you can access the word using an index number. This will make access very fast. But what's the relationship of these index numbers to the words? Given the word *morphosis*, for example, how do we find its index number?

Converting Words to Numbers

What we need is a system for turning a word into an appropriate index number. To begin, we know that computers use various schemes for representing individual characters as numbers. One such scheme is the ASCII code, in which *a* is 97, *b* is 98, and so on, up to 122 for *z*.

However, the ASCII code runs from 0 to 255, to accommodate capitals, punctuation, and so on. There are really only 26 letters in English words, so let's devise our own code, a simpler one that can potentially save memory space. Let's say *a* is 1, *b* is 2, *c* is 3, and so on up to 26 for *z*. We'll also say a blank is 0, so we have 27 characters. (Uppercase letters aren't used in this dictionary.)

How do we combine the digits from individual letters into a number that represents an entire word? There are all sorts of approaches. We'll look at two representative ones, and their advantages and disadvantages.

Adding the Digits

A simple approach to converting a word to a number might be to simply add the code numbers for each character. Say we want to convert the word *cats* to a number. First, we convert the characters to digits using our homemade code:

c = 3

a = 1

t = 20

s = 19

Then we add them:

3 + 1 + 20 + 19 = 43

Thus, in our dictionary the word *cats* would be stored in the array cell with index 43. All the other English words would likewise be assigned an array index calculated by this process.

How well would this work? For the sake of argument, let's restrict ourselves to 10-letter words. Then (remembering that a blank is 0), the first word in the dictionary, *a*, would be coded by

0 + 0 + 0 + 0 + 0 + 0 + 0 + 0 + 0 + 1 = 1

The last potential word in the dictionary would be *zzzzzzzzzz* (10 Zs). Our code obtained by adding its letters would be

26 + 26 + 26 + 26 + 26 + 26 + 26 + 26 + 26 + 26 = 260

Thus, the total range of word codes is from 1 to 260. Unfortunately, there are 50,000 words in the dictionary, so there aren't enough index numbers to go around. Each array element will need to hold about 192 words (50,000 divided by 260).

Clearly, this presents problems if we're thinking in terms of our one word-per-array element scheme. Maybe we could put a subarray or linked list of words at each array element. Unfortunately, such an approach would seriously degrade the access speed. Accessing the array element would be quick, but searching through the 192 words to find the one we wanted would be slow.

So our first attempt at converting words to numbers leaves something to be desired. Too many words have the same index. (For example, *was*, *tin*, *give*, *tend*, *moan*, *tick*, *bails*, *dredge*, and hundreds of other words add to 43, as *cats* does.) We conclude that this approach doesn't discriminate enough, so the resulting array has too few elements. We need to spread out the range of possible indices.

Multiplying by Powers

Let's try a different way to map words to numbers. If our array was too small before, let's make sure it's big enough. What would happen if we created an array in which every word, in fact every potential word, from *a* to *zzzzzzzzzz*, was guaranteed to occupy its own unique array element?

To do this, we need to be sure that every character in a word contributes in a unique way to the final number.

We'll begin by thinking about an analogous situation with numbers instead of words. Recall that in an ordinary multi-digit number, each digit-position represents a value 10 times as big as the position to its right. Thus 7,546 really means

$$7*1000 + 5*100 + 4*10 + 6*1$$

Or, writing the multipliers as powers of 10:

$$7*10^3 + 5*10^2 + 4*10^1 + 6*10^0$$

(An input routine in a computer program performs a similar series of multiplications and additions to convert a sequence of digits, entered at the keyboard, into a number stored in memory.)

In this system we break a number into its digits, multiply them by appropriate powers of 10 (because there are 10 possible digits), and add the products.

In a similar way we can decompose a word into its letters, convert the letters to their numerical equivalents, multiply them by appropriate powers of 27 (because there are 27 possible characters, including the blank), and add the results. This gives a unique number for every word.

Say we want to convert the word *cats* to a number. We convert the digits to numbers as shown earlier. Then we multiply each number by the appropriate power of 27 and add the results:

$3*27^3 + 1*27^2 + 20*27^1 + 19*27^0$

Calculating the powers gives

$3*19,683 + 1*729 + 20*27 + 19*1$

and multiplying the letter codes times the powers yields

$59,049 + 729 + 540 + 19$

which sums to 60,337.

This process does indeed generate a unique number for every potential word. We just calculated a 4-letter word. What happens with larger words? Unfortunately, the range of numbers becomes rather large. The largest 10-letter word, *zzzzzzzzzz*, translates into

$26*27^9 + 26*27^8 + 26*27^7 + 26*27^6 + 26*27^5 + 26*27^4 + 26*27^3 + 26*27^2 + 26*27^1 + 26*27^0$

Just by itself, 27^9 is more than 7,000,000,000,000, so you can see that the sum will be huge. An array stored in memory can't possibly have this many elements.

The problem is that this scheme assigns an array element to every potential word, whether it's an actual English word or not. Thus, there are cells for *aaaaaaaaaa*, *aaaaaaaaab*, *aaaaaaaaac*, and so on, up to *zzzzzzzzzz*. Only a small fraction of these cells are necessary for real words, so most array cells are empty. This situation is shown in Figure 11.2.

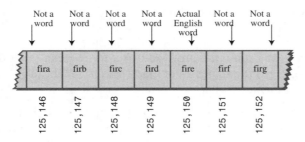

FIGURE 11.2   Index for every potential word.

Our first scheme—adding the numbers—generated too few indices. This latest scheme—adding the numbers times powers of 27—generates too many.

Hashing

What we need is a way to compress the huge range of numbers we obtain from the numbers-multiplied-by-powers system into a range that matches a reasonably sized array.

How big an array are we talking about for our English dictionary? If we have only 50,000 words, you might assume our array should have approximately this many elements. However, it turns out we're going to need an array with about twice this many cells. (It will become clear later why this is so.) So we need an array with 100,000 elements.

Thus, we look for a way to squeeze a range of 0 to more than 7,000,000,000,000 into the range 0 to 100,000. A simple approach is to use the modulo operator (%), which finds the remainder when one number is divided by another.

To see how this approach works, let's look at a smaller and more comprehensible range. Suppose we squeeze numbers in the range 0 to 199 (we'll represent them by the variable largeNumber) into the range 0 to 9 (the variable smallNumber). There are 10 numbers in the range of small numbers, so we'll say that a variable smallRange has the value 10. It doesn't really matter what the large range is (unless it overflows the program's variable size). The Java expression for the conversion is

```
smallNumber = largeNumber % smallRange;
```

The remainders when any number is divided by 10 are always in the range 0 to 9; for example, 13%10 gives 3, and 157%10 is 7. This is shown in Figure 11.3. We've squeezed the range 0–199 into the range 0–9, a 20-to-1 compression ratio.

A similar expression can be used to compress the really huge numbers that uniquely represent every English word into index numbers that fit in our dictionary array:

```
arrayIndex = hugeNumber % arraySize;
```

This is an example of a *hash function*. It *hashes* (converts) a number in a large range into a number in a smaller range. This smaller range corresponds to the index numbers in an array. An array into which data is inserted using a hash function is called a *hash table*. (We'll talk more about the design of hash functions later in the chapter.)

To review: We convert a word into a huge number by multiplying each character in the word by an appropriate power of 27.

hugeNumber = $ch0*27^9 + ch1*27^8 + ch2*27^7 + ch3*27^6 + ch4*27^5 + ch5*27^4 + ch6*27^3 + ch7*27^2 + ch8*27^1 + ch9*27^0$

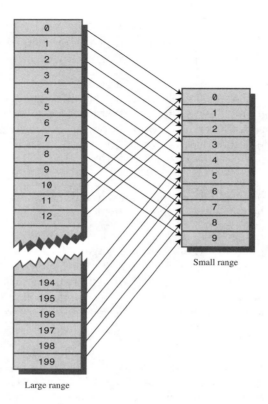

FIGURE 11.3   Range conversion.

Then, using the modulo operator (%), we squeeze the resulting huge range of numbers into a range about twice as big as the number of items we want to store. This is an example of a hash function:

```
arraySize = numberWords * 2;
arrayIndex = hugeNumber % arraySize;
```

In the huge range, each number represents a potential data item (an arrangement of letters), but few of these numbers represent actual data items (English words). A hash function transforms these large numbers into the index numbers of a much smaller array. In this array we expect that, on the average, there will be one word for every two cells. Some cells will have no words; and others, more than one.

A practical implementation of this scheme runs into trouble because hugeNumber will probably overflow its variable size, even for type long. We'll see how to deal with this problem later.

Collisions

We pay a price for squeezing a large range into a small one. There's no longer a guarantee that two words won't hash to the same array index.

This is similar to what happened when we added the letter codes, but the situation is nowhere near as bad. When we added the letters, there were only 260 possible results (for words up to 10 letters). Now we're spreading this out into 50,000 possible results.

Even so, it's impossible to avoid hashing several different words into the same array location, at least occasionally. We had hoped that we could have one data item per index number, but this turns out not to be possible. The best we can do is hope that not too many words will hash to the same index.

Perhaps you want to insert the word *melioration* into the array. You hash the word to obtain its index number but find that the cell at that number is already occupied by the word *demystify*, which happens to hash to the exact same number (for a certain size array). This situation, shown in Figure 11.4, is called a *collision*.

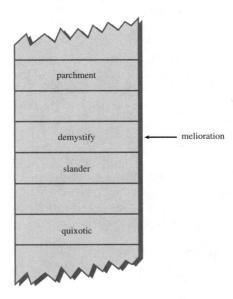

FIGURE 11.4   A collision.

It may appear that the possibility of collisions renders the hashing scheme impractical, but in fact we can work around the problem in a variety of ways.

Remember that we've specified an array with twice as many cells as data items. Thus, perhaps half the cells are empty. One approach, when a collision occurs, is to search the array in some systematic way for an empty cell and insert the new item there, instead of at the index specified by the hash function. This approach is called *open addressing*. If *cats* hashes to 5,421, but this location is already occupied by *parsnip*, then we might try to insert *cats* in 5,422, for example.

A second approach (mentioned earlier) is to create an array that consists of linked lists of words instead of the words themselves. Then, when a collision occurs, the new item is simply inserted in the list at that index. This is called *separate chaining*.

In the balance of this chapter we'll discuss open addressing and separate chaining, and then return to the question of hash functions.

So far we've focused on hashing strings. This is realistic, because many hash tables are used for storing strings. However, many other hash tables hold numbers, as in our employee-number example. In the discussion that follows, and in the Workshop applets, we use numbers—rather than strings—as keys. This makes things easier to understand and simplifies the programming examples. Keep in mind, however, that in many situations these numbers would be derived from strings.

Open Addressing

In open addressing, when a data item can't be placed at the index calculated by the hash function, another location in the array is sought. We'll explore three methods of open addressing, which vary in the method used to find the next vacant cell. These methods are *linear probing*, *quadratic probing*, and *double hashing*.

Linear Probing

In linear probing we search sequentially for vacant cells. If 5,421 is occupied when we try to insert a data item there, we go to 5,422, then 5,423, and so on, incrementing the index until we find an empty cell. This is called linear probing because it steps sequentially along the line of cells.

The Hash Workshop Applet

The Hash Workshop applet demonstrates linear probing. When you start this applet, you'll see a screen similar to Figure 11.5.

In this applet the range of keys runs from 0 to 999. The initial size of the array is 60. The hash function has to squeeze the range of keys down to match the array size. It does this with the modulo operator (%), as we've seen before:

```
arrayIndex = key % arraySize;
```

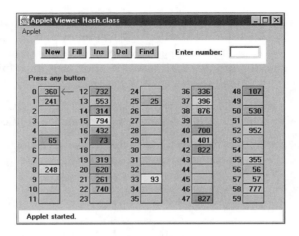

FIGURE 11.5   The Hash Workshop applet at startup.

For the initial array size of 60, this is

```
arrayIndex = key % 60;
```

This hash function is simple enough that you can solve it mentally. For a given key, keep subtracting 60 until you get a number less than 60. For example, to hash 143, subtract 60, giving 83, and then 60 again, giving 23. This is the index number where the algorithm will place 143. Thus, you can easily check that the algorithm has hashed a key to the correct address. (An array size of 10 is even easier to figure out, as a key's last digit is the index it will hash to.)

As with other applets, operations are carried out by repeatedly pressing the same button. For example, to find a data item with a specified number, click the Find button repeatedly. Remember, finish a sequence with one button before using another button. For example, don't switch from clicking Fill to some other button until the Press any key message is displayed.

All the operations require you to type a numerical value at the beginning of the sequence. The Find button requires you to type a key value, for example, while New requires the size of the new table.

The New Button   You can create a new hash table of a size you specify by using the New button. The maximum size is 60; this limitation results from the number of cells that can be viewed in the applet window. The initial size is also 60. We use this number because it makes it easy to check whether the hash values are correct, but as we'll see later, in a general-purpose hash table, the array size should be a prime number, so 59 would be a better choice.

The Fill Button   Initially, the hash table contains 30 items, so it's half full. However, you can also fill it with a specified number of data items using the Fill button. Keep clicking Fill, and when prompted, type the number of items to fill. Hash tables work best when they are not more than half or at the most two-thirds full (40 items in a 60-cell table).

You'll see that the filled cells aren't evenly distributed in the cells. Sometimes there's a sequence of several empty cells and sometimes a sequence of filled cells.

Let's call a sequence of filled cells in a hash table a *filled sequence*. As you add more and more items, the filled sequences become longer. This is called *clustering*, and is shown in Figure 11.6.

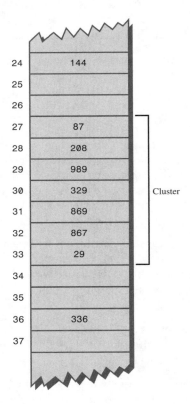

FIGURE 11.6   An example of clustering.

When you use the applet, note that it may take a long time to fill a hash table if you try to fill it too full (for example, if you try to put 59 items in a 60-cell table). You may think the program has stopped, but be patient. It's extremely inefficient at filling an almost-full array.

Also, note that if the hash table becomes completely full, the algorithms all stop working; in this applet they assume that the table has at least one empty cell.

The Find Button   The Find button starts by applying the hash function to the key value you type into the number box. This results in an array index. The cell at this index may be the key you're looking for; this is the optimum situation, and success will be reported immediately.

However, it's also possible that this cell is already occupied by a data item with some other key. This is a collision; you'll see the red arrow pointing to an occupied cell. Following a collision, the search algorithm will look at the next cell in sequence. The process of finding an appropriate cell following a collision is called a *probe*.

Following a collision, the Find algorithm simply steps along the array looking at each cell in sequence. If it encounters an empty cell before finding the key it's looking for, it knows the search has failed. There's no use looking further because the insertion algorithm would have inserted the item at this cell (if not earlier). Figure 11.7 shows successful and unsuccessful linear probes.

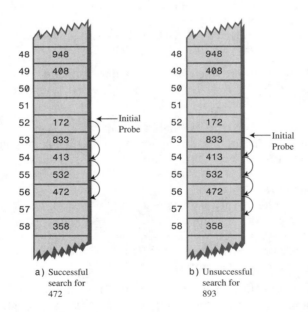

a) Successful
search for
472

b) Unsuccessful
search for
893

FIGURE 11.7   Linear probes.

The Ins Button   The Ins button inserts a data item, with a key value that you type into the number box, into the hash table. It uses the same algorithm as the Find button to locate the appropriate cell. If the original cell is occupied, it will probe linearly for a vacant cell. When it finds one, it inserts the item.

Try inserting some new data items. Type in a three-digit number and watch what happens. Most items will go into the first cell they try, but some will suffer collisions and need to step along to find an empty cell. The number of steps they take is the *probe length*. Most probe lengths are only a few cells long. Sometimes, however, you may see probe lengths of four or five cells, or even longer as the array becomes excessively full.

Notice which keys hash to the same index. If the array size is 60, the keys 7, 67, 127, 187, 247, and so on up to 967 all hash to index 7. Try inserting this sequence or a similar one. Such sequences will demonstrate the linear probe.

The Del Button   The Del button deletes an item whose key is typed by the user. Deletion isn't accomplished by simply removing a data item from a cell, leaving it empty. Why not? Remember that during insertion the probe process steps along a series of cells, looking for a vacant one. If a cell is made empty in the middle of this sequence of full cells, the Find routine will give up when it sees the empty cell, even if the desired cell can eventually be reached.

For this reason a deleted item is replaced by an item with a special key value that identifies it as deleted. In this applet we assume all legitimate key values are positive, so the deleted value is chosen as –1. Deleted items are marked with the special key *Del*.

The Insert button will insert a new item at the first available empty cell or in a *Del* item. The Find button will treat a *Del* item as an existing item for the purposes of searching for another item further along.

If there are many deletions, the hash table fills up with these ersatz *Del* data items, which makes it less efficient. For this reason many hash table implementations don't allow deletion. If it is implemented, it should be used sparingly.

Duplicates Allowed?

Can you allow data items with duplicate keys to be used in hash tables? The fill routine in the Hash Workshop applet doesn't allow duplicates, but you can insert them with the Insert button if you like. Then you'll see that only the first one can be accessed. The only way to access a second item with the same key is to delete the first one. This isn't too convenient.

You could rewrite the Find algorithm to look for all items with the same key instead of just the first one. However, it would then need to search through all the cells of every linear sequence it encountered. This wastes time for all table accesses, even when no duplicates are involved. In the majority of cases you probably want to forbid duplicates.

Clustering

Try inserting more items into the hash table in the Hash Workshop applet. As it gets more full, the clusters grow larger. Clustering can result in very long probe lengths. This means that accessing cells at the end of the sequence is very slow.

The more full the array is, the worse clustering becomes. It's not usually a problem when the array is half full, and still not too bad when it's two-thirds full. Beyond this, however, performance degrades seriously as the clusters grow larger and larger. For this reason it's critical when designing a hash table to ensure that it never becomes more than half, or at the most two-thirds, full. (We'll discuss the mathematical relationship between how full the hash table is and probe lengths at the end of this chapter.)

Java Code for a Linear Probe Hash Table

It's not hard to create methods to handle search, insertion, and deletion with linear probe hash tables. We'll show the Java code for these methods and then a complete hash.java program that puts them in context.

The find() Method

The find() method first calls hashFunc() to hash the search key to obtain the index number hashVal. The hashFunc() method applies the % operator to the search key and the array size, as we've seen before.

Next, in a while condition, find() checks whether the item at this index is empty (null). If not, it checks whether the item contains the search key. If the item does contain the key, find() returns the item. If it doesn't, find() increments hashVal and goes back to the top of the while loop to check whether the next cell is occupied. Here's the code for find():

```
public DataItem find(int key)      // find item with key
// (assumes table not full)
   {
   int hashVal = hashFunc(key);   // hash the key

   while(hashArray[hashVal] != null)  // until empty cell,
      {                               // found the key?
      if(hashArray[hashVal].getKey() == key)
         return hashArray[hashVal];   // yes, return item
      ++hashVal;                      // go to next cell
      hashVal %= arraySize;           // wrap around if necessary
      }
   return null;                       // can't find item
   }
```

As hashVal steps through the array, it eventually reaches the end. When this happens, we want it to wrap around to the beginning. We could check for this with an if statement, setting hashVal to 0 whenever it equaled the array size. However, we can accomplish the same thing by applying the % operator to hashVal and the array size.

Cautious programmers might not want to assume the table is not full, as is done here. The table should not be allowed to become full, but if it did, this method would loop forever. For simplicity we don't check for this situation.

The insert() Method

The insert() method, shown here, uses about the same algorithm as find() to locate where a data item should go. However, it's looking for an empty cell or a deleted item (key –1), rather than a specific item. When such an empty cell has been located, insert() places the new item into it.

```
public void insert(DataItem item) // insert a DataItem
// (assumes table not full)
   {
   int key = item.getKey();      // extract key
   int hashVal = hashFunc(key);  // hash the key
                                 // until empty cell or -1,
   while(hashArray[hashVal] != null &&
                      hashArray[hashVal].iData != -1)
      {
      ++hashVal;                 // go to next cell
      hashVal %= arraySize;      // wrap around if necessary
      }
   hashArray[hashVal] = item;    // insert item
   }  // end insert()
```

The delete() Method

The following delete() method finds an existing item using code similar to find(). When the item is found, delete() writes over it with the special data item nonItem, which is predefined with a key of –1.

```
public DataItem delete(int key)  // delete a DataItem
   {
   int hashVal = hashFunc(key);  // hash the key

   while(hashArray[hashVal] != null)  // until empty cell,
      {                              // found the key?
      if(hashArray[hashVal].getKey() == key)
```

```
         {
         DataItem temp = hashArray[hashVal]; // save item
         hashArray[hashVal] = nonItem;        // delete item
         return temp;                         // return item
         }
      ++hashVal;                     // go to next cell
      hashVal %= arraySize;          // wrap around if necessary
      }
   return null;                      // can't find item
   }  // end delete()
```

The hash.java **Program**

Listing 11.1 shows the complete hash.java program. In this program a DataItem object contains just one field, an integer that is its key. As in other data structures we've discussed, these objects could contain more data or a reference to an object of another class (such as employee or partNumber).

The major field in class HashTable is an array called hashArray. Other fields are the size of the array and the special nonItem object used for deletions.

LISTING 11.1   The hash.java Program

```
// hash.java
// demonstrates hash table with linear probing
// to run this program: C:>java HashTableApp
import java.io.*;
////////////////////////////////////////////////////////////////
class DataItem
   {                                 // (could have more data)
   private int iData;                // data item (key)
//-------------------------------------------------------------
   public DataItem(int ii)          // constructor
      { iData = ii; }
//-------------------------------------------------------------
   public int getKey()
      { return iData; }
//-------------------------------------------------------------
   }  // end class DataItem
////////////////////////////////////////////////////////////////
class HashTable
   {
   private DataItem[] hashArray;     // array holds hash table
   private int arraySize;
```

LISTING 11.1   Continued

```java
    private DataItem nonItem;         // for deleted items
// -------------------------------------------------------------
    public HashTable(int size)        // constructor
        {
        arraySize = size;
        hashArray = new DataItem[arraySize];
        nonItem = new DataItem(-1);    // deleted item key is -1
        }
// -------------------------------------------------------------
    public void displayTable()
        {
        System.out.print("Table: ");
        for(int j=0; j<arraySize; j++)
            {
            if(hashArray[j] != null)
                System.out.print(hashArray[j].getKey() + " ");
            else
                System.out.print("** ");
            }
        System.out.println("");
        }
// -------------------------------------------------------------
    public int hashFunc(int key)
        {
        return key % arraySize;        // hash function
        }
// -------------------------------------------------------------
    public void insert(DataItem item) // insert a DataItem
    // (assumes table not full)
        {
        int key = item.getKey();       // extract key
        int hashVal = hashFunc(key);   // hash the key
                                       // until empty cell or -1,
        while(hashArray[hashVal] != null &&
                      hashArray[hashVal].getKey() != -1)
            {
            ++hashVal;                 // go to next cell
            hashVal %= arraySize;      // wraparound if necessary
            }
        hashArray[hashVal] = item;     // insert item
        }  // end insert()
```

LISTING 11.1   Continued

```
// ------------------------------------------------------------
   public DataItem delete(int key)  // delete a DataItem
      {
      int hashVal = hashFunc(key);  // hash the key

      while(hashArray[hashVal] != null)  // until empty cell,
         {                               // found the key?
         if(hashArray[hashVal].getKey() == key)
            {
            DataItem temp = hashArray[hashVal]; // save item
            hashArray[hashVal] = nonItem;       // delete item
            return temp;                        // return item
            }
         ++hashVal;                  // go to next cell
         hashVal %= arraySize;       // wraparound if necessary
         }
      return null;                  // can't find item
      } // end delete()
// ------------------------------------------------------------
   public DataItem find(int key)    // find item with key
      {
      int hashVal = hashFunc(key);  // hash the key

      while(hashArray[hashVal] != null)  // until empty cell,
         {                               // found the key?
         if(hashArray[hashVal].getKey() == key)
            return hashArray[hashVal];   // yes, return item
         ++hashVal;                  // go to next cell
         hashVal %= arraySize;       // wraparound if necessary
         }
      return null;                  // can't find item
      }
// ------------------------------------------------------------
   } // end class HashTable
////////////////////////////////////////////////////////////////
class HashTableApp
   {
   public static void main(String[] args) throws IOException
      {
      DataItem aDataItem;
      int aKey, size, n, keysPerCell;
```

LISTING 11.1   Continued

```
                                    // get sizes
      System.out.print("Enter size of hash table: ");
      size = getInt();
      System.out.print("Enter initial number of items: ");
      n = getInt();
      keysPerCell = 10;
                                    // make table
      HashTable theHashTable = new HashTable(size);

      for(int j=0; j<n; j++)        // insert data
         {
         aKey = (int)(java.lang.Math.random() *
                                       keysPerCell * size);
         aDataItem = new DataItem(aKey);
         theHashTable.insert(aDataItem);
         }

      while(true)                   // interact with user
         {
         System.out.print("Enter first letter of ");
         System.out.print("show, insert, delete, or find: ");
         char choice = getChar();
         switch(choice)
            {
            case 's':
               theHashTable.displayTable();
               break;
            case 'i':
            System.out.print("Enter key value to insert: ");
               aKey = getInt();
               aDataItem = new DataItem(aKey);
               theHashTable.insert(aDataItem);
               break;
            case 'd':
               System.out.print("Enter key value to delete: ");
               aKey = getInt();
               theHashTable.delete(aKey);
               break;
            case 'f':
               System.out.print("Enter key value to find: ");
               aKey = getInt();
```

LISTING 11.1   Continued

```
                    aDataItem = theHashTable.find(aKey);
                    if(aDataItem != null)
                       {
                       System.out.println("Found " + aKey);
                       }
                    else
                       System.out.println("Could not find " + aKey);
                    break;
                 default:
                    System.out.print("Invalid entry\n");
                 }  // end switch
            }  // end while
       }  // end main()
//-------------------------------------------------------------
   public static String getString() throws IOException
      {
      InputStreamReader isr = new InputStreamReader(System.in);
      BufferedReader br = new BufferedReader(isr);
      String s = br.readLine();
      return s;
      }
//-------------------------------------------------------------
   public static char getChar() throws IOException
      {
      String s = getString();
      return s.charAt(0);
      }
//-------------------------------------------------------------
   public static int getInt() throws IOException
      {
      String s = getString();
      return Integer.parseInt(s);
      }
//-------------------------------------------------------------
   }  // end class HashTableApp
///////////////////////////////////////////////////////////////
```

The main() routine in the HashTableApp class contains a user interface that allows the
user to show the contents of the hash table (enter s), insert an item (i), delete an
item (d), or find an item (f).

Initially, the program asks the user to input the size of the hash table and the number of items in it. You can make it almost any size, from a few items to 10,000. (Building larger tables than this may take a little time.) Don't use the s (for show) option on tables of more than a few hundred items; they scroll off the screen and displaying them takes a long time.

A variable in main(), keysPerCell, specifies the ratio of the range of keys to the size of the array. In the listing, it's set to 10. This means that if you specify a table size of 20, the keys will range from 0 to 200.

If you want to see what's going on, it's best to create tables with fewer than about 20 items so that all the items can be displayed on one line. Here's some sample interaction with hash.java:

```
Enter size of hash table: 12
Enter initial number of items: 8

Enter first letter of show, insert, delete, or find: s
Table: 108 13 0 ** ** 113 5 66 ** 117 ** 47

Enter first letter of show, insert, delete, or find: f
Enter key value to find: 66
Found 66

Enter first letter of show, insert, delete, or find: i
Enter key value to insert: 100
Enter first letter of show, insert, delete, or find: s
Table: 108 13 0 ** 100 113 5 66 ** 117 ** 47

Enter first letter of show, insert, delete, or find: d
Enter key value to delete: 100
Enter first letter of show, insert, delete, or find: s
Table: 108 13 0 ** -1 113 5 66 ** 117 ** 47
```

Key values run from 0 to 119 (12 times 10, minus 1). The ** symbol indicates that a cell is empty. The item with key 100 is inserted at location 4 (the first item is numbered 0) because 100%12 is 4. Notice how 100 changes to –1 when this item is deleted.

Expanding the Array
One option when a hash table becomes too full is to expand its array. In Java, arrays have a fixed size and can't be expanded. Your program must create a new, larger array, and then insert the contents of the old small array into the new large one.

Remember that the hash function calculates the location of a given data item based on the array size, so items won't be located in the same place in the large array as they were in the small array. You can't therefore simply copy the items from one array to the other. You'll need to go through the old array in sequence, cell by cell, inserting each item you find into the new array with the insert() method. This is called *rehashing*. It's a time-consuming process, but necessary if the array is to be expanded.

The expanded array is usually made twice the size of the original array. Actually, because the array size should be a prime number, the new array will need to be a bit more than twice as big. Calculating the new array size is part of the rehashing process.

Here are some routines to help find the new array size (or the original array size, if you don't trust the user to pick a prime number, which is usually the case). You start off with the specified size and then look for the next prime larger than that. The getPrime() method gets the next prime larger than its argument. It calls isPrime() to check each of the numbers above the specified size.

```
private int getPrime(int min)    // returns 1st prime > min
    {
    for(int j = min+1; true; j++)    // for all j > min
       if( isPrime(j) )              // is j prime?
          return j;                  // yes, return it
    }
// ----------------------------------------------------------
   private boolean isPrime(int n)    // is n prime?
    {
    for(int j=2; (j*j <= n); j++)    // for all j
       if( n % j == 0)               // divides evenly by j?
          return false;              // yes, so not prime
    return true;                     // no, so prime
    }
```

These routines are not the ultimate in sophistication. For example, in getPrime() you could check 2 and then odd numbers from then on, instead of every number. However, such refinements don't matter much because you usually find a prime after checking only a few numbers.

Java offers a class Vector that is an array-like data structure that can be expanded. However, it's not much help because of the need to rehash all data items when the table changes size.

Quadratic Probing

We've seen that clusters can occur in the linear probe approach to open addressing. Once a cluster forms, it tends to grow larger. Items that hash to any value in the range of the cluster will step along and insert themselves at the end of the cluster, thus making it even bigger. The bigger the cluster gets, the faster it grows.

It's like the crowd that gathers when someone faints at the shopping mall. The first arrivals come because they saw the victim fall; later arrivals gather because they wondered what everyone else was looking at. The larger the crowd grows, the more people are attracted to it.

The ratio of the number of items in a table to the table's size is called the *load factor*. A table with 10,000 cells and 6,667 items has a load factor of 2/3.

```
loadFactor = nItems / arraySize;
```

Clusters can form even when the load factor isn't high. Parts of the hash table may consist of big clusters, while others are sparsely inhabited. Clusters reduce performance.

Quadratic probing is an attempt to keep clusters from forming. The idea is to probe more widely separated cells, instead of those adjacent to the primary hash site.

The Step Is the Square of the Step Number

In a linear probe, if the primary hash index is x, subsequent probes go to x+1, x+2, x+3, and so on. In quadratic probing, probes go to x+1, x+4, x+9, x+16, x+25, and so on. The distance from the initial probe is the square of the step number: $x+1^2$, $x+2^2$, $x+3^2$, $x+4^2$, $x+5^2$, and so on.

Figure 11.8 shows some quadratic probes.

It's as if a quadratic probe became increasingly desperate as its search lengthened. At first it calmly picks the adjacent cell. If that's occupied, it thinks it may be in a small cluster, so it tries something 4 cells away. If that's occupied, it becomes a little concerned, thinking it may be in a larger cluster, and tries 9 cells away. If that's occupied, it feels the first tinges of panic and jumps 16 cells away. Pretty soon, it's flying hysterically all over the place, as you can see if you try searching with the HashDouble Workshop applet when the table is almost full.

The HashDouble Applet with Quadratic Probes

The HashDouble Workshop applet allows two different kinds of collision handling: quadratic probes and double hashing. (We'll look at double hashing in the next section.) This applet generates a display much like that of the Hash Workshop applet, except that it includes radio buttons to select quadratic probing or double hashing.

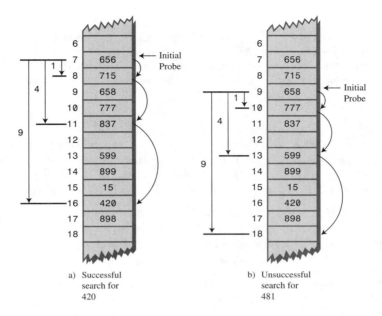

FIGURE 11.8   Quadratic probes.

To see how quadratic probes look, start up this applet and create a new hash table of 59 items using the New button. When you're asked to select double or quadratic probe, click the Quad button. After the new table is created, fill it four/fifths full using the Fill button (47 items in a 59-cell array). This is too full, but it will generate longer probes so you can study the probe algorithm.

Incidentally, if you try to fill the hash table too full, you may see the message Can't complete fill. This occurs when the probe sequences get very long. Every additional step in the probe sequence makes a bigger step size. If the sequence is too long, the step size will eventually exceed the capacity of its integer variable, so the applet shuts down the fill process before this happens.

When the table is filled, select an existing key value and use the Find key to see whether the algorithm can find it. Often the key value is located at the initial cell, or the one adjacent to it. If you're patient, however, you'll find a key that requires three or four steps, and you'll see the step size lengthen for each step. You can also use Find to search for a non-existent key; this search continues until an empty cell is encountered.

TIP

Important: Always make the array size a prime number. Use 59 instead of 60, for example. (Other primes less than 60 are 53, 47, 43, 41, 37, 31, 29, 23, 19, 17, 13, 11, 7, 5, 3, and 2.) If the array size is not prime, an endless sequence of steps may occur during a probe. If this happens during a Fill operation, the applet will be paralyzed.

The Problem with Quadratic Probes

Quadratic probes eliminate the clustering problem we saw with the linear probe, which is called *primary clustering*. However, quadratic probes suffer from a different and more subtle clustering problem. This occurs because all the keys that hash to a particular cell follow the same sequence in trying to find a vacant space.

Let's say 184, 302, 420, and 544 all hash to 7 and are inserted in this order. Then 302 will require a one-step probe, 420 will require a four-step probe, and 544 will require a nine-step probe. Each additional item with a key that hashes to 7 will require a longer probe. This phenomenon is called *secondary clustering*.

Secondary clustering is not a serious problem, but quadratic probing is not often used because there's a slightly better solution.

Double Hashing

To eliminate secondary clustering as well as primary clustering, we can use another approach: *double hashing*. Secondary clustering occurs because the algorithm that generates the sequence of steps in the quadratic probe always generates the same steps: 1, 4, 9, 16, and so on.

What we need is a way to generate probe sequences that depend on the key instead of being the same for every key. Then numbers with different keys that hash to the same index will use different probe sequences.

The solution is to hash the key a second time, using a different hash function, and use the result as the step size. For a given key the step size remains constant throughout a probe, but it's different for different keys.

Experience has shown that this secondary hash function must have certain characteristics:

- It must not be the same as the primary hash function.

- It must never output a 0 (otherwise, there would be no step; every probe would land on the same cell, and the algorithm would go into an endless loop).

Experts have discovered that functions of the following form work well:

```
stepSize = constant - (key % constant);
```

where constant is prime and smaller than the array size. For example,

```
stepSize = 5 - (key % 5);
```

This is the secondary hash function used in the HashDouble Workshop applet. Different keys may hash to the same index, but they will (most likely) generate different step sizes. With this hash function the step sizes are all in the range 1 to 5. This is shown in Figure 11.9.

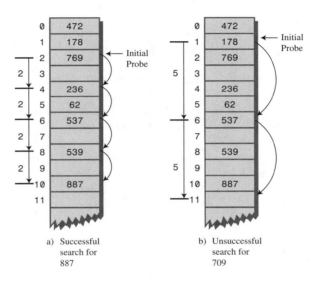

a) Successful search for 887

b) Unsuccessful search for 709

FIGURE 11.9 Double hashing.

The HashDouble Applet with Double Hashing

You can use the HashDouble Workshop applet to see how double hashing works. It starts up automatically in Double-hashing mode, but if it's in Quadratic mode, you can switch to Double by creating a new table with the New button and clicking the Double button when prompted. To best see probes at work, you'll need to fill the table rather full, say to about nine/tenths capacity or more. Even with such high load factors, most data items will be found immediately by the first hash function; only a few will require extended probe sequences.

Try finding existing keys. When one needs a probe sequence, you'll see how all the steps are the same size for a given key, but that the step size is different—between 1 and 5—for different keys.

Java Code for Double Hashing

Listing 11.2 shows the complete listing for hashDouble.java, which uses double hashing. It's similar to the hash.java program (Listing 11.1), but uses two hash functions: one for finding the initial index and the second for generating the step size. As before, the user can show the table contents, insert an item, delete an item, and find an item.

LISTING 11.2   The hashDouble.java Program

```
// hashDouble.java
// demonstrates hash table with double hashing
// to run this program: C:>java HashDoubleApp
import java.io.*;
//////////////////////////////////////////////////////////////
class DataItem
   {                                 // (could have more items)
   private int iData;                // data item (key)
//-------------------------------------------------------------
   public DataItem(int ii)           // constructor
      { iData = ii; }
//-------------------------------------------------------------
   public int getKey()
      { return iData; }
//-------------------------------------------------------------
   }  // end class DataItem
//////////////////////////////////////////////////////////////
class HashTable
   {
   private DataItem[] hashArray;     // array is the hash table
   private int arraySize;
   private DataItem nonItem;         // for deleted items
// -----------------------------------------------------------
   HashTable(int size)               // constructor
      {
      arraySize = size;
      hashArray = new DataItem[arraySize];
      nonItem = new DataItem(-1);
      }
// -----------------------------------------------------------
   public void displayTable()
      {
      System.out.print("Table: ");
      for(int j=0; j<arraySize; j++)
```

LISTING 11.2   Continued

```
         {
         if(hashArray[j] != null)
            System.out.print(hashArray[j].getKey()+ " ");
         else
            System.out.print("** ");
         }
      System.out.println("");
      }
// ------------------------------------------------------------
   public int hashFunc1(int key)
      {
      return key % arraySize;
      }
// ------------------------------------------------------------
   public int hashFunc2(int key)
      {
      // non-zero, less than array size, different from hF1
      // array size must be relatively prime to 5, 4, 3, and 2
      return 5 - key % 5;
      }
// ------------------------------------------------------------
                                  // insert a DataItem
   public void insert(int key, DataItem item)
   // (assumes table not full)
      {
      int hashVal = hashFunc1(key);   // hash the key
      int stepSize = hashFunc2(key);  // get step size
                                       // until empty cell or -1
      while(hashArray[hashVal] != null &&
                   hashArray[hashVal].getKey() != -1)
         {
         hashVal += stepSize;        // add the step
         hashVal %= arraySize;       // for wraparound
         }
      hashArray[hashVal] = item;      // insert item
      }  // end insert()
// ------------------------------------------------------------
   public DataItem delete(int key)    // delete a DataItem
      {
      int hashVal = hashFunc1(key);      // hash the key
      int stepSize = hashFunc2(key);     // get step size
```

LISTING 11.2   Continued

```
      while(hashArray[hashVal] != null)  // until empty cell,
        {                                // is correct hashVal?
        if(hashArray[hashVal].getKey() == key)
           {
           DataItem temp = hashArray[hashVal]; // save item
           hashArray[hashVal] = nonItem;       // delete item
           return temp;                        // return item
           }
        hashVal += stepSize;            // add the step
        hashVal %= arraySize;           // for wraparound
        }
      return null;                      // can't find item
      } // end delete()
// ------------------------------------------------------------
   public DataItem find(int key)     // find item with key
   // (assumes table not full)
      {
      int hashVal = hashFunc1(key);      // hash the key
      int stepSize = hashFunc2(key);     // get step size

      while(hashArray[hashVal] != null)  // until empty cell,
        {                                // is correct hashVal?
        if(hashArray[hashVal].getKey() == key)
           return hashArray[hashVal];    // yes, return item
        hashVal += stepSize;             // add the step
        hashVal %= arraySize;            // for wraparound
        }
      return null;                      // can't find item
      }
// ------------------------------------------------------------
   } // end class HashTable
/////////////////////////////////////////////////////////////////
class HashDoubleApp
   {
   public static void main(String[] args) throws IOException
      {
      int aKey;
      DataItem aDataItem;
      int size, n;
                                  // get sizes
      System.out.print("Enter size of hash table: ");
```

LISTING 11.2   Continued

```
            size = getInt();
            System.out.print("Enter initial number of items: ");
            n = getInt();
                                      // make table
            HashTable theHashTable = new HashTable(size);

            for(int j=0; j<n; j++)        // insert data
               {
               aKey = (int)(java.lang.Math.random() * 2 * size);
               aDataItem = new DataItem(aKey);
               theHashTable.insert(aKey, aDataItem);
               }

            while(true)                  // interact with user
               {
               System.out.print("Enter first letter of ");
               System.out.print("show, insert, delete, or find: ");
               char choice = getChar();
               switch(choice)
                  {
                  case 's':
                     theHashTable.displayTable();
                     break;
                  case 'i':
                     System.out.print("Enter key value to insert: ");
                     aKey = getInt();
                     aDataItem = new DataItem(aKey);
                     theHashTable.insert(aKey, aDataItem);
                     break;
                  case 'd':
                     System.out.print("Enter key value to delete: ");
                     aKey = getInt();
                     theHashTable.delete(aKey);
                     break;
                  case 'f':
                     System.out.print("Enter key value to find: ");
                     aKey = getInt();
                     aDataItem = theHashTable.find(aKey);
                     if(aDataItem != null)
                        System.out.println("Found " + aKey);
                     else
```

LISTING 11.2   Continued

```
                System.out.println("Could not find " + aKey);
             break;
          default:
             System.out.print("Invalid entry\n");
          } // end switch
       } // end while
    } // end main()
//-----------------------------------------------------------
   public static String getString() throws IOException
      {
      InputStreamReader isr = new InputStreamReader(System.in);
      BufferedReader br = new BufferedReader(isr);
      String s = br.readLine();
      return s;
      }
//-----------------------------------------------------------
   public static char getChar() throws IOException
      {
      String s = getString();
      return s.charAt(0);
      }
//-----------------------------------------------------------
   public static int getInt() throws IOException
      {
      String s = getString();
      return Integer.parseInt(s);
      }
//-----------------------------------------------------------
   } // end class HashDoubleApp
/////////////////////////////////////////////////////////////
```

Output and operation of this program are similar to those of hash.java. Table 11.1 shows what happens when 21 items are inserted into a 23-cell hash table using double hashing. The step sizes run from 1 to 5.

TABLE 11.1   Filling a 23-Cell Table Using Double Hashing

Item Number	Key	Hash Value	Step Size	Cells in Probe Sequence
1	1	1	4	
2	38	15	2	
3	37	14	3	
4	16	16	4	

TABLE 11.1   Continued

Item Number	Key	Hash Value	Step Size	Cells in Probe Sequence
5	20	20	5	
6	3	3	2	
7	11	11	4	
8	24	1	1	2
9	5	5	5	
10	16	16	4	20 1 5 9
11	10	10	5	
12	31	8	4	
13	18	18	2	
14	12	12	3	
15	30	7	5	
16	1	1	4	5 9 13
17	19	19	1	
18	36	13	4	17
19	41	18	4	22
20	15	15	5	20 2 7 12 17 22 4
21	25	2	5	7 12 17 22 4 9 14 19 1 6

The first 15 keys mostly hash to a vacant cell (the 10th one is an anomaly). After that, as the array gets more full, the probe sequences become quite long. Here's the resulting array of keys:

    ** 1 24 3 15 5 25 30 31 16 10 11 12 1 37 38 16 36 18 19 20 ** 41

Table Size a Prime Number

Double hashing requires that the size of the hash table is a prime number. To see why, imagine a situation in which the table size is not a prime number. For example, suppose the array size is 15 (indices from 0 to 14), and that a particular key hashes to an initial index of 0 and a step size of 5. The probe sequence will be 0, 5, 10, 0, 5, 10, and so on, repeating endlessly. Only these three cells are ever examined, so the algorithm will never find the empty cells that might be waiting at 1, 2, 3, and so on. The algorithm will crash and burn.

If the array size were 13, which is prime, the probe sequence will eventually visit every cell. It's 0, 5, 10, 2, 7, 12, 4, 9, 1, 6, 11, 3, and so on and on. If there is even one empty cell, the probe will find it. Using a prime number as the array size makes it impossible for any number to divide it evenly, so the probe sequence will eventually check every cell.

A similar effect occurs using the quadratic probe. In that case, however, the step size gets larger with each step and will eventually overflow the variable holding it, thus preventing an endless loop.

In general, double hashing is the probe sequence of choice when open addressing is used.

Separate Chaining

In open addressing, collisions are resolved by looking for an open cell in the hash table. A different approach is to install a linked list at each index in the hash table. A data item's key is hashed to the index in the usual way, and the item is inserted into the linked list at that index. Other items that hash to the same index are simply added to the linked list; there's no need to search for empty cells in the primary array. Figure 11.10 shows how separate chaining looks.

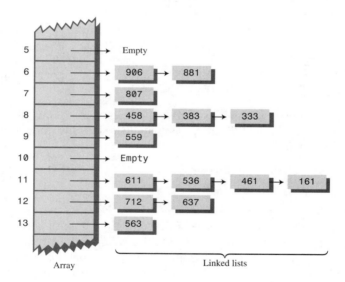

FIGURE 11.10   Example of separate chaining.

Separate chaining is conceptually somewhat simpler than the various probe schemes used in open addressing. However, the code is longer because it must include the mechanism for the linked lists, usually in the form of an additional class.

The HashChain Workshop Applet

To see how separate chaining works, start the HashChain Workshop applet. It displays an array of linked lists, as shown in Figure 11.11.

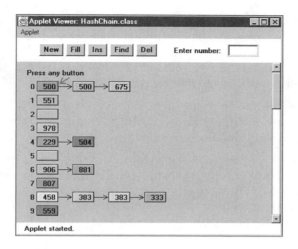

FIGURE 11.11   Separate chaining in the HashChain Workshop applet.

Each element of the array occupies one line of the display, and the linked lists extend from left to right. Initially, there are 25 cells in the array (25 lists). This is more than fits on the screen; you can move the display up and down with the scroll-bar to see the entire array. The display shows up to six items per list. You can create a hash table with up to 100 lists and use load factors up to 2.0. Higher load factors may cause the linked lists to exceed six items and run off the right edge of the screen, making it impossible to see all the items. (This may happen very occasionally even at the 2.0 load factor.)

Experiment with the HashChain applet by inserting some new items with the Ins button. You'll see how the red arrow goes immediately to the correct list and inserts the item at the beginning of the list. The lists in the HashChain applet are not sorted, so insertion does not require searching through the list. (The example program will demonstrate sorted lists.)

Try to find specified items using the Find button. During a Find operation, if there are several items on the list, the red arrow must step through the items looking for the correct one. For a successful search, half the items in the list must be examined on the average, as we discussed in Chapter 5, "Linked Lists." For an unsuccessful search, all the items must be examined.

Load Factors
The load factor (the ratio of the number of items in a hash table to its size) is typi-cally different in separate chaining than in open addressing. In separate chaining it's normal to put N or more items into an N cell array; thus, the load factor can be 1 or greater. There's no problem with this; some locations will simply contain two or more items in their lists.

Of course, if there are many items on the lists, access time is reduced because access to a specified item requires searching through an average of half the items on the list. Finding the initial cell takes fast O(1) time, but searching through a list takes time proportional to M, the average number of items on the list. This is O(M) time. Thus, we don't want the lists to become too full.

A load factor of 1, as shown in the initial Workshop applet, is common. With this load factor, roughly one-third of the cells will be empty, one-third will hold one item, and one-third will hold two or more items.

In open addressing, performance degrades badly as the load factor increases above one-half or two-thirds. In separate chaining the load factor can rise above 1 without hurting performance very much. This makes separate chaining a more robust mechanism, especially when it's hard to predict in advance how much data will be placed in the hash table.

Duplicates
Duplicates are allowed and may be generated in the Fill process. All items with the same key will be inserted in the same list, so if you need to discover all of them, you must search the entire list in both successful and unsuccessful searches. This lowers performance. The Find operation in the applet finds only the first of several duplicates.

Deletion
In separate chaining, deletion poses no special problems as it does in open addressing. The algorithm hashes to the proper list and then deletes the item from the list. Because probes aren't used, it doesn't matter if the list at a particular cell becomes empty. We've included a Del button in the Workshop applet to show how deletion works.

Table Size
With separate chaining, making the table size a prime number is not as important as it is with quadratic probes and double hashing. There are no probes in separate chaining, so we don't need to worry that a probe will go into an endless sequence because the step size divides evenly into the array size.

On the other hand, certain kinds of key distributions can cause data to cluster when the array size is not a prime number. We'll have more to say about this problem when we discuss hash functions.

Buckets
Another approach similar to separate chaining is to use an array at each location in the hash table, instead of a linked list. Such arrays are sometimes called *buckets*. This approach is not as efficient as the linked list approach, however, because of the

problem of choosing the size of the buckets. If they're too small, they may overflow, and if they're too large, they waste memory. Linked lists, which allocate memory dynamically, don't have this problem.

Java Code for Separate Chaining

The hashChain.java program includes a SortedList class and an associated Link class. Sorted lists don't speed up a successful search, but they do cut the time of an unsuccessful search in half. (As soon as an item larger than the search key is reached, which on average is half the items in a list, the search can be declared a failure.)

Deletion times are also cut in half; however, insertion times are lengthened because the new item can't just be inserted at the beginning of the list; its proper place in the ordered list must be located before it's inserted. If the lists are short, the increase in insertion times may not be important.

If many unsuccessful searches are anticipated, it may be worthwhile to use the slightly more complicated sorted list, rather than an unsorted list. However, an unsorted list is preferred if insertion speed is more important.

The hashChain.java program, shown in Listing 11.3, begins by constructing a hash table with a table size and number of items entered by the user. The user can then insert, find, and delete items, and display the list. For the entire hash table to be viewed on the screen, the size of the table must be no greater than 16 or so.

LISTING 11.3   The hashChain.java Program

```java
// hashChain.java
// demonstrates hash table with separate chaining
// to run this program: C:>java HashChainApp
import java.io.*;
////////////////////////////////////////////////////////////////
class Link
   {                                  // (could be other items)
   private int iData;                 // data item
   public Link next;                  // next link in list
// -------------------------------------------------------------
   public Link(int it)                // constructor
      { iData= it; }
// -------------------------------------------------------------
   public int getKey()
      { return iData; }
// -------------------------------------------------------------
   public void displayLink()          // display this link
      { System.out.print(iData + " "); }
```

LISTING 11.3   Continued

```
   } // end class Link
/////////////////////////////////////////////////////////////
class SortedList
   {
   private Link first;            // ref to first list item
// -------------------------------------------------------------
   public void SortedList()       // constructor
      { first = null; }
// -------------------------------------------------------------
   public void insert(Link theLink) // insert link, in order
      {
      int key = theLink.getKey();
      Link previous = null;       // start at first
      Link current = first;
                                  // until end of list,
      while( current != null && key > current.getKey() )
         {                        // or current > key,
         previous = current;
         current = current.next;  // go to next item
         }
      if(previous==null)          // if beginning of list,
         first = theLink;         //    first --> new link
      else                        // not at beginning,
         previous.next = theLink; //    prev --> new link
      theLink.next = current;     // new link --> current
      } // end insert()
// -------------------------------------------------------------
   public void delete(int key)    // delete link
      {                           // (assumes non-empty list)
      Link previous = null;       // start at first
      Link current = first;
                                  // until end of list,
      while( current != null && key != current.getKey() )
         {                        // or key == current,
         previous = current;
         current = current.next;  // go to next link
         }
                                  // disconnect link
      if(previous==null)          //    if beginning of list
         first = first.next;      //       delete first link
      else                        //    not at beginning
```

LISTING 11.3   Continued

```
            previous.next = current.next; //    delete current link
        }  // end delete()
// ------------------------------------------------------------
    public Link find(int key)          // find link
        {
        Link current = first;          // start at first
                                       // until end of list,
        while(current != null &&  current.getKey() <= key)
            {                          // or key too small,
            if(current.getKey() == key)   // is this the link?
                return current;        // found it, return link
            current = current.next;    // go to next item
            }
        return null;                   // didn't find it
        }  // end find()
// ------------------------------------------------------------
    public void displayList()
        {
        System.out.print("List (first-->last): ");
        Link current = first;       // start at beginning of list
        while(current != null)      // until end of list,
            {
            current.displayLink();  // print data
            current = current.next; // move to next link
            }
        System.out.println("");
        }
    }  // end class SortedList
////////////////////////////////////////////////////////////////
class HashTable
    {
    private SortedList[] hashArray;   // array of lists
    private int arraySize;
// ------------------------------------------------------------
    public HashTable(int size)          // constructor
        {
        arraySize = size;
        hashArray = new SortedList[arraySize];  // create array
        for(int j=0; j<arraySize; j++)          // fill array
            hashArray[j] = new SortedList();    // with lists
        }
```

LISTING 11.3   Continued

```
// --------------------------------------------------------------
   public void displayTable()
      {
      for(int j=0; j<arraySize; j++) // for each cell,
         {
         System.out.print(j + ". "); // display cell number
         hashArray[j].displayList(); // display list
         }
      }
// --------------------------------------------------------------
   public int hashFunc(int key)       // hash function
      {
      return key % arraySize;
      }
// --------------------------------------------------------------
   public void insert(Link theLink)  // insert a link
      {
      int key = theLink.getKey();
      int hashVal = hashFunc(key);    // hash the key
      hashArray[hashVal].insert(theLink); // insert at hashVal
      }  // end insert()
// --------------------------------------------------------------
   public void delete(int key)        // delete a link
      {
      int hashVal = hashFunc(key);    // hash the key
      hashArray[hashVal].delete(key); // delete link
      }  // end delete()
// --------------------------------------------------------------
   public Link find(int key)          // find link
      {
      int hashVal = hashFunc(key);    // hash the key
      Link theLink = hashArray[hashVal].find(key);  // get link
      return theLink;                 // return link
      }
// --------------------------------------------------------------
   }  // end class HashTable
////////////////////////////////////////////////////////////////
class HashChainApp
   {
   public static void main(String[] args) throws IOException
      {
```

LISTING 11.3   Continued

```java
int aKey;
Link aDataItem;
int size, n, keysPerCell = 100;
                                 // get sizes
System.out.print("Enter size of hash table: ");
size = getInt();
System.out.print("Enter initial number of items: ");
n = getInt();
                                 // make table
HashTable theHashTable = new HashTable(size);

for(int j=0; j<n; j++)           // insert data
    {
    aKey = (int)(java.lang.Math.random() *
                                keysPerCell * size);
    aDataItem = new Link(aKey);
    theHashTable.insert(aDataItem);
    }
while(true)                      // interact with user
    {
    System.out.print("Enter first letter of ");
    System.out.print("show, insert, delete, or find: ");
    char choice = getChar();
    switch(choice)
        {
        case 's':
            theHashTable.displayTable();
            break;
        case 'i':
            System.out.print("Enter key value to insert: ");
            aKey = getInt();
            aDataItem = new Link(aKey);
            theHashTable.insert(aDataItem);
            break;
        case 'd':
            System.out.print("Enter key value to delete: ");
            aKey = getInt();
            theHashTable.delete(aKey);
            break;
        case 'f':
            System.out.print("Enter key value to find: ");
```

LISTING 11.3   Continued

```
                aKey = getInt();
                aDataItem = theHashTable.find(aKey);
                if(aDataItem != null)
                   System.out.println("Found " + aKey);
                else
                   System.out.println("Could not find " + aKey);
                break;
             default:
                System.out.print("Invalid entry\n");
          }  // end switch
       }  // end while
    }  // end main()
//-------------------------------------------------------------
   public static String getString() throws IOException
      {
      InputStreamReader isr = new InputStreamReader(System.in);
      BufferedReader br = new BufferedReader(isr);
      String s = br.readLine();
      return s;
      }
//-------------------------------------------------------------
   public static char getChar() throws IOException
      {
      String s = getString();
      return s.charAt(0);
      }
//-------------------------------------------------------------
   public static int getInt() throws IOException
      {
      String s = getString();
      return Integer.parseInt(s);
      }
//-------------------------------------------------------------
   }  // end class HashChainApp
/////////////////////////////////////////////////////////////////
```

Here's the output when the user creates a table with 20 lists, inserts 20 items into it, and displays it with the s option:

```
Enter size of hash table: 20
Enter initial number of items: 20
Enter first letter of show, insert, delete, or find: s
```

```
0. List (first-->last): 240 1160
1. List (first-->last):
2. List (first-->last):
3. List (first-->last): 143
4. List (first-->last): 1004
5. List (first-->last): 1485 1585
6. List (first-->last):
7. List (first-->last): 87 1407
8. List (first-->last):
9. List (first-->last): 309
10. List (first-->last): 490
11. List (first-->last):
12. List (first-->last): 872
13. List (first-->last): 1073
14. List (first-->last): 594 954
15. List (first-->last): 335
16. List (first-->last): 1216
17. List (first-->last): 1057 1357
18. List (first-->last): 938 1818
19. List (first-->last):
```

If you insert more items into this table, you'll see the lists grow longer but maintain their sorted order. You can delete items as well.

We'll return to the question of when to use separate chaining when we discuss hash table efficiency later in this chapter.

Hash Functions

In this section we'll explore the issue of what makes a good hash function and see whether we can improve the approach to hashing strings mentioned at the beginning of this chapter.

Quick Computation

A good hash function is simple, so it can be computed quickly. The major advantage of hash tables is their speed. If the hash function is slow, this speed will be degraded. A hash function with many multiplications and divisions is not a good idea. (The bit-manipulation facilities of Java or C++, such as shifting bits right to divide a number by a multiple of 2, can sometimes be used to good advantage.)

The purpose of a hash function is to take a range of key values and transform them into index values in such a way that the key values are distributed randomly across all the indices of the hash table. Keys may be completely random or not so random.

Random Keys

A so-called *perfect* hash function maps every key into a different table location. This is only possible for keys that are unusually well behaved and whose range is small enough to be used directly as array indices (as in the employee-number example at the beginning of this chapter).

In most cases neither of these situations exists, and the hash function will need to compress a larger range of keys into a smaller range of index numbers.

The distribution of key values in a particular database determines what the hash function needs to do. In this chapter we've assumed that the data was randomly distributed over its entire range. In this situation the hash function

```
index = key % arraySize;
```

is satisfactory. It involves only one mathematical operation, and if the keys are truly random, the resulting indices will be random too, and therefore well distributed.

Non-Random Keys

However, data is often distributed non-randomly. Imagine a database that uses car-part numbers as keys. Perhaps these numbers are of the form

033-400-03-94-05-0-535

This is interpreted as follows:

- Digits 0–2: Supplier number (1 to 999, currently up to 70)
- Digits 3–5: Category code (100, 150, 200, 250, up to 850)
- Digits 6–7: Month of introduction (1 to 12)
- Digits 8–9: Year of introduction (00 to 99)
- Digits 10–11: Serial number (1 to 99, but never exceeds 100)
- Digit 12: Toxic risk flag (0 or 1)
- Digits 13–15: Checksum (sum of other fields, modulo 100)

The key used for the part number shown would be 0,334,000,394,050,535. However, such keys are not randomly distributed. The majority of numbers from 0 to 9,999,999,999,999,999 can't actually occur (for example, supplier numbers higher than 70, category codes that aren't multiples of 50, and months from 13 to 99). Also, the checksum is not independent of the other numbers. Some work should be done to these part numbers to help ensure that they form a range of more truly random numbers.

Don't Use Non-Data

The key fields should be squeezed down until every bit counts. For example, the category codes should be changed to run from 0 to 15. Also, the checksum should be removed because it doesn't add any additional information; it's deliberately redundant. Various bit-twiddling techniques are appropriate for compressing the various fields in the key.

Use All the Data

Every part of the key (except non-data, as just described) should contribute to the hash function. Don't just use the first four digits or some such expurgation. The more data that contributes to the key, the more likely it is that the keys will hash evenly into the entire range of indices.

Sometimes the range of keys is so large it overflows type int or type long variables. We'll see how to handle overflow when we talk about hashing strings in a moment.

To summarize: The trick is to find a hash function that's simple and fast, yet excludes the non-data parts of the key and uses all the data.

Use a Prime Number for the Modulo Base

Often the hash function involves using the modulo operator (%) with the table size. We've already seen that it's important for the table size to be a prime number when using a quadratic probe or double hashing. However, if the keys themselves may not be randomly distributed, it's important for the table size to be a prime number no matter what hashing system is used.

This is true because, if many keys share a divisor with the array size, they may tend to hash to the same location, causing clustering. Using a prime table size eliminates this possibility. For example, if the table size is a multiple of 50 in our car-part example, the category codes will all hash to index numbers that are multiples of 50. However, with a prime number such as 53, you are guaranteed that no keys will divide evenly into the table size.

The moral is to examine your keys carefully and tailor your hash algorithm to remove any irregularity in the distribution of the keys.

Hashing Strings

We saw at the beginning of this chapter how to convert short strings to key numbers by multiplying digit codes by powers of a constant. In particular, we saw that the four-letter word *cats* could turn into a number by calculating

```
key = 3*27³ + 1*27² + 20*27¹ + 19*27⁰
```

This approach has the desirable attribute of involving all the characters in the input string. The calculated key value can then be hashed into an array index in the usual way:

```
index = (key) % arraySize;
```

Here's a Java method that finds the key value of a string:

```
public static int hashFunc1(String key)
    {
    int hashVal = 0;
    int pow27 = 1;                          // 1, 27, 27*27, etc

    for(int j=key.length()-1; j>=0; j--) // right to left
        {
        int letter = key.charAt(j) - 96;  // get char code
        hashVal += pow27 * letter;         // times power of 27
        pow27 *= 27;                       // next power of 27
        }
    return hashVal % arraySize;
    }  // end hashFunc1()
```

The loop starts at the rightmost letter in the word. If there are N letters, this is N-1. The numerical equivalent of the letter, according to the code we devised at the beginning of this chapter (a=1 and so on), is placed in `letter`. This is then multiplied by a power of 27, which is 1 for the letter at N-1, 27 for the letter at N-2, and so on.

The `hashFunc1()` method is not as efficient as it might be. Aside from the character conversion, there are two multiplications and an addition inside the loop. We can eliminate a multiplication by taking advantage of a mathematical identity called Horner's method. (Horner was an English mathematician, 1773–1827.) This states that an expression like

```
var4*n⁴ + var3*n³ + var2*n² + var1*n¹ + var0*n⁰
```

can be written as

```
(((var4*n + var3)*n + var2)*n + var1)*n + var0
```

To evaluate this equation, we can start inside the innermost parentheses and work outward. If we translate this to a Java method, we have the following code:

```
public static int hashFunc2(String key)
    {
    int hashVal = key.charAt(0) - 96;
```

```
    for(int j=1; j<key.length(); j++)    // left to right
       {
       int letter = key.charAt(j) - 96;  // get char code
       hashVal = hashVal * 27 + letter;  // multiply and add
       }
    return hashVal % arraySize;          // mod
    } // end hashFunc2()
```

Here we start with the leftmost letter of the word (which is somewhat more natural than starting on the right), and we have only one multiplication and one addition each time through the loop (aside from extracting the character from the string).

The hashFunc2() method unfortunately can't handle strings longer than about seven letters. Longer strings cause the value of hashVal to exceed the size of type int. (If we used type long, the same problem would still arise for somewhat longer strings.)

Can we modify this basic approach so we don't overflow any variables? Notice that the key we eventually end up with is always less than the array size because we apply the modulo operator. It's not the final index that's too big; it's the intermediate key values.

It turns out that with Horner's formulation we can apply the modulo operator (%) at each step in the calculation. This gives the same result as applying the modulo operator once at the end but avoids overflow. (It does add an operation inside the loop.) The hashFunc3() method shows how this looks:

```
public static int hashFunc3(String key)
   {
   int hashVal = 0;
   for(int j=0; j<key.length(); j++)     // left to right
      {
      int letter = key.charAt(j) - 96;   // get char code
      hashVal = (hashVal * 27 + letter) % arraySize; // mod
      }
   return hashVal;                        // no mod
   } // end hashFunc3()
```

This approach or something like it is normally taken to hash a string. Various bit-manipulation tricks can be played as well, such as using a base of 32 (or a larger power of 2) instead of 27, so that multiplication can be effected using the shift operator (>>), which is faster than the modulo operator (%).

You can use an approach similar to this to convert any kind of string to a number suitable for hashing. The strings can be words, names, or any other concatenation of characters.

Folding

Another reasonable hash function involves breaking the key into groups of digits and adding the groups. This ensures that all the digits influence the hash value. The number of digits in a group should correspond to the size of the array. That is, for an array of 1,000 items, use groups of three digits each.

For example, suppose you want to hash nine-digit Social Security numbers for linear probing. If the array size is 1,000, you would divide the nine-digit number into three groups of three digits. If a particular SSN was 123-45-6789, you would calculate a key value of 123+456+789 = 1368. You can use the % operator to trim such sums so the highest index is 999. In this case, 1368%1000 = 368. If the array size is 100, you would need to break the nine-digit key into four two-digit numbers and one one-digit number: 12+34+56+78+9 = 189, and 189%100 = 89.

It's easier to imagine how this works when the array size is a multiple of 10. However, for best results it should be a prime number, as we've seen for other hash functions. We'll leave an implementation of this scheme as an exercise.

Hashing Efficiency

We've noted that insertion and searching in hash tables can approach O(1) time. If no collision occurs, only a call to the hash function and a single array reference are necessary to insert a new item or find an existing item. This is the minimum access time.

If collisions occur, access times become dependent on the resulting probe lengths. Each cell accessed during a probe adds another time increment to the search for a vacant cell (for insertion) or for an existing cell. During an access, a cell must be checked to see whether it's empty, and—in the case of searching or deletion—whether it contains the desired item.

Thus, an individual search or insertion time is proportional to the length of the probe. This is in addition to a constant time for the hash function.

The average probe length (and therefore the average access time) is dependent on the load factor (the ratio of items in the table to the size of the table). As the load factor increases, probe lengths grow longer.

We'll look at the relationship between probe lengths and load factors for the various kinds of hash tables we've studied.

Open Addressing

The loss of efficiency with high load factors is more serious for the various open addressing schemes than for separate chaining.

In open addressing, unsuccessful searches generally take longer than successful searches. During a probe sequence, the algorithm can stop as soon as it finds the desired item, which is, on the average, halfway through the probe sequence. On the other hand, it must go all the way to the end of the sequence before it's sure it can't find an item.

Linear Probing

The following equations show the relationship between probe length (P) and load factor (L) for linear probing. For a successful search it's

$$P = (1 + 1 / (1 - L)^2) / 2$$

and for an unsuccessful search it's

$$P = (1 + 1 / (1 - L)) / 2$$

These formulas are from Knuth (see Appendix B, "Further Reading"), and their derivation is quite complicated. Figure 11.12 shows the result of graphing these equations.

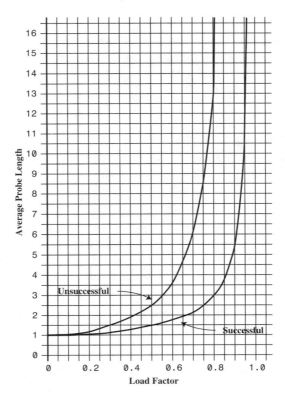

FIGURE 11.12   Linear probe performance.

At a load factor of 1/2, a successful search takes 1.5 comparisons and an unsuccessful search takes 2.5. At a load factor of 2/3, the numbers are 2.0 and 5.0. At higher load factors the numbers become very large.

The moral, as you can see, is that the load factor must be kept under 2/3 and preferably under 1/2. On the other hand, the lower the load factor, the more memory is needed for a given amount of data. The optimum load factor in a particular situation depends on the trade-off between memory efficiency, which decreases with lower load factors, and speed, which increases.

Quadratic Probing and Double Hashing

Quadratic probing and double hashing share their performance equations. These equations indicate a modest superiority over linear probing. For a successful search, the formula (again from Knuth) is

```
-log₂(1-loadFactor) / loadFactor
```

For an unsuccessful search it is

```
1 / (1-loadFactor)
```

Figure 11.13 shows graphs of these formulas. At a load factor of 0.5, successful and unsuccessful searches both require an average of two probes. At a 2/3 load factor, the numbers are 2.37 and 3.0, and at 0.8 they're 2.90 and 5.0. Thus, somewhat higher load factors can be tolerated for quadratic probing and double hashing than for linear probing.

Separate Chaining

The efficiency analysis for separate chaining is different, and generally easier, than for open addressing.

We want to know how long it takes to search for or insert an item into a separate-chaining hash table. We'll assume that the most time-consuming part of these operations is comparing the search key of the item with the keys of other items in the list. We'll also assume that the time required to hash to the appropriate list and to determine when the end of a list has been reached is equivalent to one key comparison. Thus, all operations require 1+nComps time, where nComps is the number of key comparisons.

Let's say that the hash table consists of arraySize elements, each of which holds a list, and that N data items have been inserted in the table. Then, on the average, each list will hold N divided by arraySize items:

```
AverageListLength = N / arraySize
```

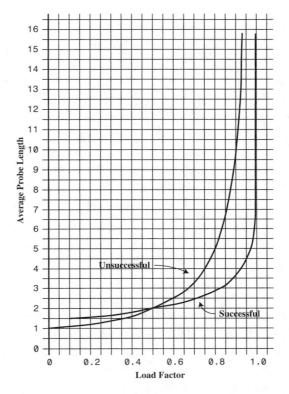

FIGURE 11.13   Quadratic-probe and double-hashing performance.

This is the same as the definition of the load factor:

```
loadFactor = N / arraySize
```

so the average list length equals the load factor.

Searching

In a successful search, the algorithm hashes to the appropriate list and then searches along the list for the item. On the average, half the items must be examined before the correct one is located. Thus, the search time is

```
1 + loadFactor / 2
```

This is true whether the lists are ordered or not. In an unsuccessful search, if the lists are unordered, all the items must be searched, so the time is

```
1 + loadFactor
```

These formulas are graphed in Figure 11.14.

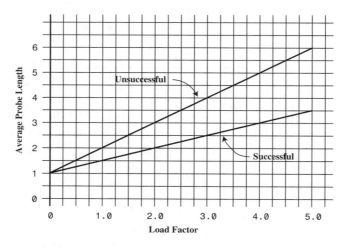

FIGURE 11.14   Separate-chaining performance.

For an ordered list, only half the items must be examined in an unsuccessful search, so the time is the same as for a successful search.

In separate chaining it's typical to use a load factor of about 1.0 (the number of data items equals the array size). Smaller load factors don't improve performance significantly, but the time for all operations increases linearly with load factor, so going beyond 2 or so is generally a bad idea.

Insertion
If the lists are not ordered, insertion is always immediate, in the sense that no comparisons are necessary. The hash function must still be computed, so let's call the insertion time 1.

If the lists are ordered, then, as with an unsuccessful search, an average of half the items in each list must be examined, so the insertion time is

```
1 + loadFactor / 2
```

Open Addressing Versus Separate Chaining

If open addressing is to be used, double hashing seems to be the preferred system by a small margin over quadratic probing. The exception is the situation in which plenty of memory is available and the data won't expand after the table is created; in this case linear probing is somewhat simpler to implement and, if load factors below 0.5 are used, causes little performance penalty.

If the number of items that will be inserted in a hash table isn't known when the table is created, separate chaining is preferable to open addressing. Increasing the load factor causes major performance penalties in open addressing, but performance degrades only linearly in separate chaining.

When in doubt, use separate chaining. Its drawback is the need for a linked list class, but the payoff is that adding more data than you anticipated won't cause performance to slow to a crawl.

Hashing and External Storage

At the end of Chapter 10, "2-3-4 Trees and External Storage," we discussed using B-trees as data structures for external (disk-based) storage. Let's look briefly at the use of hash tables for external storage.

Recall from Chapter 10 that a disk file is divided into blocks containing many records, and that the time to access a block is much larger than any internal processing on data in main memory. For these reasons the overriding consideration in devising an external storage strategy is minimizing the number of block accesses.

On the other hand, external storage is not expensive per byte, so it may be acceptable to use large amounts of it, more than is strictly required to hold the data, if by so doing we can speed up access time. This is possible using hash tables.

Table of File Pointers

The central feature in external hashing is a hash table containing block numbers, which refer to blocks in external storage. The hash table is sometimes called an *index* (in the sense of a book's index). It can be stored in main memory or, if it is too large, stored externally on disk, with only part of it being read into main memory at a time. Even if it fits entirely in main memory, a copy will probably be maintained on the disk and read into memory when the file is opened.

Non-Full Blocks

Let's reuse the example from Chapter 10 in which the block size is 8,192 bytes, and a record is 512 bytes. Thus, a block can hold 16 records. Every entry in the hash table points to one of these blocks. Let's say there are 100 blocks in a particular file.

The index (hash table) in main memory holds pointers to the file blocks, which start at 0 at the beginning of the file and run up to 99.

In external hashing it's important that blocks don't become full. Thus, we might store an average of 8 records per block. Some blocks would have more records, and some fewer. There would be about 800 records in the file. This arrangement is shown in Figure 11.15.

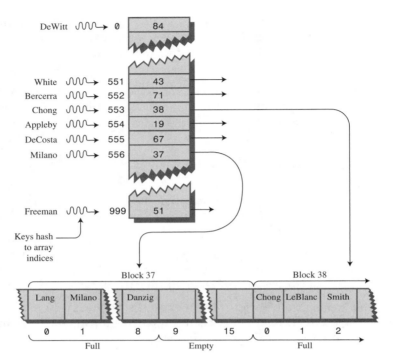

FIGURE 11.15   External hashing.

All records with keys that hash to the same value are located in the same block. To find a record with a particular key, the search algorithm hashes the key, uses the hash value as an index to the hash table, gets the block number at that index, and reads the block.

This process is efficient because only one block access is necessary to locate a given item. The downside is that considerable disk space is wasted because the blocks are, by design, not full.

To implement this scheme, we must choose the hash function and the size of the hash table with some care so that a limited number of keys hash to the same value. In our example, we want only eight records per key, on the average.

Full Blocks

Even with a good hash function, a block will occasionally become full. This situation can be handled using variations of the collision-resolution schemes discussed for internal hash tables: open addressing and separate chaining.

In open addressing, if, during insertion, one block is found to be full, the algorithm inserts the new record in a neighboring block. In linear probing this is the next block, but it could also be selected using a quadratic probe or double hashing. In separate chaining, special overflow blocks are made available; when a primary block is found to be full, the new record is inserted in the overflow block.

Full blocks are undesirable because an additional disk access is necessary for the second block; this doubles the access time. However, this is acceptable if it happens rarely.

We've discussed only the simplest hash table implementation for external storage. There are many more complex approaches that are beyond the scope of this book.

Summary

- A hash table is based on an array.

- The range of key values is usually greater than the size of the array.

- A key value is hashed to an array index by a hash function.

- An English-language dictionary is a typical example of a database that can be efficiently handled with a hash table.

- The hashing of a key to an already-filled array cell is called a collision.

- Collisions can be handled in two major ways: open addressing and separate chaining.

- In open addressing, data items that hash to a full array cell are placed in another cell in the array.

- In separate chaining, each array element consists of a linked list. All data items hashing to a given array index are inserted in that list.

- We discussed three kinds of open addressing: linear probing, quadratic probing, and double hashing.

- In linear probing the step size is always 1, so if x is the array index calculated by the hash function, the probe goes to x, x+1, x+2, x+3, and so on.

- The number of such steps required to find a specified item is called the probe length.

- In linear probing, contiguous sequences of filled cells appear. They are called primary clusters, and they reduce performance.

- In quadratic probing the offset from x is the square of the step number, so the probe goes to x, x+1, x+4, x+9, x+16, and so on.

- Quadratic probing eliminates primary clustering but suffers from the less severe secondary clustering.

- Secondary clustering occurs because all the keys that hash to the same value follow the same sequence of steps during a probe.

- All keys that hash to the same value follow the same probe sequence because the step size does not depend on the key, but only on the hash value.

- In double hashing the step size depends on the key and is obtained from a secondary hash function.

- If the secondary hash function returns a value s in double hashing, the probe goes to x, x+s, x+2s, x+3s, x+4s, and so on, where s depends on the key but remains constant during the probe.

- The load factor is the ratio of data items in a hash table to the array size.

- The maximum load factor in open addressing should be around 0.5. For double hashing at this load factor, searches will have an average probe length of 2.

- Search times go to infinity as load factors approach 1.0 in open addressing.

- It's crucial that an open-addressing hash table does not become too full.

- A load factor of 1.0 is appropriate for separate chaining.

- At this load factor a successful search has an average probe length of 1.5, and an unsuccessful search, 2.0.

- Probe lengths in separate chaining increase linearly with load factor.

- A string can be hashed by multiplying each character by a different power of a constant, adding the products, and using the modulo operator (%) to reduce the result to the size of the hash table.

- To avoid overflow, we can apply the modulo operator at each step in the process, if the polynomial is expressed using Horner's method.

- Hash table sizes should generally be prime numbers. This is especially important in quadratic probing and separate chaining.

- Hash tables can be used for external storage. One way to do this is to have the elements in the hash table contain disk-file block numbers.

Questions

These questions are intended as a self-test for readers. Answers may be found in Appendix C.

1. Using big O notation, say how long it takes (ideally) to find an item in a hash table.

2. A _____ transforms a range of key values into a range of index values.

3. Open addressing refers to

    a. keeping many of the cells in the array unoccupied.

    b. keeping an open mind about which address to use.

    c. probing at cell x+1, x+2, and so on until an empty cell is found.

    d. looking for another location in the array when the one you want is occupied.

4. Using the next available position after an unsuccessful probe is called _____.

5. What are the first five step sizes in quadratic probing?

6. Secondary clustering occurs because

    a. many keys hash to the same location.

    b. the sequence of step lengths is always the same.

    c. too many items with the same key are inserted.

    d. the hash function is not perfect.

7. Separate chaining involves the use of a _____ at each location.

8. A reasonable load factor in separate chaining is _____.

9. True or False: A possible hash function for strings involves multiplying each character by an ever-increasing power.

10. The best technique when the amount of data is not well known is

    a. linear probing.

    b. quadratic probing.

    c. double hashing.

    d. separate chaining.

11. If digit folding is used in a hash function, the number of digits in each group should reflect _____.

12. True or False: In linear probing an unsuccessful search takes longer than a successful search.

13. In separate chaining the time to insert a new item

    a. increases linearly with the load factor.

    b. is proportional to the number of items in the table.

    c. is proportional to the number of lists.

    d. is proportional to the percentage of full cells in the array.

14. True or False: In external hashing, it's important that the records don't become full.

15. In external hashing, all records with keys that hash to the same value are located in _____.

Experiments

Carrying out these experiments will help to provide insights into the topics covered in the chapter. No programming is involved.

1. In linear probing, the time for an unsuccessful search is related to the cluster size. Using the Hash workshop applet, find the average cluster size for 30 items filled into 60 cells, with a load factor of 0.5. Consider an isolated cell (that is, with empty cells on both sides) to be a cluster of size 1. To find the average, you could count the number of cells in each cluster and divide by the number of clusters, but there's an easier way. What is it? Repeat this experiment for a half-dozen 30-item fills and average the cluster sizes. Repeat the entire process for load factors of 0.6, 0.7, 0.8, and 0.9. Do your results agree with the chart in Figure 11.12?

2. With the HashDouble Workshop applet, make a small quadratic hash table, with a size that is *not* a prime number, say 24. Fill it very full, say 16 items. Now search for non-existent key values. Try different keys until you find one that causes the quadratic probe to go into an unending sequence. This happens because the quadratic step size, modulo a non-prime array size, forms a repeating series. The moral: Make your array size a prime number.

3. With the HashChain applet, create an array with 25 cells, and then fill it with 50 items, with a load factor of 2.0. Inspect the linked lists that are displayed. Add the lengths of all these linked lists and divide by the number of lists to find the average list length. On the average, you'll need to search this length in an unsuccessful search. (Actually, there's a quicker way to find this average length. What is it?)

Programming Projects

Writing programs to solve the Programming Projects helps to solidify your understanding of the material and demonstrates how the chapter's concepts are applied. (As noted in the Introduction, qualified instructors may obtain completed solutions to the Programming Projects on the publisher's Web site.)

11.1 Modify the hash.java program (Listing 11.1) to use quadratic probing.

11.2 Implement a linear probe hash table that stores strings. You'll need a hash function that converts a string to an index number; see the section "Hashing Strings" in this chapter. Assume the strings will be lowercase words, so 26 characters will suffice.

11.3 Write a hash function to implement a digit-folding approach in the hash function (as described in the "Hash Functions" section of this chapter). Your program should work for any array size and any key length. Use linear probing. Accessing a group of digits in a number may be easier than you think. Does it matter if the array size is not a multiple of 10?

11.4 Write a rehash() method for the hash.java program. It should be called by insert() to move the entire hash table to an array about twice as large whenever the load factor exceeds 0.5. The new array size should be a prime number. Refer to the section "Expanding the Array" in this chapter. Don't forget you'll need to handle items that have been "deleted," that is, written over with –1.

11.5 Instead of using a linked list to resolve collisions, as in separate chaining, use a binary search tree. That is, create a hash table that is an array of trees. You can use the hashChain.java program (Listing 11.3) as a starting point and the Tree class from the tree.java program (Listing 8.1) in Chapter 8. To display a small tree-based hash table, you could use an inorder traversal of each tree.

The advantage of a tree over a linked list is that it can be searched in O(logN) instead of O(N) time. This time savings can be a significant advantage if very high load factors are encountered. Checking 15 items takes a maximum of 15 comparisons in a list but only 4 in a tree.

Duplicates can present problems in both trees and hash tables, so add some code that prevents a duplicate key from being inserted in the hash table. (Beware: The find() method in Tree assumes a non-empty tree.) To shorten the listing for this program, you can forget about deletion, which for trees requires a lot of code.

12

Heaps

We saw in Chapter 4, "Stacks and Queues," that a priority queue is a data structure that offers convenient access to the data item with the smallest (or largest) key.

Priority queues may be used for task scheduling in computers, where some programs and activities should be executed sooner than others and are therefore given a higher priority.

Another example is in weapons systems, say in a navy cruiser. Numerous threats—airplanes, missiles, submarines, and so on—are detected and must be prioritized. For example, a missile that's a short distance from the cruiser is assigned a higher priority than an aircraft a long distance away so that countermeasures (surface-to-air missiles, for example) can deal with it first.

Priority queues are also used internally in other computer algorithms. In Chapter 14, "Weighted Graphs," we'll see priority queues used in graph algorithms, such as Dijkstra's algorithm.

A priority queue is an Abstract Data Type (ADT) offering methods that allow removal of the item with the maximum (or minimum) key value, insertion, and sometimes other operations. As with other ADTs, priority queues can be implemented using a variety of underlying structures. In Chapter 4 we saw a priority queue implemented as an ordered array. The trouble with that approach is that, even though removal of the largest item is accomplished in fast $O(1)$ time, insertion requires slow $O(N)$ time, because an average of half the items in the array must be moved to insert the new one in order.

In this chapter we'll describe another structure that can be used to implement a priority queue: the heap. A heap is a kind of tree. It offers both insertion and deletion in O(logN) time. Thus, it's not quite as fast for deletion, but much faster for insertion. It's the method of choice for implementing priority queues where speed is important and there will be many insertions.

NOTE

Don't confuse the term *heap*, used here for a special kind of binary tree, with the same term used to mean the portion of computer memory available to a programmer with new in languages like Java and C++.

Introduction to Heaps

A heap is a binary tree with these characteristics:

- It's complete. This means it's completely filled in, reading from left to right across each row, although the last row need not be full. Figure 12.1 shows complete and incomplete trees.

- It's (usually) implemented as an array. We described in Chapter 8, "Binary Trees," how binary trees can be stored in arrays, rather than using references to connect the nodes.

- Each node in a heap satisfies the *heap condition*, which states that every node's key is larger than (or equal to) the keys of its children.

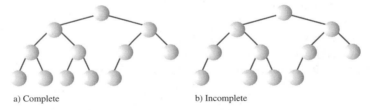

a) Complete   b) Incomplete

FIGURE 12.1   Complete and incomplete binary trees.

Figure 12.2 shows a heap and its relationship to the array used to implement it. The array is what's stored in memory; the heap is only a conceptual representation. Notice that the tree is complete and that the heap condition is satisfied for all the nodes.

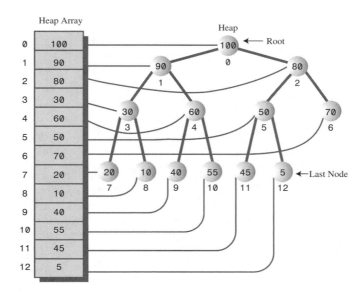

FIGURE 12.2   A heap and its underlying array.

The fact that a heap is a complete binary tree implies that there are no "holes" in the array used to represent it. Every cell is filled, from 0 to N-1. (N is 13 in Figure 12.2.)

We'll assume in this chapter that the maximum key (rather than the minimum) is in the root. A priority queue based on such a heap is a *descending-priority* queue. (We discussed ascending-priority queues in Chapter 4.)

Priority Queues, Heaps, and ADTs

We'll be talking about heaps in this chapter, although heaps are mostly used to implement priority queues. However, there's a very close relationship between a priority queue and the heap used to implement it. This relationship is demonstrated in the following abbreviated code:

```
class Heap
   {
   private Node heapArray[];

   public void insert(Node nd)
      { }
   public Node remove()
      { }
   }
```

```
class priorityQueue
   {
   private Heap theHeap;

   public void insert(Node nd)
      { theHeap.insert(nd); }
   public Node remove()
      ( return theHeap.remove() }
   }
```

The methods for the priorityQueue class are simply wrapped around the methods for the underlying Heap class; they have the same functionality. This example makes it conceptually clear that a priority queue is an ADT that can be implemented in a variety of ways, while a heap is a more fundamental kind of data structure. In this chapter, for simplicity, we'll simply show the heap's methods without the priority-queue wrapping.

Weakly Ordered

A heap is weakly ordered compared with a binary search tree, in which all a node's left descendants have keys less than all its right descendants. This implies, as we saw, that in a binary search tree you can traverse the nodes in order by following a simple algorithm.

In a heap, traversing the nodes in order is difficult because the organizing principle (the heap condition) is not as strong as the organizing principle in a tree. All you can say about a heap is that, along every path from the root to a leaf, the nodes are arranged in descending order. As you can see in Figure 12.2, the nodes to the left or right of a given node, or on higher or lower levels—provided they're not on the same path—can have keys larger or smaller than the node's key. Except where they share the same nodes, paths are independent of each other.

Because heaps are weakly ordered, some operations are difficult or impossible. Besides its failure to support traversal, a heap also does not allow convenient searching for a specified key. This is because there's not enough information to decide which of a node's two children to pick in trying to descend to a lower level during the search. It follows that a node with a specified key can't be deleted, at least in O(logN) time, because there's no way to find it. (These operations can be carried out, by looking at every cell of the array in sequence, but this is only possible in slow O(N) time.)

Thus, the organization of a heap may seem dangerously close to randomness. Nevertheless, the ordering is just sufficient to allow fast removal of the maximum node and fast insertion of new nodes. These operations are all that's needed to use a heap as a priority queue. We'll discuss briefly how these operations are carried out and then see them in action in a Workshop applet.

Removal

Removal means removing the node with the maximum key. This node is always the root, so removing it is easy. The root is always at index 0 of the heap array:

```
maxNode = heapArray[0];
```

The problem is that once the root is gone, the tree is no longer complete; there's an empty cell. This "hole" must be filled in. We could shift all the elements in the array down one cell, but there's a much faster approach. Here are the steps for removing the maximum node:

1. Remove the root.

2. Move the last node into the root.

3. Trickle the last node down until it's below a larger node and above a smaller one.

The *last* node is the rightmost node in the lowest occupied level of the tree. This corresponds to the last filled cell in the array. (See the node at index 12, with the value 5, in Figure 12.2.) To copy this node into the root is straightforward:

```
heapArray[0] = heapArray[N-1];
N--;
```

The removal of the root decreases the size of the array by one.

To *trickle* (the terms *bubble* or *percolate* are also used) a node up or down means to move it along a path step by step, swapping it with the node ahead of it, checking at each step to see whether it's in its proper position. In step 3 the node at the root is too small for that position, so it's trickled down the heap into its proper place. We'll see the code for this later.

Step 2 restores the completeness characteristic of the heap (no holes), and step 3 restores the heap condition (every node larger than its children). The removal process is shown in Figure 12.3.

In part a) of this figure the last node (30) is copied to the root, which is removed. In parts b), c), and d), the last node is trickled down to its appropriate position, which happens to be on the bottom row. (This isn't always the case; the trickle-down process may stop at a middle row as well.) Part e) shows the node in its correct position.

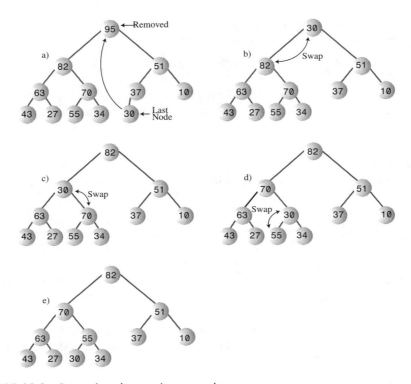

FIGURE 12.3 Removing the maximum node.

At each position of the target node the trickle-down algorithm checks which child is larger. It then swaps the target node with the larger child. If it tried to swap with the smaller child, that child would become the parent of a larger child, which violates the heap condition. Correct and incorrect swaps are shown in Figure 12.4.

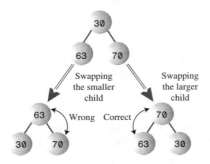

FIGURE 12.4 Which child to swap?

Insertion

Inserting a node is also easy. Insertion uses trickle up, rather than trickle down. Initially, the node to be inserted is placed in the first open position at the end of the array, increasing the array size by one:

```
heapArray[N] = newNode;
N++;
```

The problem is that it's likely that this will destroy the heap condition. This happens if the new node's key is larger than its newly acquired parent. Because this parent is on the bottom of the heap, it's likely to be small, so the new node is likely to be larger. Thus, the new node will usually need to be trickled upward until it's below a node with a larger key and above a node with a smaller key. The insertion process is shown in Figure 12.5.

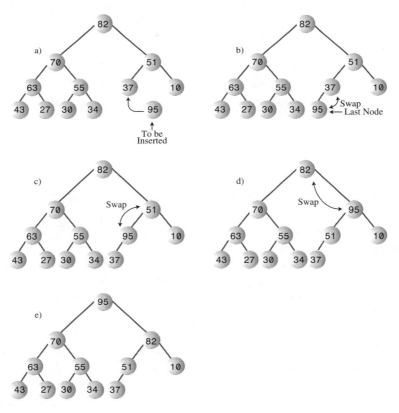

FIGURE 12.5   Inserting a node.

The trickle-up algorithm is somewhat simpler than trickling down because two children don't need to be compared. A node has only one parent, and the target node is simply swapped with its parent. In the figure the final correct position for the new node happens to be the root, but a new node can also end up at an intermediate level.

By comparing Figures 12.4 and 12.5, you can see that if you remove a node and then insert the same node the result is not necessarily the restoration of the original heap. A given set of nodes can be arranged in many valid heaps, depending on the order in which nodes are inserted.

Not Really Swapped

In Figures 12.4 and 12.5 we showed nodes being swapped in the trickle-down and trickle-up processes. Swapping is conceptually the easiest way to understand insertion and deletion, and indeed some heap implementations actually use swaps. Figure 12.6a shows a simplified version of swaps used in the trickle-down process. After three swaps, node A will end up in position D, and nodes B, C, and D will each move up one level.

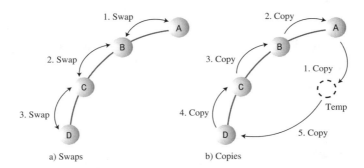

a) Swaps   b) Copies

FIGURE 12.6   Trickling with swaps and copies.

However, a swap requires three copies, so the three swaps shown in Figure 12.6a take nine copies. We can reduce the total number of copies necessary in a trickle algorithm by substituting copies for swaps.

Figure 12.6b shows how five copies do the work of three swaps. First, node A is saved temporarily. Then B is copied over A, C is copied over B, and D is copied over C. Finally, A is copied back from temporary storage onto position D. We have reduced the number of copies from nine to five.

In the figure we're moving node A three levels. The savings in copy time grow larger as the number of levels increases because the two copies from and to temporary storage account for less of the total. For a large number of levels the savings in the number of copies approach a factor of three.

Another way to visualize trickle-up and trickle-down processes being carried out with copies is to think of a "hole"—the absence of a node—moving down in a trickle up and up in a trickle down. For example, in Figure 12.6b, copying A to Temp creates a "hole" at A. The "hole" actually consists of the earlier copy of a node that will be moved; it's still there but it's irrelevant. Copying B to A moves the "hole" from A to B, in the opposite direction from the node. Step by step the "hole" trickles downward.

The Heap Workshop Applet

The Heap Workshop applet demonstrates the operations we discussed in the preceding section: It allows you to insert new items into a heap and remove the largest item. In addition, you can change the priority of a given item.

When you start up the Heap Workshop applet, you'll see a display similar to Figure 12.7.

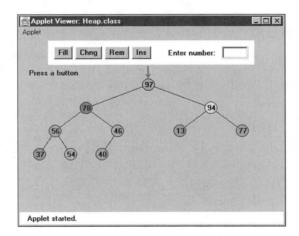

FIGURE 12.7   The Heap Workshop applet.

There are four buttons: Fill, Chng, Rem, and Ins, for fill, change, remove, and insert. Let's see how they work.

The Fill Button

The heap contains 10 nodes when the applet is first started. Using the Fill button, you can create a new heap with any number of nodes from 1 to 31. Press Fill repeatedly, and type in the desired number when prompted.

The Change Button

It's possible to change the priority of an existing node. This procedure is useful in many situations. For example, in our cruiser example, a threat such as an approaching airplane may reverse course away from the carrier; its priority should be lowered to reflect this new development, although the aircraft would remain in the priority queue until it was out of radar range.

To change the priority of a node, repeatedly press the Chng button. When prompted, click on the node with the mouse. This will position the red arrow on the node. Then, when prompted, type in the node's new priority.

If the node's priority is raised, it will trickle upward to a new position. If the priority is lowered, the node will trickle downward.

The Remove Button

Repeatedly pressing the Rem button causes the node with the highest key, located at the root, to be removed. You'll see it disappear, and then be replaced by the last (rightmost) node on the bottom row. Finally, this node will trickle down until it reaches the position that reestablishes the heap order.

The Insert Button

A new node is always inserted initially in the first available array cell, just to the right of the last node on the bottom row of the heap. From there it trickles up to the appropriate position. Pressing the Ins button repeatedly carries out this operation.

Java Code for Heaps

The complete code for heap.java is shown later in this section. Before we get to it, we'll focus on the individual operations of insertion, removal, and change.

Here are some points to remember from Chapter 8 about representing a tree as an array. For a node at index x in the array,

- Its parent is (x-1) / 2.

- Its left child is 2*x + 1.

- Its right child is 2*x + 2.

These relationships can be seen in Figure 12.2.

NOTE

Remember that the / symbol, when applied to integers, performs integer division, in which the answer is rounded to the lowest integer.

Insertion

We place the trickle-up algorithm in its own method. The insert() method, which includes a call to this trickleUp() method, is straightforward:

```
public boolean insert(int key)
   {
   if(currentSize==maxSize)          // if array is full,
      return false;                  //    failure
   Node newNode = new Node(key);     // make a new node
   heapArray[currentSize] = newNode; // put it at the end
   trickleUp(currentSize++);         // trickle it up
   return true;                      // success
   }  // end insert()
```

We check to make sure the array isn't full and then make a new node using the key value passed as an argument. This node is inserted at the end of the array. Finally, the trickleUp() routine is called to move this node up to its proper position.

In trickleUp() (shown below) the argument is the index of the newly inserted item. We find the parent of this position and then save the node in a variable called bottom. Inside the while loop, the variable index will trickle up the path toward the root, pointing to each node in turn. The while loop runs as long as we haven't reached the root (index>0), and the key (iData) of index's parent is less than the new node.

The body of the while loop executes one step of the trickle-up process. It first copies the parent node into index, moving the node down. (This has the effect of moving the "hole" upward.) Then it moves index upward by giving it its parent's index, and giving its parent *its* parent's index.

```
public void trickleUp(int index)
   {
   int parent = (index-1) / 2;
   Node bottom = heapArray[index];

   while( index > 0 &&
           heapArray[parent].getKey() < bottom.getKey() )
      {
      heapArray[index] = heapArray[parent]; // move node down
      index = parent;                       // move index up
      parent = (parent-1) / 2;        // parent <- its parent
      } // end while
   heapArray[index] = bottom;
   }  // end trickleUp()
```

Finally, when the loop has exited, the newly inserted node, which has been temporarily stored in `bottom`, is inserted into the cell pointed to by `index`. This is the first location where it's not larger than its parent, so inserting it here satisfies the heap condition.

Removal

The removal algorithm is also not complicated if we subsume the trickle-down algorithm into its own routine. We save the node from the root, copy the last node (at index `currentSize-1`) into the root, and call `trickleDown()` to place this node in its appropriate location.

```
public Node remove()           // delete item with max key
   {                           // (assumes non-empty list)
   Node root = heapArray[0];   // save the root
   heapArray[0] = heapArray[--currentSize];  // root <- last
   trickleDown(0);             // trickle down the root
   return root;                // return removed node
   }  // end remove()
```

This method returns the node that was removed; the user of the heap usually needs to process it in some way.

The `trickleDown()` routine is more complicated than `trickleUp()` because we must determine which of the two children is larger. First, we save the node at `index` in a variable called `top`. If `trickleDown()` has been called from `remove()`, `index` is the root, but, as we'll see, it can be called from other routines as well.

The `while` loop will run as long as `index` is not on the bottom row—that is, as long as it has at least one child. Within the loop we check if there is a right child (there may be only a left), and if so, compare the children's keys, setting `largerChild` appropriately.

Then we check if the key of the original node (now in `top`) is greater than that of `largerChild`; if so, the trickle-down process is complete and we exit the loop.

```
public void trickleDown(int index)
   {
   int largerChild;
   Node top = heapArray[index];      // save root
   while(index < currentSize/2)      // while node has at
      {                              //    least one child,
      int leftChild = 2*index+1;
      int rightChild = leftChild+1;

                                     // find larger child
```

```
        if( rightChild < currentSize &&   // (rightChild exists?)
                        heapArray[leftChild].getKey() <
                        heapArray[rightChild].getKey() )
            largerChild = rightChild;
        else
            largerChild = leftChild;
                                        // top >= largerChild?
        if(top.getKey() >= heapArray[largerChild].getKey())
            break;
                                        // shift child up
        heapArray[index] = heapArray[largerChild];
        index = largerChild;            // go down
        } // end while
    heapArray[index] = top;             // index <- root
    } // end trickleDown()
```

On exiting the loop we need only restore the node stored in top to its appropriate position, pointed to by index.

Key Change

After we've created the trickleDown() and trickleUp() methods, we can easily implement an algorithm to change the priority (the key) of a node and then trickle it up or down to its proper position. The change() method accomplishes this:

```
public boolean change(int index, int newValue)
    {
    if(index<0 || index>=currentSize)
        return false;
    int oldValue = heapArray[index].getKey(); // remember old
    heapArray[index].setKey(newValue);  // change to new

    if(oldValue < newValue)             // if raised,
        trickleUp(index);               // trickle it up
    else                                // if lowered,
        trickleDown(index);             // trickle it down
    return true;
    } // end change()
```

This routine first checks that the index given in the first argument is valid, and if so, changes the iData field of the node at that index to the value specified as the second argument.

Then, if the priority has been raised, the node is trickled up; if it's been lowered, it's trickled down.

Actually, the difficult part of changing a node's priority is not shown in this routine: finding the node you want to change. In the change() method just shown, we supply the index as an argument, and in the Heap Workshop applet, the user simply clicks on the selected node. In a real-world application a mechanism would be needed to find the appropriate node; as we've seen, the only node to which we normally have convenient access in a heap is the one with the largest key.

The problem can be solved in linear O(N) time by searching the array sequentially. Or, a separate data structure (perhaps a hash table) could be updated with the new index value whenever a node was moved in the priority queue. This would allow quick access to any node. Of course, keeping a second structure updated would itself be time-consuming.

The Array Size

We should note that the array size, equivalent to the number of nodes in the heap, is a vital piece of information about the heap's state and a critical field in the Heap class. Nodes copied from the last position aren't erased, so the only way for algorithms to know the location of the last occupied cell is to refer to the current size of the array.

The heap.java Program

The heap.java program, shown in Listing 12.1, uses a Node class whose only field is the iData variable that serves as the node's key. As usual, this class would hold many other fields in a useful program. The Heap class contains the methods we discussed, plus isEmpty() and displayHeap(), which outputs a crude but comprehensible character-based representation of the heap.

LISTING 12.1   The heap.java Program

```
// heap.java
// demonstrates heaps
// to run this program: C>java HeapApp
import java.io.*;
////////////////////////////////////////////////////////////////
class Node
   {
   private int iData;              // data item (key)
// -------------------------------------------------------------
   public Node(int key)            // constructor
```

LISTING 12.1   Continued

```java
      { iData = key; }
// ------------------------------------------------------------
   public int getKey()
      { return iData; }
// ------------------------------------------------------------
   public void setKey(int id)
      { iData = id; }
// ------------------------------------------------------------
   }  // end class Node
////////////////////////////////////////////////////////////////
class Heap
   {
   private Node[] heapArray;
   private int maxSize;            // size of array
   private int currentSize;       // number of nodes in array
// ------------------------------------------------------------
   public Heap(int mx)            // constructor
      {
      maxSize = mx;
      currentSize = 0;
      heapArray = new Node[maxSize];  // create array
      }
// ------------------------------------------------------------
   public boolean isEmpty()
      { return currentSize==0; }
// ------------------------------------------------------------
   public boolean insert(int key)
      {
      if(currentSize==maxSize)
         return false;
      Node newNode = new Node(key);
      heapArray[currentSize] = newNode;
      trickleUp(currentSize++);
      return true;
      }  // end insert()
// ------------------------------------------------------------
   public void trickleUp(int index)
      {
      int parent = (index-1) / 2;
      Node bottom = heapArray[index];
```

LISTING 12.1   Continued

```
        while( index > 0 &&
              heapArray[parent].getKey() < bottom.getKey() )
           {
           heapArray[index] = heapArray[parent];  // move it down
           index = parent;
           parent = (parent-1) / 2;
           }  // end while
        heapArray[index] = bottom;
        }  // end trickleUp()
// -----------------------------------------------------------
    public Node remove()            // delete item with max key
        {                           // (assumes non-empty list)
        Node root = heapArray[0];
        heapArray[0] = heapArray[--currentSize];
        trickleDown(0);
        return root;
        }  // end remove()
// -----------------------------------------------------------
    public void trickleDown(int index)
        {
        int largerChild;
        Node top = heapArray[index];        // save root
        while(index < currentSize/2)        // while node has at
           {                                //    least one child,
           int leftChild = 2*index+1;
           int rightChild = leftChild+1;
                                            // find larger child
           if(rightChild < currentSize &&  // (rightChild exists?)
                            heapArray[leftChild].getKey() <
                            heapArray[rightChild].getKey())
              largerChild = rightChild;
           else
              largerChild = leftChild;
                                            // top >= largerChild?
           if( top.getKey() >= heapArray[largerChild].getKey() )
              break;
                                            // shift child up
           heapArray[index] = heapArray[largerChild];
           index = largerChild;             // go down
           }  // end while
        heapArray[index] = top;             // root to index
```

LISTING 12.1   Continued

```
      } // end trickleDown()
// ------------------------------------------------------------
   public boolean change(int index, int newValue)
      {
      if(index<0 || index>=currentSize)
         return false;
      int oldValue = heapArray[index].getKey(); // remember old
      heapArray[index].setKey(newValue);  // change to new

      if(oldValue < newValue)              // if raised,
         trickleUp(index);                 // trickle it up
      else                                 // if lowered,
         trickleDown(index);               // trickle it down
      return true;
      } // end change()
// ------------------------------------------------------------
   public void displayHeap()
      {
      System.out.print("heapArray: ");    // array format
      for(int m=0; m<currentSize; m++)
         if(heapArray[m] != null)
            System.out.print( heapArray[m].getKey() + " ");
         else
            System.out.print( "-- ");
      System.out.println();
                                           // heap format
      int nBlanks = 32;
      int itemsPerRow = 1;
      int column = 0;
      int j = 0;                           // current item
      String dots = "...............................";
      System.out.println(dots+dots);       // dotted top line

      while(currentSize > 0)               // for each heap item
         {
         if(column == 0)                   // first item in row?
            for(int k=0; k<nBlanks; k++)   // preceding blanks
               System.out.print(' ');
                                           // display item
         System.out.print(heapArray[j].getKey());
```

LISTING 12.1   Continued

```
            if(++j == currentSize)             // done?
               break;

            if(++column==itemsPerRow)          // end of row?
               {
               nBlanks /= 2;                    // half the blanks
               itemsPerRow *= 2;                // twice the items
               column = 0;                      // start over on
               System.out.println();            //    new row
               }
            else                                // next item on row
               for(int k=0; k<nBlanks*2-2; k++)
                  System.out.print(' ');        // interim blanks
            }  // end for
         System.out.println("\n"+dots+dots); // dotted bottom line
         }  // end displayHeap()
// --------------------------------------------------------------
   }  // end class Heap
////////////////////////////////////////////////////////////////
class HeapApp
   {
   public static void main(String[] args) throws IOException
      {
      int value, value2;
      Heap theHeap = new Heap(31);  // make a Heap; max size 31
      boolean success;

      theHeap.insert(70);               // insert 10 items
      theHeap.insert(40);
      theHeap.insert(50);
      theHeap.insert(20);
      theHeap.insert(60);
      theHeap.insert(100);
      theHeap.insert(80);
      theHeap.insert(30);
      theHeap.insert(10);
      theHeap.insert(90);

      while(true)                        // until [Ctrl]-[C]
         {
         System.out.print("Enter first letter of ");
```

LISTING 12.1 Continued

```java
                System.out.print("show, insert, remove, change: ");
                int choice = getChar();
                switch(choice)
                    {
                    case 's':                          // show
                        theHeap.displayHeap();
                        break;
                    case 'i':                          // insert
                        System.out.print("Enter value to insert: ");
                        value = getInt();
                        success = theHeap.insert(value);
                        if( !success )
                            System.out.println("Can't insert; heap full");
                        break;
                    case 'r':                          // remove
                        if( !theHeap.isEmpty() )
                            theHeap.remove();
                        else
                            System.out.println("Can't remove; heap empty");
                        break;
                    case 'c':                          // change
                        System.out.print("Enter current index of item: ");
                        value = getInt();
                        System.out.print("Enter new key: ");
                        value2 = getInt();
                        success = theHeap.change(value, value2);
                        if( !success )
                            System.out.println("Invalid index");
                        break;
                    default:
                        System.out.println("Invalid entry\n");
                    }  // end switch
                }  // end while
            }  // end main()
//-------------------------------------------------------------
    public static String getString() throws IOException
        {
        InputStreamReader isr = new InputStreamReader(System.in);
        BufferedReader br = new BufferedReader(isr);
        String s = br.readLine();
        return s;
```

LISTING 12.1   Continued

```
      }
//------------------------------------------------------------
   public static char getChar() throws IOException
      {
      String s = getString();
      return s.charAt(0);
      }
//------------------------------------------------------------
   public static int getInt() throws IOException
      {
      String s = getString();
      return Integer.parseInt(s);
      }
//------------------------------------------------------------
   }  // end class HeapApp
////////////////////////////////////////////////////////////////
```

The array places the heap's root at index 0. Some heap implementations start the array with the root at 1, using position 0 as a sentinel value with the largest possible key. This saves an instruction in some of the algorithms but complicates things conceptually.

The main() routine in HeapApp creates a heap with a maximum size of 31 (dictated by the limitations of the display routine) and inserts into it 10 nodes with random keys. Then it enters a loop in which the user can enter s, i, r, or c, for show, insert, remove, or change.

Here's some sample interaction with the program:

```
Enter first letter of show, insert, remove, change: s
heapArray: 100 90 80 30 60 50 70 20 10 40

..............................................................
                              100
               90                              80
        30             60             50              70
     20     10      40
..............................................................
Enter first letter of show, insert, remove, change: i
Enter value to insert: 53
Enter first letter of show, insert, remove, change: s
heapArray: 100 90 80 30 60 50 70 20 10 40 53

..............................................................
```

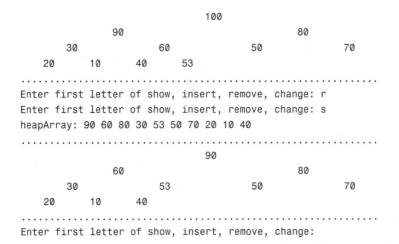

```
                              100
            90                            80
        30              60            50          70
     20      10      40      53
.........................................................
Enter first letter of show, insert, remove, change: r
Enter first letter of show, insert, remove, change: s
heapArray: 90 60 80 30 53 50 70 20 10 40
.........................................................
                              90
            60                            80
        30              53            50          70
     20      10      40
.........................................................
Enter first letter of show, insert, remove, change:
```

The user displays the heap, adds an item with a key of 53, shows the heap again, removes the item with the greatest key, and shows the heap a third time. The show() routine displays both the array and the tree versions of the heap. You'll need to use your imagination to fill in the connections between nodes.

Expanding the Heap Array

What happens if, while a program is running, too many items are inserted for the size of the heap array? A new array can be created, and the data from the old array copied into it. (Unlike the situation with hash tables, changing the size of a heap doesn't require reordering the data.) The copying operation takes linear time, but enlarging the array size shouldn't be necessary very often, especially if the array size is increased substantially each time it's expanded (by doubling it, for example).

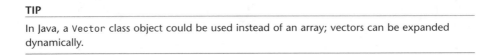

TIP

In Java, a Vector class object could be used instead of an array; vectors can be expanded dynamically.

Efficiency of Heap Operations

For a heap with a substantial number of items, the trickle-up and trickle-down algorithms are the most time-consuming part of the operations we've seen. These algorithms spend time in a loop, repeatedly moving nodes up or down along a path. The number of copies necessary is bounded by the height of the heap; if there are five levels, four copies will carry the "hole" from the top to the bottom. (We'll ignore the two moves used to transfer the end node to and from temporary storage; they're always necessary, so they require constant time.)

The `trickleUp()` method has only one major operation in its loop: comparing the key of the new node with the node at the current location. The `trickleDown()` method needs two comparisons: one to find the largest child and a second to compare this child with the "last" node. They must both copy a node from top to bottom or bottom to top to complete the operation.

A heap is a special kind of binary tree, and as we saw in Chapter 8, the number of levels L in a binary tree equals $\log_2(N+1)$, where N is the number of nodes. The `trickleUp()` and `trickleDown()` routines cycle through their loops L-1 times, so the first takes time proportional to $\log_2 N$, and the second somewhat more because of the extra comparison. Thus, the heap operations we've talked about here all operate in O(logN) time.

A Tree-based Heap

In the figures in this chapter we've shown heaps as if they were trees because it's easier to visualize them that way, but the implementation has been based on an array. However, it's possible to use an actual tree-based implementation. The tree will be a binary tree, but it won't be a search tree because, as we've seen, the ordering principle is not that strong. It will also be a complete tree, with no missing nodes. Let's call such a tree a *tree heap*.

One problem with tree heaps is finding the last node. You need to find this node to remove the maximum item, because it's the last node that's inserted in place of the deleted root (and then trickled down). You also need to find the first empty node, because that's where you insert a new node (and then trickle it up). You can't search for these nodes because you don't know their values, and anyway it's not a search tree. However, these locations are not hard to find in a complete tree if you keep track of the number of nodes in the tree.

As we saw in the discussion of the Huffman tree in Chapter 8, you can represent the path from root to leaf as a binary number, with the binary digits indicating the path from each parent to its child: 0 for left and 1 for right.

It turns out there's a simple relationship between the number of nodes in the tree and the binary number that codes the path to the last node. Assume the root is numbered 1; the next row has nodes 2 and 3; the third row has nodes 4, 5, 6, and 7; and so on. Start with the node number you want to find. This will be the last node or the first null node. Convert the node number to binary. For example, say there are 29 nodes in the tree and you want to find the last node. The number 29 decimal is 11101 binary. Remove the initial 1, leaving 1101. This is the path from the root to node 29: right, right, left, right. The first available null node is 30, which (after removing the initial 1) is 1110 binary: right, right, right, left.

To carry out the calculation, you can repeatedly use the % operator to find the remainder (0 or 1) when the node number n is divided by 2 and then use the /

operator to actually divide n by 2. When n is less than 1, you're done. The sequence of remainders, which you can save in an array or string, is the binary number. (The least significant bits correspond to the lower end of the path.) Here's how you might calculate it:

```
while(n >= 1)
   {
   path[j++] = n % 2;
   n = n / 2;
   }
```

You could also use a recursive approach in which the remainders are calculated each time the function calls itself and the appropriate direction is taken each time it returns.

After the appropriate node (or null child) is found, the heap operations are fairly straightforward. When trickling up or down, the structure of the tree doesn't change, so you don't need to move the actual nodes around. You can simply copy the data from one node to the next. This way, you don't need to connect and disconnect all the children and parents for a simple move. The Node class will need a field for the parent node because you'll need to access the parent when you trickle up. We'll leave the implementation of the tree heap as a programming project.

The tree heap operates in O(logN) time. As in the array-based heap the time is mostly spent doing the trickle-up and trickle-down operations, which take time proportional to the height of the tree.

Heapsort

The efficiency of the heap data structure lends itself to a surprisingly simple and very efficient sorting algorithm called *heapsort*.

The basic idea is to insert all the unordered items into a heap using the normal insert() routine. Repeated application of the remove() routine will then remove the items in sorted order. Here's how that might look:

```
for(j=0; j<size; j++)
   theHeap.insert( anArray[j] );    // from unsorted array
for(j=0; j<size; j++)
   anArray[j] = theHeap.remove();  // to sorted array
```

Because insert() and remove() operate in O(logN) time, and each must be applied N times, the entire sort requires O(N*logN) time, which is the same as quicksort. However, it's not quite as fast as quicksort, partly because there are more operations in the inner while loop in trickleDown() than in the inner loop in quicksort.

However, several tricks can make heapsort more efficient. The first saves time, and the second saves memory.

Trickling Down in Place

If we insert N new items into a heap, we apply the `trickleUp()` method N times. However, all the items can be placed in random locations in the array and then rearranged into a heap with only N/2 applications of `trickleDown()`. This offers a small speed advantage.

Two Correct Subheaps Make a Correct Heap

To see how this approach works, you should know that `trickleDown()` will create a correct heap if, when an out-of-order item is placed at the root, both the child subheaps of this root are correct heaps. (The root can itself be the root of a subheap as well as of the entire heap.) This is shown in Figure 12.8.

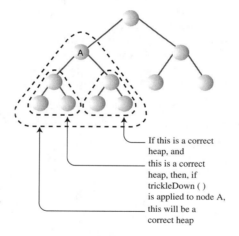

If this is a correct heap, and this is a correct heap, then, if trickleDown ( ) is applied to node A, this will be a correct heap

FIGURE 12.8   Both subheaps must be correct.

This suggests a way to transform an unordered array into a heap. We can apply `trickleDown()` to the nodes on the bottom of the (potential) heap—that is, at the end of the array—and work our way upward to the root at index 0. At each step the subheaps below us will already be correct heaps because we already applied `trickleDown()` to them. After we apply `trickleDown()` to the root, the unordered array will have been transformed into a heap.

Notice, however, that the nodes on the bottom row—those with no children—are already correct heaps, because they are trees with only one node; they have no relationships to be out of order. Therefore, we don't need to apply `trickleDown()` to these nodes. We can start at node N/2-1, the rightmost node with children, instead of N-1, the last node. Thus, we need only half as many trickle operations as we would using insert() N times. Figure 12.9 shows the order in which the trickle-down algorithm is applied, starting at node 6 in a 15-node heap.

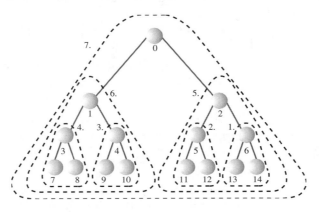

FIGURE 12.9   Order of applying `trickleDown()`.

The following code fragment applies `trickleDown()` to all nodes, except those on the bottom row, starting at N/2-1 and working back to the root:

```
for(j=size/2-1; j >=0; j--)
   theHeap.trickleDown(j);
```

A Recursive Approach

A recursive approach can also be used to form a heap from an array. A `heapify()` method is applied to the root. It calls itself for the root's two children, then for each of these children's two children, and so on. Eventually, it works its way down to the bottom row, where it returns immediately whenever it finds a node with no children.

After it has called itself for two child subtrees, `heapify()` then applies `trickleDown()` to the root of the subtree. This ensures that the subtree is a correct heap. Then `heapify()` returns and works on the subtree one level higher.

```
heapify(int index)        // transform array into heap
   {
   if(index > N/2-1)      // if node has no children,
      return;             //    return
   heapify(index*2+2);    // turn right subtree into heap
   heapify(index*2+1);    // turn left subtree into heap
   trickleDown(index);    // apply trickle-down to this node
   }
```

This recursive approach is probably not quite as efficient as the simple loop.

Using the Same Array

Our initial code fragment showed unordered data in an array. This data was then inserted into a heap, and finally removed from the heap and written back to the array in sorted order. In this procedure two size-N arrays are required: the initial array and the array used by the heap.

In fact, the same array can be used both for the heap and for the initial array. This cuts in half the amount of memory needed for heapsort; no memory beyond the initial array is necessary.

We've already seen how `trickleDown()` can be applied to half the elements of an array to transform them into a heap. We transform the unordered array data into a heap in place; only one array is necessary for this task. Thus, the first step in heapsort requires only one array.

However, the situation becomes more complicated when we apply `remove()` repeatedly to the heap. Where are we going to put the items that are removed?

Each time an item is removed from the heap, an element at the end of the heap array becomes empty; the heap shrinks by one. We can put the recently removed item in this newly freed cell. As more items are removed, the heap array becomes smaller and smaller, while the array of ordered data becomes larger and larger. Thus, with a little planning, it's possible for the ordered array and the heap array to share the same space. This is shown in Figure 12.10.

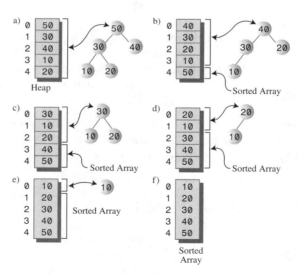

FIGURE 12.10   Dual-purpose array.

The `heapSort.java` Program

We can put these two tricks—applying `trickleDown()` without using `insert()`, and using the same array for the initial data and the heap—together in a program that performs heapsort. Listing 12.2 shows the complete `heapSort.java` program.

LISTING 12.2   The `heapSort.java` Program

```
// heapSort.java
// demonstrates heap sort
// to run this program: C>java HeapSortApp
import java.io.*;
////////////////////////////////////////////////////////////////
class Node
   {
   private int iData;             // data item (key)
// -------------------------------------------------------------
   public Node(int key)           // constructor
      { iData = key; }
// -------------------------------------------------------------
   public int getKey()
      { return iData; }
// -------------------------------------------------------------
   }  // end class Node
////////////////////////////////////////////////////////////////
class Heap
   {
   private Node[] heapArray;
   private int maxSize;           // size of array
   private int currentSize;       // number of items in array
// -------------------------------------------------------------
   public Heap(int mx)            // constructor
      {
      maxSize = mx;
      currentSize = 0;
      heapArray = new Node[maxSize];
      }
// -------------------------------------------------------------
   public Node remove()           // delete item with max key
      {                           // (assumes non-empty list)
      Node root = heapArray[0];
      heapArray[0] = heapArray[--currentSize];
      trickleDown(0);
```

LISTING 12.2   Continued

```
          return root;
          }  // end remove()
//  ---------------------------------------------------------
      public void trickleDown(int index)
         {
         int largerChild;
         Node top = heapArray[index];       // save root
         while(index < currentSize/2)       // not on bottom row
            {
            int leftChild = 2*index+1;
            int rightChild = leftChild+1;
                                            // find larger child
            if(rightChild < currentSize &&  // right ch exists?
                          heapArray[leftChild].getKey() <
                          heapArray[rightChild].getKey())
               largerChild = rightChild;
            else
               largerChild = leftChild;
                                            // top >= largerChild?
            if(top.getKey() >= heapArray[largerChild].getKey())
               break;
                                            // shift child up
            heapArray[index] = heapArray[largerChild];
            index = largerChild;            // go down
            }  // end while
         heapArray[index] = top;            // root to index
         }  // end trickleDown()
//  ---------------------------------------------------------
      public void displayHeap()
         {
         int nBlanks = 32;
         int itemsPerRow = 1;
         int column = 0;
         int j = 0;                         // current item
         String dots = "...............................";
         System.out.println(dots+dots);     // dotted top line

         while(currentSize > 0)             // for each heap item
            {
            if(column == 0)                 // first item in row?
               for(int k=0; k<nBlanks; k++) // preceding blanks
```

LISTING 12.2   Continued

```
                System.out.print(' ');
                                           // display item
        System.out.print(heapArray[j].getKey());

        if(++j == currentSize)              // done?
           break;

        if(++column==itemsPerRow)           // end of row?
           {
           nBlanks /= 2;                    // half the blanks
           itemsPerRow *= 2;                // twice the items
           column = 0;                      // start over on
           System.out.println();            //    new row
           }
        else                                // next item on row
           for(int k=0; k<nBlanks*2-2; k++)
              System.out.print(' ');        // interim blanks
        }  // end for
     System.out.println("\n"+dots+dots); // dotted bottom line
     }  // end displayHeap()
// -------------------------------------------------------------
   public void displayArray()
      {
      for(int j=0; j<maxSize; j++)
         System.out.print(heapArray[j].getKey() + " ");
      System.out.println("");
      }
// -------------------------------------------------------------
   public void insertAt(int index, Node newNode)
      { heapArray[index] = newNode; }
// -------------------------------------------------------------
   public void incrementSize()
      { currentSize++; }
// -------------------------------------------------------------
   }  // end class Heap
/////////////////////////////////////////////////////////////////
class HeapSortApp
   {
   public static void main(String[] args) throws IOException
      {
      int size, j;
```

LISTING 12.2   Continued

```
System.out.print("Enter number of items: ");
size = getInt();
Heap theHeap = new Heap(size);

for(j=0; j<size; j++)        // fill array with
   {                         //    random nodes
   int random = (int)(java.lang.Math.random()*100);
   Node newNode = new Node(random);
   theHeap.insertAt(j, newNode);
   theHeap.incrementSize();
   }

System.out.print("Random: ");
   theHeap.displayArray();  // display random array

for(j=size/2-1; j>=0; j--)  // make random array into heap
   theHeap.trickleDown(j);

System.out.print("Heap:   ");
theHeap.displayArray();      // display heap array
theHeap.displayHeap();       // display heap

for(j=size-1; j>=0; j--)    // remove from heap and
   {                         //    store at array end
   Node biggestNode = theHeap.remove();
   theHeap.insertAt(j, biggestNode);
   }
System.out.print("Sorted: ");
theHeap.displayArray();       // display sorted array
   }  // end main()
// ------------------------------------------------------------
   public static String getString() throws IOException
      {
      InputStreamReader isr = new InputStreamReader(System.in);
      BufferedReader br = new BufferedReader(isr);
      String s = br.readLine();
      return s;
      }
//------------------------------------------------------------
   public static int getInt() throws IOException
      {
```

LISTING 12.2   Continued

```
        String s = getString();
        return Integer.parseInt(s);
        }
// ------------------------------------------------------------
    }  // end class HeapSortApp
```

The Heap class is much the same as in the heap.java program (Listing 12.1), except that to save space we've removed the trickleUp() and insert() methods, which aren't necessary for heapsort. We've also added an insertAt() method, which allows direct insertion into the heap's array.

Notice that this addition is not in the spirit of object-oriented programming. The Heap class interface is supposed to shield class users from the underlying implementation of the heap. The underlying array should be invisible, but insertAt() allows direct access to it. In this situation we accept the violation of OOP principles because the array is so closely tied to the heap architecture.

An incrementSize() method is another addition to the heap class. It might seem as though we could combine this with insertAt(), but when we're inserting into the array in its role as an ordered array, we don't want to increase the heap size, so we keep these functions separate.

The main() routine in the HeapSortApp class does the following:

1. Gets the array size from the user.

2. Fills the array with random data.

3. Turns the array into a heap with N/2 applications of trickleDown().

4. Removes the items from the heap and writes them back at the end of the array.

After each step the array contents are displayed. The user selects the array size. Here's some sample output from heapSort.java:

```
Enter number of items: 10
Random: 81 6 23 38 95 71 72 39 34 53
Heap:   95 81 72 39 53 71 23 38 34 6
.....................................................
                         95
              81                    72
         39        53          71        23
      38    34    6
.....................................................
Sorted: 6 23 34 38 39 53 71 72 81 95
```

The Efficiency of Heapsort

As we noted, heapsort runs in O(N*logN) time. Although it may be slightly slower than quicksort, an advantage over quicksort is that it is less sensitive to the initial distribution of data. Certain arrangements of key values can reduce quicksort to slow O(N^2) time, whereas heapsort runs in O(N*logN) time no matter how the data is distributed.

Summary

- In an ascending-priority queue the item with the largest key is said to have the highest priority. (It's the smallest item in a descending queue.)

- A priority queue is an Abstract Data Type (ADT) that offers methods for insertion of data and removal of the largest (or smallest) item.

- A heap is an efficient implementation of an ADT priority queue.

- A heap offers removal of the largest item, and insertion, in O(N*logN) time.

- The largest item is always in the root.

- Heaps do not support ordered traversal of the data, locating an item with a specific key, or deletion.

- A heap is usually implemented as an array representing a complete binary tree. The root is at index 0 and the last item at index N-1.

- Each node has a key less than its parents and greater than its children.

- An item to be inserted is always placed in the first vacant cell of the array and then trickled up to its appropriate position.

- When an item is removed from the root, it's replaced by the last item in the array, which is then trickled down to its appropriate position.

- The trickle-up and trickle-down processes can be thought of as a sequence of swaps, but are more efficiently implemented as a sequence of copies.

- The priority of an arbitrary item can be changed. First, its key is changed. Then, if the key was increased, the item is trickled up, but if the key was decreased, the item is trickled down.

- A heap can be based on a binary tree (not a search tree) that mirrors the heap structure; this is called a treeheap.

- Algorithms exist to find the last occupied node or the first free node in a treeheap.

- Heapsort is an efficient sorting procedure that requires O(N*logN) time.

- Conceptually, heapsort consists of making N insertions into a heap, followed by N removals.

- Heapsort can be made to run faster by applying the trickle-down algorithm directly to N/2 items in the unsorted array, rather than inserting N items.

- The same array can be used for the initial unordered data, for the heap array, and for the final sorted data. Thus, heapsort requires no extra memory.

Questions

These questions are intended as a self-test for readers. Answers may be found in Appendix C.

1. What does the term *complete* mean when applied to binary trees?

    a. All the necessary data has been inserted.

    b. All the rows are filled with nodes, except possibly the bottom one.

    c. All existing nodes contain data.

    d. The node arrangement satisfies the heap condition.

2. What does the term *weakly ordered* mean when applied to heaps?

3. A node is always removed from the _____.

4. To "trickle up" a node in a descending heap means

    a. to repeatedly exchange it with its parent until it's larger than its parent.

    b. to repeatedly exchange it with its child until it's larger than its child.

    c. to repeatedly exchange it with its child until it's smaller than its child.

    d. to repeatedly exchange it with its parent until it's smaller than its parent.

5. A heap can be represented by an array because a heap

    a. is complete.

    b. is weakly ordered.

    c. is a binary tree.

    d. satisfies the heap condition.

6. The last node in a heap is

    a. always a left child.

    b. always a right child.

    c. always on the bottom row.

    d. never less than its sibling.

7. A heap is to a priority queue as a(n) _____ is to a stack.

8. Insertion into a descending heap involves trickle _____.

9. Heapsort involves

    a. removing data from a heap and then inserting it again.

    b. inserting data into a heap and then removing it.

    c. copying data from one heap to another.

    d. copying data from the array representing a heap to the heap.

10. How many arrays, each big enough to hold all the data, does it take to sort a heap?

Experiments

Carrying out these experiments will help to provide insights into the topics covered in the chapter. No programming is involved.

1. Does the order in which data is inserted in a heap affect the arrangement of the heap? Use the Heap Workshop applet to find out.

2. Use the Workshop applet's Ins button to insert 10 items in ascending order into an empty heap. If you remove these items with the Rem button, will they come off in the reverse order?

3. Insert some items with equal keys. Then remove them. Can you tell from this whether heapsort is stable? The color of the nodes is the secondary data item.

Programming Projects

Writing programs to solve the Programming Projects helps to solidify your understanding of the material and demonstrates how the chapter's concepts are applied. (As noted in the Introduction, qualified instructors may obtain completed solutions to the Programming Projects on the publisher's Web site.)

12.1. Convert the heap.java program (Listing 12.1) so the heap is an ascending, rather than a descending, heap. (That is, the node at the root is the smallest rather than the largest.) Make sure all operations work correctly.

12.2. In the heap.java program the insert() method inserts a new node in the heap and ensures the heap condition is preserved. Write a toss() method that places a new node in the heap array without attempting to maintain the heap condition. (Perhaps each new item can simply be placed at the end of the array.) Then write a restoreHeap() method that restores the heap condition throughout the entire heap. Using toss() repeatedly followed by a single restoreHeap() is more efficient than using insert() repeatedly when a large amount of data must be inserted at one time. See the description of heapsort for clues. To test your program, insert a few items, toss in some more, and then restore the heap.

12.3. Implement the PriorityQ class in the priorityQ.java program (Listing 4.6) using a heap instead of an array. You should be able to use the Heap class in the heap.java program (Listing 12.1) without modification. Make it a descending queue (largest item is removed).

12.4. One problem with implementing a priority queue with an array-based heap is the fixed size of the array. If your data outgrows the array, you'll need to expand the array, as we did for hash tables in Programming Project 11.4 in Chapter 11, "Hash Tables." You can avoid this problem by implementing a priority queue with an ordinary binary search tree rather than a heap. A tree can grow as large as it wants (except for system-memory constraints).

Start with the Tree class from the tree.java program (Listing 8.1). Modify this class so it supports priority queues by adding a removeMax() method that removes the largest item. In a heap this is easy, but in a tree it's slightly harder. How do you find the largest item in a tree? Do you need to worry about both its children when you delete it? Implementing change() is optional. It's easily handled in a binary search tree by deleting the old item and inserting a new one with a different key.

The application should relate to a PriorityQ class; the Tree class should be invisible to main() (except perhaps for displaying the tree while you're debugging). Insertion and removeMax() will operate in O(logN) time.

12.5. Write a program that implements the tree heap (the tree-based implementation of the heap) discussed in the text. Make sure you can remove the largest item, insert items, and change an item's key.

13
Graphs

Graphs are one of the most versatile structures used in computer programming. The sorts of problems that graphs can help to solve are generally quite different from those we've dealt with thus far in this book. If you're dealing with general kinds of data storage problems, you probably won't need a graph, but for some problems—and they tend to be interesting ones—a graph is indispensable.

Our discussion of graphs is divided into two chapters. In this chapter we'll cover the algorithms associated with unweighted graphs, show some algorithms that these graphs can represent, and present two Workshop applets to model them. In the next chapter we'll look at the more complicated algorithms associated with weighted graphs.

Introduction to Graphs

Graphs are data structures rather like trees. In fact, in a mathematical sense, a tree is a kind of graph. In computer programming, however, graphs are used in different ways than trees.

The data structures examined previously in this book have an architecture dictated by the algorithms used on them. For example, a binary tree is shaped the way it is because that shape makes it easy to search for data and insert new data. The edges in a tree represent quick ways to get from node to node.

Graphs, on the other hand, often have a shape dictated by a physical or abstract problem. For example, nodes in a graph may represent cities, while edges may represent airline flight routes between the cities. Another more abstract example is a graph representing the individual tasks necessary to complete a project. In the graph, nodes may represent tasks, while directed edges (with an arrow at one end) indicate which task must be completed before another. In both cases, the shape of the graph arises from the specific real-world situation.

Before going further, we must mention that, when discussing graphs, nodes are traditionally called *vertices* (the singular is *vertex*). This is probably because the nomenclature for graphs is older than that for trees, having arisen in mathematics centuries ago. Trees are more closely associated with computer science. However, both terms are used more or less interchangeably.

Definitions

Figure 13.1a shows a simplified map of the freeways in the vicinity of San Jose, California. Figure 13.1b shows a graph that models these freeways.

In the graph, circles represent freeway interchanges, and straight lines connecting the circles represent freeway segments. The circles are *vertices*, and the lines are *edges*. The vertices are usually labeled in some way—often, as shown here, with letters of the alphabet. Each edge is bounded by the two vertices at its ends.

The graph doesn't attempt to reflect the geographical positions shown on the map; it shows only the relationships of the vertices and the edges—that is, which edges are connected to which vertex. It doesn't concern itself with physical distances or directions. Also, one edge may represent several different route numbers, as in the case of the edge from I to H, which involves routes 101, 84, and 280. It's the *connectedness* (or lack of it) of one intersection to another that's important, not the actual routes.

Adjacency

Two vertices are said to be *adjacent* to one another if they are connected by a single edge. Thus, in Figure 13.1, vertices I and G are adjacent, but vertices I and F are not. The vertices adjacent to a given vertex are sometimes said to be its *neighbors*. For example, the neighbors of G are I, H, and F.

Paths

A *path* is a sequence of edges. Figure 13.1 shows a path from vertex B to vertex J that passes through vertices A and E. We can call this path BAEJ. There can be more than one path between two vertices; another path from B to J is BCDJ.

Connected Graphs

A graph is said to be *connected* if there is at least one path from every vertex to every other vertex, as in the graph in Figure 13.2a. However, if "You can't get there from here" (as Vermont farmers traditionally tell city slickers who stop to ask for directions), the graph is not connected, as in Figure 13.2b.

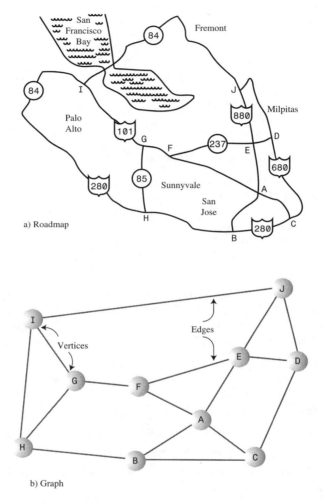

FIGURE 13.1 Roadmap and graph.

A non-connected graph consists of several *connected components*. In Figure 13.2b, A and B are one connected component, and C and D are another.

For simplicity, the algorithms we'll be discussing in this chapter are written to apply to connected graphs, or to one connected component of a non-connected graph. If appropriate, small modifications will usually enable them to work with non-connected graphs as well.

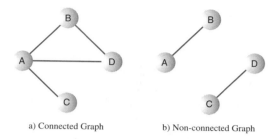

a) Connected Graph   b) Non-connected Graph

FIGURE 13.2   Connected and non-connected graphs.

Directed and Weighted Graphs

The graphs in Figures 13.1 and 13.2 are *non-directed* graphs. That means that the edges don't have a *direction*; you can go either way on them. Thus, you can go from vertex A to vertex B, or from vertex B to vertex A, with equal ease. (Non-directed graphs model freeways appropriately, because you can usually go either way on a freeway.)

However, graphs are often used to model situations in which you can go in only one direction along an edge—from A to B but not from B to A, as on a one-way street. Such a graph is said to be *directed*. The allowed direction is typically shown with an arrowhead at the end of the edge.

In some graphs, edges are given a *weight*, a number that can represent the physical distance between two vertices, or the time it takes to get from one vertex to another, or how much it costs to travel from vertex to vertex (on airline routes, for example). Such graphs are called *weighted* graphs. We'll explore them in the next chapter.

We're going to begin this chapter by discussing simple undirected, unweighted graphs; later we'll explore directed unweighted graphs.

We have by no means covered all the definitions that apply to graphs; we'll introduce more as we go along.

Historical Note

One of the first mathematicians to work with graphs was Leonhard Euler in the early eighteenth century. He solved a famous problem dealing with the bridges in the town of Königsberg, Poland. This town included an island and seven bridges, as shown in Figure 13.3a.

The problem, much discussed by the townsfolk, was to find a way to walk across all seven bridges without recrossing any of them. We won't recount Euler's solution to the problem; it turns out that there is no such path. However, the key to his solution was to represent the problem as a graph, with land areas as vertices and bridges as edges, as shown in Figure 13.3b. This is perhaps the first example of a graph being used to represent a problem in the real world.

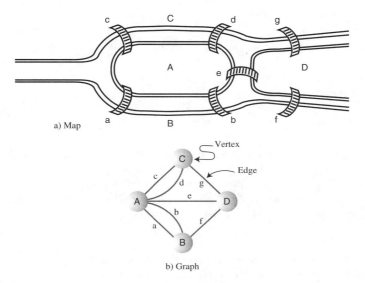

a) Map

b) Graph

FIGURE 13.3   The bridges of Königsberg.

Representing a Graph in a Program

It's all very well to think about graphs in the abstract, as Euler and other mathematicians did until the invention of the computer, but we want to represent graphs by using a computer. What sort of software structures are appropriate to model a graph? We'll look at vertices first and then at edges.

Vertices

In a very abstract graph program you could simply number the vertices 0 to N-1 (where N is the number of vertices). You wouldn't need any sort of variable to hold the vertices because their usefulness would result from their relationships with other vertices.

In most situations, however, a vertex represents some real-world object, and the object must be described using data items. If a vertex represents a city in an airline route simulation, for example, it may need to store the name of the city, its altitude, its location, and other such information. Thus, it's usually convenient to represent a vertex by an object of a vertex class. Our example programs store only a letter (like A), used as a label for identifying the vertex, and a flag for use in search algorithms, as we'll see later. Here's how the Vertex class looks:

```
class Vertex
   {
   public char label;          // label (e.g. 'A')
   public boolean wasVisited;

   public Vertex(char lab)     // constructor
      {
      label = lab;
      wasVisited = false;
      }
   } // end class Vertex
```

Vertex objects can be placed in an array and referred to using their index number. In our examples we'll store them in an array called vertexList. The vertices might also be placed in a list or some other data structure. Whatever structure is used, this storage is for convenience only. It has no relevance to how the vertices are connected by edges. For this, we need another mechanism.

Edges

In Chapter 9, "Red-Black Trees," we saw that a computer program can represent trees in several ways. Mostly we examined trees in which each node contained references to its children, but we also mentioned that an array could be used, with a node's position in the array indicating its relationship to other nodes. Chapter 12, "Heaps," described arrays used to represent a kind of tree called a *heap*.

A graph, however, doesn't usually have the same kind of fixed organization as a tree. In a binary tree, each node has a maximum of two children, but in a graph each vertex may be connected to an arbitrary number of other vertices. For example, in Figure 13.2a, vertex A is connected to three other vertices, whereas C is connected to only one.

To model this sort of free-form organization, a different approach to representing edges is preferable to that used for trees. Two methods are commonly used for graphs: the *adjacency matrix* and the *adjacency list*. (Remember that one vertex is said to be *adjacent* to another if they're connected by a single edge.)

The Adjacency Matrix

An adjacency matrix is a two-dimensional array in which the elements indicate whether an edge is present between two vertices. If a graph has N vertices, the adjacency matrix is an NxN array. Table 13.1 shows the adjacency matrix for the graph in Figure 13.2a.

TABLE 13.1   Adjacency Matrix

	A	B	C	D
A	0	1	1	1
B	1	0	0	1
C	1	0	0	0
D	1	1	0	0

The vertices are used as headings for both rows and columns. An edge between two vertices is indicated by a 1; the absence of an edge is a 0. (You could also use Boolean true/false values.) As you can see, vertex A is adjacent to all three other vertices, B is adjacent to A and D, C is adjacent only to A, and D is adjacent to A and B. In this example, the "connection" of a vertex to itself is indicated by 0, so the diagonal from upper left to lower right, A-A to D-D, which is called the *identity diagonal*, is all 0s. The entries on the identity diagonal don't convey any real information, so you can equally well put 1s along it, if that's more convenient in your program.

Note that the triangular-shaped part of the matrix above the identity diagonal is a mirror image of the part below; both triangles contain the same information. This redundancy may seem inefficient, but there's no convenient way to create a triangular array in most computer languages, so it's simpler to accept the redundancy. Consequently, when you add an edge to the graph, you must make two entries in the adjacency matrix rather than one.

The Adjacency List

The other way to represent edges is with an adjacency list. The *list* in *adjacency list* refers to a linked list of the kind we examined in Chapter 5, "Linked Lists." Actually, an adjacency list is an array of lists (or sometimes a list of lists). Each individual list shows what vertices a given vertex is adjacent to. Table 13.2 shows the adjacency lists for the graph of Figure 13.2a.

TABLE 13.2   Adjacency Lists

Vertex	List Containing Adjacent Vertices
A	B—>C—>D
B	A—>D
C	A
D	A—>B

In this table, the —> symbol indicates a link in a linked list. Each link in the list is a vertex. Here the vertices are arranged in alphabetical order in each list, although that's not really necessary. Don't confuse the contents of adjacency lists with paths. The adjacency list shows which vertices are adjacent to—that is, one edge away from—a given vertex, not paths from vertex to vertex.

Later we'll discuss when to use an adjacency matrix as opposed to an adjacency list. The Workshop applets shown in this chapter all use the adjacency matrix approach, but sometimes the list approach is more efficient.

Adding Vertices and Edges to a Graph

To add a vertex to a graph, you make a new vertex object with `new` and insert it into your vertex array, `vertexList`. In a real-world program a vertex might contain many data items, but for simplicity we'll assume that it contains only a single character. Thus, the creation of a vertex looks something like this:

```
vertexList[nVerts++] = new Vertex('F');
```

This inserts a vertex F, where `nVerts` is the number of vertices currently in the graph.

How you add an edge to a graph depends on whether you're using an adjacency matrix or adjacency lists to represent the graph. Let's say that you're using an adjacency matrix and want to add an edge between vertices 1 and 3. These numbers correspond to the array indices in `vertexList` where the vertices are stored. When you first created the adjacency matrix `adjMat`, you filled it with 0s. To insert the edge, you say

```
adjMat[1][3] = 1;
adjMat[3][1] = 1;
```

If you were using an adjacency list, you would add a 1 to the list for 3, and a 3 to the list for 1.

The `Graph` Class

Let's look at a class `Graph` that contains methods for creating a vertex list and an adjacency matrix, and for adding vertices and edges to a `Graph` object:

```
class Graph
   {
   private final int MAX_VERTS = 20;
   private Vertex vertexList[]; // array of vertices
   private int adjMat[][];      // adjacency matrix
   private int nVerts;          // current number of vertices
// -------------------------------------------------------------
   public Graph()               // constructor
      {
      vertexList = new Vertex[MAX_VERTS];
                                       // adjacency matrix
      adjMat = new int[MAX_VERTS][MAX_VERTS];
```

```
      nVerts = 0;
      for(int j=0; j<MAX_VERTS; j++)        // set adjacency
         for(int k=0; k<MAX_VERTS; k++)     //    matrix to 0
            adjMat[j][k] = 0;
      }  // end constructor
// -------------------------------------------------------------
   public void addVertex(char lab)    // argument is label
      {
      vertexList[nVerts++] = new Vertex(lab);
      }
// -------------------------------------------------------------
   public void addEdge(int start, int end)
      {
      adjMat[start][end] = 1;
      adjMat[end][start] = 1;
      }
// -------------------------------------------------------------
   public void displayVertex(int v)
      {
      System.out.print(vertexList[v].label);
      }
// -------------------------------------------------------------
   }  // end class Graph
```

Within the Graph class, vertices are identified by their index number in vertexList.

We've already discussed most of the methods shown here. To display a vertex, we simply print out its one-character label.

The adjacency matrix (or the adjacency list) provides information that is local to a given vertex. Specifically, it tells you which vertices are connected by a single edge to a given vertex. To answer more global questions about the arrangement of the vertices, we must resort to various algorithms. We'll begin with searches.

Searches

One of the most fundamental operations to perform on a graph is finding which vertices can be reached from a specified vertex. For example, imagine trying to find out how many towns in the United States can be reached by passenger train from Kansas City (assuming that you don't mind changing trains). Some towns could be reached. Others couldn't be reached because they didn't have passenger rail service. Possibly others couldn't be reached, even though they had rail service, because their rail system (the narrow-gauge Hayfork-Hicksville RR, for example) didn't connect with the standard-gauge line you started on or any of the lines that could be reached from your line.

Here's another situation in which you might need to find all the vertices reachable from a specified vertex. Imagine that you're designing a printed circuit board, like the ones inside your computer. (Open it up and take a look!) Various components—mostly integrated circuits (ICs)—are placed on the board, with pins from the ICs protruding through holes in the board. The ICs are soldered in place, and their pins are electrically connected to other pins by *traces*—thin metal lines applied to the surface of the circuit board, as shown in Figure 13.4. (No, you don't need to worry about the details of this figure.)

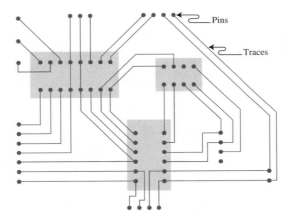

FIGURE 13.4   Pins and traces on a circuit board.

In a graph, each pin might be represented by a vertex, and each trace by an edge. On a circuit board there are many electrical circuits that aren't connected to each other, so the graph is by no means a connected one. During the design process, therefore, it may be genuinely useful to create a graph and use it to find which pins are connected to the same electrical circuit.

Assume that you've created such a graph. Now you need an algorithm that provides a systematic way to start at a specified vertex and then move along edges to other vertices in such a way that, when it's done, you are guaranteed that it has *visited* every vertex that's connected to the starting vertex. Here, as it did in Chapter 8, "Binary Trees," where we discussed binary trees, *visit* means to perform some operation on the vertex, such as displaying it.

There are two common approaches to searching a graph: *depth-first search (DFS)* and *breadth-first search (BFS)*. Both will eventually reach all connected vertices. The depth-first search is implemented with a stack, whereas the breadth-first search is implemented with a queue. These mechanisms result, as we'll see, in the graph being searched in different ways.

Depth-First Search

The depth-first search uses a stack to remember where it should go when it reaches a dead end. We'll show an example, encourage you to try similar examples with the GraphN Workshop applet, and then finally show some code that carries out the search.

An Example

We'll discuss the idea behind the depth-first search in relation to Figure 13.5. The numbers in this figure show the order in which the vertices are visited.

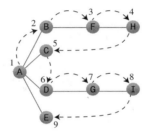

FIGURE 13.5   Depth-first search.

To carry out the depth-first search, you pick a starting point—in this case, vertex A. You then do three things: visit this vertex, push it onto a stack so you can remember it, and mark it so you won't visit it again.

Next, you go to any vertex adjacent to A that hasn't yet been visited. We'll assume the vertices are selected in alphabetical order, so that brings up B. You visit B, mark it, and push it on the stack.

Now what? You're at B, and you do the same thing as before: go to an adjacent vertex that hasn't been visited. This leads you to F. We can call this process Rule 1.

RULE 1
_____

If possible, visit an adjacent unvisited vertex, mark it, and push it on the stack.

Applying Rule 1 again leads you to H. At this point, however, you need to do something else because there are no unvisited vertices adjacent to H. Here's where Rule 2 comes in.

If you can't follow Rule 1, then, if possible, pop a vertex off the stack.

Following this rule, you pop H off the stack, which brings you back to F. F has no unvisited adjacent vertices, so you pop it. Ditto B. Now only A is left on the stack.

A, however, does have unvisited adjacent vertices, so you visit the next one, C. But C is the end of the line again, so you pop it and you're back to A. You visit D, G, and I, and then pop them all when you reach the dead end at I. Now you're back to A. You visit E, and again you're back to A.

This time, however, A has no unvisited neighbors, so we pop it off the stack. But now there's nothing left to pop, which brings up Rule 3.

RULE 3

If you can't follow Rule 1 or Rule 2, you're done.

Table 13.3 shows how the stack looks in the various stages of this process, as applied to Figure 13.5.

TABLE 13.3   Stack Contents During Depth-First Search

Event	Stack
Visit A	A
Visit B	AB
Visit F	ABF
Visit H	ABFH
Pop H	ABF
Pop F	AB
Pop B	A
Visit C	AC
Pop C	A
Visit D	AD
Visit G	ADG
Visit I	ADGI
Pop I	ADG
Pop G	AD
Pop D	A
Visit E	AE
Pop E	A
Pop A	
Done	

The contents of the stack is the route you took from the starting vertex to get where you are. As you move away from the starting vertex, you push vertices as you go. As you move back toward the starting vertex, you pop them. The order in which you visit the vertices is ABFHCDGIE.

You might say that the depth-first search algorithm likes to get as far away from the starting point as quickly as possible and returns only when it reaches a dead end. If you use the term *depth* to mean the distance from the starting point, you can see where the name *depth-first search* comes from.

An Analogy

An analogy you might think about in relation to depth-first search is a maze. The maze—perhaps one of the people-size ones made of hedges, popular in England—consists of narrow passages (think of edges) and intersections where passages meet (vertices).

Suppose that someone is lost in the maze. She knows there's an exit and plans to traverse the maze systematically to find it. Fortunately, she has a ball of string and a marker pen. She starts at some intersection and goes down a randomly chosen passage, unreeling the string. At the next intersection, she goes down another randomly chosen passage, and so on, until finally she reaches a dead end.

At the dead end she retraces her path, reeling in the string, until she reaches the previous intersection. Here she marks the path she's been down so she won't take it again, and tries another path. When she's marked all the paths leading from that intersection, she returns to the previous intersection and repeats the process.

The string represents the stack: It "remembers" the path taken to reach a certain point.

The GraphN Workshop Applet and DFS

You can try out the depth-first search with the DFS button in the GraphN workshop applet. (The N is for *not directed, not weighted*.)

Start the applet. At the beginning, there are no vertices or edges, just an empty rectangle. You create vertices by double-clicking the desired location. The first vertex is automatically labeled A, the second one is B, and so on. They're colored randomly.

To make an edge, drag from one vertex to another. Figure 13.6 shows the graph of Figure 13.5 as it looks when created using the applet.

There's no way to delete individual edges or vertices, so if you make a mistake, you'll need to start over by clicking the New button, which erases all existing vertices and edges. (It warns you before it does this.) Clicking the View button switches you to the adjacency matrix for the graph you've made, as shown in Figure 13.7. Clicking View again switches you back to the graph.

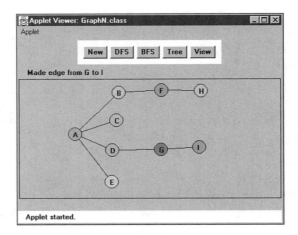

FIGURE 13.6   The GraphN workshop applet.

Applet Viewer: GraphN.class
Applet

| New | DFS | BFS | Tree | View |

Press View button again to show graph

	A	B	C	D	E	F	G	H	I
A	0	1	1	1	1	0	0	0	0
B	1	0	0	0	0	1	0	0	0
C	1	0	0	0	0	0	0	0	0
D	1	0	0	0	0	0	1	0	0
E	1	0	0	0	0	0	0	0	0
F	0	1	0	0	0	0	0	1	0
G	0	0	0	1	0	0	0	0	1
H	0	0	0	0	0	1	0	0	0
I	0	0	0	0	0	0	1	0	0

Applet started.

FIGURE 13.7   Adjacency matrix view in GraphN.

To run the depth-first search algorithm, click the DFS button repeatedly. You'll be prompted to click (*not* double-click) the starting vertex at the beginning of the process.

You can re-create the graph of Figure 13.6, or you can create simpler or more complex ones of your own. After you play with it a while, you can predict what the algorithm will do next (unless the graph is too weird).

If you use the algorithm on an unconnected graph, it will find only those vertices that are connected to the starting vertex.

Java Code

A key to the DFS algorithm is being able to find the vertices that are unvisited and adjacent to a specified vertex. How do you do this? The adjacency matrix is the key. By going to the row for the specified vertex and stepping across the columns, you can pick out the columns with a 1; the column number is the number of an adjacent vertex. You can then check whether this vertex is unvisited. If so, you've found what you want—the next vertex to visit. If no vertices on the row are simultaneously 1 (adjacent) and also unvisited, there are no unvisited vertices adjacent to the specified vertex. We put the code for this process in the getAdjUnvisitedVertex() method:

```
// returns an unvisited vertex adjacent to v
public int getAdjUnvisitedVertex(int v)
   {
   for(int j=0; j<nVerts; j++)
      if(adjMat[v][j]==1 && vertexList[j].wasVisited==false)
         return j;                 // return first such vertex
   return -1;                      // no such vertices
   }  // end getAdjUnvisitedVertex()
```

Now we're ready for the dfs() method of the Graph class, which actually carries out the depth-first search. You can see how this code embodies the three rules listed earlier. It loops until the stack is empty. Within the loop, it does four things:

1. It examines the vertex at the top of the stack, using peek().

2. It tries to find an unvisited neighbor of this vertex.

3. If it doesn't find one, it pops the stack.

4. If it finds such a vertex, it visits that vertex and pushes it onto the stack.

Here's the code for the dfs() method:

```
public void dfs()  // depth-first search
   {                                  // begin at vertex 0
   vertexList[0].wasVisited = true;   // mark it
   displayVertex(0);                  // display it
   theStack.push(0);                  // push it

   while( !theStack.isEmpty() )       // until stack empty,
      {
      // get an unvisited vertex adjacent to stack top
      int v = getAdjUnvisitedVertex( theStack.peek() );
      if(v == -1)                     // if no such vertex,
         theStack.pop();              //    pop a new one
```

```
    else                        // if it exists,
       {
       vertexList[v].wasVisited = true;  // mark it
       displayVertex(v);                 // display it
       theStack.push(v);                 // push it
       }
    }  // end while

    // stack is empty, so we're done
    for(int j=0; j<nVerts; j++)    // reset flags
       vertexList[j].wasVisited = false;
    }  // end dfs
```

At the end of dfs(), we reset all the wasVisited flags so we'll be ready to run dfs()
again later. The stack should already be empty, so it doesn't need to be reset.

Now we have all the pieces of the Graph class we need. Here's some code that creates
a graph object, adds some vertices and edges to it, and then performs a depth-first
search:

```
Graph theGraph = new Graph();
theGraph.addVertex('A');    // 0  (start for dfs)
theGraph.addVertex('B');    // 1
theGraph.addVertex('C');    // 2
theGraph.addVertex('D');    // 3
theGraph.addVertex('E');    // 4

theGraph.addEdge(0, 1);     // AB
theGraph.addEdge(1, 2);     // BC
theGraph.addEdge(0, 3);     // AD
theGraph.addEdge(3, 4);     // DE

System.out.print("Visits: ");
theGraph.dfs();             // depth-first search
System.out.println();
```

Figure 13.8 shows the graph created by this code. Here's the output:

```
Visits: ABCDE
```

You can modify this code to create the graph of your choice, and then run it to see it
carry out the depth-first search.

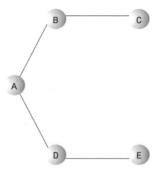

FIGURE 13.8   Graph used by dfs.java and bfs.java.

The dfs.java **Program**

Listing 13.1 shows the dfs.java program, which includes the dfs() method. It includes a version of the StackX class from Chapter 4, "Stacks and Queues."

LISTING 13.1   The dfs.java Program

```java
// dfs.java
// demonstrates depth-first search
// to run this program: C>java DFSApp
//////////////////////////////////////////////////////////////
class StackX
   {
   private final int SIZE = 20;
   private int[] st;
   private int top;
// --------------------------------------------------------------
   public StackX()              // constructor
      {
      st = new int[SIZE];      // make array
      top = -1;
      }
// --------------------------------------------------------------
   public void push(int j)   // put item on stack
      { st[++top] = j; }
// --------------------------------------------------------------
   public int pop()             // take item off stack
      { return st[top--]; }
// --------------------------------------------------------------
   public int peek()            // peek at top of stack
```

LISTING 13.1   Continued

```
      { return st[top]; }
// -----------------------------------------------------------
   public boolean isEmpty()  // true if nothing on stack-
      { return (top == -1); }
// -----------------------------------------------------------
   }  // end class StackX
/////////////////////////////////////////////////////////////////
class Vertex
   {
   public char label;        // label (e.g. 'A')
   public boolean wasVisited;
// -----------------------------------------------------------
   public Vertex(char lab)   // constructor
      {
      label = lab;
      wasVisited = false;
      }
// -----------------------------------------------------------
   }  // end class Vertex
/////////////////////////////////////////////////////////////////
class Graph
   {
   private final int MAX_VERTS = 20;
   private Vertex vertexList[]; // list of vertices
   private int adjMat[][];       // adjacency matrix
   private int nVerts;           // current number of vertices
   private StackX theStack;
// -----------------------------------------------------------
   public Graph()                // constructor
      {
      vertexList = new Vertex[MAX_VERTS];
                                        // adjacency matrix
      adjMat = new int[MAX_VERTS][MAX_VERTS];
      nVerts = 0;
      for(int j=0; j<MAX_VERTS; j++)    // set adjacency
         for(int k=0; k<MAX_VERTS; k++) //    matrix to 0
            adjMat[j][k] = 0;
      theStack = new StackX();
      }  // end constructor
// -----------------------------------------------------------
   public void addVertex(char lab)
```

LISTING 13.1   Continued

```
      {
      vertexList[nVerts++] = new Vertex(lab);
      }
// ------------------------------------------------------------
   public void addEdge(int start, int end)
      {
      adjMat[start][end] = 1;
      adjMat[end][start] = 1;
      }
// ------------------------------------------------------------
   public void displayVertex(int v)
      {
      System.out.print(vertexList[v].label);
      }
// ------------------------------------------------------------
   public void dfs()   // depth-first search
      {                                     // begin at vertex 0
      vertexList[0].wasVisited = true;  // mark it
      displayVertex(0);                 // display it
      theStack.push(0);                 // push it

      while( !theStack.isEmpty() )        // until stack empty,
         {
         // get an unvisited vertex adjacent to stack top
         int v = getAdjUnvisitedVertex( theStack.peek() );
         if(v == -1)                    // if no such vertex,
            theStack.pop();
         else                           // if it exists,
            {
            vertexList[v].wasVisited = true;  // mark it
            displayVertex(v);                 // display it
            theStack.push(v);                 // push it
            }
         }  // end while

      // stack is empty, so we're done
      for(int j=0; j<nVerts; j++)           // reset flags
         vertexList[j].wasVisited = false;
      }  // end dfs
// ------------------------------------------------------------
   // returns an unvisited vertex adj to v
```

LISTING 13.1   Continued

```
   public int getAdjUnvisitedVertex(int v)
      {
      for(int j=0; j<nVerts; j++)
         if(adjMat[v][j]==1 && vertexList[j].wasVisited==false)
            return j;
      return -1;
      }   // end getAdjUnvisitedVertex()
// ------------------------------------------------------------
   }  // end class Graph
////////////////////////////////////////////////////////////////
class DFSApp
   {
   public static void main(String[] args)
      {
      Graph theGraph = new Graph();
      theGraph.addVertex('A');      // 0  (start for dfs)
      theGraph.addVertex('B');      // 1
      theGraph.addVertex('C');      // 2
      theGraph.addVertex('D');      // 3
      theGraph.addVertex('E');      // 4

      theGraph.addEdge(0, 1);       // AB
      theGraph.addEdge(1, 2);       // BC
      theGraph.addEdge(0, 3);       // AD
      theGraph.addEdge(3, 4);       // DE

      System.out.print("Visits: ");
      theGraph.dfs();               // depth-first search
      System.out.println();
      }  // end main()
   }  // end class DFSApp
////////////////////////////////////////////////////////////////
```

Depth-First Search and Game Simulations

Depth-first searches are often used in simulations of games (and game-like situations in the real world). In a typical game you can choose one of several possible actions. Each choice leads to further choices, each of which leads to further choices, and so on into an ever-expanding tree-shaped graph of possibilities. A choice point corresponds to a vertex, and the specific choice taken corresponds to an edge, which leads to another choice-point vertex.

Imagine a game of tic-tac-toe. If you go first, you can make one of nine possible moves. Your opponent can counter with one of eight possible moves, and so on. Each move leads to another group of choices by your opponent, which leads to another series of choices for you, until the last square is filled.

When you are deciding what move to make, one approach is to mentally imagine a move, then your opponent's possible responses, then your responses, and so on. You can decide what to do by seeing which move leads to the best outcome. In simple games like tic-tac-toe the number of possible moves is sufficiently limited that it's possible to follow each path to the end of the game. After you've analyzed the paths completely, you know which move to make first. This can be represented by a graph with one node representing your first move, which is connected to eight nodes representing your opponent's possible responses, each of which is connected to seven nodes representing your responses, and so on. All these paths from the beginning node to an end node include nine nodes. For a complete analysis you'll need to draw nine graphs, one for each starting move.

Even in this simple game the number of paths is surprisingly large. If we ignore simplifications from symmetry, there are 9*8*7*6*5*4*3*2*1 paths in the nine graphs. This is 9 factorial (9!) or 362,880. In a game like chess where the number of possible moves is much greater, even the most powerful computers (like IBM's "Deep Blue") cannot "see" to the end of the game. They can only follow a path to a certain depth and then evaluate the board to see if it appears more favorable than other choices.

The natural way to examine such situations in a computer program is to use a depth-first search. At each node you decide what move to make next, as is done in the getAdjUnvisitedVertex() method in the dfs.java program (Listing 13.1). If there are still unvisited nodes (choice points), you push the current one on the stack and go on to the next. If you find you can't make a move (getAdjUnvisitedVertex() returns –1) in a certain situation, you backtrack by popping a node off the stack (which corresponds to taking back a move) and see if the resulting position has any unexplored choices.

You can think of the sequences of moves in a game as a tree, with nodes representing moves. The first move is the root. In tic-tac-toe, after the first move there are eight possible second moves, each represented by a node connected to the root. After each of these eight second moves, there are seven possible third moves represented by nodes connected to the second-move nodes. You end up with a tree with 9! possible paths from the root to the leaves. This is called the *game tree*.

Actually, the number of branches in the game tree is reduced because the game is often won before all the squares are filled. However, the tic-tac-toe game tree is still very large and complex, and this is a simple game compared with many others, such as chess.

Only some paths in a game tree lead to a successful conclusion. For example, some lead to a win by your opponent. When you reach such an ending, you must back up, or *backtrack*, to a previous node and try a different path. In this way you explore the tree until you find a path with a successful conclusion. Then you make the first move along this path.

Breadth-First Search

As we saw in the depth-first search, the algorithm acts as though it wants to get as far away from the starting point as quickly as possible. In the breadth-first search, on the other hand, the algorithm likes to stay as close as possible to the starting point. It visits all the vertices adjacent to the starting vertex, and only then goes further afield. This kind of search is implemented using a queue instead of a stack.

An Example

Figure 13.9 shows the same graph as Figure 13.5, but here the breadth-first search is used. Again, the numbers indicate the order in which the vertices are visited.

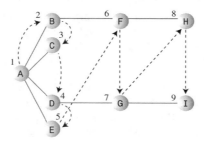

FIGURE 13.9   Breadth-first search.

A is the starting vertex, so you visit it and make it the current vertex. Then you follow these rules:

RULE 1

Visit the next unvisited vertex (if there is one) that's adjacent to the current vertex, mark it, and insert it into the queue.

RULE 2

If you can't carry out Rule 1 because there are no more unvisited vertices, remove a vertex from the queue (if possible) and make it the current vertex.

RULE 3

If you can't carry out Rule 2 because the queue is empty, you're done.

Thus, you first visit all the vertices adjacent to A, inserting each one into the queue as you visit it. Now you've visited A, B, C, D, and E. At this point the queue (from front to rear) contains BCDE.

There are no more unvisited vertices adjacent to A, so you remove B from the queue and look for vertices adjacent to it. You find F, so you insert it in the queue. There are no more unvisited vertices adjacent to B, so you remove C from the queue. It has no adjacent unvisited vertices, so you remove D and visit G. D has no more adjacent unvisited vertices, so you remove E. Now the queue is FG. You remove F and visit H, and then you remove G and visit I.

Now the queue is HI, but when you've removed each of these and found no adjacent unvisited vertices, the queue is empty, so you're done. Table 13.4 shows this sequence.

TABLE 13.4   Queue Contents During Breadth-First Search

Event	Queue (Front to Rear)
Visit A	
Visit B	B
Visit C	BC
Visit D	BCD
Visit E	BCDE
Remove B	CDE
Visit F	CDEF
Remove C	DEF
Remove D	EF
Visit G	EFG
Remove E	FG
Remove F	G
Visit H	GH
Remove G	H
Visit I	HI
Remove H	I
Remove I	
Done	

At each moment, the queue contains the vertices that have been visited but whose neighbors have not yet been fully explored. (Contrast this breadth-first search with the depth-first search, where the contents of the stack is the route you took from the starting point to the current vertex.) The nodes are visited in the order ABCDEFGHI.

The GraphN Workshop Applet and BFS

Use the GraphN workshop applet to try out a breadth-first search using the BFS button. Again, you can experiment with the graph of Figure 13.9, or you can make up your own.

Notice the similarities and the differences of the breadth-first search compared with the depth-first search.

You can think of the breadth-first search as proceeding like ripples widening when you drop a stone in water—or, for those of you who enjoy epidemiology, as the influenza virus carried by air travelers from city to city. First, all the vertices one edge (plane flight) away from the starting point are visited, then all the vertices two edges away are visited, and so on.

Java Code

The bfs() method of the Graph class is similar to the dfs() method, except that it uses a queue instead of a stack and features nested loops instead of a single loop. The outer loop waits for the queue to be empty, whereas the inner one looks in turn at each unvisited neighbor of the current vertex. Here's the code:

```java
public void bfs()                    // breadth-first search
   {                                 // begin at vertex 0
   vertexList[0].wasVisited = true;  // mark it
   displayVertex(0);                 // display it
   theQueue.insert(0);               // insert at tail
   int v2;

   while( !theQueue.isEmpty() )      // until queue empty,
      {
      int v1 = theQueue.remove();    // remove vertex at head
      // until it has no unvisited neighbors
      while( (v2=getAdjUnvisitedVertex(v1)) != -1 )
         {                                // get one,
         vertexList[v2].wasVisited = true; // mark it
         displayVertex(v2);               // display it
         theQueue.insert(v2);             // insert it
         }  // end while(unvisited neighbors)
      }  // end while(queue not empty)

   // queue is empty, so we're done
   for(int j=0; j<nVerts; j++)           // reset flags
      vertexList[j].wasVisited = false;
   }  // end bfs()
```

Given the same graph as in dfs.java (shown earlier in Figure 13.8), the output from bfs.java is now

```
Visits: ABDCE
```

The bfs.java Program

The bfs.java program, shown in Listing 13.2, is similar to dfs.java except for the inclusion of a Queue class (modified from the version in Chapter 4) instead of a StackX class, and a bfs() method instead of a dfs() method.

LISTING 13.2   The bfs.java Program

```java
// bfs.java
// demonstrates breadth-first search
// to run this program: C>java BFSApp
/////////////////////////////////////////////////////////////////
class Queue
   {
   private final int SIZE = 20;
   private int[] queArray;
   private int front;
   private int rear;
// -------------------------------------------------------------
   public Queue()              // constructor
      {
      queArray = new int[SIZE];
      front = 0;
      rear = -1;
      }
// -------------------------------------------------------------
   public void insert(int j) // put item at rear of queue
      {
      if(rear == SIZE-1)
         rear = -1;
      queArray[++rear] = j;
      }
// -------------------------------------------------------------
   public int remove()         // take item from front of queue
      {
      int temp = queArray[front++];
      if(front == SIZE)
         front = 0;
      return temp;
```

LISTING 13.2   Continued

```
      }
// ------------------------------------------------------------
   public boolean isEmpty()  // true if queue is empty
      {
      return ( rear+1==front || (front+SIZE-1==rear) );
      }
// ------------------------------------------------------------
   }  // end class Queue
////////////////////////////////////////////////////////////////
class Vertex
   {
   public char label;        // label (e.g. 'A')
   public boolean wasVisited;
// ------------------------------------------------------------
   public Vertex(char lab)   // constructor
      {
      label = lab;
      wasVisited = false;
      }
// ------------------------------------------------------------
   }  // end class Vertex
////////////////////////////////////////////////////////////////
class Graph
   {
   private final int MAX_VERTS = 20;
   private Vertex vertexList[]; // list of vertices
   private int adjMat[][];      // adjacency matrix
   private int nVerts;          // current number of vertices
   private Queue theQueue;
// ------------------
   public Graph()               // constructor
      {
      vertexList = new Vertex[MAX_VERTS];
                                       // adjacency matrix
      adjMat = new int[MAX_VERTS][MAX_VERTS];
      nVerts = 0;
      for(int j=0; j<MAX_VERTS; j++)       // set adjacency
         for(int k=0; k<MAX_VERTS; k++)   //    matrix to 0
            adjMat[j][k] = 0;
      theQueue = new Queue();
      }  // end constructor
```

LISTING 13.2   Continued

```
// ---------------------------------------------------------------
   public void addVertex(char lab)
      {
      vertexList[nVerts++] = new Vertex(lab);
      }
// ---------------------------------------------------------------
   public void addEdge(int start, int end)
      {
      adjMat[start][end] = 1;
      adjMat[end][start] = 1;
      }
// ---------------------------------------------------------------
   public void displayVertex(int v)
      {
      System.out.print(vertexList[v].label);
      }
// ---------------------------------------------------------------
   public void bfs()                      // breadth-first search
      {                                    // begin at vertex 0
      vertexList[0].wasVisited = true;  // mark it
      displayVertex(0);                     // display it
      theQueue.insert(0);                   // insert at tail
      int v2;

      while( !theQueue.isEmpty() )     // until queue empty,
         {
         int v1 = theQueue.remove();   // remove vertex at head
         // until it has no unvisited neighbors
         while( (v2=getAdjUnvisitedVertex(v1)) != -1 )
            {                                    // get one,
            vertexList[v2].wasVisited = true;  // mark it
            displayVertex(v2);                   // display it
            theQueue.insert(v2);                 // insert it
            }   // end while
         }  // end while(queue not empty)

      // queue is empty, so we're done
      for(int j=0; j<nVerts; j++)                // reset flags
         vertexList[j].wasVisited = false;
      }  // end bfs()
// ---------------------------------------------------------------
```

LISTING 13.2   Continued

```java
   // returns an unvisited vertex adj to v
   public int getAdjUnvisitedVertex(int v)
      {
      for(int j=0; j<nVerts; j++)
         if(adjMat[v][j]==1 && vertexList[j].wasVisited==false)
            return j;
      return -1;
      } // end getAdjUnvisitedVertex()
// --------------------------------------------------------------
   } // end class Graph
////////////////////////////////////////////////////////////////
class BFSApp
   {
   public static void main(String[] args)
      {
      Graph theGraph = new Graph();
      theGraph.addVertex('A');    // 0  (start for dfs)
      theGraph.addVertex('B');    // 1
      theGraph.addVertex('C');    // 2
      theGraph.addVertex('D');    // 3
      theGraph.addVertex('E');    // 4

      theGraph.addEdge(0, 1);     // AB
      theGraph.addEdge(1, 2);     // BC
      theGraph.addEdge(0, 3);     // AD
      theGraph.addEdge(3, 4);     // DE

      System.out.print("Visits: ");
      theGraph.bfs();             // breadth-first search
      System.out.println();
      } // end main()
   } // end class BFSApp
////////////////////////////////////////////////////////////////
```

The breadth-first search has an interesting property: It first finds all the vertices that are one edge away from the starting point, then all the vertices that are two edges away, and so on. This is useful if you're trying to find the shortest path from the starting vertex to a given vertex. You start a BFS, and when you find the specified vertex, you know the path you've traced so far is the shortest path to the node. If there were a shorter path, the BFS would have found it already.

Minimum Spanning Trees

Suppose that you've designed a printed circuit board like the one shown in Figure 13.4, and you want to be sure you've used the minimum number of traces. That is, you don't want any extra connections between pins; such extra connections would take up extra room and make other circuits more difficult to lay out.

It would be nice to have an algorithm that, for any connected set of pins and traces (vertices and edges, in graph terminology), would remove any extra traces. The result would be a graph with the minimum number of edges necessary to connect the vertices. For example, Figure 13.10a shows five vertices with an excessive number of edges, while Figure 13.10b shows the same vertices with the minimum number of edges necessary to connect them. This constitutes a *minimum spanning tree (MST)*.

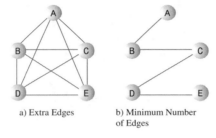

a) Extra Edges   b) Minimum Number
                            of Edges

FIGURE 13.10   Minimum spanning tree.

There are many possible minimum spanning trees for a given set of vertices. Figure 13.10b shows edges AB, BC, CD, and DE, but edges AC, CE, ED, and DB would do just as well. The arithmetically inclined will note that the number of edges E in a minimum spanning tree is always one less than the number of vertices V:

E = V – 1

Remember that we're not worried here about the length of the edges. We're not trying to find a minimum physical length, just the minimum number of edges. (This will change when we talk about weighted graphs in the next chapter.)

The algorithm for creating the minimum spanning tree is almost identical to that used for searching. It can be based on either the depth-first search or the breadth-first search. In our example we'll use the depth-first search.

Perhaps surprisingly, by executing the depth-first search and recording the edges you've traveled to make the search, you automatically create a minimum spanning tree. The only difference between the minimum spanning tree method mst(), which we'll see in a moment, and the depth-first search method dfs(), which we saw earlier, is that mst() must somehow record the edges traveled.

GraphN Workshop Applet

Repeatedly clicking the Tree button in the GraphN workshop algorithm will create a minimum spanning tree for any graph you create. Try it out with various graphs. You'll see that the algorithm follows the same steps as when using the DFS button to do a search. When you use Tree, however, the appropriate edge is darkened when the algorithm assigns it to the minimum spanning tree. When the algorithm is finished, the applet removes all the non-darkened lines, leaving only the minimum spanning tree. A final button press restores the original graph, in case you want to use it again.

Java Code for the Minimum Spanning Tree

Here's the code for the mst() method:

```
while( !theStack.isEmpty() )         // until stack empty
   {                                 // get stack top
   int currentVertex = theStack.peek();
   // get next unvisited neighbor
   int v = getAdjUnvisitedVertex(currentVertex);
   if(v == -1)                       // if no more neighbors
      theStack.pop();                //    pop it away
   else                              // got a neighbor
      {
      vertexList[v].wasVisited = true;  // mark it
      theStack.push(v);                 // push it
                                        // display edge
      displayVertex(currentVertex);     // from currentV
      displayVertex(v);                 // to v
      System.out.print(" ");
      }
   }   // end while(stack not empty)

   // stack is empty, so we're done
   for(int j=0; j<nVerts; j++)       // reset flags
      vertexList[j].wasVisited = false;
}  // end mst()
```

As you can see, this code is very similar to dfs(). In the else statement, however, the current vertex and its next unvisited neighbor are displayed. These two vertices define the edge that the algorithm is currently traveling to get to a new vertex, and it's these edges that make up the minimum spanning tree.

In the main() part of the mst.java program, we create a graph by using these statements:

```
Graph theGraph = new Graph();
theGraph.addVertex('A');      // 0  (start for mst)
theGraph.addVertex('B');      // 1
theGraph.addVertex('C');      // 2
theGraph.addVertex('D');      // 3
theGraph.addVertex('E');      // 4

theGraph.addEdge(0, 1);       // AB
theGraph.addEdge(0, 2);       // AC
theGraph.addEdge(0, 3);       // AD
theGraph.addEdge(0, 4);       // AE
theGraph.addEdge(1, 2);       // BC
theGraph.addEdge(1, 3);       // BD
theGraph.addEdge(1, 4);       // BE
theGraph.addEdge(2, 3);       // CD
theGraph.addEdge(2, 4);       // CE
theGraph.addEdge(3, 4);       // DE
```

The graph that results is the one shown in Figure 13.10a. When the mst() method has done its work, only four edges are left, as shown in Figure 13.10b. Here's the output from the mst.java program:

```
Minimum spanning tree: AB BC CD DE
```

As we noted, this is only one of many possible minimum scanning trees that can be created from this graph. Using a different starting vertex, for example, would result in a different tree. So would small variations in the code, such as starting at the end of the vertexList[] instead of the beginning in the getAdjUnvisitedVertex() method.

The minimum spanning tree is easily derived from the depth-first search because the DFS visits all the nodes, but only once. It never goes to a node that has already been visited. When it looks down an edge that has a visited node at the end, it doesn't follow that edge. It never travels any edges that aren't necessary. Thus, the path of the DFS algorithm through the graph must be a minimum spanning tree.

The mst.java Program

Listing 13.3 shows the mst.java program. It's similar to dfs.java, except for the mst() method and the graph created in main().

LISTING 13.3   The mst.java Program

```
// mst.java
// demonstrates minimum spanning tree
// to run this program: C>java MSTApp
//////////////////////////////////////////////////////////////////
```

LISTING 13.3   Continued

```
class StackX
   {
   private final int SIZE = 20;
   private int[] st;
   private int top;
// ------------------------------------------------------------
   public StackX()              // constructor
      {
      st = new int[SIZE];     // make array
      top = -1;
      }
// ------------------------------------------------------------
   public void push(int j)   // put item on stack
      { st[++top] = j; }
// ------------------------------------------------------------
   public int pop()            // take item off stack
      { return st[top--]; }
// ------------------------------------------------------------
   public int peek()           // peek at top of stack
      { return st[top]; }
// ------------------------------------------------------------
   public boolean isEmpty()  // true if nothing on stack
      { return (top == -1); }
// ------------------------------------------------------------
   }  // end class StackX
////////////////////////////////////////////////////////////////
class Vertex
   {
   public char label;         // label (e.g. 'A')
   public boolean wasVisited;
// ------------------------------------------------------------
   public Vertex(char lab)   // constructor
      {
      label = lab;
      wasVisited = false;
      }
// ------------------------------------------------------------
   }  // end class Vertex
////////////////////////////////////////////////////////////////
class Graph
   {
```

LISTING 13.3   Continued

```java
    private final int MAX_VERTS = 20;
    private Vertex vertexList[]; // list of vertices
    private int adjMat[][];       // adjacency matrix
    private int nVerts;           // current number of vertices
    private StackX theStack;
// -------------------------------------------------------------
    public Graph()                // constructor
        {
        vertexList = new Vertex[MAX_VERTS];
                                              // adjacency matrix
        adjMat = new int[MAX_VERTS][MAX_VERTS];
        nVerts = 0;
        for(int j=0; j<MAX_VERTS; j++)      // set adjacency
           for(int k=0; k<MAX_VERTS; k++)   //    matrix to 0
              adjMat[j][k] = 0;
        theStack = new StackX();
        }   // end constructor
// -------------------------------------------------------------
    public void addVertex(char lab)
        {
        vertexList[nVerts++] = new Vertex(lab);
        }
// -------------------------------------------------------------
    public void addEdge(int start, int end)
        {
        adjMat[start][end] = 1;
        adjMat[end][start] = 1;
        }
// -------------------------------------------------------------
    public void displayVertex(int v)
        {
        System.out.print(vertexList[v].label);
        }
// -------------------------------------------------------------
    public void mst()  // minimum spanning tree (depth first)
        {                                     // start at 0
        vertexList[0].wasVisited = true;   // mark it
        theStack.push(0);                  // push it

        while( !theStack.isEmpty() )       // until stack empty
           {                               // get stack top
```

LISTING 13.3   Continued

```
            int currentVertex = theStack.peek();
            // get next unvisited neighbor
            int v = getAdjUnvisitedVertex(currentVertex);
            if(v == -1)                    // if no more neighbors
               theStack.pop();             //    pop it away
            else                           // got a neighbor
               {
               vertexList[v].wasVisited = true;  // mark it
               theStack.push(v);                 // push it
                                                 // display edge
               displayVertex(currentVertex);   // from currentV
               displayVertex(v);               // to v
               System.out.print(" ");
               }
            }  // end while(stack not empty)

            // stack is empty, so we're done
            for(int j=0; j<nVerts; j++)          // reset flags
               vertexList[j].wasVisited = false;
         }  // end tree
// ------------------------------------------------------------
   // returns an unvisited vertex adj to v
   public int getAdjUnvisitedVertex(int v)
      {
      for(int j=0; j<nVerts; j++)
         if(adjMat[v][j]==1 && vertexList[j].wasVisited==false)
            return j;
      return -1;
      }  // end getAdjUnvisitedVertex()
// ------------------------------------------------------------
   }  // end class Graph
////////////////////////////////////////////////////////////////
class MSTApp
   {
   public static void main(String[] args)
      {
      Graph theGraph = new Graph();
      theGraph.addVertex('A');     // 0  (start for mst)
      theGraph.addVertex('B');     // 1
      theGraph.addVertex('C');     // 2
      theGraph.addVertex('D');     // 3
```

LISTING 13.3   Continued

```
        theGraph.addVertex('E');      // 4

        theGraph.addEdge(0, 1);       // AB
        theGraph.addEdge(0, 2);       // AC
        theGraph.addEdge(0, 3);       // AD
        theGraph.addEdge(0, 4);       // AE
        theGraph.addEdge(1, 2);       // BC
        theGraph.addEdge(1, 3);       // BD
        theGraph.addEdge(1, 4);       // BE
        theGraph.addEdge(2, 3);       // CD
        theGraph.addEdge(2, 4);       // CE
        theGraph.addEdge(3, 4);       // DE

        System.out.print("Minimum spanning tree: ");
        theGraph.mst();               // minimum spanning tree
        System.out.println();
        }  // end main()
    }  // end class MSTApp
//////////////////////////////////////////////////////////////////
```

The statements in main() form a graph that can be visualized as a five-pointed star with every node connected to every other node. The output is

```
Minimum spanning tree: AB BC CD DE
```

Topological Sorting with Directed Graphs

Topological sorting is another operation that can be modeled with graphs. It's useful in situations in which items or events must be arranged in a specific order. Let's look at an example.

An Example: Course Prerequisites

In high school and college, students find (sometimes to their dismay) that they can't take just any course they want. Some courses have prerequisites—other courses that must be taken first. Indeed, taking certain courses may be a prerequisite to obtaining a degree in a certain field. Figure 13.11 shows a somewhat fanciful arrangement of courses necessary for graduating with a degree in mathematics.

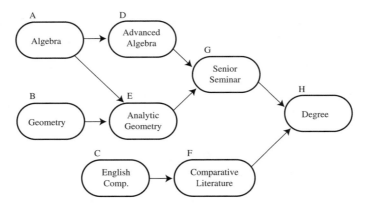

FIGURE 13.11   Course prerequisites.

To obtain your degree, you must complete the Senior Seminar and (because of pressure from the English Department) Comparative Literature. But you can't take Senior Seminar without having already taken Advanced Algebra and Analytic Geometry, and you can't take Comparative Literature without taking English Composition. Also, you need Geometry for Analytic Geometry, and Algebra for both Advanced Algebra and Analytic Geometry.

Directed Graphs

As Figure 13.11 shows, a graph can represent this sort of arrangement. However, the graph needs a feature we haven't seen before: The edges need to have a *direction*. When this is the case, the graph is called a *directed* graph. In a directed graph you can proceed only one way along an edge. The arrows in the figure show the direction of the edges.

In a program, the difference between a non-directed graph and a directed graph is that an edge in a directed graph has only one entry in the adjacency matrix. Figure 13.12 shows a small directed graph; Table 13.5 shows its adjacency matrix.

FIGURE 13.12   A small directed graph.

TABLE 13.5 Adjacency Matrix for a Small Directed Graph

	A	B	C
A	0	1	0
B	0	0	1
C	0	0	0

Each edge is represented by a single 1. The row labels show where the edge starts, and the column labels show where it ends. Thus, the edge from A to B is represented by a single 1 at row A column B. If the directed edge were reversed so that it went from B to A, there would be a 1 at row B column A instead.

For a non-directed graph, as we noted earlier, half of the adjacency matrix mirrors the other half, so half the cells are redundant. However, for a weighted graph, every cell in the adjacency matrix conveys unique information. The halves are not mirror images.

For a directed graph, the method that adds an edge thus needs only a single statement,

```
public void addEdge(int start, int end)  // directed graph
   {
   adjMat[start][end] = 1;
   }
```

instead of the two statements required in a non-directed graph.

If you use the adjacency-list approach to represent your graph, then A has B in its list but—unlike a non-directed graph—B does not have A in its list.

Topological Sorting

Imagine that you make a list of all the courses necessary for your degree, using Figure 13.11 as your input data. You then arrange the courses in the order you need to take them. Obtaining your degree is the last item on the list, which might look like this:

BAEDGCFH

Arranged this way, the graph is said to be *topologically sorted*. Any course you must take before some other course occurs before it in the list.

Actually, many possible orderings would satisfy the course prerequisites. You could take the English courses C and F first, for example:

CFBAEDGH

This also satisfies all the prerequisites. There are many other possible orderings as well. When you use an algorithm to generate a topological sort, the approach you take and the details of the code determine which of various valid sortings are generated.

Topological sorting can model other situations besides course prerequisites. Job scheduling is an important example. If you're building a car, you want to arrange things so that brakes are installed before the wheels, and the engine is assembled before it's bolted onto the chassis. Car manufacturers use graphs to model the thousands of operations in the manufacturing process, to ensure that everything is done in the proper order.

Modeling job schedules with graphs is called *critical path analysis*. Although we don't show it here, a weighted graph (discussed in the next chapter) can be used, which allows the graph to include the time necessary to complete different tasks in a project. The graph can then tell you such things as the minimum time necessary to complete the entire project.

The GraphD Workshop Applet

The GraphD workshop applet models directed graphs. This applet operates in much the same way as GraphN but provides a dot near one end of each edge to show which direction the edge is pointing. Be careful: The direction you drag the mouse to create the edge determines the direction of the edge. Figure 13.13 shows the GraphD workshop applet used to model the course-prerequisite situation of Figure 13.11.

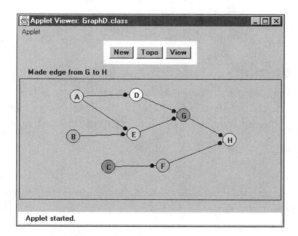

FIGURE 13.13   The GraphD workshop applet.

The idea behind the topological sorting algorithm is unusual but simple. Two steps are necessary:

STEP 1

Find a vertex that has no successors.

The successors to a vertex are those vertices that are directly "downstream" from it— that is, connected to it by an edge that points in their direction. If there is an edge pointing from A to B, then B is a successor to A. In Figure 13.11, the only vertex with no successors is H.

STEP 2

Delete this vertex from the graph, and insert its label at the beginning of a list.

Steps 1 and 2 are repeated until all the vertices are gone. At this point, the list shows the vertices arranged in topological order.

You can see the process at work by using the GraphD applet. Construct the graph of Figure 13.11 (or any other graph, if you prefer) by double-clicking to make vertices and dragging to make edges. Then repeatedly click the Topo button. As each vertex is removed, its label is placed at the beginning of the list below the graph.

Deleting a vertex may seem like a drastic step, but it's the heart of the algorithm. The algorithm can't figure out the second vertex to remove until the first vertex is gone. If you need to, you can save the graph's data (the vertex list and the adjacency matrix) elsewhere and restore it when the sort is completed, as we do in the GraphD applet.

The algorithm works because if a vertex has no successors, it must be the last one in the topological ordering. As soon as it's removed, one of the remaining vertices must have no successors, so it will be the next-to-last one in the ordering, and so on.

The topological sorting algorithm works on unconnected graphs as well as connected graphs. This models the situation in which you have two unrelated goals, such as getting a degree in mathematics and at the same time obtaining a certificate in first aid.

Cycles and Trees

One kind of graph the topological-sort algorithm cannot handle is a graph with *cycles*. What's a cycle? It's a path that ends up where it started. In Figure 13.14 the path B-C-D-B forms a cycle. (Notice that A-B-C-A is not a cycle because you can't go from C to A.)

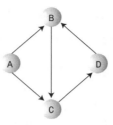

FIGURE 13.14   Graph with a cycle.

A cycle models the Catch-22 situation (which some students claim to have actually encountered at certain institutions), in which course B is a prerequisite for course C, C is a prerequisite for D, and D is a prerequisite for B.

A graph with no cycles is called a *tree*. The binary and multiway trees we saw earlier in this book are trees in this sense. However, the trees that arise in graphs are more general than binary and multiway trees, which have a fixed maximum number of child nodes. In a graph, a vertex in a tree can be connected to any number of other vertices, provided that no cycles are created.

It's easy to figure out if a non-directed graph has cycles. If a graph with N nodes has more than N-1 edges, it must have cycles. You can make this clear to yourself by trying to draw a graph with N nodes and N edges that does not have any cycles.

A topological sort must be carried out on a directed graph with no cycles. Such a graph is called a *directed acyclic graph*, often abbreviated DAG.

Java Code

Here's the Java code for the topo() method, which carries out the topological sort:

```
public void topo()              // topological sort
   {
   int orig_nVerts = nVerts;  // remember how many verts

   while(nVerts > 0)            // while vertices remain,
      {
      // get a vertex with no successors, or -1
      int currentVertex = noSuccessors();
      if(currentVertex == -1)        // must be a cycle
         {
         System.out.println("ERROR: Graph has cycles");
         return;
         }
      }
```

```
      // insert vertex label in sorted array (start at end)
      sortedArray[nVerts-1] = vertexList[currentVertex].label;

      deleteVertex(currentVertex);  // delete vertex
      }  // end while

   // vertices all gone; display sortedArray
   System.out.print("Topologically sorted order: ");
   for(int j=0; j<orig_nVerts; j++)
      System.out.print( sortedArray[j] );
   System.out.println("");
   }  // end topo
```

The work is done in the while loop, which continues until the number of vertices is reduced to 0. Here are the steps involved:

1. Call noSuccessors() to find any vertex with no successors.

2. If such a vertex is found, put the vertex label at the end of sortedArray[] and delete the vertex from graph.

3. If an appropriate vertex isn't found, the graph must have a cycle.

The last vertex to be removed appears first on the list, so the vertex label is placed in sortedArray starting at the end and working toward the beginning, as nVerts (the number of vertices in the graph) gets smaller.

If vertices remain in the graph but all of them have successors, the graph must have a cycle, and the algorithm displays a message and quits. If there are no cycles, the while loop exits, and the list from sortedArray is displayed, with the vertices in topologically sorted order.

The noSuccessors() method uses the adjacency matrix to find a vertex with no successors. In the outer for loop, it goes down the rows, looking at each vertex. For each vertex, it scans across the columns in the inner for loop, looking for a 1. If it finds one, it knows that that vertex has a successor, because there's an edge from that vertex to another one. When it finds a 1, it bails out of the inner loop so that the next vertex can be investigated.

Only if an entire row is found with no 1s do we know we have a vertex with no successors; in this case, its row number is returned. If no such vertex is found, −1 is returned. Here's the noSuccessors() method:

```
public int noSuccessors()  // returns vert with no successors
   {                        // (or -1 if no such verts)
```

```
      boolean isEdge;  // edge from row to column in adjMat

      for(int row=0; row<nVerts; row++)  // for each vertex,
         {
         isEdge = false;                 // check edges
         for(int col=0; col<nVerts; col++)
            {
            if( adjMat[row][col] > 0 )   // if edge to
               {                         // another,
               isEdge = true;
               break;                    // this vertex
               }                         //    has a successor
            }                            //    try another
         if( !isEdge )                   // if no edges,
            return row;                  //    has no successors
         }
      return -1;                         // no such vertex
      }  // end noSuccessors()
```

Deleting a vertex is straightforward except for a few details. The vertex is removed from the vertexList[] array, and the vertices above it are moved down to fill up the vacant position. Likewise, the row and column for the vertex are removed from the adjacency matrix, and the rows and columns above and to the right are moved down and to the left to fill the vacancies. These tasks are carried out by the deleteVertex(), moveRowUp(), and moveColLeft() methods, which you can examine in the complete listing for topo.java (Listing 13.4). It's actually more efficient to use the adjacency-list representation of the graph for this algorithm, but that would take us too far afield.

The main() routine in this program calls on methods, similar to those we saw earlier, to create the same graph shown in Figure 13.10. The addEdge() method, as we noted, inserts a single number into the adjacency matrix because this is a directed graph. Here's the code for main():

```
public static void main(String[] args)
   {
   Graph theGraph = new Graph();
   theGraph.addVertex('A');    // 0
   theGraph.addVertex('B');    // 1
   theGraph.addVertex('C');    // 2
   theGraph.addVertex('D');    // 3
   theGraph.addVertex('E');    // 4
   theGraph.addVertex('F');    // 5
```

```
    theGraph.addVertex('G');    // 6
    theGraph.addVertex('H');    // 7

    theGraph.addEdge(0, 3);     // AD
    theGraph.addEdge(0, 4);     // AE
    theGraph.addEdge(1, 4);     // BE
    theGraph.addEdge(2, 5);     // CF
    theGraph.addEdge(3, 6);     // DG
    theGraph.addEdge(4, 6);     // EG
    theGraph.addEdge(5, 7);     // FH
    theGraph.addEdge(6, 7);     // GH

    theGraph.topo();            // do the sort
    }  // end main()
```

After the graph is created, main() calls topo() to sort the graph and display the result. Here's the output:

```
Topologically sorted order: BAEDGCFH
```

Of course, you can rewrite main() to generate other graphs.

The Complete topo.java Program

You've seen most of the routines in topo.java already. Listing 13.4 shows the complete program.

LISTING 13.4 The topo.java Program

```
// topo.java
// demonstrates topological sorting
// to run this program: C>java TopoApp
////////////////////////////////////////////////////////////////
class Vertex
    {
    public char label;        // label (e.g. 'A')
// --------------------------------------------------------------
    public Vertex(char lab)   // constructor
        { label = lab; }
    }  // end class Vertex
////////////////////////////////////////////////////////////////
class Graph
    {
    private final int MAX_VERTS = 20;
```

LISTING 13.4   Continued

```java
   private Vertex vertexList[]; // list of vertices
   private int adjMat[][];      // adjacency matrix
   private int nVerts;          // current number of vertices
   private char sortedArray[];
// ------------------------------------------------------------
   public Graph()               // constructor
      {
      vertexList = new Vertex[MAX_VERTS];
                                        // adjacency matrix
      adjMat = new int[MAX_VERTS][MAX_VERTS];
      nVerts = 0;
      for(int j=0; j<MAX_VERTS; j++)      // set adjacency
         for(int k=0; k<MAX_VERTS; k++)   //    matrix to 0
            adjMat[j][k] = 0;
      sortedArray = new char[MAX_VERTS];  // sorted vert labels
      }  // end constructor
// ------------------------------------------------------------
   public void addVertex(char lab)
      {
      vertexList[nVerts++] = new Vertex(lab);
      }
// ------------------------------------------------------------
   public void addEdge(int start, int end)
      {
      adjMat[start][end] = 1;
      }
// ------------------------------------------------------------
   public void displayVertex(int v)
      {
      System.out.print(vertexList[v].label);
      }
// ------------------------------------------------------------
   public void topo()  // topological sort
      {
      int orig_nVerts = nVerts;  // remember how many verts

      while(nVerts > 0)  // while vertices remain,
         {
         // get a vertex with no successors, or -1
         int currentVertex = noSuccessors();
         if(currentVertex == -1)       // must be a cycle
```

LISTING 13.4   Continued

```
            {
            System.out.println("ERROR: Graph has cycles");
            return;
            }
         // insert vertex label in sorted array (start at end)
         sortedArray[nVerts-1] = vertexList[currentVertex].label;

         deleteVertex(currentVertex);  // delete vertex
         }  // end while

      // vertices all gone; display sortedArray
      System.out.print("Topologically sorted order: ");
      for(int j=0; j<orig_nVerts; j++)
         System.out.print( sortedArray[j] );
      System.out.println("");
      }  // end topo
// --------------------------------------------------------------
   public int noSuccessors()  // returns vert with no successors
      {                       // (or -1 if no such verts)
      boolean isEdge;  // edge from row to column in adjMat

      for(int row=0; row<nVerts; row++)  // for each vertex,
         {
         isEdge = false;                 // check edges
         for(int col=0; col<nVerts; col++)
            {
            if( adjMat[row][col] > 0 )   // if edge to
               {                         // another,
               isEdge = true;
               break;                    // this vertex
               }                         //     has a successor
            }                            //     try another
         if( !isEdge )                   // if no edges,
            return row;                  //     has no successors
         }
      return -1;                         // no such vertex
      }  // end noSuccessors()
// --------------------------------------------------------------
   public void deleteVertex(int delVert)
      {
      if(delVert != nVerts-1)        // if not last vertex,
```

LISTING 13.4   Continued

```
         {                                   // delete from vertexList
         for(int j=delVert; j<nVerts-1; j++)
            vertexList[j] = vertexList[j+1];
                                             // delete row from adjMat
         for(int row=delVert; row<nVerts-1; row++)
            moveRowUp(row, nVerts);
                                             // delete col from adjMat
         for(int col=delVert; col<nVerts-1; col++)
            moveColLeft(col, nVerts-1);
         }
      nVerts--;                              // one less vertex
      }  // end deleteVertex
// -------------------------------------------------------------
   private void moveRowUp(int row, int length)
      {
      for(int col=0; col<length; col++)
         adjMat[row][col] = adjMat[row+1][col];
      }
// -------------------------------------------------------------
   private void moveColLeft(int col, int length)
      {
      for(int row=0; row<length; row++)
         adjMat[row][col] = adjMat[row][col+1];
      }
// -------------------------------------------------------------
   }  // end class Graph
////////////////////////////////////////////////////////////////
class TopoApp
   {
   public static void main(String[] args)
      {
      Graph theGraph = new Graph();
      theGraph.addVertex('A');     // 0
      theGraph.addVertex('B');     // 1
      theGraph.addVertex('C');     // 2
      theGraph.addVertex('D');     // 3
      theGraph.addVertex('E');     // 4
      theGraph.addVertex('F');     // 5
      theGraph.addVertex('G');     // 6
      theGraph.addVertex('H');     // 7
```

LISTING 13.4   Continued

```
    theGraph.addEdge(0, 3);      // AD
    theGraph.addEdge(0, 4);      // AE
    theGraph.addEdge(1, 4);      // BE
    theGraph.addEdge(2, 5);      // CF
    theGraph.addEdge(3, 6);      // DG
    theGraph.addEdge(4, 6);      // EG
    theGraph.addEdge(5, 7);      // FH
    theGraph.addEdge(6, 7);      // GH

    theGraph.topo();             // do the sort
    }  // end main()
  }  // end class TopoApp
//////////////////////////////////////////////////////////////////
```

In the next chapter, we'll see what happens when edges are given a weight as well as a direction.

Connectivity in Directed Graphs

We've seen how in a non-directed graph you can find all the vertices that are connected by doing a depth-first or breadth-first search. When we try to find all the connected vertices in a directed graph, things get more complicated. You can't just start from a randomly selected vertex and expect to reach all the other connected vertices.

Consider the graph in Figure 13.15. If you start on A, you can get to C but not to any of the other vertices. If you start on B, you can't get to D, and if you start on C, you can't get anywhere. The meaningful question about connectivity is: What vertices can you reach if you start on a particular vertex?

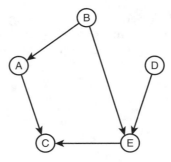

FIGURE 13.15   A directed graph.

The Connectivity Table

You can easily modify the dfs.java program (Listing 13.1) to start the search on each vertex in turn. For the graph of Figure 13.15 the output will look something like this:

```
AC
BACE
C
DEC
EC
```

This is the connectivity table for the directed graph. The first letter is the starting vertex and subsequent letters show the vertices that can be reached (either directly or via other vertices) from the starting vertex.

Warshall's Algorithm

In some applications it's important to find out quickly whether one vertex is reachable from another vertex. Perhaps you want to fly from Athens to Murmansk on Hubris Airlines and you don't care how many intermediate stops you need to make. Is this trip possible?

You could examine the connectivity table, but then you would need to look through all the entries on a given row, which would take O(N) time (where N is the average number of vertices reachable from a given vertex). But you're in a hurry; is there a faster way?

It's possible to construct a table that will tell you instantly (that is, in O(1) time) whether one vertex is reachable from another. Such a table can be obtained by systematically modifying a graph's adjacency matrix. The graph represented by this revised adjacency matrix is called the *transitive closure* of the original graph.

Remember that in an ordinary adjacency matrix the row number indicates where an edge starts and the column number indicates where it ends. (This is similar to the arrangement in the connectivity table.) A 1 at the intersection of row C and column D means there's an edge from vertex C to vertex D. You can get from one vertex to the other in one step. (Of course, in a directed graph it does not follow that you can go the other way, from D to C.) Table 13.6 shows the adjacency matrix for the graph of Figure 13.15.

TABLE 13.6   Adjacency Matrix

	A	B	C	D	E
A	0	0	1	0	0
B	1	0	0	0	1
C	0	0	0	0	0

TABLE 13.6   Continued

	A	B	C	D	E
D	0	0	0	0	1
E	0	0	1	0	0

We can use *Warshall's algorithm* to change the adjacency matrix into the transitive closure of the graph. This algorithm does a lot in a few lines of code. It's based on a simple idea:

> If you can get from vertex L to vertex M, and you can get from M to N, then you can get from L to N.

We've derived a two-step path from two one-step paths. The adjacency matrix shows all possible one-step paths, so it's a good starting place to apply this rule.

You might wonder if this algorithm can find paths of more than two edges. After all, the rule only talks about combining two one-edge paths into one two-edge path. As it turns out, the algorithm will build on previously discovered multi-edge paths to create paths of arbitrary length. The implementation we will describe guarantees this result, but a proof is beyond the scope of this book.

Here's how it works. We'll use Table 13.6 as an example. We're going to examine every cell in the adjacency matrix, one row at a time.

Row A
We start with row A. There's nothing in columns A and B, but there's a 1 at column C, so we stop there.

Now the 1 at this location says there is a path from A to C. If we knew there was a path from some vertex X to A, then we would know there was a path from X to C. Where are the edges (if any) that end at A? They're in column A. So we examine all the cells in column A. In Table 13.6 there's only one 1 in column A: at row B. It says there's an edge from B to A. So we know there's an edge from B to A, and another (the one we started with) from A to C. From this we infer that we can get from B to C in two steps. You can verify this is true by looking at the graph in Figure 13.15.

To record this result, we put a 1 at the intersection of row B and column C. The result is shown in Figure 13.16a.

The remaining cells of row A are blank.

Rows B, C, and D
We go to row B. The first cell, at column A, has a 1, indicating an edge from B to A. Are there any edges that end at B? We look in column B, but it's empty, so we know that none of the 1s we find in row B will result in finding longer paths because no edges end at B.

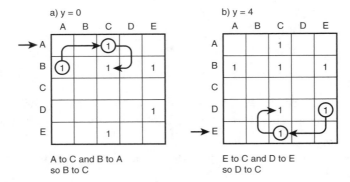

FIGURE 13.16   Steps in Warshall's algorithm.

Row C has no 1s at all, so we go to row D. Here we find an edge from D to E. However, column D is empty, so there are no edges that end on D.

Row E
In row E we see there's an edge from E to C. Looking in column E we see the first entry is for the edge B to E, so with B to E and E to C we infer there's a path from B to C. However, it's already been discovered, as indicated by the 1 at that location.

There's another 1 in column E, at row D. This edge from D to E plus the one from E to C imply a path from D to C, so we insert a 1 in that cell. The result is shown in Figure 13.16b.

Warshall's algorithm is now complete. We've added two 1s to the adjacency matrix, which now shows which nodes are reachable from another node in any number of steps. If we drew a graph based on this new matrix, it would be the transitive closure of the graph in Figure 13.15.

Implementation of Warshall's Algorithm

One way to implement Warshall's algorithm is with three nested loops (as suggested by Sedgewick; see Appendix B, "Further Reading"). The outer loop looks at each row; let's call its variable y. The loop inside that looks at each cell in the row; it uses variable x. If a 1 is found in cell (x, y), there's an edge from y to x, and the third (innermost) loop is activated; it uses variable z.

The third loop examines the cells in column y, looking for an edge that ends at y. (Note that y is used for rows in the first loop but for the column in the third loop.) If there's a 1 in column y at row z, then there's an edge from z to y. With one edge from z to y and another from y to x, it follows that there's a path from z to x, so you can put a 1 at (x, z). We'll leave the details as an exercise.

Summary

- Graphs consist of vertices connected by edges.

- Graphs can represent many real-world entities, including airline routes, electrical circuits, and job scheduling.

- Search algorithms allow you to visit each vertex in a graph in a systematic way. Searches are the basis of several other activities.

- The two main search algorithms are depth-first search (DFS) and breadth-first search (BFS).

- The depth-first search algorithm can be based on a stack; the breadth-first search algorithm can be based on a queue.

- A minimum spanning tree (MST) consists of the minimum number of edges necessary to connect all a graph's vertices.

- A slight modification of the depth-first search algorithm on an unweighted graph yields its minimum spanning tree.

- In a directed graph, edges have a direction (often indicated by an arrow).

- A topological sorting algorithm creates a list of vertices arranged so that a vertex A precedes a vertex B in the list if there's a path from A to B.

- A topological sort can be carried out only on a DAG, a directed acyclic (no cycles) graph.

- Topological sorting is typically used for scheduling complex projects that consist of tasks contingent on other tasks.

- Warshall's algorithm finds whether there is a connection, of either one or multiple edges, from any vertex to any other vertex.

Questions

These questions are intended as a self-test for readers. Answers may be found in Appendix C.

1. In a graph, an _____ connects two _____ .

2. How do you tell, by looking at its adjacency matrix, how many edges there are in an undirected graph?

3. In a game simulation, what graph entity corresponds to a choice about what move to make?

4. A directed graph is one in which

   a. you must follow the minimum spanning tree.

   b. you must go from vertex A to vertex B to vertex C and so on.

   c. you can go in only one direction from one given vertex to another.

   d. you can go in only one direction on any given path.

5. If an adjacency matrix has rows {0,1,0,0}, {1,0,1,1}, {0,1,0,0}, and {0,1,0,0}, what is the corresponding adjacency list?

6. A minimum spanning tree is a graph in which

   a. the number of edges connecting all the vertices is as small as possible.

   b. the number of edges is equal to the number of vertices.

   c. all unnecessary vertices have been removed.

   d. every combination of two vertices is connected by the minimum number of edges.

7. How many different minimum spanning trees are there in an undirected graph of three vertices and three edges?

8. An undirected graph must have a cycle if

   a. any vertex can be reached from some other vertex.

   b. the number of paths is greater than the number of vertices.

   c. the number of edges is equal to the number of vertices.

   d. the number of paths is less than the number of edges.

9. A _____ is a graph with no cycles.

10. Can a minimum spanning tree for an undirected graph have cycles?

11. True or False: There may be many correct topological sorts for a given graph.

12. Topological sorting results in

   a. vertices arranged so the directed edges all go in the same direction.

   b. vertices listed in order of increasing number of edges from the beginning vertex.

   c. vertices arranged so A precedes B, which precedes C, and so on.

   d. vertices listed so the ones later in the list are downstream from the ones earlier.

13. What's a DAG?

14. Can a tree have cycles?

15. What evidence does the `topo.java` program (Listing 13.4) use to deduce that a graph has a cycle?

Experiments

Carrying out these experiments will help to provide insights into the topics covered in the chapter. No programming is involved.

1. Using the GraphN workshop applet, draw a graph with five vertices and seven edges. Then, without using the View button, write down the adjacency matrix for the graph. When you're done, push the View button to see if you got it right.

2. A normal tic-tac-toe game is played on a 3×3 board, but for simplicity, think of a 2×2 tic-tac-toe game, in which a player needs only 2 Xs or 2 Os to win. Use the GraphN applet to create a graph corresponding to such a 2×2 game. Do you really need a depth of 4?

3. Create a five-vertex adjacency matrix and insert 0s and 1s randomly. Don't worry about symmetry. Now, without using the View button, create the corresponding directed graph using the GraphD workshop applet. When you're done, push the View button to see if the graph corresponds to your adjacency matrix.

4. In the GraphD workshop applet, see if you can create a graph with a cycle that the Topo routine cannot identify.

Programming Projects

Writing programs to solve the Programming Projects helps to solidify your understanding of the material and demonstrates how the chapter's concepts are applied. (As noted in the Introduction, qualified instructors may obtain completed solutions to the Programming Projects on the publisher's Web site.)

13.1 Modify the `bfs.java` program (Listing 13.2) to find the minimum spanning tree using a breadth-first search, rather than the depth-first search shown in `mst.java` (Listing 13.3). In main(), create a graph with 9 vertices and 12 edges, and find its minimum spanning tree.

13.2 Modify the `dfs.java` program (Listing 13.1) to use adjacency lists rather than an adjacency matrix. You can obtain a list by adapting the Link and LinkList classes from the `linkList2.java` program (Listing 5.2) in Chapter 5. Modify the find() routine from LinkList to search for an unvisited vertex rather than for a key value.

13.3   Modify the dfs.java program (Listing 13.1) to display a connectivity table for a directed graph, as described in the section "Connectivity in Directed Graphs."

13.4   Implement Warshall's algorithm to find the transitive closure for a graph. You could start with the code from Programming Project 13.3. It's useful to be able to display the adjacency matrix at various stages of the algorithm's operation.

13.5   The Knight's Tour is an ancient and famous chess puzzle. The object is to move a knight from one square to another on an otherwise empty chess board until it has visited every square exactly once. Write a program that solves this puzzle using a depth-first search. It's best to make the board size variable so that you can attempt solutions for smaller boards. The regular 8×8 board can take years to solve on a desktop computer, but a 5×5 board takes only a minute or so.

Refer to the section "Depth-First Search and Game Simulations" in this chapter. It may be easier to think of a new knight being created and remaining on the new square when a move is made. This way, a knight corresponds to a vertex, and a sequence of knights can be pushed onto the stack. When the board is completely filled with knights (the stack is full), you win. In this problem the board is traditionally numbered sequentially, from 1 at the upper-left corner to 64 at the lower-right corner (or 1 to 25 on a 5×5 board). When looking for its next move, a knight must not only make a legal knight's move, it must also not move off the board or onto an already-occupied (visited) square. If you make the program display the board and wait for a keypress after every move, you can watch the progress of the algorithm as it places more and more knights on the board, and then, when it gets boxed in, backtracks by removing some knights and trying a different series of moves. We'll have more to say about the complexity of this problem in the next chapter.

14

Weighted Graphs

In the preceding chapter we saw that a graph's edges can have direction. In this chapter we'll explore another edge feature: weight. For example, if vertices in a weighted graph represent cities, the weight of the edges might represent distances between the cities, or costs to fly between them, or the number of automobile trips made annually between them (a figure of interest to highway engineers).

When we include weight as a feature of a graph's edges, some interesting and complex questions arise. What is the minimum spanning tree for a weighted graph? What is the shortest (or cheapest) distance from one vertex to another? Such questions have important applications in the real world.

We'll first examine a weighted but non-directed graph and its minimum spanning tree. In the second half of this chapter we'll examine graphs that are both directed and weighted, in connection with the famous Dijkstra's algorithm, used to find the shortest path from one vertex to another.

Minimum Spanning Tree with Weighted Graphs

To introduce weighted graphs, we'll return to the question of the minimum spanning tree. Creating such a tree is a bit more complicated with a weighted graph than with an unweighted one. When all edges are the same weight, it's fairly straightforward—as we saw in Chapter 13, "Graphs"—for the algorithm to choose one to add to the minimum spanning tree. But when edges can have different weights, some arithmetic is needed to choose the right one.

An Example: Cable TV in the Jungle

Suppose we want to install a cable television line that connects six towns in the mythical country of Magnaguena. Five links will connect the six cities, but which five links should they be? The cost of connecting each pair of cities varies, so we must pick the route carefully to minimize the overall cost.

Figure 14.1 shows a weighted graph with six vertices, representing the towns Ajo, Bordo, Colina, Danza, Erizo, and Flor. Each edge has a weight, shown by a number alongside the edge. Imagine that these numbers represent the cost, in millions of Magnaguenian dollars, of installing a cable link between two cities. (Notice that some links are impractical because of distance or terrain; for example, we will assume that it's too far from Ajo to Colina or from Danza to Flor, so these links don't need to be considered and don't appear on the graph.)

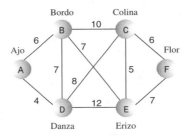

FIGURE 14.1   A weighted graph.

How can we pick a route that minimizes the cost of installing the cable system? The answer is to calculate a minimum spanning tree. It will have five links (one fewer than the number of towns), it will connect all six towns, and it will minimize the total cost of building these links. Can you figure out this route by looking at the graph in Figure 14.1? If not, you can solve the problem with the GraphW Workshop applet.

The GraphW Workshop Applet

The GraphW Workshop applet is similar to GraphN and GraphD, but it creates weighted, undirected graphs. Before you drag from vertex to vertex to create an edge, you must type the weight of the edge into the text box in the upper-right corner.

This applet carries out only one algorithm: When you repeatedly click the Tree button, it finds the minimum spanning tree for whatever graph you have created. The New and View buttons work as in previous graph applets to erase an old graph and to view the adjacency matrix.

Try out this applet by creating some small graphs and finding their minimum spanning trees. (For some configurations you'll need to be careful positioning the vertices so that the weight numbers don't fall on top of each other.)

As you step through the algorithm, you'll see that vertices acquire red borders and edges are made thicker when they're added to the minimum spanning tree. Vertices that are in the tree are also listed below the graph, on the left. On the right, the contents of a priority queue (PQ) are shown. The items in the priority queue are edges. For instance, the entry AB6 in the queue is the edge from A to B, which has a weight of 6. We'll explain what the priority queue does after we've shown an example of the algorithm.

Use the GraphW Workshop applet to construct the graph of Figure 14.1. The result should resemble Figure 14.2.

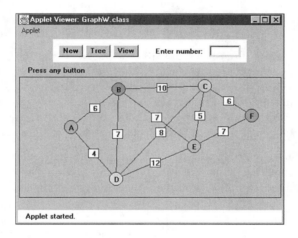

FIGURE 14.2   The GraphW Workshop applet.

Now find this graph's minimum spanning tree by stepping through the algorithm with the Tree button. The result should be the minimum spanning tree shown in Figure 14.3.

The applet should discover that the minimum spanning tree consists of the edges AD, AB, BE, EC, and CF, for a total edge weight of 28. The order in which the edges are specified is unimportant. If you start at a different vertex, you will create a tree with the same edges, but in a different order.

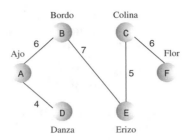

FIGURE 14.3   The minimum spanning tree.

Send Out the Surveyors

The algorithm for constructing the minimum spanning tree is a little involved, so we're going to introduce it using an analogy involving cable TV employees. You are one employee—a manager, of course—and there are also various surveyors.

A computer algorithm (unless perhaps it's a neural network) doesn't "know" about all the data in a given problem at once; it can't deal with the big picture. It must acquire the data little by little, modifying its view of things as it goes along. With graphs, algorithms tend to start at some vertex and work outward, acquiring data about nearby vertices before finding out about vertices farther away. We saw examples of this in the depth-first and breadth-first searches in the preceding chapter.

In a similar way, we're going to assume that you don't start out knowing the costs of installing the cable TV line between all the pairs of towns in Magnaguena. Acquiring this information takes time. That's where the surveyors come in.

Starting in Ajo

You start by setting up an office in Ajo. (You could start in any town, but Ajo has the best restaurants.) Only two towns are reachable from Ajo: Bordo and Danza (see Figure 14.1). You hire two tough, jungle-savvy surveyors and send them out along the dangerous wilderness trails, one to Bordo and one to Danza. Their job is to determine the cost of installing cable along these routes.

The first surveyor arrives in Bordo, having completed her survey, and calls you on her cell phone; she says it will cost 6 million dollars to install the cable link between Ajo and Bordo. The second surveyor, who has had some trouble with crocodiles, reports a little later from Danza that the Ajo–Danza link, which crosses more level country, will cost only 4 million dollars. You make a list:

- Ajo–Danza, $4 million

- Ajo–Bordo, $6 million

You always list the links in order of increasing cost; we'll see why this is a good idea soon.

Building the Ajo–Danza Link

At this point you figure you can send out the construction crew to actually install the cable from Ajo to Danza. How can you be sure the Ajo–Danza route will eventually be part of the cheapest solution (the minimum spanning tree)? So far, you know the cost of only two links in the system. Don't you need more information?

To get a feel for this situation, try to imagine some other route linking Ajo to Danza that would be cheaper than the direct link. If it doesn't go directly to Danza, this other route must go through Bordo and circle back to Danza, possibly via one or more other towns. But you already know the link to Bordo is more expensive, at 6 million dollars, than the link to Danza, at 4. So even if the remaining links in this hypothetical circle route are cheap, as shown in Figure 14.4, it will still be more expensive to get to Danza by going through Bordo. Also, it will be more expensive to get to towns on the circle route, like X, by going through Bordo than by going through Danza.

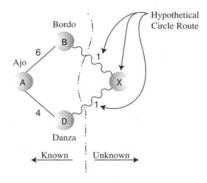

FIGURE 14.4 Hypothetical circle route.

We conclude that the Ajo–Danza route will be part of the minimum spanning tree. This isn't a formal proof (which is beyond the scope of this book), but it does suggest your best bet is to pick the cheapest link. So you build the Ajo–Danza link and install an office in Danza.

Why do you need an office? Due to a Magnaguena government regulation, you must install an office in a town before you can send out surveyors from that town to adjacent towns. In graph terms, you must add a vertex to the tree before you can learn the weight of the edges leading away from that vertex. All towns with offices are connected by cable with each other; towns with no offices are not yet connected.

Building the Ajo–Bordo Link

After you've completed the Ajo–Danza link and built your office in Danza, you can send out surveyors from Danza to all the towns reachable from there. These are Bordo, Colina, and Erizo. The surveyors reach their destinations and report back costs of 7, 8, and 12 million dollars, respectively. (Of course, you don't send a surveyor to Ajo because you've already surveyed the Ajo–Danza route and installed its cable.)

Now you know the costs of four links from towns with offices to towns with no offices:

- Ajo–Bordo, $6 million

- Danza–Bordo, $7 million

- Danza–Colina, $8 million

- Danza–Erizo, $12 million

Why isn't the Ajo–Danza link still on the list? Because you've already installed the cable there; there's no point giving any further consideration to this link. The route on which a cable has just been installed is always removed from the list.

At this point it may not be obvious what to do next. There are many potential links to choose from. What do you imagine is the best strategy now? Here's the rule:

RULE

From the list, always pick the cheapest edge.

Actually, you already followed this rule when you chose which route to follow from Ajo; the Ajo–Danza edge was the cheapest. Here the cheapest edge is Ajo–Bordo, so you install a cable link from Ajo to Bordo for a cost of 6 million dollars, and build an office in Bordo.

Let's pause for a moment and make a general observation. At a given time in the cable system construction, there are three kinds of towns:

1. Towns that have offices and are linked by cable. (In graph terms they're in the minimum spanning tree.)

2. Towns that aren't linked yet and have no office, but for which you know the cost to link them to at least one town with an office. We can call these "fringe" towns.

3. Towns you don't know anything about.

At this stage, Ajo, Danza, and Bordo are in category 1, Colina and Erizo are in category 2, and Flor is in category 3, as shown in Figure 14.5. As we work our way through the algorithm, towns move from category 3 to 2, and from 2 to 1.

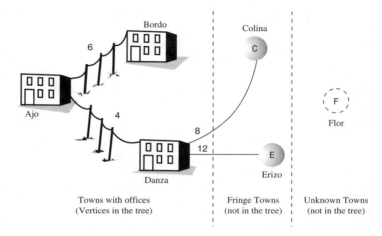

FIGURE 14.5   Partway through the minimum spanning tree algorithm.

Building the Bordo–Erizo Link

At this point, Ajo, Danza, and Bordo are connected to the cable system and have offices. You already know the costs from Ajo and Danza to towns in category 2, but you don't know these costs from Bordo. So from Bordo you send out surveyors to Colina and Erizo. They report back costs of 10 million dollars to Colina and 7 to Erizo. Here's the new list:

- Bordo–Erizo, $7 million

- Danza–Colina, $8 million

- Bordo–Colina, $10 million

- Danza–Erizo, $12 million

The Danza–Bordo link was on the previous list but is not on this one because, as we noted, there's no point in considering links to towns that are already connected, even by an indirect route.

From this list we can see that the cheapest route is Bordo–Erizo, at 7 million dollars. You send out the crew to install this cable link, and you build an office in Erizo (see Figure 14.3).

Building the Erizo–Colina Link

From Erizo the surveyors report back costs of 5 million dollars to Colina and 7 to Flor. The Danza–Erizo link from the previous list must be removed because Erizo is now a connected town. Your new list is

- Erizo–Colina, $5 million

- Erizo–Flor, $7 million

- Danza–Colina, $8 million

- Bordo–Colina, $10 million

The cheapest of these links is Erizo–Colina, so you build this link and install an office in Colina.

And, Finally, the Colina–Flor Link

The choices are narrowing. After you remove already-linked towns, your list now shows only

- Colina–Flor, $6 million

- Erizo–Flor, $7 million

You install the last link of cable from Colina to Flor, build an office in Flor, and you're done. You know you're done because there's now an office in every town. You've constructed the cable route Ajo–Danza, Ajo–Bordo, Bordo–Erizo, Erizo–Colina, and Colina–Flor, as shown earlier in Figure 14.3. This is the cheapest possible route linking the six towns of Magnaguena.

Creating the Algorithm

Using the somewhat fanciful idea of installing a cable TV system, we've shown the main ideas behind the minimum spanning tree for weighted graphs. Now let's see how we'd go about creating the algorithm for this process.

The Priority Queue

The key activity in carrying out the algorithm, as described in the cable TV example, was maintaining a list of the costs of links between pairs of cities. We decided where to build the next link by selecting the minimum of these costs.

A list in which we repeatedly select the minimum value suggests a priority queue as an appropriate data structure, and in fact this turns out to be an efficient way to handle the minimum spanning tree problem. Instead of a list or array, we use a priority queue. In a serious program this priority queue might be based on a heap, as described in Chapter 12, "Heaps." This would speed up operations on large priority queues. However, in our demonstration program we'll use a priority queue based on a simple array.

Outline of the Algorithm

Let's restate the algorithm in graph terms (as opposed to cable TV terms):

Start with a vertex, and put it in the tree. Then repeatedly do the following:

1. Find all the edges from the newest vertex to other vertices that aren't in the tree. Put these edges in the priority queue.

2. Pick the edge with the lowest weight, and add this edge and its destination vertex to the tree.

Repeat these steps until all the vertices are in the tree. At that point, you're done.

In step 1, *newest* means most recently installed in the tree. The edges for this step can be found in the adjacency matrix. After step 1, the list will contain all the edges from vertices in the tree to vertices on the fringe.

Extraneous Edges

In maintaining the list of links, we went to some trouble to remove links that led to a town that had recently become connected. If we didn't do this, we would have ended up installing unnecessary cable links.

In a programming algorithm we must likewise make sure that we don't have any edges in the priority queue that lead to vertices that are already in the tree. We could go through the queue looking for and removing any such edges each time we added a new vertex to the tree. As it turns out, it is easier to keep only one edge from the tree to a given fringe vertex in the priority queue at any given time. That is, the queue should contain only one edge to each category 2 vertex.

You'll see that this is what happens in the GraphW Workshop applet. There are fewer edges in the priority queue than you might expect—just one entry for each category 2 vertex. Step through the minimum spanning tree for Figure 14.1 and verify that this is what happens. Table 14.1 shows how edges with duplicate destinations have been removed from the priority queue.

TABLE 14.1   Edge Pruning

Step Number List	Unpruned Edge List	Pruned Edge (in Priority Queue)	Duplicate Removed from Priority Queue
1	AB6, AD4	AB6, AD4	
2	DE12, DC8, DB7, AB6	DE12, DC8, AB6	DB7(AB6)
3	DE12, BC10, DC8, BE7	DC8, BE7	DE12(BE7), BC10(DC8)
4	BC10, DC8, EF7, EC5	EF7, EC5	BC10(EC5), DC8(EC5)
5	EF7, CF6	CF6	EF7

Remember that an edge consists of a letter for the source (starting) vertex of the edge, a letter for the destination (ending vertex), and a number for the weight. The second column in this table corresponds to the lists you kept when constructing the cable TV system. It shows all edges from category 1 vertices (those in the tree) to category 2 vertices (those with at least one known edge from a category 1 vertex).

The third column is what you see in the priority queue when you run the GraphW applet. Any edge with the same destination vertex as another edge, and which has a greater weight, has been removed.

The fourth column shows the edges that have been removed and, in parentheses, the edge with the smaller weight that superseded it and remains in the queue. Remember that as you go from step to step the last entry on the list is always removed because this edge is added to the tree.

Looking for Duplicates in the Priority Queue

How do we make sure there is only one edge per category 2 vertex? Each time we add an edge to the queue, we make sure there's no other edge going to the same destination. If there is, we keep only the one with the smallest weight.

This necessitates looking through the priority queue item by item, to see if there's such a duplicate edge. Priority queues are not designed for random access, so this is not an efficient activity. However, violating the spirit of the priority queue is necessary in this situation.

Java Code

The method that creates the minimum spanning tree for a weighted graph, mstw(), follows the algorithm outlined earlier. As in our other graph programs, it assumes there's a list of vertices in vertexList[], and that it will start with the vertex at index 0. The currentVert variable represents the vertex most recently added to the tree. Here's the code for mstw():

```
public void mstw()              // minimum spanning tree
   {
   currentVert = 0;             // start at 0

   while(nTree < nVerts-1)    // while not all verts in tree
      {                         // put currentVert in tree
      vertexList[currentVert].isInTree = true;
      nTree++;

      // insert edges adjacent to currentVert into PQ
      for(int j=0; j<nVerts; j++)   // for each vertex,
         {
         if(j==currentVert)          // skip if it's us
            continue;
         if(vertexList[j].isInTree) // skip if in the tree
            continue;
         int distance = adjMat[currentVert][j];
         if( distance == INFINITY)  // skip if no edge
            continue;
         putInPQ(j, distance);       // put it in PQ (maybe)
         }
      if(thePQ.size()==0)             // no vertices in PQ?
            {
            System.out.println(" GRAPH NOT CONNECTED");
            return;
            }
      // remove edge with minimum distance, from PQ
      Edge theEdge = thePQ.removeMin();
      int sourceVert = theEdge.srcVert;
      currentVert = theEdge.destVert;

      // display edge from source to current
      System.out.print( vertexList[sourceVert].label );
      System.out.print( vertexList[currentVert].label );
      System.out.print(" ");
      }  // end while(not all verts in tree)

   // mst is complete
   for(int j=0; j<nVerts; j++)      // unmark vertices
      vertexList[j].isInTree = false;
   }  // end mstw()
```

The algorithm is carried out in the while loop, which terminates when all vertices are in the tree. Within this loop the following activities take place:

1. The current vertex is placed in the tree.

2. The edges adjacent to this vertex are placed in the priority queue (if appropriate).

3. The edge with the minimum weight is removed from the priority queue. The destination vertex of this edge becomes the current vertex.

Let's look at these steps in more detail. In step 1, the currentVert is placed in the tree by marking its isInTree field.

In step 2, the edges adjacent to this vertex are considered for insertion in the priority queue. The edges are examined by scanning across the row whose number is currentVert in the adjacency matrix. An edge is placed in the queue unless one of these conditions is true:

• The source and destination vertices are the same.

• The destination vertex is in the tree.

• There is no edge to this destination.

If none of these conditions are true, the putInPQ() method is called to put the edge in the priority queue. Actually, this routine doesn't always put the edge in the queue either, as we'll see in a moment.

In step 3, the edge with the minimum weight is removed from the priority queue. This edge and its destination vertex are added to the tree, and the source vertex (currentVert) and destination vertex are displayed.

At the end of mstw(), the vertices are removed from the tree by resetting their isInTree variables. That isn't strictly necessary in this program because only one tree is created from the data. However, it's good housekeeping to restore the data to its original form when you finish with it.

As we noted, the priority queue should contain only one edge with a given destination vertex. The putInPQ() method makes sure this is true. It calls the find() method of the PriorityQ class, which has been doctored to find the edge with a specified destination vertex. If there is no such vertex, and find() therefore returns –1, then putInPQ() simply inserts the edge into the priority queue. However, if such an edge does exist, putInPQ() checks to see whether the existing edge or the new proposed edge has the lower weight. If it's the old edge, no change is necessary. If the new one has a lower weight, the old edge is removed from the queue and the new one is installed. Here's the code for putInPQ():

```
public void putInPQ(int newVert, int newDist)
   {
   // is there another edge with the same destination vertex?
   int queueIndex = thePQ.find(newVert);  // got edge's index
   if(queueIndex != -1)                    // if there is one,
      {                                    // get edge
      Edge tempEdge = thePQ.peekN(queueIndex);
      int oldDist = tempEdge.distance;
      if(oldDist > newDist)               // if new edge shorter,
         {
         thePQ.removeN(queueIndex);  // remove old edge
         Edge theEdge = new Edge(currentVert, newVert, newDist);
         thePQ.insert(theEdge);       // insert new edge
         }
      // else no action; just leave the old vertex there
      } // end if
   else  // no edge with same destination vertex
      {                                    // so insert new one
      Edge theEdge = new Edge(currentVert, newVert, newDist);
      thePQ.insert(theEdge);
      }
   } // end putInPQ()
```

The `mstw.java` Program

The `PriorityQ` class uses an array to hold the members. As we noted, in a program dealing with large graphs, a heap would be more appropriate than the array shown here. The `PriorityQ` class has been augmented with various methods. It can, as we've seen, find an edge with a given destination vertex with `find()`. It can also peek at an arbitrary member with `peekN()`, and remove an arbitrary member with `removeN()`. Most of the rest of this program you've seen before. Listing 14.1 shows the complete `mstw.java` program.

LISTING 14.1   The `mstw.java` Program

```
// mstw.java
// demonstrates minimum spanning tree with weighted graphs
// to run this program: C>java MSTWApp
/////////////////////////////////////////////////////////////////
class Edge
   {
   public int srcVert;   // index of a vertex starting edge
   public int destVert;  // index of a vertex ending edge
```

LISTING 14.1   Continued

```
   public int distance;  // distance from src to dest
// -------------------------------------------------------------
   public Edge(int sv, int dv, int d)  // constructor
      {
      srcVert = sv;
      destVert = dv;
      distance = d;
      }
// -------------------------------------------------------------
   }  // end class Edge
/////////////////////////////////////////////////////////////////
class PriorityQ
   {
   // array in sorted order, from max at 0 to min at size-1
   private final int SIZE = 20;
   private Edge[] queArray;
   private int size;
// -------------------------------------------------------------
   public PriorityQ()             // constructor
      {
      queArray = new Edge[SIZE];
      size = 0;
      }
// -------------------------------------------------------------
   public void insert(Edge item)  // insert item in sorted order
      {
      int j;

      for(j=0; j<size; j++)            // find place to insert
         if( item.distance >= queArray[j].distance )
            break;

      for(int k=size-1; k>=j; k--)     // move items up
         queArray[k+1] = queArray[k];

      queArray[j] = item;              // insert item
      size++;
      }
// -------------------------------------------------------------
   public Edge removeMin()            // remove minimum item
      { return queArray[--size]; }
```

LISTING 14.1   Continued

```
// ---------------------------------------------------------
   public void removeN(int n)        // remove item at n
      {
      for(int j=n; j<size-1; j++)     // move items down
         queArray[j] = queArray[j+1];
      size--;
      }
// ---------------------------------------------------------
   public Edge peekMin()        // peek at minimum item
      { return queArray[size-1]; }
// ---------------------------------------------------------
   public int size()            // return number of items
      { return size; }
// ---------------------------------------------------------
   public boolean isEmpty()      // true if queue is empty
      { return (size==0); }
// ---------------------------------------------------------
   public Edge peekN(int n)      // peek at item n
      { return queArray[n]; }
// ---------------------------------------------------------
   public int find(int findDex)  // find item with specified
      {                          // destVert value
      for(int j=0; j<size; j++)
         if(queArray[j].destVert == findDex)
            return j;
      return -1;
      }
// ---------------------------------------------------------
   } // end class PriorityQ
/////////////////////////////////////////////////////////////
class Vertex
   {
   public char label;        // label (e.g. 'A')
   public boolean isInTree;
// ---------------------------------------------------------
   public Vertex(char lab)   // constructor
      {
      label = lab;
      isInTree = false;
      }
// ---------------------------------------------------------
```

LISTING 14.1 Continued

```
   }  // end class Vertex
///////////////////////////////////////////////////////////////
class Graph
   {
   private final int MAX_VERTS = 20;
   private final int INFINITY = 1000000;
   private Vertex vertexList[]; // list of vertices
   private int adjMat[][];       // adjacency matrix
   private int nVerts;           // current number of vertices
   private int currentVert;
   private PriorityQ thePQ;
   private int nTree;            // number of verts in tree
// -------------------------------------------------------------
   public Graph()                // constructor
      {
      vertexList = new Vertex[MAX_VERTS];
                                           // adjacency matrix
      adjMat = new int[MAX_VERTS][MAX_VERTS];
      nVerts = 0;
      for(int j=0; j<MAX_VERTS; j++)     // set adjacency
         for(int k=0; k<MAX_VERTS; k++)  //    matrix to 0
            adjMat[j][k] = INFINITY;
      thePQ = new PriorityQ();
      }  // end constructor
// -------------------------------------------------------------
   public void addVertex(char lab)
      {
      vertexList[nVerts++] = new Vertex(lab);
      }
// -------------------------------------------------------------
   public void addEdge(int start, int end, int weight)
      {
      adjMat[start][end] = weight;
      adjMat[end][start] = weight;
      }
// -------------------------------------------------------------
   public void displayVertex(int v)
      {
      System.out.print(vertexList[v].label);
      }
// -------------------------------------------------------------
```

LISTING 14.1   Continued

```
public void mstw()              // minimum spanning tree
   {
   currentVert = 0;             // start at 0

   while(nTree < nVerts-1)   // while not all verts in tree
      {                         // put currentVert in tree
      vertexList[currentVert].isInTree = true;
      nTree++;

      // insert edges adjacent to currentVert into PQ
      for(int j=0; j<nVerts; j++)   // for each vertex,
         {
         if(j==currentVert)          // skip if it's us
            continue;
         if(vertexList[j].isInTree) // skip if in the tree
            continue;
         int distance = adjMat[currentVert][j];
         if( distance == INFINITY)  // skip if no edge
            continue;
         putInPQ(j, distance);       // put it in PQ (maybe)
         }
      if(thePQ.size()==0)             // no vertices in PQ?
         {
         System.out.println(" GRAPH NOT CONNECTED");
         return;
         }
      // remove edge with minimum distance, from PQ
      Edge theEdge = thePQ.removeMin();
      int sourceVert = theEdge.srcVert;
      currentVert = theEdge.destVert;

      // display edge from source to current
      System.out.print( vertexList[sourceVert].label );
      System.out.print( vertexList[currentVert].label );
      System.out.print(" ");
      }  // end while(not all verts in tree)

   // mst is complete
   for(int j=0; j<nVerts; j++)     // unmark vertices
      vertexList[j].isInTree = false;
   }  // end mstw
```

LISTING 14.1   Continued

```
// --------------------------------------------------------------
   public void putInPQ(int newVert, int newDist)
      {
      // is there another edge with the same destination vertex?
      int queueIndex = thePQ.find(newVert);
      if(queueIndex != -1)                  // got edge's index
         {
         Edge tempEdge = thePQ.peekN(queueIndex);  // get edge
         int oldDist = tempEdge.distance;
         if(oldDist > newDist)             // if new edge shorter,
            {
            thePQ.removeN(queueIndex);  // remove old edge
            Edge theEdge =
                      new Edge(currentVert, newVert, newDist);
            thePQ.insert(theEdge);       // insert new edge
            }
         // else no action; just leave the old vertex there
         } // end if
      else  // no edge with same destination vertex
         {                                  // so insert new one
         Edge theEdge = new Edge(currentVert, newVert, newDist);
         thePQ.insert(theEdge);
         }
      } // end putInPQ()
// --------------------------------------------------------------
   } // end class Graph
////////////////////////////////////////////////////////////////
class MSTWApp
   {
   public static void main(String[] args)
      {
      Graph theGraph = new Graph();
      theGraph.addVertex('A');     // 0  (start for mst)
      theGraph.addVertex('B');     // 1
      theGraph.addVertex('C');     // 2
      theGraph.addVertex('D');     // 3
      theGraph.addVertex('E');     // 4
      theGraph.addVertex('F');     // 5

      theGraph.addEdge(0, 1, 6);  // AB   6
      theGraph.addEdge(0, 3, 4);  // AD   4
```

LISTING 14.1   Continued

```
        theGraph.addEdge(1, 2, 10); // BC 10
        theGraph.addEdge(1, 3, 7);  // BD  7
        theGraph.addEdge(1, 4, 7);  // BE  7
        theGraph.addEdge(2, 3, 8);  // CD  8
        theGraph.addEdge(2, 4, 5);  // CE  5
        theGraph.addEdge(2, 5, 6);  // CF  6
        theGraph.addEdge(3, 4, 12); // DE 12
        theGraph.addEdge(4, 5, 7);  // EF  7

        System.out.print("Minimum spanning tree: ");
        theGraph.mstw();               // minimum spanning tree
        System.out.println();
        }  // end main()
    }  // end class MSTWApp
//////////////////////////////////////////////////////////////
```

The main() routine in class MSTWApp creates the tree in Figure 14.1. Here's the output:

Minimum spanning tree: AD AB BE EC CF

The Shortest-Path Problem

Perhaps the most commonly encountered problem associated with weighted graphs is that of finding the shortest path between two given vertices. The solution to this problem is applicable to a wide variety of real-world situations, from the layout of printed circuit boards to project scheduling. It is a more complex problem than we've seen before, so let's start by looking at a (somewhat) real-world scenario in the same mythical country of Magnaguena introduced in the preceding section.

The Railroad Line

This time we're concerned with railroads rather than cable TV. However, this project is not as ambitious as the last one. We're not going to build the railroad; it already exists. We just want to find the cheapest route from one city to another.

The railroad charges passengers a fixed fare to travel between any two towns. These fares are shown in Figure 14.6. That is, the fare from Ajo to Bordo is $50, from Bordo to Danza is $90, and so on. These rates are the same whether the ride between two towns is part of a longer itinerary or not (unlike the situation with today's airline fares).

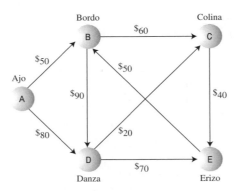

FIGURE 14.6   Train fares in Magnaguena.

The edges in Figure 14.6 are *directed*. They represent single-track railroad lines, on which (in the interest of safety) travel is permitted in only one direction. For example, you can go directly from Ajo to Bordo, but not from Bordo to Ajo.

Although in this situation we're interested in the cheapest fares, the graph problem is nevertheless always referred to as the *shortest-path problem (SPP)*. Here *shortest* doesn't necessarily mean shortest in terms of distance; it can also mean cheapest, fastest, or best route by some other measure.

Cheapest Fares

There are several possible routes between any two towns. For example, to take the train from Ajo to Erizo, you could go through Danza, or you could go through Bordo and Colina, or through Danza and Colina, or you could take several other routes. (It's not possible to reach the town of Flor by rail because it lies beyond the rugged Sierra Descaro range, so it doesn't appear on the graph. This is fortunate, because it reduces the size of certain lists we'll need to make.)

The shortest-path problem is this: For a given starting point and destination, what's the cheapest route? In Figure 14.6, you can see (with a little mental effort) that the cheapest route from Ajo to Erizo passes through Danza and Colina; it will cost you $140.

A Directed, Weighted Graph

As we noted, our railroad has only single-track lines, so you can go in only one direction between any two cities. This corresponds to a directed graph. We could have portrayed the more realistic situation in which you can go either way between two cities for the same price; this would correspond to a non-directed graph. However, the shortest-path problem is similar in these cases, so for variety we'll show how it looks in a directed graph.

Dijkstra's Algorithm

The solution we'll show for the shortest-path problem is called Dijkstra's algorithm, after Edsger Dijkstra, who first described it in 1959. This algorithm is based on the adjacency matrix representation of a graph. Somewhat surprisingly, it finds not only the shortest path from one specified vertex to another, but also the shortest paths from the specified vertex to all the other vertices.

Agents and Train Rides

To see how Dijkstra's algorithm works, imagine that you want to find the cheapest way to travel from Ajo to all the other towns in Magnaguena. You (and various agents you will hire) are going to play the role of the computer program carrying out Dijkstra's algorithm. Of course, in real life you could probably obtain a schedule from the railroad with all the fares. The algorithm, however, must look at one piece of information at a time, so (as in the preceding section) we'll assume that you are similarly unable to see the big picture.

At each town, the stationmaster can tell you how much it will cost to travel to the other towns that you can reach directly (that is, in a single ride, without passing through another town). Alas, he cannot tell you the fares to towns further than one ride away. You keep a notebook, with a column for each town. You hope to end up with each column showing the cheapest route from your starting point to that town.

The First Agent: In Ajo

Eventually, you're going to place an agent in every town; this agent's job is to obtain information about ticket costs to other towns. You yourself are the agent in Ajo.

All the stationmaster in Ajo can tell you is that it will cost $50 to ride to Bordo and $80 to ride to Danza. You write this information in your notebook, as shown in Table 14.2.

TABLE 14.2   Step 1: An Agent at Ajo

From Ajo to→	Bordo	Colina	Danza	Erizo
Step 1	50 (via Ajo)	inf	80 (via Ajo)	inf

The entry "inf" is short for "infinity," and means that you can't get from Ajo to the town shown in the column head, or at least that you don't yet know how to get there. (In the algorithm infinity will be represented by a very large number, which will help with calculations, as we'll see.) The table entries in parentheses show the last town visited before you arrive at the various destinations. We'll see later why this is good to know. What do you do now? Here's the rule you'll follow:

> **RULE**
>
> Always send an agent to the town whose overall fare from the starting point (Ajo) is the cheapest.

You don't consider towns that already have an agent. Notice that this is not the same rule as that used in the minimum spanning tree problem (the cable TV installation). There, you picked the least expensive single *link* (edge) from the connected towns to an unconnected town. Here, you pick the least expensive *total route* from Ajo to a town with no agent. In this particular point in your investigation these two approaches amount to the same thing, because all known routes from Ajo consist of only one edge, but as you send agents to more towns, the routes from Ajo will become the sum of several direct edges.

The Second Agent: In Bordo

The cheapest fare from Ajo is to Bordo, at $50. So you hire a passerby and send him to Bordo, where he'll be your agent. When he's there, he calls you by telephone and tells you that the Bordo stationmaster says it costs $60 to ride to Colina and $90 to Danza.

Doing some quick arithmetic, you figure it must be $50 plus $60, or $110 to go from Ajo to Colina via Bordo, so you modify the entry for Colina. You also can see that, going via Bordo, it must be $50 plus $90, or $140, from Ajo to Danza. However—and this is a key point—you already know it's only $80 going directly from Ajo to Danza. You care only about the *cheapest* route from Ajo, so you ignore the more expensive route, leaving this entry as it was. The resulting notebook entries are shown as the last row in Table 14.3. Figure 14.7 shows the situation geographically.

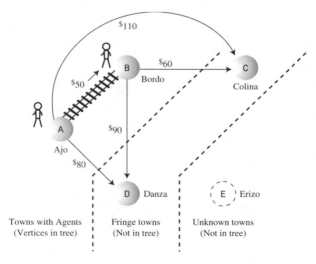

FIGURE 14.7   Following step 2 in the shortest-path algorithm.

TABLE 14.3   Step 2: Agents at Ajo and Bordo

From Ajo to→	Bordo	Colina	Danza	Erizo
Step 1	50 (via Ajo)	inf	80 (via Ajo)	inf
Step 2	50 (via Ajo)*	110 (via Bordo)	80 (via Ajo)	inf

After we've installed an agent in a town, we can be sure that the route taken by the agent to get to that town is the cheapest route. Why? Consider the present case. If there were a cheaper route than the direct one from Ajo to Bordo, it would need to go through some other town. But the only other way out of Ajo is to Danza, and that ride is already more expensive than the direct route to Bordo. Adding additional fares to get from Danza to Bordo would make the Danza route still more expensive.

From this we decide that from now on we won't need to update the entry for the cheapest fare from Ajo to Bordo. This fare will not change, no matter what we find out about other towns. We'll put a * next to it to show that there's an agent in the town and that the cheapest fare to it is fixed.

Three Kinds of Towns

As in the minimum spanning tree algorithm, we're dividing the towns into three categories:

1. Towns in which we've installed an agent; they're in the tree.

2. Towns with known fares from towns with an agent; they're on the fringe.

3. Unknown towns.

At this point Ajo and Bordo are category 1 towns because they have agents there. Category 1 towns form a tree consisting of paths that all begin at the starting vertex and that each end on a different destination vertex. (This is not the same tree, of course, as a minimum spanning tree.)

Some other towns have no agents, but you know the fares to them because you have agents in adjacent category 1 towns. You know the fare from Ajo to Danza is $80 and from Bordo to Colina is $60. Because the fares to them are known, Danza and Colina are category 2 (fringe) towns.

You don't know anything yet about Erizo, it's an "unknown" town. Figure 14.7 shows these categories at the current point in the algorithm.

As in the minimum spanning tree algorithm, this algorithm moves towns from the unknown category to the fringe category, and from the fringe category to the tree, as it goes along.

The Third Agent: In Danza
At this point, the cheapest route you know that goes from Ajo to any town without
an agent is $80, the direct route from Ajo to Danza. Both the Ajo–Bordo–Colina
route at $110 and the Ajo–Bordo–Danza route at $140 are more expensive.

You hire another passerby and send her to Danza, with an $80 ticket. She reports
that from Danza it's $20 to Colina and $70 to Erizo. Now you can modify your entry
for Colina. Before, it was $110 from Ajo, going via Bordo. Now you see you can
reach Colina for only $100, going via Danza. Also, you now know a fare from Ajo to
the previously unknown Erizo: it's $150, via Danza. You note these changes, as
shown in Table 14.4 and Figure 14.8.

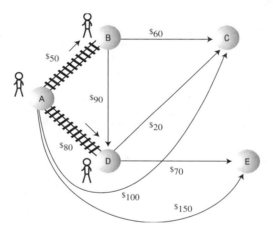

FIGURE 14.8   Following step 3 in the shortest-path algorithm.

TABLE 14.4   Step 3: Agents at Ajo, Bordo, and Danza

From Ajo to→	Bordo	Colina	Danza	Erizo
Step 1	50 (via Ajo)	inf	80 (via Ajo)	inf
Step 2	50 (via Ajo)*	110 (via Bordo)	80 (via Ajo)	inf
Step 3	50 (via Ajo)*	100 (via Danza)	80 (via Ajo)*	150 (via Danza)

The Fourth Agent: In Colina
Now the cheapest path to any town without an agent is the $100 trip from Ajo to
Colina, going via Danza. Accordingly, you dispatch an agent over this route to
Colina. He reports that it's $40 from there to Erizo. Now you can calculate that, since
Colina is $100 from Ajo (via Danza), and Erizo is $40 from Colina, you can reduce
the minimum Ajo-to-Erizo fare from $150 (the Ajo–Danza–Erizo route) to $140 (the

Ajo–Danza–Colina–Erizo route). You update your notebook accordingly, as shown in Table 14.5 and Figure 14.9.

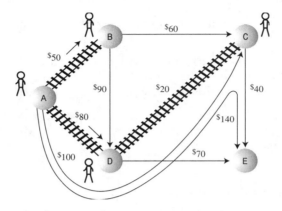

FIGURE 14.9 Following step 4 in the shortest-path algorithm.

TABLE 14.5 Step 4: Agents in Ajo, Bordo, Danza, and Colina

From Ajo to→	Bordo	Colina	Danza	Erizo
Step 1	50 (via Ajo)	inf	80 (via Ajo)	inf
Step 2	50 (via Ajo)*	110 (via Bordo)	80 (via Ajo)	inf
Step 3	50 (via Ajo)*	100 (via Danza)	80 (via Ajo)*	150 (via Danza)
Step 4	50 (via Ajo)*	100 (via Danza)*	80 (via Ajo)*	140 (via Colina)

The Last Agent: In Erizo

The cheapest path from Ajo to any town you know about that doesn't have an agent is now $140 to Erizo, via Danza and Colina. You dispatch an agent to Erizo, but she reports that there are no routes from Erizo to towns without agents. (There's a route to Bordo, but Bordo has an agent.) Table 14.6 shows the final line in your notebook; all you've done is add a star to the Erizo entry to show that an agent is there.

TABLE 14.6 Step 5: Agents in Ajo, Bordo, Danza, Colina, and Erizo

From Ajo to→	Bordo	Colina	Danza	Erizo
Step 1	50 (via Ajo)	inf	80 (via Ajo)	inf
Step 2	50 (via Ajo)*	110 (via Bordo)	80 (via Ajo)	inf
Step 3	50 (via Ajo)*	100 (via Danza)	80 (via Ajo)*	150 (via Danza)
Step 4	50 (via Ajo)*	100 (via Danza)*	80 (via Ajo)*	140 (via Colina)
Step 5	50 (via Ajo)*	100 (via Danza)*	80 (via Ajo)*	140 (via Colina)*

When there's an agent in every town, you know the fares from Ajo to every other town. So you're done. With no further calculations, the last line in your notebook shows the cheapest routes from Ajo to all other towns.

This narrative has demonstrated the essentials of Dijkstra's algorithm. The key points are

- Each time you send an agent to a new town, you use the new information provided by that agent to revise your list of fares. Only the cheapest fare (that you know about) from the starting point to a given town is retained.

- You always send the new agent to the town that has the cheapest path from the starting point (not the cheapest edge from any town with an agent, as in the minimum spanning tree).

Using the GraphDW Workshop Applet

Let's see how Dijkstra's algorithm looks using the GraphDW (for Directed and Weighted) Workshop applet. Use the applet to create the graph from Figure 14.6. The result should look something like Figure 14.10. (We'll see how to make the table appear below the graph in a moment.) This is a weighted, directed graph, so to make an edge, you must type a number before dragging, and you must drag in the correct direction, from the start to the destination.

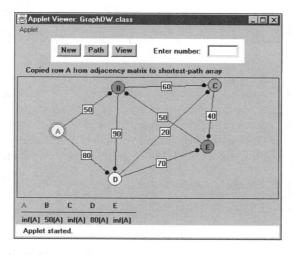

FIGURE 14.10   The railroad scenario in GraphDW.

When the graph is complete, click the Path button, and when prompted, click the A vertex. A few more clicks on Path will place A in the tree, shown with a red circle around A.

The Shortest-Path Array

An additional click of the Path button will install a table under the graph, as you can see in Figure 14.10. The corresponding message is `Copied row A from adjacency matrix to shortest-path array`. Dijkstra's algorithm starts by copying the appropriate row of the adjacency matrix (that is, the row for the starting vertex) to an array. (Remember that you can examine the adjacency matrix at any time by pressing the View button.)

This array is called the "shortest-path" array. It corresponds to the most recent row of notebook entries you made while determining the cheapest train fares in Magnaguena. This array will hold the current versions of the shortest paths to the other vertices, which we can call the *destination* vertices. These destination vertices are represented by the column heads in Table 14.7.

TABLE 14.7   Step 1: The Shortest-Path Array

A	B	C	D	E
inf(A)	50(A)	inf(A)	80(A)	inf(A)

In the applet, the shortest-path figures in the array are followed by the *parent vertex* enclosed in parentheses. The parent is the vertex you reached just before you reached the destination vertex. In this case the parents are all A because we've moved only one edge away from A.

If a fare is unknown (or meaningless, as from A to A), it's shown as infinity, represented by "inf," as in the rail-fare notebook entries. Notice that the column heads of those vertices that have already been added to the tree are shown in red. The entries for these columns won't change.

Minimum Distance

Initially, the algorithm knows the distances from A to other vertices that are exactly one edge from A. Only B and D are adjacent to A, so they're the only ones whose distances are shown. The algorithm picks the minimum distance. Another click on Path will show you the message

`Minimum distance from A is 50, to vertex B`

The algorithm adds this vertex to the tree, so the next click will show you

`Added vertex B to tree`

Now B is circled in the graph, and the B column head is in red. The edge from A to B is made darker to show it's also part of the tree.

Column by Column in the Shortest-Path Array

Now the algorithm knows not only all the edges from A, but the edges from B as well. So it goes through the shortest-path array, column by column, checking whether a shorter path than that shown can be calculated using this new information. Vertices that are already in the tree, here A and B, are skipped. First, column C is examined. You'll see the message

```
To C: A to B (50) plus edge BC (60) less than A to C (inf)
```

The algorithm has found a shorter path to C than that shown in the array. The array shows infinity in the C column. But from A to B is 50 (which the algorithm finds in the B column in the shortest-path array) and from B to C is 60 (which it finds in row B column C in the adjacency matrix). The sum is 110. The 110 distance is less than infinity, so the algorithm updates the shortest-path array for column C, inserting 110. This is followed by a B in parentheses, because that's the last vertex before reaching C; B is the parent of C.

Next, the D column is examined. You'll see the message

```
To D: A to B (50) plus edge BD (90) greater than or equal to A to D (80)
```

The algorithm is comparing the previously shown distance from A to D, which is 80 (the direct route), with a possible route via B (that is, A–B–D). But path A–B is 50 and edge BD is 90, so the sum is 140. This is bigger than 80, so 80 is not changed.

For column E, the message is

```
To E: A to B (50) plus edge BE (inf) greater than or equal to A to E (inf)
```

The newly calculated route from A to E via B (50 plus infinity) is still greater than or equal to the current one in the array (infinity), so the E column is not changed. The shortest-path array now looks like Table 14.8.

TABLE 14.8   Step 2: The Shortest-Path Array

A	B	C	D	E
inf(A)	50(A)	110(B)	80(A)	inf(A)

Now we can see more clearly the role played by the parent vertex shown in parentheses after each distance. Each column shows the distance from A to an ending vertex. The parent is the immediate predecessor of the ending vertex along the path from A. In column C, the parent vertex is B, meaning that the shortest path from A to C passes through B just before it gets to C. This information is used by the algorithm to place the appropriate edge in the tree. (When the distance is infinity, the parent vertex is meaningless and is shown as A.)

New Minimum Distance
Now that the shortest-path array has been updated, the algorithm finds the shortest distance in the array, as you will see with another Path keypress. The message is

```
Minimum distance from A is 80, to vertex D
```

Accordingly, the message

```
Added vertex D to tree
```

appears and the new vertex and edge AC are added to the tree.

Do It Again and Again
Now the algorithm goes through the shortest-path array again, checking and updating the distances for destination vertices not in the tree; only C and E are still in this category. Column C and E are both updated. The result is shown in Table 14.9.

TABLE 14.9   Step 3: The Shortest-Path Array

A	B	C	D	E
inf(A)	50(A)	100(D)	80(A)	150(D)

The shortest path from A to a non-tree vertex is 100, to vertex C, so C is added to the tree.

Next time through the shortest-path array, only the distance to E is considered. It can be shortened by going via C, so we have the entries shown in Table 14.10.

TABLE 14.10   Step 4: The Shortest-Path Array

A	B	C	D	E
inf(A)	50(A)	100(D)	80(A)	140(C)

Now the last vertex, E, is added to the tree, and you're done. The shortest-path array shows the shortest distances from A to all the other vertices. The tree consists of all the vertices and the edges AB, AD, DC, and CE, shown with thick lines.

You can work backward to reconstruct the sequence of vertices along the shortest path to any vertex. For the shortest path to E, for example, the parent of E, shown in the array in parentheses, is C. The predecessor of C, again from the array, is D, and the predecessor of D is A. So the shortest path from A to E follows the route A–D–C–E.

Experiment with other graphs using GraphDW, starting with small ones. You'll find that after a while you can predict what the algorithm is going to do, and you'll be on your way to understanding Dijkstra's algorithm.

Java Code

The code for the shortest-path algorithm is among the most complex in this book, but even so it's not beyond mere mortals. We'll look first at a helper class and then at the chief method that executes the algorithm, path(), and finally at two methods called by path() to carry out specialized tasks.

The sPath Array and the DistPar Class

As we've seen, the key data structure in the shortest-path algorithm is an array that keeps track of the minimum distances from the starting vertex to the other vertices (destination vertices). During the execution of the algorithm, these distances are changed, until at the end they hold the actual shortest distances from the start. In the example code, this array is called sPath[] (for shortest paths).

As we've seen, it's important to record not only the minimum distance from the starting vertex to each destination vertex, but also the path taken. Fortunately, the entire path need not be explicitly stored. It's only necessary to store the parent of the destination vertex. The parent is the vertex reached just before the destination. We've seen this in the Workshop applet, where, if 100(D) appears in the C column, it means that the cheapest path from A to C is 100, and D is the last vertex before C on this path.

There are several ways to keep track of the parent vertex, but we choose to combine the parent with the distance and put the resulting object into the sPath[] array. We call this class of objects DistPar (for distance-parent).

```
class DistPar                 // distance and parent
   {                          // items stored in sPath array
   public int distance;    // distance from start to this vertex
   public int parentVert; // current parent of this vertex

   public DistPar(int pv, int d)   // constructor
      {
      distance = d;
      parentVert = pv;
      }
   }
```

The path() method

The path() method carries out the actual shortest-path algorithm. It uses the DistPar class and the Vertex class, which we saw in the mstw.java program (Listing 14.1). The path() method is a member of the Graph class, which we also saw in mstw.java in a somewhat different version.

```java
public void path()                  // find all shortest paths
   {
   int startTree = 0;               // start at vertex 0
   vertexList[startTree].isInTree = true;
   nTree = 1;                       // put it in tree

   // transfer row of distances from adjMat to sPath
   for(int j=0; j<nVerts; j++)
      {
      int tempDist = adjMat[startTree][j];
      sPath[j] = new DistPar(startTree, tempDist);
      }

   // until all vertices are in the tree
   while(nTree < nVerts)
      {
      int indexMin = getMin();      // get minimum from sPath
      int minDist = sPath[indexMin].distance;

      if(minDist == INFINITY)       // if all infinite
         {                          // or in tree,
         System.out.println("There are unreachable vertices");
         break;                     // sPath is complete
         }
      else
         {                          // reset currentVert
         currentVert = indexMin;    // to closest vert
         startToCurrent = sPath[indexMin].distance;
         // minimum distance from startTree is
         // to currentVert, and is startToCurrent
         }
      // put current vertex in tree
      vertexList[currentVert].isInTree = true;
      nTree++;
      adjust_sPath();               // update sPath[] array
      }  // end while(nTree<nVerts)

   displayPaths();                  // display sPath[] contents

   nTree = 0;                       // clear tree
   for(int j=0; j<nVerts; j++)
      vertexList[j].isInTree = false;
   }  // end path()
```

The starting vertex is always at index 0 of the vertexList[] array. The first task in path() is to put this vertex into the tree. As the algorithm proceeds, we'll be moving other vertices into the tree as well. The Vertex class contains a flag that indicates whether a vertex object is in the tree. Putting a vertex in the tree consists of setting this flag and incrementing nTree, which counts how many vertices are in the tree.

Second, path() copies the distances from the appropriate row of the adjacency matrix to sPath[]. This is always row 0, because for simplicity we assume 0 is the index of the starting vertex. Initially, the parent field of all the sPath[] entries is A, the starting vertex.

We now enter the main while loop of the algorithm. This loop terminates after all the vertices have been placed in the tree. There are basically three actions in this loop:

1. Choose the sPath[] entry with the minimum distance.

2. Put the corresponding vertex (the column head for this entry) in the tree. This becomes the "current vertex," currentVert.

3. Update all the sPath[] entries to reflect distances from currentVert.

If path() finds that the minimum distance is infinity, it knows that some vertices are unreachable from the starting point. Why? Because not all the vertices are in the tree (the while loop hasn't terminated), and yet there's no way to get to these extra vertices; if there were, there would be a non-infinite distance.

Before returning, path() displays the final contents of sPath[] by calling the displayPaths() method. This is the only output from the program. Also, path() sets nTree to 0 and removes the isInTree flags from all the vertices, in case they might be used again by another algorithm (although they aren't in this program).

Finding the Minimum Distance with getMin()
To find the sPath[] entry with the minimum distance, path() calls the getMin() method. This routine is straightforward; it steps across the sPath[] entries and returns with the column number (the array index) of the entry with the minimum distance.

```
public int getMin()              // get entry from sPath
   {                             //    with minimum distance
   int minDist = INFINITY;       // assume large minimum
   int indexMin = 0;
   for(int j=1; j<nVerts; j++)   // for each vertex,
      {                          // if it's in tree and
      if( !vertexList[j].isInTree &&  // smaller than old one
                      sPath[j].distance < minDist )
```

```
        {
        minDist = sPath[j].distance;
        indexMin = j;               // update minimum
        }
    }  // end for
    return indexMin;                // return index of minimum
    }  // end getMin()
```

We could have used a priority queue as the basis for the shortest-path algorithm, as we did in the previous section to find the minimum spanning tree. If we had, the getMin() method would not have been necessary; the minimum-weight edge would have appeared automatically at the front of the queue. However, the array approach shown makes it easier to see what's going on.

Updating sPath[] **with** adjust_sPath()
The adjust_sPath() method is used to update the sPath[] entries to reflect new information obtained from the vertex just inserted in the tree. When this routine is called, currentVert has just been placed in the tree, and startToCurrent is the current entry in sPath[] for this vertex. The adjust_sPath() method now examines each vertex entry in sPath[], using the loop counter column to point to each vertex in turn. For each sPath[] entry, provided the vertex is not in the tree, it does three things:

1. It adds the distance to the current vertex (already calculated and now in startToCurrent) to the edge distance from currentVert to the column vertex. We call the result startToFringe.

2. It compares startToFringe with the current entry in sPath[].

3. If startToFringe is less, it replaces the entry in sPath[].

This is the heart of Dijkstra's algorithm. It keeps sPath[] updated with the shortest distances to all the vertices that are currently known. Here's the code for adjust_sPath():

```
public void adjust_sPath()
    {
    // adjust values in shortest-path array sPath
    int column = 1;                 // skip starting vertex
    while(column < nVerts)          // go across columns
        {
        // if this column's vertex already in tree, skip it
        if( vertexList[column].isInTree )
            {
            column++;
            continue;
```

```
         }
      // calculate distance for one sPath entry
                 // get edge from currentVert to column
      int currentToFringe = adjMat[currentVert][column];
                 // add distance from start
      int startToFringe = startToCurrent + currentToFringe;
                 // get distance of current sPath entry
      int sPathDist = sPath[column].distance;

      // compare distance from start with sPath entry
      if(startToFringe < sPathDist)  // if shorter,
         {                                 // update sPath
         sPath[column].parentVert = currentVert;
         sPath[column].distance = startToFringe;
         }
      column++;
      }  // end while(column < nVerts)
   }  // end adjust_sPath()
```

The main() routine in the path.java program creates the tree of Figure 14.6 and displays its shortest-path array. Here's the code:

```
public static void main(String[] args)
   {
   Graph theGraph = new Graph();
   theGraph.addVertex('A');       // 0  (start)
   theGraph.addVertex('B');       // 1
   theGraph.addVertex('C');       // 2
   theGraph.addVertex('D');       // 3
   theGraph.addVertex('E');       // 4

   theGraph.addEdge(0, 1, 50);  // AB 50
   theGraph.addEdge(0, 3, 80);  // AD 80
   theGraph.addEdge(1, 2, 60);  // BC 60
   theGraph.addEdge(1, 3, 90);  // BD 90
   theGraph.addEdge(2, 4, 40);  // CE 40
   theGraph.addEdge(3, 2, 20);  // DC 20
   theGraph.addEdge(3, 4, 70);  // DE 70
   theGraph.addEdge(4, 1, 50);  // EB 50

   System.out.println("Shortest paths");
   theGraph.path();               // shortest paths
   System.out.println();
   }  // end main()
```

The output of this program is

```
A=inf(A)  B=50(A)  C=100(D)  D=80(A)  E=140(C)
```

The path.java Program

Listing 14.2 shows the complete code for the path.java program. Its various components were all discussed earlier.

LISTING 14.2   The path.java Program

```
// path.java
// demonstrates shortest path with weighted, directed graphs
// to run this program: C>java PathApp
////////////////////////////////////////////////////////////////
class DistPar                 // distance and parent
   {                          // items stored in sPath array
   public int distance;       // distance from start to this vertex
   public int parentVert;     // current parent of this vertex
// -----------------------------------------------------------
   public DistPar(int pv, int d)   // constructor
      {
      distance = d;
      parentVert = pv;
      }
   }  // end class DistPar
////////////////////////////////////////////////////////////////
class Vertex
   {
   public char label;         // label (e.g. 'A')
   public boolean isInTree;
// -----------------------------------------------------------
   public Vertex(char lab)    // constructor
      {
      label = lab;
      isInTree = false;
      }
// -----------------------------------------------------------
   }  // end class Vertex
////////////////////////////////////////////////////////////////
class Graph
   {
   private final int MAX_VERTS = 20;
```

LISTING 14.2   Continued

```
private final int INFINITY = 1000000;
private Vertex vertexList[]; // list of vertices
private int adjMat[][];       // adjacency matrix
private int nVerts;           // current number of vertices
private int nTree;            // number of verts in tree
private DistPar sPath[];      // array for shortest-path data
private int currentVert;      // current vertex
private int startToCurrent;   // distance to currentVert
// ------------------------------------------------------------
public Graph()                // constructor
   {
   vertexList = new Vertex[MAX_VERTS];
                                    // adjacency matrix
   adjMat = new int[MAX_VERTS][MAX_VERTS];
   nVerts = 0;
   nTree = 0;
   for(int j=0; j<MAX_VERTS; j++)    // set adjacency
      for(int k=0; k<MAX_VERTS; k++) //     matrix
         adjMat[j][k] = INFINITY;    //     to infinity
   sPath = new DistPar[MAX_VERTS];   // shortest paths
   }  // end constructor
// ------------------------------------------------------------
public void addVertex(char lab)
   {
   vertexList[nVerts++] = new Vertex(lab);
   }
// ------------------------------------------------------------
public void addEdge(int start, int end, int weight)
   {
   adjMat[start][end] = weight;  // (directed)
   }
// ------------------------------------------------------------
public void path()                 // find all shortest paths
   {
   int startTree = 0;              // start at vertex 0
   vertexList[startTree].isInTree = true;
   nTree = 1;                      // put it in tree

   // transfer row of distances from adjMat to sPath
   for(int j=0; j<nVerts; j++)
      {
```

LISTING 14.2   Continued

```
            int tempDist = adjMat[startTree][j];
            sPath[j] = new DistPar(startTree, tempDist);
            }

      // until all vertices are in the tree
      while(nTree < nVerts)
         {
         int indexMin = getMin();    // get minimum from sPath
         int minDist = sPath[indexMin].distance;

         if(minDist == INFINITY)      // if all infinite
            {                         // or in tree,
            System.out.println("There are unreachable vertices");
            break;                    // sPath is complete
            }
         else
            {                         // reset currentVert
            currentVert = indexMin;   // to closest vert
            startToCurrent = sPath[indexMin].distance;
            // minimum distance from startTree is
            // to currentVert, and is startToCurrent
            }
         // put current vertex in tree
         vertexList[currentVert].isInTree = true;
         nTree++;
         adjust_sPath();              // update sPath[] array
         }   // end while(nTree<nVerts)

      displayPaths();                 // display sPath[] contents

      nTree = 0;                      // clear tree
      for(int j=0; j<nVerts; j++)
         vertexList[j].isInTree = false;
      }   // end path()
// ------------------------------------------------------------
   public int getMin()              // get entry from sPath
      {                             //    with minimum distance
      int minDist = INFINITY;       // assume minimum
      int indexMin = 0;
      for(int j=1; j<nVerts; j++)   // for each vertex,
         {                          // if it's in tree and
```

LISTING 14.2   Continued

```
          if( !vertexList[j].isInTree &&   // smaller than old one
                              sPath[j].distance < minDist )
             {
             minDist = sPath[j].distance;
             indexMin = j;                // update minimum
             }
          }  // end for
       return indexMin;                   // return index of minimum
       }  // end getMin()
// -----------------------------------------------------------------
   public void adjust_sPath()
       {
       // adjust values in shortest-path array sPath
       int column = 1;                    // skip starting vertex
       while(column < nVerts)             // go across columns
          {
          // if this column's vertex already in tree, skip it
          if( vertexList[column].isInTree )
             {
             column++;
             continue;
             }
          // calculate distance for one sPath entry
                       // get edge from currentVert to column
          int currentToFringe = adjMat[currentVert][column];
                       // add distance from start
          int startToFringe = startToCurrent + currentToFringe;
                       // get distance of current sPath entry
          int sPathDist = sPath[column].distance;

          // compare distance from start with sPath entry
          if(startToFringe < sPathDist)  // if shorter,
             {                                // update sPath
             sPath[column].parentVert = currentVert;
             sPath[column].distance = startToFringe;
             }
          column++;
          }  // end while(column < nVerts)
       }  // end adjust_sPath()
// -----------------------------------------------------------------
   public void displayPaths()
```

LISTING 14.2   Continued

```
      {
      for(int j=0; j<nVerts; j++) // display contents of sPath[]
         {
         System.out.print(vertexList[j].label + "=");   // B=
         if(sPath[j].distance == INFINITY)
            System.out.print("inf");                     // inf
         else
            System.out.print(sPath[j].distance);         // 50
         char parent = vertexList[ sPath[j].parentVert ].label;
         System.out.print("(" + parent + ") ");          // (A)
         }
      System.out.println("");
      }
// ------------------------------------------------------------
   } // end class Graph
////////////////////////////////////////////////////////////////
class PathApp
   {
   public static void main(String[] args)
      {
      Graph theGraph = new Graph();
      theGraph.addVertex('A');      // 0  (start)
      theGraph.addVertex('C');      // 2
      theGraph.addVertex('B');      // 1
      theGraph.addVertex('D');      // 3
      theGraph.addVertex('E');      // 4

      theGraph.addEdge(0, 1, 50);  // AB 50
      theGraph.addEdge(0, 3, 80);  // AD 80
      theGraph.addEdge(1, 2, 60);  // BC 60
      theGraph.addEdge(1, 3, 90);  // BD 90
      theGraph.addEdge(2, 4, 40);  // CE 40
      theGraph.addEdge(3, 2, 20);  // DC 20
      theGraph.addEdge(3, 4, 70);  // DE 70
      theGraph.addEdge(4, 1, 50);  // EB 50

      System.out.println("Shortest paths");
      theGraph.path();            // shortest paths
      System.out.println();
      } // end main()
   } // end class PathApp
////////////////////////////////////////////////////////////////
```

The All-Pairs Shortest-Path Problem

In discussing connectivity in Chapter 13, we wanted to know whether it was possible to fly from Athens to Murmansk if we didn't care how many stops we made. With weighted graphs we can answer the second question that might occur to you as you wait at the Hubris Airlines ticket counter: How much will the journey cost?

To find whether a trip was possible, we created a connectivity table. With weighted graphs we want a table that gives the minimum cost from any vertex to any other vertex using multiple edges. This is called the *all-pairs shortest-path* problem.

You can create such a table by running the path.java program using each vertex in turn as the starting vertex. This will yield something like Table 14.11.

TABLE 14.11   All-Pairs Shortest-Path Table

	A	B	C	D	E
A	—	50	100	80	140
B	—	—	60	90	100
C	—	90	—	180	40
D	—	110	20	—	60
E	—	50	110	140	—

In the preceding chapter we found that Warshall's algorithm was a quicker way to create a table showing which vertices could be reached from a given vertex using one or many steps. An analogous approach for weighted graphs uses Floyd's algorithm, discovered by Robert Floyd in 1962. This is another way to create the kind of table shown in Table 14.11.

Let's discuss Floyd's algorithm with a simpler graph. Figure 14.11 shows a weighted directed graph and its adjacency matrix.

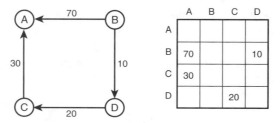

FIGURE 14.11   A weighted graph and its adjacency matrix.

The adjacency matrix shows the cost of all the one-edge paths. We want to extend this matrix to show the cost of all paths regardless of length. For example, it's clear from Figure 14.11 that we can go from B to C at a cost of 30 (10 from B to D plus 20 from D to C).

As in Warshall's algorithm we systematically modify the adjacency matrix. We examine every cell in every row. If there's a non-zero weight (say a 30 at row C column A), we then look in column C (because C is the row where the 30 is). If we find an entry in column C (say a 40 at row D), we know there is a path from C to A with a weight of 30 and a path from D to C with a weight of 40. From this, we can deduce that there's a two-edge path from D to A with a weight of 70. Figure 14.12 shows the steps when Floyd's algorithm is applied to the graph in Figure 14.11.

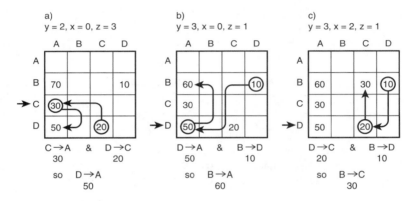

FIGURE 14.12 Floyd's algorithm.

Row A is empty, so there's nothing to do there. In row B there's a 70 in column A and a 10 in column D, but there's nothing in column B, so the entries in row B can't be combined with any edges ending on B.

In row C, however, we find a 30 at column A. Looking in column C, we find a 20 at row D. Now we have C to A with a weight of 30 and D to C with a weight of 20, so we have D to A with a weight of 50.

Row D shows an interesting situation: We will lower an existing cost. There's a 50 in column A. There's also 10 in row B of column D, so we know there's a path from B to A with a cost of 60. However, there's already a cost of 70 in this cell. What do we do? Because 60 is less than 70, we replace the 70 with 60. In the case of multiple paths from one vertex to another, we want the table to reflect the path with the lowest cost.

The implementation of Floyd's algorithm is similar to that for Warshall's algorithm. However, instead of simply inserting a 1 into the table when a two-edge path is found, we add the costs of the two one-edge paths and insert the sum. We'll leave the details as an exercise.

Efficiency

So far we haven't discussed the efficiency of the various graph algorithms. The issue is complicated by the two ways of representing graphs: the adjacency matrix and adjacency lists.

If an adjacency matrix is used, the algorithms we've discussed mostly require $O(V^2)$ time, where V is the number of vertices. Why? If you analyze the algorithms, you'll see that they involve examining each vertex once, and for that vertex going across its row in the adjacency matrix, looking at each edge in turn. In other words, each cell of the adjacency matrix, which has V^2 cells, is examined.

For large matrices $O(V^2)$ isn't very good performance. If the graph is dense, there isn't much we can do about improving this performance. (As we noted earlier, by *dense* we mean a graph that has many edges—one in which many or most of the cells in the adjacency matrix are filled.)

However, many graphs are *sparse*, the opposite of dense. There's no clear-cut definition of how many edges a graph must have to be described as sparse or dense, but if each vertex in a large graph is connected by only a few edges, the graph would normally be described as sparse.

In a sparse graph, running times can be improved by using the adjacency-list representation rather than the adjacency matrix. This is easy to understand: You don't waste time examining adjacency-matrix cells that don't hold edges.

For unweighted graphs the depth-first search with adjacency lists requires $O(V+E)$ time, where V is the number of vertices and E is the number of edges. For weighted graphs, both the minimum spanning tree and the shortest-path algorithm require $O((E+V)logV)$ time. In large, sparse graphs these times can represent dramatic improvements over the adjacency matrix approach. However, the algorithms are somewhat more complicated, which is why we've used the adjacency-matrix approach throughout this chapter. You can consult Sedgewick (see Appendix B, "Further Reading") and other writers for examples of graph algorithms using the adjacency-list approach.

Warshall's and Floyd's algorithms are slower than the other algorithms we've discussed so far in this book. They both operate in $O(V^3)$ time. This is the result of the three nested loops used in their implementation.

Intractable Problems

In this book we've seen big O values ranging from $O(1)$, through $O(N)$, $O(N*logN)$, $O(N^2)$, up to (for Warshall's and Floyd's algorithms) $O(N^3)$. Even $O(N^3)$ can be solved in a reasonable length of time for values N in the thousands. Algorithms with these big O values can be used to find solutions to most practical problems.

However, some algorithms have big O values that are so large that they can be used only for relatively small values of N. Many real-world problems that require such algorithms simply cannot be solved in a reasonable length of time. Such problems are said to be *intractable*. (Another term used for such problems is *NP complete*, where NP means non-deterministic polynomial. An explanation of what this means is beyond the scope of this book.)

The Knight's Tour

The Knight's Tour (Programming Project 13.5 in Chapter 13) is an example of an intractable problem because the number of possible moves is so large. The total number of possible move sequences is difficult to calculate, but we can approximate it. Each move can end on a maximum of eight squares. This number is reduced by moves that would be off the edge of the board and moves that would end on a square that was already visited. In the early stages of a tour, there will be closer to eight moves, but this number will gradually decrease as the board fills up. Let's assume (conservatively) an average of only two possible moves from each position. After the initial square, the knight can visit 63 more squares. Thus, there is a total of 2^{63} possible moves. This is about 10^{19}. Assume a computer can make a million moves a second (10^6). There are roughly 10^7 seconds in a year, so the computer can make 10^{13} moves in a year. Solving the puzzle by brute force can therefore be expected to take 10^6 or around a million years.

This particular problem can be made more tractable if strategies are used to "prune" the game tree. One is Warnsdorff's heuristic (H.C. von Warnsdorff, 1823), which specifies that you always move to the square that has the fewest possible exit moves.

The Traveling Salesman Problem

Here's another famous intractable problem. Suppose you're a salesperson and you need to drive to all the cities where you have clients. You would like to minimize the number of miles you travel. You know the distance from each city to every other city. You want to start in your home city, visit each client city once and only once, and return to your home city. In what sequence should you visit these cities to minimize the total miles traveled? In graph theory this is called the *traveling salesman problem*, often abbreviated TSP.

Figure 14.13 shows an arrangement of cities and distances. What's the shortest way to travel from A through each of the other cities and back to A? Notice that it's not necessary that every pair of cities be connected by an edge. Because of geography it may be impossible to drive from Washington, D.C., to New York without going through Philadelphia, for example.

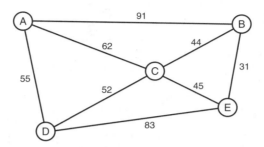

FIGURE 14.13   Cities and distances.

To find the shortest route, you list all the possible permutations of cities (Boston-Seattle-Miami, Boston-Miami-Seattle, Miami-Boston-Seattle, and so on) and calculate the total distance for each permutation. The route ABCEDA has a total length of 318. The route ABCDEA is impossible because there is no edge from E back to A.

Unfortunately, the number of permutations can be very large: It's the factorial of the number of cities (not counting your home city). If there are 6 cities to visit, there are 6 choices for the first city, 5 for the second, 4 for the third, and so on; a total of 6*5*4*3*2*1 or 720 possible routes. The problem is impractical to solve for even 50 cities. Again, there are strategies to reduce the number of sequences that must be checked, but this helps only a little. A weighted graph is used to implement the problem, with weights representing miles and vertices representing the cities. The graph can be non-directed if the distance is the same going from A to B as from B to A, as it usually is when driving. If the weights represent airfares, they may be different in different directions, in which case a directed graph is used.

Hamiltonian Cycles

A problem that's similar to the TSP but more abstract is that of finding the Hamiltonian cycle of a graph. As we noted earlier, a cycle is a path that starts and ends on the same vertex. A Hamiltonian cycle is one that visits every other vertex in the graph exactly once. Unlike we did with the TSP, we don't care about distances; all we want to know is whether such a cycle exists. In Figure 14.13 the route ABCEDA is a Hamiltonian cycle, while ABCDEA is not. The Knight's Tour problem is an example of a Hamiltonian cycle (if you assume the knight returns to its starting square).

Finding a Hamiltonian cycle takes the same $O(N!)$ time as the TSP. You'll see the term *exponential time* used for big O values such as 2^N and $N!$ (which grows even more rapidly than the exponential 2^N).

Summary

- In a weighted graph, edges have an associated number called the weight, which might represent distances, costs, times, or other quantities.

- The minimum spanning tree in a weighted graph minimizes the weights of the edges necessary to connect all the vertices.

- An algorithm using a priority queue can be used to find the minimum spanning tree of a weighted graph.

- The minimum spanning tree of a weighted graph models real-world situations such as installing utility cables between cities.

- The shortest-path problem in a non-weighted graph involves finding the minimum number of edges between two vertices.

- Solving the shortest-path problem for weighted graphs yields the path with the minimum total edge weight.

- The shortest-path problem for weighted graphs can be solved with Dijkstra's algorithm.

- The algorithms for large, sparse graphs generally run much faster if the adjacency-list representation of the graph is used rather than the adjacency matrix.

- The all-pairs shortest-path problem is to find the total weight of the edges between every pair of vertices in a graph. Floyd's algorithm can be used to solve this problem.

- Some graph algorithms take exponential time and are therefore not practical for graphs with more than a few vertices.

Questions

These questions are intended as a self-test for readers. Answers may be found in Appendix C.

1. The weight in a weighted graph is a property of the graph's _____.

2. In a weighted graph, the minimum spanning tree (MST) tries to minimize

    a. the number of edges from the starting vertex to a specified vertex.

    b. the number of edges connecting all the vertices.

    c. the total weight of the edges from the starting vertex to a specified vertex.

    d. the total weight of edges connecting all the vertices.

3. True or False: The weight of the MST depends on the starting vertex.

4. In the MST algorithm, what is removed from the priority queue?

5. In the cable TV example, each edge added to the MST connects

   a. the starting vertex to an adjacent vertex.

   b. an already-connected city to an unconnected city.

   c. the current vertex to an adjacent vertex.

   d. two cities with offices.

6. The MST algorithm "prunes" an edge from the list when the edge leads to a vertex that _____ .

7. True or False: The shortest-path problem (SPP) must be carried out on a directed graph.

8. Dijkstra's algorithm finds the shortest path

   a. from one specified vertex to all other vertices.

   b. from one specified vertex to another specified vertex.

   c. from all vertices to all other vertices that can be reached along one edge.

   d. from all vertices to all other vertices that can be reached along multiple edges.

9. True or False: The rule in Dijkstra's algorithm is to always put in the tree the vertex that is closest to the starting vertex.

10. In the railroad fares example, a fringe town is one

    a. to which the distance is known, but from which no distances are known.

    b. which is in the tree.

    c. to which the distance is known and which is in the tree.

    d. which is completely unknown.

11. The all-pairs shortest-path problem involves finding the shortest path

    a. from the starting vertex to every other vertex.

    b. from every vertex to every other vertex.

    c. from the starting vertex to every vertex that is one edge away.

    d. from every vertex to every other vertex that is one or more edges away.

12. Floyd's algorithm is to weighted graphs what _____ is to unweighted graphs.

13. Floyd's algorithm uses the _____ representation of a graph.

14. What is an approximate big O time for an attempt to solve the knight's tour?

15. In Figure 14.13, is the route ABCEDA the minimum solution for the traveling salesman problem?

Experiments

Carrying out these experiments will help to provide insights into the topics covered in the chapter. No programming is involved.

1. Use the GraphW Workshop applet to find the minimum spanning tree of the graph shown in Figure 14.6, "Train fares in Magnaguena." Consider the graph to be undirected; that is, ignore the arrows.

2. Use the GraphDW Workshop applet to solve the shortest-path problem for the graph in Figure 14.6, "Train fares in Magnaguena," but derive new weights for all the edges by subtracting those shown in the figure from 100.

3. Draw a graph with five vertices and five edges. Then use pencil and paper to implement Djikstra's algorithm for this graph. Show the tree and the shortest-path array at each step.

Programming Projects

Writing programs to solve the Programming Projects helps to solidify your understanding of the material and demonstrates how the chapter's concepts are applied. (As noted in the Introduction, qualified instructors may obtain completed solutions to the Programming Projects on the publisher's Web site.)

14.1 Modify the path.java program (Listing 14.2) to print a table of the minimum costs to get from any vertex to any other vertex. This exercise will require some fiddling with routines that assume the starting vertex is always A.

14.2 So far we've implemented graphs as adjacency matrices or adjacency lists. Another approach is to use Java references to represent edges, so that a Vertex object contains a list of references to other vertices that it's connected to. In a directed graph a reference used this way is especially intuitive because it "points" from one vertex to another. Write a program that implements this scheme. The main() method should be similar to main() in the path.java program (Listing 14.2) so that it creates the graph shown in Figure 14.6 using the same addVertex() and addEdge() calls. It should then display a connectivity table of the graph to prove that the graph is constructed properly. You'll need to store the weight of each edge somewhere. One approach is to use an Edge class, which stores its weight and the vertex on which it ends. Each vertex then keeps a list of Edge objects—that is, edges that start on that vertex.

14.3 Implement Floyd's algorithm. You can start with the path.java program (Listing 14.2) and modify it as appropriate. For instance, you can delete all the shortest-path code. Keep the infinity representation for unreachable vertices. By doing this, you will avoid the need to check for 0 when comparing an existing cost with a newly derived cost. The costs on all possible routes will be less than infinity. You should be able to enter graphs of arbitrary complexity into main().

14.4 Implement the traveling salesman problem described in the "Intractable Problems" section in this chapter. In spite of its intractability, it will have no trouble solving the problem for small N, say 10 cities or fewer. Try a non-directed graph. Use the brute-force approach of testing every possible sequence of cities. For a way to permute the sequence of cities, see the anagram.java program (Listing 6.2) in Chapter 6, "Recursion." Use infinity to represent non-existent edges. That way, you won't need to abort the calculation of a sequence when it turns out that an edge from one city to the next does not exist; any total greater than infinity is an impossible route. Also, don't worry about eliminating symmetrical routes. Display both ABCDEA and AEDCBA, for example.

14.5 Write a program that discovers and displays all the Hamiltonian cycles of a weighted, non-directed graph.

15

When to Use What

In this chapter we briefly summarize what we've learned so far, with an eye toward deciding what data structure or algorithm to use in a particular situation.

This chapter comes with the usual caveats. Of necessity, it's very general. Every real-world situation is unique, so what we say here may not be the right answer to your problem. This chapter is divided into these somewhat arbitrary sections:

- General-purpose data structures: arrays, linked lists, trees, hash tables
- Specialized data structures: stacks, queues, priority queues, graphs
- Sorting: insertion sort, Shellsort, quicksort, mergesort, heapsort
- Graphs: adjacency matrix, adjacency list
- External storage: sequential storage, indexed files, B-trees, hashing

---

NOTE

For detailed information on these topics, refer to the individual chapters in this book.

General-Purpose Data Structures

If you need to store real-world data such as personnel records, inventories, contact lists, or sales data, you need a general-purpose data structure. The structures of this type that we've discussed in this book are arrays, linked lists, trees, and hash tables. We call these general-purpose data structures because they are used to store and retrieve data using key values. This works for general-purpose database programs (as opposed to specialized structures such as stacks, which allow access to only certain data items).

Which of these general-purpose data structures is appropriate for a given problem? Figure 15.1 shows a first approximation to this question. However, there are many factors besides those shown in the figure. For more detail, we'll explore some general considerations first and then zero in on the individual structures.

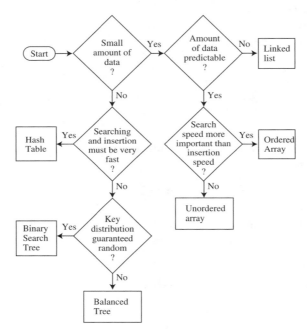

FIGURE 15.1   Relationship of general-purpose data structures.

Speed and Algorithms

The general-purpose data structures can be roughly arranged in terms of speed: Arrays and linked lists are slow, trees are fairly fast, and hash tables are very fast.

However, don't draw the conclusion from Figure 15.1 that it's always best to use the fastest structures. There's a penalty for using them. First, they are—in varying degrees—more complex to program than the array and linked list. Also, hash tables require you to know in advance about how much data can be stored, and they don't use memory very efficiently. Ordinary binary trees will revert to slow O(N) operation for ordered data, and balanced trees, which avoid this problem, are difficult to program.

Processor Speed: A Moving Target

The fast structures come with penalties, and another development makes the slow structures more attractive. Every year there's an increase in the CPU and memory-access speed of the latest computers. Moore's Law (postulated by Gordon Moore in 1965) specifies that CPU performance will double every 18 months. This adds up to an astonishing difference in performance between the earliest computers and those available today, and there's no reason to think this increase will slow down any time soon.

Suppose a computer a few years ago handled an array of 100 objects in acceptable time. Now, computers are much faster, so an array with 10,000 objects might run at the same speed. Many writers provide estimates of the maximum size you can make a data structure before it becomes too slow. Don't trust these estimates (including those in this book). Today's estimate doesn't apply to tomorrow.

Instead, start by considering the simple data structures. Unless it's obvious they'll be too slow, code a simple version of an array or linked list and see what happens. If it runs in acceptable time, look no further. Why slave away on a balanced tree when no one would ever notice if you used an array instead? Even if you must deal with thousands or tens of thousands of items, it's still worthwhile to see how well an array or linked list will handle them. Only when experimentation shows their performance to be too slow should you revert to more sophisticated data structures.

Advantages of Java References

Java has an advantage over some languages in the speed with which objects can be manipulated, because, in most data structures, Java stores only references, not actual objects. Therefore, most algorithms will run faster than in languages where actual objects occupy space in a data structure. In analyzing the algorithms, it's not the case, as when objects themselves are stored, that the time to "move" an object depends on the size of the object. Because only a reference is moved, it doesn't matter how large the object is.

Of course, in other languages, such as C++, pointers to objects can be stored instead of the objects themselves; this has the same effect as using references, but the syntax is more complicated.

Libraries

Libraries of data structures are available commercially in all major programming languages. Languages themselves may have some structures built in. Java, for example, includes Vector, Stack, and Hashtable classes. C++ includes the Standard Template Library (STL), which contains classes for many data structures and algorithms.

Using a commercial library may eliminate or at least reduce the programming necessary to create the data structures described in this book. When that's the case, using a complex structure such as a balanced tree, or a delicate algorithm such as quicksort, becomes a more attractive possibility. However, you must ensure that the class can be adapted to your particular situation.

Arrays

In many situations the array is the first kind of structure you should consider when storing and manipulating data. Arrays are useful when

- The amount of data is reasonably small.

- The amount of data is predictable in advance.

If you have plenty of memory, you can relax the second condition; just make the array big enough to handle any foreseeable influx of data.

If insertion speed is important, use an unordered array. If search speed is important, use an ordered array with a binary search. Deletion is always slow in arrays because an average of half the items must be moved to fill in the newly vacated cell. Traversal is fast in an ordered array but not supported in an unordered array.

Vectors, such as the Vector class supplied with Java, are arrays that expand themselves when they become too full. Vectors may work when the amount of data isn't known in advance. However, there may periodically be a significant pause while they enlarge themselves by copying the old data into the new space.

Linked Lists

Consider a linked list whenever the amount of data to be stored cannot be predicted in advance or when data will frequently be inserted and deleted. The linked list obtains whatever storage it needs as new items are added, so it can expand to fill all of available memory; and there is no need to fill "holes" during deletion, as there is in arrays.

Insertion is fast in an unordered list. Searching and deletion are slow (although deletion is faster than in an array), so, like arrays, linked lists are best used when the amount of data is comparatively small.

A linked list is somewhat more complicated to program than an array, but is simple compared with a tree or hash table.

Binary Search Trees

A binary tree is the first structure to consider when arrays and linked lists prove too slow. A tree provides fast O(logN) insertion, searching, and deletion. Traversal is

O(N), which is the maximum for any data structure (by definition, you must visit every item). You can also find the minimum and maximum quickly and traverse a range of items.

An unbalanced binary tree is much easier to program than a balanced tree, but unfortunately ordered data can reduce its performance to O(N) time, no better than a linked list. However, if you're sure the data will arrive in random order, there's no point using a balanced tree.

Balanced Trees

Of the various kinds of balanced trees, we discussed red-black trees and 2-3-4 trees. They are both balanced trees, and thus guarantee O(logN) performance whether the input data is ordered or not. However, these balanced trees are challenging to program, with the red-black tree being the more difficult. They also impose additional memory overhead, which may or may not be significant.

The problem of complex programming may be reduced if a commercial class can be used for a tree. In some cases a hash table may be a better choice than a balanced tree. Hash-table performance doesn't degrade when the data is ordered.

There are other kinds of balanced trees, including AVL trees, splay trees, 2-3 trees, and so on, but they are not as commonly used as the red-black tree.

Hash Tables

Hash tables are the fastest data storage structure. This makes them a necessity for situations in which a computer program, rather than a human, is interacting with the data. Hash tables are typically used in spelling checkers and as symbol tables in computer language compilers, where a program must check thousands of words or symbols in a fraction of a second.

Hash tables may also be useful when a person, as opposed to a computer, initiates data-access operations. As noted earlier, hash tables are not sensitive to the order in which data is inserted, and so can take the place of a balanced tree. Programming is much simpler than for balanced trees.

Hash tables require additional memory, especially for open addressing. Also, the amount of data to be stored must be known fairly accurately in advance, because an array is used as the underlying structure.

A hash table with separate chaining is the most robust implementation, unless the amount of data is known accurately in advance, in which case open addressing offers simpler programming because no linked list class is required.

Hash tables don't support any kind of ordered traversal, or access to the minimum or maximum items. If these capabilities are important, the binary search tree is a better choice.

Comparing the General-Purpose Storage Structures

Table 15.1 summarizes the speeds of the various general-purpose data storage structures using Big O notation.

TABLE 15.1   General-Purpose Data Storage Structures

Data Structure	Search	Insertion	Deletion	Traversal
Array	O(N)	O(1)	O(N)	—
Ordered array	O(logN)	O(N)	O(N)	O(N)
Linked list	O(N)	O(1)	O(N)	—
Ordered linked list	O(N)	O(N)	O(N)	O(N)
Binary tree (average)	O(logN)	O(logN)	O(logN)	O(N)
Binary tree (worst case)	O(N)	O(N)	O(N)	O(N)
Balanced tree (average and worst case)	O(logN)	O(logN)	O(logN)	O(N)
Hash table	O(1)	O(1)	O(1)	—

Insertion in an unordered array is assumed to be at the end of the array. The ordered array uses a binary search, which is fast, but insertion and deletion require moving half the items on the average, which is slow. Traversal implies visiting the data items in order of ascending or descending keys; the — means this operation is not supported.

Special-Purpose Data Structures

The special-purpose data structures discussed in this book are the stack, the queue, and the priority queue. These structures, rather than supporting a database of user-accessible data, are usually used by a computer program to aid in carrying out some algorithm. We've seen examples of this throughout this book, such as in Chapters 13, "Graphs," and 14, "Weighted Graphs," where stacks, queues, and priority queues are all used in graph algorithms.

Stacks, queues, and priority queues are Abstract Data Types (ADTs) that are implemented by a more fundamental structure such as an array, linked list, or (in the case of the priority queue) a heap. These ADTs present a simple interface to the user, typically allowing only insertion and the ability to access or delete only one data item. These items are

- For stacks: the last item inserted

- For queues: the first item inserted

- For priority queues: the item with the highest priority

These ADTs can be seen as conceptual aids. Their functionality could be obtained using the underlying structure (such as an array) directly, but the reduced interface they offer simplifies many problems.

These ADTs can't be conveniently searched for an item by key value or traversed.

Stack

A stack is used when you want access only to the last data item inserted; it's a Last-In-First-Out (LIFO) structure.

A stack is often implemented as an array or a linked list. The array implementation is efficient because the most recently inserted item is placed at the end of the array, where it's also easy to delete. Stack overflow can occur, but is not likely if the array is reasonably sized, because stacks seldom contain huge amounts of data.

If the stack will contain a lot of data and the amount can't be predicted accurately in advance (as when recursion is implemented as a stack), a linked list is a better choice than an array. A linked list is efficient because items can be inserted and deleted quickly from the head of the list. Stack overflow can't occur (unless the entire memory is full). A linked list is slightly slower than an array because memory allocation is necessary to create a new link for insertion, and deallocation of the link is necessary at some point following removal of an item from the list.

Queue

A queue is used when you want access only to the first data item inserted; it's a First-In-First-Out (FIFO) structure.

Like stacks, queues can be implemented as arrays or linked lists. Both are efficient. The array requires additional programming to handle the situation in which the queue wraps around at the end of the array. A linked list must be double-ended, to allow insertions at one end and deletions at the other.

As with stacks, the choice between an array implementation and a linked list implementation is determined by how well the amount of data can be predicted. Use the array if you know about how much data there will be; otherwise, use a linked list.

Priority Queue

A priority queue is used when the only access desired is to the data item with the highest priority. This is the item with the largest (or sometimes the smallest) key.

Priority queues can be implemented as an ordered array or as a heap. Insertion into an ordered array is slow, but deletion is fast. With the heap implementation, both insertion and deletion take $O(logN)$ time.

Use an array or a double-ended linked list if insertion speed is not a problem. The array works when the amount of data to be stored can be predicted in advance; the linked list when the amount of data is unknown. If speed is important, a heap is a better choice.

Comparison of Special-Purpose Structures

Table 15.2 shows the Big O times for stacks, queues, and priority queues. These structures don't support searching or traversal.

TABLE 15.2 Special-Purpose Data Storage Structures

Data Structure	Insertion	Deletion	Comment
Stack (array or linked list)	O(1)	O(1)	Deletes most recently inserted item
Queue (array or linked list)	O(1)	O(1)	Deletes least recently inserted item
Priority queue (ordered array)	O(N)	O(1)	Deletes highest-priority item
Priority queue (heap)	O(logN)	O(logN)	Deletes highest-priority item

Sorting

As with the choice of data structures, it's worthwhile initially to try a slow but simple sort, such as the insertion sort. It may be that the fast processing speeds available in modern computers will allow sorting of your data in reasonable time. (As a wild guess, the slow sort might be appropriate for fewer than 1,000 items.)

Insertion sort is also good for almost-sorted files, operating in about O(N) time if not too many items are out of place. This is typically the case where a few new items are added to an already-sorted file.

If the insertion sort proves too slow, then the Shellsort is the next candidate. It's fairly easy to implement, and not very temperamental. Sedgewick estimates it to be useful up to 5,000 items.

Only when the Shellsort proves too slow should you use one of the more complex but faster sorts: mergesort, heapsort, or quicksort. Mergesort requires extra memory, heapsort requires a heap data structure, and both are somewhat slower than quicksort, so quicksort is the usual choice when the fastest sorting time is necessary.

However, quicksort is suspect if there's a danger that the data may not be random, in which case it may deteriorate to $O(N^2)$ performance. For potentially non-random data, heapsort is better. Quicksort is also prone to subtle errors if it is not implemented correctly. Small mistakes in coding can make it work poorly for certain arrangements of data, a situation that may be hard to diagnose.

Table 15.3 summarizes the running time for various sorting algorithms. The column labeled Comparison attempts to estimate the minor speed differences between algorithms with the same average Big O times. (There's no entry for Shellsort because there are no other algorithms with the same Big O performance.)

TABLE 15.3   Comparison of Sorting Algorithms

Sort	Average	Worst	Comparison	Extra Memory
Bubble	$O(N^2)$	$O(N^2)$	Poor	No
Selection	$O(N^2)$	$O(N^2)$	Fair	No
Insertion	$O(N^2)$	$O(N^2)$	Good	No
Shellsort	$O(N^{3/2})$	$O(N^{3/2})$	—	No
Quicksort	$O(N*logN)$	$O(N^2)$	Good	No
Mergesort	$O(N*logN)$	$O(N*logN)$	Fair	Yes
Heapsort	$O(N*logN)$	$O(N*logN)$	Fair	No

Graphs

Graphs are unique in the pantheon of data storage structures. They don't store general-purpose data, and they don't act as programmer's tools for use in other algorithms. Instead, they directly model real-world situations. The structure of the graph reflects the structure of the problem.

When you need a graph, nothing else will do, so there's no decision to be made about when to use one. The primary choice is how to represent the graph: using an adjacency matrix or adjacency lists. Your choice depends on whether the graph is full, when the adjacency matrix is preferred, or sparse, when the adjacency list should be used.

The depth-first search and breadth-first search run in $O(V^2)$ time, where V is the number of vertices, for adjacency matrix representation. They run in $O(V+E)$ time, where E is the number of edges, for adjacency list representation. Minimum spanning trees and shortest paths run in $O(V^2)$ time using an adjacency matrix and $O((E+V)logV)$ time using adjacency lists. You'll need to estimate V and E for your graph and do the arithmetic to see which representation is appropriate.

External Storage

In the previous discussion we assumed that data was stored in main memory. However, amounts of data too large to store in memory must be stored in external storage, which generally means disk files. We discussed external storage in the second parts of Chapter 10, "2-3-4 Trees and External Storage," and Chapter 11, "Hash Tables."

We assumed that data is stored in a disk file in fixed-size units called blocks, each of which holds a number of records. (A record in a disk file holds the same sort of data as an object in main memory.) Like an object, a record has a key value used to access it.

We also assumed that reading and writing operations always involve a single block, and these read and write operations are far more time-consuming than any processing of data in main memory. Thus, for fast operation the number of disk accesses must be minimized.

Sequential Storage

The simplest approach is to store records randomly and read them sequentially when searching for one with a particular key. New records can simply be inserted at the end of the file. Deleted records can be marked as deleted, or records can be shifted down (as in an array) to fill in the gap.

On the average, searching and deletion will involve reading half the blocks, so sequential storage is not very fast, operating in O(N) time. Still, it might be satisfactory for a small number of records.

Indexed Files

Speed is increased dramatically when indexed files are used. In this scheme an index of keys and corresponding block numbers is kept in main memory. To access a record with a specified key, the index is consulted. It supplies the block number for the key, and only one block needs to be read, taking O(1) time.

Several indices with different kinds of keys can be used (one for last names, one for Social Security numbers, and so on). This scheme works well until the index becomes too large to fit in memory.

Typically, the index files are themselves stored on disk and read into memory as needed.

The disadvantage of indexed files is that at some point the index must be created. This probably involves reading through the file sequentially, so creating the index is slow. Also, the index will need to be updated when items are added to the file.

B-trees

B-trees are multiway trees, commonly used in external storage, in which nodes correspond to blocks on the disk. As in other trees, the algorithms find their way down the tree, reading one block at each level. B-trees provide searching, insertion, and deletion of records in O(logN) time. This is quite fast and works even for very large files. However, the programming is not trivial.

Hashing

If it's acceptable to use about twice as much external storage as a file would normally take, then external hashing might be a good choice. It has the same access time as indexed files, O(1), but can handle larger files.

Figure 15.2 shows, rather impressionistically, these choices for external storage structures.

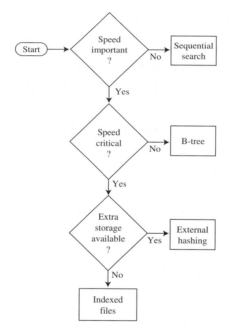

FIGURE 15.2   Relationship of external storage choices.

Virtual Memory

Sometimes you can let your operating system's virtual memory capabilities (if it has them) solve disk access problems with little programming effort on your part.

If you read a file that's too big to fit in main memory, the virtual memory system will read in that part of the file that fits and store the rest on the disk. As you access different parts of the file, they will be read from the disk automatically and placed in memory.

You can apply internal algorithms to the entire file just as if it was all in memory at the same time, and let the operating system worry about reading the appropriate part of the file if it isn't in memory already.

Of course, operation will be much slower than when the entire file is in memory, but this would also be true if you dealt with the file block by block using one of the external-storage algorithms. It may be worth simply ignoring the fact that a file doesn't fit in memory and seeing how well your algorithms work with the help of virtual memory. Especially for files that aren't much larger than the available memory, this may be an easy solution.

Onward

We've come to the end of our survey of data structures and algorithms. The subject is large and complex, so no one book can make you an expert, but we hope this book has made it easy for you to learn about the fundamentals. Appendix B, "Further Reading," contains suggestions for further study.

A

Running the Workshop Applets and Example Programs

In this appendix we discuss the details of running the Workshop applets and the example programs.

- The Workshop applets are graphics-based demonstration programs that show what trees and other data structures look like.

- The example programs, whose code is shown in the text, present runnable Java code.

We also discuss the Sun Microsystems Java 2 Standard Edition (J2SE) Software Development Kit (SDK), which you can use not only to run the applets and example programs in this book but to modify the example programs and to write your own programs.

Downloadable versions of this book's applets and example programs are available on the Sams Web site: www.samspublishing.com. Log on, and use the book's International Standard Book Number (ISBN) to access the book's Web page, where you'll find a link to the downloads.

The Workshop Applets

An *applet* is a special kind of Java program that is easy to send over the Internet's World Wide Web. Because Java applets are designed for the Internet, they can run on any computer platform that has an appropriate Web browser or applet viewer.

In this book, the Workshop applets provide dynamic, interactive graphics-based demonstrations of the concepts discussed in the text. For example, Chapter 8, "Binary Trees," includes a Workshop applet that shows a tree in the applet window. Clicking the applet's buttons will show the steps involved in inserting a new node into the tree, deleting an existing node, traversing the tree, and so on. Other chapters include appropriate Workshop applets.

You can run the Workshop applets immediately after downloading them, using most popular Web browsers. This includes the current versions of Microsoft Internet Explorer and Netscape Communicator. Commercial Java development products also have an applet viewer utility that runs applets. You can also run the Workshop applets with the `appletviewer` utility included with the SDK.

Here's how to run the applets with a typical Web browser. Working offline, select Open from the File menu and navigate to the appropriate directory. Each Workshop applet consists of a subdirectory containing several files with the `.class` extension and one file with the `.html` extension. Open the `.html` file. The applet should appear on the screen.

The Example Programs

The example programs are intended to show as simply as possible how the data structures and algorithms discussed in this book can be implemented in Java. These example programs consist of Java *applications* (as opposed to applets). Java applications are not meant to be sent over the Web, but instead run as normal programs on a specific machine.

Java applications can run in either *console mode* or graphics mode. For simplicity, our example programs run in console mode, which means that output is displayed as text and input is performed by the user typing at the keyboard. In the Windows environment the console mode runs in an MS-DOS box. There is no graphics display in console mode.

The source code for the example programs is presented in the text of the book. Source files, consisting of the same text as in the book, can be downloaded from the Sams Web site.

The Sun Microsystem's Software Development Kit

Both the Workshop applets and the example programs can be executed using utility programs that are part of Sun's SDK. The SDK can be downloaded from Sun's Web site: www.sun.com. Look for the Java 2 Standard Edition (J2SE) Software Development Kit. This is a large download, but it gives you everything you need not only to run the applets and programs in this book, but to develop your own Java applets and applications.

Command-line Programs

The SDK operates in text mode, using the command line to launch its various programs. In Windows, you'll need to open an MS-DOS box to obtain this command line. Click the Start button, and find the program called MS-DOS Prompt. It may be in the Accessories folder, and it may be called something else, like Command Prompt.

Then, in MS-DOS, use the cd (for Change Directory) command to move to the appropriate subdirectory on your hard disk, where either a Workshop applet or an example program is stored. Then execute the applet or program using the appropriate SDK utility as detailed below.

Setting the Path

In Windows, the location of the SDK utility programs should be specified in a PATH statement in the autoexec.bat file so they can be accessed conveniently from within any subdirectory. This PATH statement may be placed automatically in your autoexec.bat file when you run the setup program for the SDK. Otherwise, use the Notepad utility to insert the line

```
SET PATH=C:\JDK1.4.0\BIN;
```

into the autoexec.bat file, following any other SET PATH commands. You'll find autoexec.bat in your root directory. Close the MS-DOS box and open a new one to activate this new path. (Modify the version number and directory name as necessary.)

Viewing the Workshop Applets

To use the SDK to run the Workshop applets, first use the cd command in MS-DOS to navigate to the desired subdirectory. For example, to execute the Array workshop applet from Chapter 2, "Arrays," move to its directory:

```
C:\>cd javaapps
C:\javaapps>cd chap02
C:\javaapps\chap02>cd Array
```

Then use the appletviewer utility from the SDK to execute the applet's .html file:

```
C:\javaapps\chap02\Array>appletviewer Array.html
```

The applet should start running. (Sometimes an applet takes a while to load, so be patient.) The applet's appearance should be close to the screen shots shown in the text. It won't look exactly the same because every applet viewer and browser interprets HTML and Java format somewhat differently.

As we noted, you can also use most Web browsers to execute the applets.

Operating the Workshop Applets

Each chapter gives instructions for operating specific Workshop applets. In general, remember that in most cases you'll need to repeatedly click a single button to carry out an operation. Each press of the Ins button in the Array Workshop applet, for example, causes one step of the insertion process to be carried out. Generally, a message is displayed telling what's happening at each step.

You should complete each operation—that is, each sequence of button clicks—before clicking a different button to start a different operation. For example, keep clicking the Find button until the item with the specified key is located, and you see the message Press any button. Only then should you switch to another operation involving another button, such as inserting a new item with the Ins button.

The sorting applets from Chapter 3, "Simple Sorting," and Chapter 7, "Advanced Sorting," have a Step button with which you can view the sorting process one step at a time. They also have a Run mode in which the sort runs at high speed without additional button clicks. Just click the Run button once and watch the bars sort themselves. To pause, you can click the Step button at any time. Running can be resumed by clicking the Run button again.

It's not intended that you study the code for the Workshop applets, which is mostly concerned with the graphic presentation. Hence, source listings are not provided.

Running the Example Programs

Each example program consists of a subdirectory containing a .java file and a number of .class files. The .java file is the source file, which also appears in the text. It must be compiled before you can run it. The .class files are compiled and ready to run if you have a Java interpreter.

You can use the Java interpreter from Sun's SDK to run the example programs directly from the .class files. For each program, one .class file ends with the letters App, for application. It's this file that must be invoked with java.

From an MS-DOS prompt, go to the appropriate subdirectory (using the cd command) and find this App file. For example, for the insertSort program of Chapter 3, go to the InsertSort subdirectory for Chapter 3. (Don't confuse the directory holding the applets with the directory holding the example programs.) You'll find a .java file and several .class files. One of these is insertSortApp.class. To execute the program, enter

```
C:\chap03\InsertSort>java insertSortApp
```

Don't type a file extension after the filename. The insertSort program should run, and you'll see a text display of unsorted and sorted data. In some example programs you'll see a prompt inviting you to enter input, which you type at the keyboard.

Compiling the Example Programs

You can experiment with the example programs by modifying them and then compiling and running the modified versions. You can also write your own applications from scratch, compile them, and run them. To compile a Java application, you use the javac program, invoking the example's .java file. For example, to compile the insertSort program, you would go to the insertSort directory and enter

```
C:\chap03\insertSort>javac insertSort.java
```

This time you do need to add the .java file extension. This command will compile the .java file into as many .class files as there are classes in the program. If there are errors in the source code, you'll see them displayed on the screen.

Editing the Source Code

Many text editors are appropriate for modifying the .java source files or writing new ones. For example, you can invoke an MS-DOS editor called edit from the DOS command line, and Windows includes the Notepad editor. Many commercial text editors are available as well.

Don't use a fancy word processor, such as Microsoft Word, for editing source files. Word processors typically generate output files with strange characters and formatting information, which the Java interpreter won't understand.

Terminating the Example Programs

You can terminate any running console-mode program, including any of the example programs, by pressing the [control]-[c] key combination (the Control key and the C key pressed at the same time). Some example programs have a termination procedure that's mentioned in the text, such as pressing [Enter←] at the beginning of a line, but for the others you must press [control]-[c] .

Multiple Class Files

Often several Workshop applets, or several example programs, will use .class files with the same names. Note, however, that these files may not be identical. The applet or example program may not work if the wrong class file is used with it, even if the file has the correct name.

Invoking the wrong file should not normally be a problem because all the files for a given program are placed in the same subdirectory. However, if you move files by hand, be careful not to inadvertently copy a file to the wrong directory. Doing this may cause problems that are hard to trace.

Other Development Systems

There are many other Java development systems besides Sun's SDK. Products are available from Symantec, Microsoft, Borland, and so on. Sun itself has a Java development system called Sun ONE Studio 4 (it was formerly called Forte). These products are generally faster and more convenient to use than the SDK. They typically combine all functions—editing, compiling, and execution—in a single window.

For use with the example programs in this book, such development systems should be able to handle Java version 1.4.0 or later. Many example programs (specifically, those that include user input) cannot be compiled with products designed for earlier versions of Java. (However, with minor modifications, some of which are mentioned in Chapter 1, "Overview," the .java files can be made to compile with older development systems.)

B

Further Reading

In this appendix we'll mention some books on various aspects of software development, including data structures and algorithms. This is a subjective list; there are many other excellent titles on all the topics mentioned.

Data Structures and Algorithms

The definitive reference for any study of data structures and algorithms is *The Art of Computer Programming* by Donald E. Knuth, of Stanford University (Addison Wesley, 1998). This seminal work, originally published in the 1970s, is now in its third edition. It consists of three volumes (recently made available in a boxed set): *Volume 1: Fundamental Algorithms*, *Volume 2: Seminumerical Algorithms*, and *Volume 3: Sorting and Searching*. Of these, the last is the most relevant to the topics in this book. This work is highly mathematical and does not make for easy reading, but it is the bible for anyone contemplating serious research in the field.

A somewhat more accessible text is Robert Sedgewick's *Algorithms in C++* (Addison Wesley, 1998). This book is adapted from the earlier *Algorithms* (Addison Wesley, 1988) in which the code examples were written in Pascal. It is comprehensive and authoritative. The text and code examples are quite compact, and require close reading. *Algorithms in Java* by Robert Sedgewick and Michael Schidlowsky (Addison Wesley, 2002) covers the ground in Java.

A good text for an undergraduate course in data structures and algorithms is *Data Abstraction and Problem Solving with C++: Walls and Mirrors* by Janet J. Prichard and Frank M. Carrano (Benjamin Cummings, 2001). There are many illustrations, and the chapters end with exercises and projects. The Java version is *Data Abstraction and Problem Solving with Java: Walls and Mirrors* by Frank M. Carrano and Janet L. Prichard.

Practical Algorithms in C++ by Bryan Flamig (John Wiley and Sons, 1995) covers many of the usual topics in addition to some not frequently covered by other books, such as algorithm generators and string searching.

Programming Pearls by Jon Louis Bentley (Addison Wesley, 1999) was originally written in 1986 but is nevertheless stuffed full of great advice for the programmer. Much of the material deals with data structures and algorithms.

Some other worthwhile texts on data structures and algorithms are *Classic Data Structures in C++* by Timothy A. Budd (Addison Wesley, 2000); *Data Structures and Problem Solving Using C++* by Mark Allen Weiss (Addison Wesley, 1999); and *Data Structures Using C and C++* by Yedidyah Langsam et al. (Prentice Hall, 1996).

Object-Oriented Programming Languages

For an accessible and thorough introduction to Java and object-oriented programming, try *Object-Oriented Programming in Java* by Stephen Gilbert and Bill McCarty (Waite Group Press, 1997).

If you're interested in C++, try *Object-Oriented Programming in C++, Fourth Edition*, by Robert Lafore (Sams Publishing, 2001).

The Java Programming Language, Third Edition, by Ken Arnold, James Gosling, and David Holmes (Addison Wesley, 2000) deals with Java syntax and is certainly authoritative (although briefer than many books): Gosling, who works at Sun Microsystems, is the creator of Java.

Core Java 2, Fifth Edition, by Cay S. Horstmann and Gary Cornell (Prentice Hall, 2000) is a multivolume series that covers in depth but very accessibly almost everything you want to know about programming in Java.

Object-Oriented Design (OOD) and Software Engineering

For an easy, non-academic introduction to software engineering, try *The Object Primer: The Application Developer's Guide to Object-Orientation, Second Edition,* by Scott W. Ambler (Cambridge University Press, 2001). This short book explains in plain language how to design a large software application. The title is a bit of a misnomer; it goes way beyond mere OO concepts.

Object-Oriented Design in Java by Stephen Gilbert and Bill McCarty (Waite Group Press, 1998) is an unusually accessible text.

A classic in the field of OOD is *Object-Oriented Analysis and Design with Applications* by Grady Booch (Addison Wesley, 1994). The author is one of the pioneers in this field and the creator of the Booch notation for depicting class relationships. This book isn't easy for beginners, but is essential for more advanced readers.

An early book on OOD is *The Mythical Man-Month* by Frederick P. Brooks, Jr. (Addison Wesley, 1975, reprinted in 1995), which explains in a very clear and literate way some of the reasons why good software design is necessary. It is said to have sold more copies than any other computer book.

Other good texts on OOD are *An Introduction to Object-Oriented Programming, Third Edition*, by Timothy Budd (Addison Wesley, 2002); *Object-Oriented Design Heuristics* by Arthur J. Riel (Addison Wesley, 1996); and *Design Patterns: Elements of Reusable Object-Oriented Software* by Erich Gamma et al. (Addison Wesley, 1995).

C
Answers to Questions

Chapter 1, Overview

Answers to Questions

1. insert, search for, delete
2. sorting
3. c
4. search key
5. b
6. a
7. d
8. method
9. dot
10. data types

Chapter 2, Arrays

Answers to Questions

1. d
2. True
3. b
4. False
5. new
6. d

7. interface

8. d

9. raising to a power

10. 3

11. 8

12. 6

13. False

14. a

15. constant

16. objects

Chapter 3, Simple Sorting

Answers to Questions

1. d

2. comparing and swapping (or copying)

3. False

4. a

5. False

6. b

7. False

8. three

9. Items with indices less than or equal to outer are sorted.

10. c

11. d

12. copies

13. b

14. Items with indices less than <u>outer</u> are partially sorted.

15. b

Chapter 4, Stacks and Queues

Answers to Questions

1. 10
2. b
3. Last-In-First-Out; and First-In-First-Out
4. False. It's the other way around.
5. b
6. It doesn't move at all.
7. 45
8. False. They take O(1) time.
9. c
10. O(N)
11. c
12. True
13. b
14. Yes, you would need a method to find the minimum value.
15. a

Chapter 5, Linked Lists

Answers to Questions

1. b
2. first
3. d
4. 2
5. 1
6. c
7. `current.next=null;`
8. Java's garbage collection process destroys it.

9. a

10. empty

11. a linked list

12. once, if the links include a `previous` reference

13. a double-ended list

14. b

15. Usually, the list. They both do `push()` and `pop()` in O(1) time, but the list uses memory more efficiently.

Chapter 6, Recursion

Answers to Questions

1. 10

2. d

3. 2

4. 10

5. false

6. `"ed"`

7. b

8. c

9. divide-and-conquer

10. the range of cells to search

11. the number of disks to transfer

12. c

13. b

14. b

15. stack

Chapter 7, Advanced Sorting

Answers to Questions

1. c

2. 40

3. d

4. false

5. $O(N*logN)$, $O(N^2)$

6. a

7. pivot

8. d

9. true

10. c

11. partitioning the resulting subarrays

12. b

13. pivot

14. $log_2 N$

15. true

Chapter 8, Binary Trees

Answers to Questions

1. $O(logN)$

2. b

3. True

4. 5

5. c

6. node, tree

7. a

8. c

9. finding

10. A, A's left-child descendents

11. d

12. 2*n+1

13. False

14. compress

15. c

Chapter 9, Red-Black Trees

Answers to Questions

1. in order (or inverse order)

2. b

3. False

4. d

5. b

6. rotations, changing the colors of nodes

7. red

8. a

9. left child, right child

10. d

11. a node, its two children

12. b

13. True

14. a

15. True

Chapter 10, 2-3-4 Trees and External Storage

Answers to Questions

1. b

2. balanced

3. 2

4. False

5. b

6. the root is split

7. a

8. 2

9. color flip

10. b

11. O(logN)

12. d

13. many

14. True

15. a

Chapter 11, Hash Tables

Answers to Questions

1. O(1)

2. hash function

3. d

4. linear probing

5. 1, 4, 9, 16, 25

6. b

7. linked list

8. 1.0

9. True

10. d

11. the array size

12. False

13. a

14. False

15. the same block

Chapter 12, Heaps

Answers to Questions

1. b

2. Both the right and left children have keys less than (or equal to) the parent.

3. root

4. a

5. a

6. c

7. array (or linked list)

8. up

9. b

10. one

Chapter 13, Graphs

Answers to Questions

1. edge, nodes (or vertices)

2. Count the number of 1s and divide by 2 (assuming the identity diagonal is all 0s).

3. node

4. d

5. A:B, B:A—>C—>D, C:B, D:B

6. a

7. 3

8. c

9. tree

10. No

11. True

12. d

13. A directed acyclic graph

14. No, by definition

15. Some vertices remain, but none have no successors.

Chapter 14, Weighted Graphs

Answers to Questions

1. edges

2. d

3. False

4. the lowest-weight (cheapest) edge

5. b

6. is already the destination of an edge with a lower weight

7. False

8. a

9. True

10. a

11. b

12. Warshall's algorithm

13. adjacency matrix

14. 2^N, where N is the number of squares on the board minus 1

15. No

Index

Symbols

D

How can we make this index more useful? Email us at indexes@samspublishing.com

O

objects, 16

accessing methods, 17

arrays, 40

classes, 16

comp to nodes, 367

creating, 17

sorting, 103

Java code, 104

lexicographical comparisons, 107

stability, 107

storing, 64

classDataArray.java, 65, 69-70

Person class, 65

OOP (object-oriented programming), 14

bank.java, 18-20

inheritance, 21

objects, 16-17

polymorphism, 21

procedural languages, 14-15

open addressing, 528

double hashing, 544

HashDouble applet, 545

Java code, 546, 550

hashing efficiency, 566-568

linear probe Java code

array expansion, 540

delete() method, 534

find() method, 533

hash.java, 535, 539-540

insert() method, 534

linear probing, 528

clustering, 533

duplicates, 532

Hash workshop applet, 528-532

quadratic probing, 542

HashDouble applet, 542

step, 542

operators

assignment (=), 23

dot (.), 17

equality (==), 25

new, 24

arrays, 40

overloaded, 25

saving on stack, 158

OrdArray class, 58

ordered arrays, 52

advantages, 61

binary search, 54

binary trees, 365

find() method, 56, 58

linear search, 53

OrdArray class, 58

Ordered Workshop applet, 52

binary search, Guess-a-Number game, 54

linear search, 53

OrderedArray.java, 59-61, 66

output, 27

outside grandchild, rotations, 454

overloading operators, 25

P

package access, 186

Params class, 295

parent (binary trees), 369

parent vertices, 695

parentheses (), 127

How can we make this index more useful? Email us at indexes@samspublishing.com

How can we make this index more useful? Email us at indexes@samspublishing.com

How can we make this index more useful? Email us at indexes@samspublishing.com

X-Y-Z

if   OP thie > OP TOP
            push OP thie